Ethnopharmacology and OMICS Advances in Medicinal Plants Volume 1

AF212732

Mukesh Nandave • Rohit Joshi •
Jyoti Upadhyay
Editors

Ethnopharmacology and OMICS Advances in Medicinal Plants Volume 1

Uncovering Diversity and Ethnopharmacological Aspects

Springer

Editors
Mukesh Nandave
Pharmacology and Pharmaceutical Biotech.
DPSRU
New Delhi, Delhi, India

Rohit Joshi
Biotechnology Division
Inst. of Himalayan Bioresource Tech.
Palampur, Himachal Pradesh, India

Jyoti Upadhyay
Health Sciences and Technology
UPES University
Dehradun, Uttarakhand, India

ISBN 978-981-97-2369-0 ISBN 978-981-97-2367-6 (eBook)
https://doi.org/10.1007/978-981-97-2367-6

© The Editor(s) (if applicable) and The Author(s), under exclusive license to Springer Nature Singapore Pte Ltd. 2024
This work is subject to copyright. All rights are solely and exclusively licensed by the Publisher, whether the whole or part of the material is concerned, specifically the rights of translation, reprinting, reuse of illustrations, recitation, broadcasting, reproduction on microfilms or in any other physical way, and transmission or information storage and retrieval, electronic adaptation, computer software, or by similar or dissimilar methodology now known or hereafter developed.
The use of general descriptive names, registered names, trademarks, service marks, etc. in this publication does not imply, even in the absence of a specific statement, that such names are exempt from the relevant protective laws and regulations and therefore free for general use.
The publisher, the authors and the editors are safe to assume that the advice and information in this book are believed to be true and accurate at the date of publication. Neither the publisher nor the authors or the editors give a warranty, expressed or implied, with respect to the material contained herein or for any errors or omissions that may have been made. The publisher remains neutral with regard to jurisdictional claims in published maps and institutional affiliations.

This Springer imprint is published by the registered company Springer Nature Singapore Pte Ltd.
The registered company address is: 152 Beach Road, #21-01/04 Gateway East, Singapore 189721, Singapore

If disposing of this product, please recycle the paper.

Foreword

Ancient works, like Ayurveda, have listed the medicinal values of many plants. Validating this information and *Uncovering Diversity and Ethnopharmacological Insights of Medicinal Plants* has been an ongoing activity in many laboratories in India and abroad. Last couple of decades has brought out quality information through genomics and other omics approaches and also through high-end spectrophotometric techniques about our understanding of the various metabolic pathways and presence of different compounds and also finding new uses of compounds in medicine. Some of this information has been collated in the present volume entitled *Ethnopharmacology and OMICS Advances in Medicinal Plants Volume 1: Uncovering Diversity and Ethnopharmacological Insights of Medicinal Plants*. The comprehensive volume stands as a testament to the profound interplay between biodiversity and the therapeutic properties of plants, synthesizing a wealth of knowledge that integrates traditional wisdom with rigorous scientific inquiry and conservation efforts.

The chapters in this book cover diverse topics, each offering an up to date information and a unique perspective on the role of medicinal plants in both human life and ecosystems. They range from the biodiversity of medicinal plants to the endophytic fungal diversity that thrives within them, the intricate web of life, revealing the interconnections within ecosystems. In addition, exploration of genome-based barcoding techniques illuminates the genetic blueprints that underpin the diverse array of medicinal plants, offering valuable insights into their classification and conservation. A special focus is placed on the cultural and scientific dimensions of ethnopharmacology in India, showcasing the rich heritage of traditional medicine in modern context. The case study of *Dolichandrone falcata* in the Western Ghats offers a compelling example of how ethnobotanical and conservational studies can provide valuable insights into the ecological and cultural significance of medicinal plants within specific regions.

This book also presents nutraceutical formulations derived from medicinal plants, highlighting their potential for holistic well-being and at the same time addresses the cutting-edge applications of nanobiotechnology in the field of medicinal plants, offering a glimpse into the future of drug delivery and therapeutic applications. The chapters dedicated to ethnopharmacognosy and traditional knowledge databases provide valuable overview on the relationships between humans and plants, showcasing the wealth of knowledge accumulated over generations and its significance for the conservation and sustainable use of medicinal plants.

In summary, this comprehensive volume stands as a testament to the enduring importance of medicinal plants in both our lives and ecosystems. These contents of this book will inspire students, faculty, and researchers to further explore the intricate world of nature's pharmacy. My sincere congratulations go to the editors whose efforts have brought out this book in the present format.

Sudhir K. Sopory
International Centre for Genetic Engineering
and Biotechnology (ICGEB)
New Delhi, India
Jawaharlal Nehru University
New Delhi, India
ICGEB
New Delhi, India

Preface

Nature's pharmacy has been a timeless healer, with medicinal plants and traditional remedies standing as ancient guardians of our well-being, offering remedies for centuries to ailments that afflict the body, mind, and spirit.

As we delve into the world of medicinal plants and biodiversity, our book entitled *Ethnopharmacology and OMICS Advances in Medicinal Plants Volume 1: Uncovering Diversity and Ethnopharmacological Insights of Medicinal Plants* embarks on a multidimensional journey, encompassing diverse facets of traditional wisdom, botanical exploration, and scientific inquiry. The richness of biodiversity, the hidden treasures within endophytic fungi, and the genomic secrets encoded in the DNA of medicinal plants all contribute to the content of this holistic book.

Our journey begins with a meticulous exploration of the biodiversity of medicinal plants and endophytic fungal diversity, unravelling the symbiotic relationships between plants and these hidden fungi that often hold the keys to untold therapeutic potentials. The pages that follow delve into the innovative realm of genome-based barcoding, where the language of DNA becomes a powerful tool for plant identification, classification, and conservation.

Ethnopharmacology and ethnobotanical studies take us on a cultural expedition, where traditional knowledge meets rigorous scientific scrutiny. Here, the ancient wisdom of communities around the world converges with modern research methodologies to unearth the secrets of plant-based medicines.

In the evolving landscape of healthcare, nutraceutical formulations play a pivotal role. This book sheds light on the fusion of nutrition and pharmaceuticals, exploring the potential of medicinal plants in promoting holistic well-being. The exploration extends further into the realm of ethnopharmacognosy, unravelling the intricate relationships between people and plants, deeply embedded in the cultural fabric of societies.

The significance of traditional knowledge finds a modern expression in the creation of a comprehensive database, safeguarding the wealth of information amassed over generations. Nanotechnology, with its infinitesimal scale, offers new possibilities for drug delivery and therapeutic applications, bringing us to the forefront of cutting-edge research on medicinal plants.

As we navigate through the chapters, mutagenesis and agronomical approaches for conservation become crucial strategies in preserving the delicate balance of biodiversity. Bioassays of secondary metabolites offer insights into the potency of medicinal plants, contributing to the development of effective pharmaceuticals.

Intellectual property rights emerge as a paramount concern in our exploration. Traditional knowledge on medicinal plants and creating IPR out of them requires careful legal and ethical considerations. Biotic stress challenges the resilience of medicinal plants, emphasizing the urgent need for sustainable conservation practices.

The journey culminates in the application of metabolomics, offering a comprehensive view of the intricate biochemical pathways that define the therapeutic potential of medicinal plants.

In this compendium, we invite you to traverse the landscapes of knowledge, tradition, and innovation. *Ethnopharmacology and OMICS Advances in Medicinal Plants Volume 1: Uncovering Diversity and Ethnopharmacological Insights of Medicinal Plants* seeks to be a guiding light in the exploration of nature's pharmacy, inspiring both seasoned researchers and curious minds alike.

We extend our heartfelt thanks to Dr. Naren Aggarwal, Ms. Swati Sharma, and Ms. Vijayalakshmi Dayalan from Springer Nature for their exceptional management in producing this book under challenging circumstances; their valuable contribution is highly appreciated. We would also like to thank all contributors for sharing their expertise and to our readers for joining us on this enlightening journey.

New Delhi, Delhi, India Mukesh Nandave
Palampur, Himachal Pradesh, India Rohit Joshi
Dehradun, Uttarakhand, India Jyoti Upadhyay

Contents

1 **Factors Affecting Biodiversity and Conservation of Medicinal Plants** 1
Acharya Balkrishna, Priya Yadav, Priyanka Yadav, Aakanksha Saini, Brijesh Kumar, Netrapal Singh, Dushyant Sharma, and Ved Priya Arya

2 **Assessment of the Medicinal Flora for Treating Skin Disorders in Indonesia** 17
Inggit Puji Astuti, Ria Cahyaningsih, R. Syamsul Hidayat, Titi Juhaeti, Iin P. Husaini, Endah Suhaendah, R. Vitri Garvita, Dewi Maharani, Hary Wawangningrum, Saniyatun Mar'atus Solihah, and Rusmanto

3 **Endophytic Fungal Diversity of Medicinal Plants and Their Role During Stress Response** 57
Ankita Chandel, Harleen Kaur, Ravneet Kaur, Swapnil Singh, Neetu Jagota, and Ashish Sharma

4 **Traditional Knowledge Databases of Medicinal Plants** 75
Acharya Balkrishna, Aakanksha Saini, Priya Yadav, Priyanka Yadav, Brijesh Kumar, Dushyant Sharma, Netrapal Singh, and Ved Priya Arya

5 **Ethnopharmacognosy of Medicinal Plants: Challenges and Biotechnological Advancements** 93
Gaurav Upadhyay and Priya Thakur

6 **Ethnobotanical and Conservational Studies of Medicinal Plants: A Case Study of *Dolichandrone falcata* (Wall. Ex DC.) Seem. in Western Ghats, India** 103
Suhas Daswadikar, Mandar Muley, Satyendra Prasad, and Prakash Itankar

7 **Bridging Cultures and Medicine: Quantitative Insights in Ethnopharmacology** 115
Varsha Srivastava, Areeba Insaf, and Sayeed Ahmad

8 Unveiling Plant-Based Healing Wisdom Through Ethnobotany and Medicinal Ethnopharmacology 149
Niharika, Mukta Satsangi, Shahid Umar, Akbar Ali, Bushra Parveen, and Sayeed Ahmad

9 A Current Perspective of Ethnopharmacology in India 173
Amrita Mishra and Arun K. Mishra

10 Databases of Medicinal Plants: An Update 191
Gandhar Pradhan and Yogesh A. Kulkarni

11 Role of Genome-Based Barcoding in Quality Control and Standardization of Medicinal Plants 203
S. Shreedevasena and Krishnan P. Abhijith

12 Future Directions and Emerging Trends in Medicinal Plant DNA Barcoding 219
Aftab Alam, Mohammad Usman, Bushra Parveen, and Sayeed Ahmad

13 Conservation of High-Value Medicinal Plants Through Agronomical Approaches 239
Pooja Oli, Kuldeep Joshi, Shailaja Punetha, Manish Sharma, Vineet Kaswan, and Mithilesh Singh

14 Understanding the Interplay: Medicinal Plants and Biotic Stress 259
S. Shreedevasena, N. Kavya, Namburi Karunakar Reddy, and Pooja S. Patel

15 Metabolomics in Medicinal Plants: Unraveling the Therapeutic Potential of Natural Compounds 285
Naveen Reddy Penumallu, Sageer Abass, Rabea Parveen, and Sayeed Ahmed

16 Exploring the Therapeutic Potential of Secondary Metabolites in Medicinal Plants 313
Monalisha Samal, Abdur Rahman, and Sayeed Ahmad

17 Integrative Approaches for Enhanced Secondary Metabolite Production 331
Pradeep Kumar Shukla, Revanth Kishore Pulamolu, and Pragati Shukla Nee Misra

18 Nutraceuticals Derived from Medicinal Plants: A Pediatric Perspective 373
Yogesh Nikam, Mandar Muley, Satyendra Prasad, and Prakash Itankar

19 Nutraceutical Formulations from Medicinal Plants: A Potential Therapeutic Agent 391
Rohan Sarkar, Swarajya Laxmi Nayak, Manish Kumar Suthar, and Manish Das

20 Nanobiotechnology for Medicinal Plants: A Comprehensive Review 419
Shreya Agrawal, Neha Bhandari, and Pooja Saklani

21 Genetic Enhancement Strategies for Medicinal Plants: Exploring Mutagenesis 445
Manish Sharma, Shailaja Punetha, B. K. Rajkumar, Vineet Kaswan, and Pooja Oli

22 Intellectual Property Rights Related to Medicinal Plants 457
Irmanida Batubara, Shadila Fira Asoka, I Ketut Mudite Adnyane, and Wisnu Ananta Kusuma

Editors and Contributors

About the Editors

Mukesh Nandave is an associate professor and head of the Department of Pharmacology and Pharmaceutical Biotechnology at the Delhi Pharmaceutical Sciences and Research University, Government of NCT of Delhi, New Delhi, India. He has earlier served as associate professor and head of the Department of Pharmacology at NMIMS University, Mumbai (2010–2017). He has also worked as research scientist in the Medical Affairs and Clinical Research Department of Ranbaxy Research Laboratories (currently known as Sun Pharmaceutical Industries Limited, Gurugram). Dr. Nandave earned his Ph.D. in pharmacology from AIIMS, Delhi, and received his postdoctoral training from the Ohio State University Medical Center, Columbus, USA.

For over 20 years, Dr. Nandave has been actively engaged in investigating the role of nutraceuticals, herbomineral formulations, and phytoconstituents in mitigating myocardial ischemia and reperfusion injury, diabetes, obesity, and pain management. He has published more than 80 papers in peer-reviewed national and international journals and book chapters.

Dr. Nandave has received numerous awards, including the Distinguished Leadership Award from the International Academy of Cardiovascular Sciences; Prof. S. C. Lahiri Oration and G. Achari Gold Medal from the Indian Pharmacological Society; Young Pharmacy Teacher of the Year Award from the Association of Pharmaceutical Teachers of India (APTI); Early Investigator Award from the International Society for Heart Research; Prof. Duggirala Visweswaram and Prof. Sreemantula Satyanarayana Prize; Best Research Output of the Year for 2014–2015, 2013–2014, and 2012–2013 Award from SVKM's NMIMS University.

Dr. Nandave is the secretary general of the International Academy of Cardiovascular Sciences (IACS), India, and treasurer of the Society for the Promotion and Research of Cardiovascular Sciences (SPARCS). He is a life member of various professional bodies, including the International Society for Heart Research (ISHR), the International Academy of Cardiovascular Sciences (IACS), the Indian Pharmacological Society (IPS), the Indian Pharmaceutical Association (IPA), the Association of Physiologists and Pharmacologists of India (APPI),

Association of Pharmaceutical Teachers of India (APTI), and Society for Ethnopharmacology.

Rohit Joshi is currently working as a senior scientist and assistant professor, Biotechnology Division, at CSIR-Institute of Himalayan Bioresource Technology, Palampur, Himachal Pradesh, India. Dr. Joshi earned his Ph.D. with major in plant physiology and minor in molecular biology and biotechnology from G.B. Pant University of Agriculture and Technology, Pantnagar, Uttarakhand, India, and received his postdoctoral training from Jawaharlal Nehru University, New Delhi, India; Indian Agricultural Research Institute, New Delhi, India; International Centre for Genetic Engineering and Biotechnology, New Delhi, India; and School of Plant, Environmental, and Soil Sciences, Louisiana State University, Baton Rouge, LA, USA.

For over 15 years, Dr. Joshi has been engaged in investigating the molecular mechanism and signalling response of various plant species under different abiotic stresses using various OMICS approaches. He has published 58 articles in peer-reviewed journals and 24 book chapters. Dr. Joshi has received numerous awards, including the R.D. Asana Gold Medal Award, Young Scientist Award, Third Prize in the Lightning Talk session of India-EMBO symposium, Young Investigator talk at ICGEB, Fast Track Young Scientist (SERB), and Dr. D.S. Kothari post-doctoral fellowship (UGC). He is elected as a member of the National Academy of Sciences, India (NASI), and Plant Tissue Culture Association of India, PTCA(I).

Jyoti Upadhyay is an associate professor at the Department of Pharmaceutical Sciences, School of Health Sciences and Technology, UPES, Dehradun, Uttarakhand, India. She is also the coordinator of Pharmacovigilance Centre, UPES recognized by the National Coordination Centre (NCC), Indian Pharmacopoeia Commission (IPC), Ghaziabad. She completed her Ph.D. in pharmaceutical sciences from Kumaun University, Nainital, India. During her Ph.D., she worked on the assessment of persistent organic pollutants on the risk of birth defects. Her research interests are pharmacology and toxicology, maternal and child health, and pharmacovigilance. She has received academic excellence awards and research awards at UPES, Dehradun. She has published more than 45 research and review articles. She is a life member of the Indian Pharmacological Society and Society of Biological Chemists.

Contributors

Sageer Abass Centre of Excellence in Unani Medicine (Pharmacognosy and Pharmacology), Bioactive Natural Product Laboratory, School of Pharmaceutical Education and Research, Jamia Hamdard, New Delhi, India

K. P. Abhijith School of Crop Improvement, ICAR-IARI Assam, Gogamukh, Dhemaji, Assam, India

I. Ketut Mudite Adnyane School of Veterinary Medicine and Biomedical Sciences, IPB University, Dramaga, Indonesia

Shreya Agrawal Department of Biotechnology, Hemvati Nandan Bahuguna Garhwal University (A Central University), Srinagar Garhwal, Uttarakhand, India

Sayeed Ahmad Centre of Excellence in Unani Medicine (Pharmacognosy and Pharmacology), Bioactive Natural Product Laboratory (BNPL), Department of Pharmacognosy and Phytochemistry, School of Pharmaceutical Education and Research, Jamia Hamdard, New Delhi, India

Aftab Alam Centre of Excellence in Unani Medicine (Pharmacognosy and Pharmacology) and Bioactive Natural Product Laboratory, SPER, Jamia Hamdard, New Delhi, India

Department of Pharmacology, SPER, Jamia Hamdard, New Delhi, India

Akbar Ali Department of Botany, Dayalbagh Educational Institute, Dayalbagh, Agra, Uttar Pradesh, India

Ved Priya Arya Patanjali Herbal Research Department, Patanjali Research Institute, Haridwar, Uttarakhand, India

University of Patanjali, Haridwar, Uttarakhand, India

Centre of Excellence, Patanjali Ayurved Hospital, Haridwar, Uttarakhand, India

Shadila Fira Asoka Tropical Biopharmaca Research Center—IPB University, Dramaga, Indonesia

Department of Chemistry, Faculty of Mathematics and Natural Sciences, IPB University, Dramaga, Indonesia

Inggit Puji Astuti Research Center for Plant Conservation, Botanical Gardens, and Forestry, National Research and Innovation Agency (BRIN), Kawasan Sains dan Teknologi (KST) Soekarno, Bogor, West Java, Indonesia

Acharya Balkrishna Patanjali Herbal Research Department, Patanjali Research Institute, Haridwar, Uttarakhand, India

University of Patanjali, Haridwar, Uttarakhand, India

Irmanida Batubara Tropical Biopharmaca Research Center—IPB University, Dramaga, Indonesia

Department of Chemistry, Faculty of Mathematics and Natural Sciences, IPB University, Dramaga, Indonesia

Neha Bhandari Department of Biotechnology, Hemvati Nandan Bahuguna Garhwal University (A Central University), Srinagar Garhwal, Uttarakhand, India

Ria Cahyaningsih Research Center for Plant Conservation, Botanical Gardens, and Forestry, National Research and Innovation Agency (BRIN), Kawasan Sains dan Teknologi (KST) Soekarno, Bogor, West Java, Indonesia

Ankita Chandel Department of Botany and Environment Science, DAV University, Jalandhar, Punjab, India

Manish Das ICAR-Directorate of Medicinal and Aromatic Plants Research, Boriavi, Anand, Gujarat, India

Suhas Daswadikar Department of Pharmaceutical Sciences, Rashtrasant Tukadoji Maharaj Nagpur University, Nagpur, Maharashtra, India

R. Vitri Garvita Research Center for Plant Conservation, Botanical Gardens, and Forestry, National Research and Innovation Agency (BRIN), Kawasan Sains dan Teknologi (KST) Soekarno, Bogor, West Java, Indonesia

R. Syamsul Hidayat Research Center for Plant Conservation, Botanical Gardens, and Forestry, National Research and Innovation Agency (BRIN), Kawasan Sains dan Teknologi (KST) Soekarno, Bogor, West Java, Indonesia

Iin P. Husaini Research Center for Plant Conservation, Botanical Gardens, and Forestry, National Research and Innovation Agency (BRIN), Kawasan Sains dan Teknologi (KST) Soekarno, Bogor, West Java, Indonesia

Areeba Insaf Centre of Excellence in Unani Medicine (Pharmacognosy and Phytochemistry), Bioactive Natural Product Laboratory, Department of Pharmacognosy and Phytochemistry, School of Pharmaceutical Education and Research, Jamia Hamdard, New Delhi, India

Prakash Itankar Department of Pharmaceutical Sciences, Rashtrasant Tukadoji Maharaj Nagpur University, Nagpur, Maharashtra, India

Neetu Jagota Department of Botany and Environment Science, DAV University, Jalandhar, Punjab, India

Kuldeep Joshi G.B. Pant National Institute of Himalayan Environment, Kosi-Katarmal, Uttarakhand, India

Titi Juhaeti Research Center for Plant Conservation, Botanical Gardens, and Forestry, National Research and Innovation Agency (BRIN), Kawasan Sains dan Teknologi (KST) Soekarno, Bogor, West Java, Indonesia

Vineet Kaswan College of Basic Science and Humanities, Sardarkrushinagar Dantiwada Agricultural University, Sardarkrushinagar, Gujarat, India

Harleen Kaur Department of Botany and Environment Science, DAV University, Jalandhar, Punjab, India

Ravneet Kaur Department of Botany and Environment Science, DAV University, Jalandhar, Punjab, India

N. Kavya University of Agricultural Sciences, GKVK, Bangalore, India

Mohammad Ahmed Khan Department of Pharmacology, School of Pharmaceutical Education and Research, Jamia Hamdard, New Delhi, India

Yogesh A. Kulkarni Shobhaben Pratapbhai Patel School of Pharmacy and Technology Management, SVKM's Narsee Monjee Institute of Management Studies (NMIMS) Deemed to be University, Mumbai, India

Brijesh Kumar Patanjali Herbal Research Department, Patanjali Research Institute, Haridwar, Uttarakhand, India

Wisnu Ananta Kusuma Faculty of Mathematics and Natural Sciences, IPB University, Dramaga, Indonesia

Dewi Maharani Research Center for Plant Conservation, Botanical Gardens, and Forestry, National Research and Innovation Agency (BRIN), Kawasan Sains dan Teknologi (KST) Soekarno, Bogor, West Java, Indonesia

Amrita Mishra Department of B.Pharm (Ayu), Delhi Pharmaceutical Sciences and Research University, New Delhi, India

Arun K. Mishra SOS School of Pharmacy, IFTM University, Moradabad, India

Pragati Shukla Nee Misra Centre for Tissue Culture Technology, Jacob Institute of Biotechnology and Bioengineering, Sam Higginbottom University of Agriculture Technology and Sciences, Prayagraj, Uttar Pradesh, India

Mandar Muley Department of Pharmaceutical Sciences, Rashtrasant Tukadoji Maharaj Nagpur University, Nagpur, Maharashtra, India

Swarajyalaxmi Nayak ICAR-Directorate of Medicinal and Aromatic Plants Research, Boriavi, Anand, Gujarat, India

Niharika Department of Botany, School of Chemical and Life Sciences, Jamia Hamdard, New Delhi, India

Centre of Excellence in Unani Medicine (Pharmacognosy and Pharmacology), Bioactive Natural Product Laboratory (BNPL), Department of Pharmacognosy and Phytochemistry, School of Pharmaceutical Education and Research, Jamia Hamdard, New Delhi, India

Yogesh Nikam Department of Pharmaceutical Sciences, Rashtrasant Tukadoji Maharaj Nagpur University, Nagpur, Maharashtra, India

Pooja Oli G.B. Pant National Institute of Himalayan Environment, Kosi-Katarmal, Uttarakhand, India

Bushra Parveen Department of Pharmacology, School of Pharmaceutical Education and Research, Jamia Hamdard, New Delhi, India

Rabea Parveen Department of Pharmaceutics, School of Pharmaceutical Education and Research, Jamia Hamdard, New Delhi, India

Pooja S. Patel University of Agricultural Sciences, GKVK, Bangalore, India

Gandhar Pradhan Shobhaben Pratapbhai Patel School of Pharmacy and Technology Management, SVKM's Narsee Monjee Institute of Management Studies (NMIMS) Deemed to be University, Mumbai, India

Satyendra Prasad Department of Pharmaceutical Sciences, Rashtrasant Tukadoji Maharaj Nagpur University, Nagpur, Maharashtra, India

Revanth Kishore Pulamolu Naini Institute of Agriculture, Sam Higginbottom University of Agriculture Technology and Sciences, Prayagraj, Uttar Pradesh, India

Shailaja Punetha G.B. Pant National Institute of Himalayan Environment, Kosi-Katarmal, Uttarakhand, India

Abdur Rahman Centre of Excellence in Unani Medicine (Pharmacognosy and Pharmacology), Bioactive Natural Product Laboratory, School of Pharmaceutical Education and Research, Jamia Hamdard, New Delhi, India

B. K. Rajkumar Main Cotton Research Station, Navsari Agricultural University, Surat, Gujarat, India

Namburi Karunakar Reddy University of Agricultural Sciences, GKVK, Bangalore, India

Naveen Reddy Centre of Excellence in Unani Medicine (Pharmacognosy and Pharmacology), Bioactive Natural Product Laboratory, School of Pharmaceutical Education and Research, Jamia Hamdard, New Delhi, India

Rusmanto Research Center for Plant Conservation, Botanical Gardens, and Forestry, National Research and Innovation Agency (BRIN), Kawasan Sains dan Teknologi (KST) Soekarno, Bogor, West Java, Indonesia

Aakanksha Saini Patanjali Herbal Research Department, Patanjali Research Institute, Haridwar, Uttarakhand, India

Pooja Saklani Department of Biotechnology, Hemvati Nandan Bahuguna Garhwal University (A Central University), Srinagar Garhwal, Uttarakhand, India

Monalisha Samal Centre of Excellence in Unani Medicine (Pharmacognosy and Pharmacology), Bioactive Natural Product Laboratory, School of Pharmaceutical Education and Research, Jamia Hamdard, New Delhi, India

Rohan Sarkar ICAR-Directorate of Medicinal and Aromatic Plants Research, Boriavi, Anand, Gujarat, India

Mukta Satsangi Centre of Excellence in Unani Medicine (Pharmacognosy and Pharmacology), Bioactive Natural Product Laboratory (BNPL), Department of Pharmacognosy and Phytochemistry, School of Pharmaceutical Education and Research, Jamia Hamdard, New Delhi, India

Department of Botany, Dayalbagh Educational Institute, Dayalbagh, Agra, Uttar Pradesh, India

Ashish Sharma Department of Botany and Environment Science, DAV University, Jalandhar, Punjab, India

Dushyant Sharma Patanjali Herbal Research Department, Patanjali Research Institute, Haridwar, Uttarakhand, India

Manish Sharma Pulses Research Station, Sardarkrushinagar Dantiwada Agricultural University, Sardarkrushinagar, Gujarat, India

S. Shreedevasena ICAR-Directorate of Medicinal and Aromatic Plant Research, Anand, Gujarat, India

Pradeep Kumar Shukla Department of Biological Sciences, Faculty of Science, Sam Higginbottom University of Agriculture Technology and Sciences, Prayagraj, Uttar Pradesh, India

Mithilesh Singh G.B. Pant National Institute of Himalayan Environment, Kosi-Katarmal, Uttarakhand, India

Netrapal Singh Patanjali Herbal Research Department, Patanjali Research Institute, Haridwar, Uttarakhand, India

Swapnil Singh Department of Botany and Environment Science, DAV University, Jalandhar, Punjab, India

Saniyatun Mar'atus Solihah Conservation Department, Mitra Natura Raya, Bogor Botanic Gardens, Bogor, West Java, Indonesia

Varsha Srivastava Centre of Excellence in Unani Medicine (Pharmacognosy and Phytochemistry), Bioactive Natural Product Laboratory, Department of Pharmacognosy and Phytochemistry, School of Pharmaceutical Education and Research, Jamia Hamdard, New Delhi, India

Endah Suhaendah Research Center for Plant Conservation, Botanical Gardens, and Forestry, National Research and Innovation Agency (BRIN), Kawasan Sains dan Teknologi (KST) Soekarno, Bogor, West Java, Indonesia

Manish Kumar Suthar ICAR-Directorate of Medicinal and Aromatic Plants Research, Boriavi, Anand, Gujarat, India

Priya Thakur MET Faculty of Pharmacy, MIT Campus, Ram Ganga Vihar, Moradabad, Uttar Pradesh, India

Shahid Umar Department of Botany, School of Chemical and Life Sciences, Jamia Hamdard, New Delhi, India

Gaurav Upadhyay Ram-Eesh Institute of Vocational and Technical Education, Greater Noida, Uttar Pradesh, India

Mohammad Usman Department of Pharmacognosy and Phytochemistry, SPER, Jamia Hamdard, New Delhi, India

Universal Biotech, Khari Baoli, Delhi, India

Hary Wawangningrum Research Center for Plant Conservation, Botanical Gardens, and Forestry, National Research and Innovation Agency (BRIN), Kawasan Sains dan Teknologi (KST) Soekarno, Bogor, West Java, Indonesia

Priya Yadav Patanjali Herbal Research Department, Patanjali Research Institute, Haridwar, Uttarakhand, India

Priyanka Yadav Patanjali Herbal Research Department, Patanjali Research Institute, Haridwar, Uttarakhand, India

Abbreviations

ABTS	2,2′-Azino-bis(3-ethylbenzothiazoline-6-sulfonic acid)
ACF	Ag-deposited activated carbon fibre
ADHD	Attention deficit hyperactivity disorder
ADME	Absorption, distribution, metabolism, excretion
AFLP	Amplified fragment length polymorphism
AFM	Atomic force microscopy
AHAP	2-Amino-6N-hydroxyl amino purine
ANOVA	Analysis of variance
API	Ayurvedic Pharmacopoeia of India
AR	Augmented reality
ARIMA	Autoregressive integrated moving average
ASSURED	Affordable, sensitive, specific, user-friendly, rapid, equipment-free, and delivered
ASTA	American Spice Trade Association
ASU	Ayurveda, Siddha, Unani
AYUSH	Ayurveda, Yoga & Naturopathy, Unani, Siddha, and Homeopathy
BDU	Bromo-deoxy-uridine
BINs	Barcode Index Numbers
BOLD	Barcode of Life Database
BRM	Biological Reference Material
BU	5-Bromo-uracil
C3G	Cyanidin-3-glucoside
CAGR	Compound annual growth rate
CAM	Complementary and alternative medicine
CBD	Convention on Biological Diversity
CBOL	Consortium for the Barcode of Life
CCIM	Central Council of Indian Medicine
CCRUM	Central Council for Research in Unani Medicine
CDSCO	Central Drugs Standard Control Organisation
CI	Combination Index
CIRISM	Central Institute for Research in Indigenous System of Medicine
CITES	Convention on International Trade in Endangered Species
COI	Cytochrome-C oxidase subunit 1
COX	Cyclooxygenase

CPMs	Chinese medicinal plants
CRISPR	Clustered regularly interspaced short palindrome repeats
CSIR	Council of Scientific and Industrial Research
CThTV	Curvularia thermal tolerance virus
CUTS	Consumer Unity and Trust Society
DDBJ	DNA Data Bank Of Japan
DES	Diethyl sulfate
DLS	Dynamic light scattering
DMBA	7,12-Dimethylbenzanthracene
DMS	Dimethyl sulfate
DPPH	1,1-Diphenyl-2-picrylhydrazyl
EDX	Energy-dispersive X-ray spectroscopy
EGCG	Epigallocatechin-3-gallate
EMS	Ethyl methanesulfonate
ENA	European Nucleotide Archive
ENNG	1-Ethyl-2-nitro-1-nitrosoguanidine
ENU	1-Ethyl-1-nitrosourea
EPM	Elevated plus maze
ESA	European Spice Association
FDI	Foreign direct investment
FESEM	Field emission scanning electron microscopy
FICI	Fractional inhibitory concentration index
FSSAI	Food Safety and Standards Authority of India
FTIR	Fourier-transform infrared spectroscopy
FYM	Farmyard manure
GACP	Good agricultural and collection practices
GAP	Good agricultural practices
GBIF	Global Biodiversity Information Facility
GMOs	Genetically modified organisms
GMP	Good manufacturing practices
HAP	6-N-Hydroxylaminopurine
HDCC	Herbal Drugs Consultative Committee
HDI	Human development index
HF	Hydrogen fluoride
HPLC	High-performance liquid chromatography
HP-TLC	High-performance-thin layer chromatography
HRM	High-resolution melting
HR-MAS	High-resolution magic angle spinning
iBOL	International Barcode of Life Project
IBSD	Institute of Bioresources and Sustainable Development
ICDS	Integrated Child Development Scheme
ICMR	Indian Council of Medical Research
ICP-MS	Inductively coupled plasma-mass spectrometry
IORA	Indian Ocean Rim Association
IPCC	Intergovernmental Panel on Climate Change

IPR	Intellectual property rights
ISM	Indian Systems of Medicine
ISMH	Indian Systems of Medicine and Homeopathy
ISSRs	Inter simple sequence repeats
IUCN	International Union for Conservation of Nature
LAMP	Loop-mediated isothermal amplification
LC-MS	Liquid chromatography-mass spectrometry
LOX	Lipoxygenase
MALDI-MS	Matrix-assisted laser desorption/ionization mass spectrometry
MAPs	Medicinal and aromatic plants
MBT	Marble burying test
MDM	Mid-day meal
MDR	Multi-drug resistance
MH	Maleic hydrazide
MIC	Minimum inhibitory concentration
MMDBD	Medicinal Materials DNA Barcode Database
MMS	Methyl methanesulfonate
MNNG	1-Methyl-2-nitro-1-nitrosoguanidine
MNU	N-Methyl-N-nitrosourea
MOA	Mode of action
MPs	Medicinal plants
MRSA	Methicillin-resistant *S. aureus*
NABARD	National Bank for Agriculture and Rural Development
NBT	New breeding techniques
NCBI	National Center for Biotechnology Information
NGS	Next-generation sequencing
NHM	National Health Mission
NMPB	National Medicinal Plants Board
NMR	Nuclear magnetic resonance
NRHM	National Rural Health Mission
NUHM	National Urban Health Mission
ODM	Oligonucleotide-directed mutagenesis
OECD	Organization for Economic Co-operation and Development
OFSP	Orange-fleshed sweet potatoes
OTUs	Operational taxonomic units
PASS	Prediction of Activity Spectra for Substances
PCA	Principal component analysis
PCR	Polymerase chain reaction
POWO	Plants of the World
PROBE	Public Report on Basic Education
QTL	Quantitative trait loci
RAPD	Random amplified polymorphic DNA
RFLP	Restriction fragment length polymorphism
RIS	Revenue Information System
RNS	Reactive nitrogen species

ROS	Reactive oxygen species
SAED	Selected area electron diffraction
SARS-CoV	Severe acute respiratory syndrome-corona virus
SC	Stratum corneum
SCC-25	Human oral squamous carcinoma cell line
SCD	Sickle cell disease
SDN	Site-directed nuclease
SEM	Scanning electron microscopy
SNP	Single nucleotide polymorphism
SOD	Superoxide dismutase
SSR	Simple sequence repeat
SWHD	San Wu Huangqin Decoction
TALEN	Transcription activator-like effector nuclease
TCM	Traditional Chinese Medicine
TCM-ID	Traditional Chinese Medicine Information Database
TEM	Transmission electron microscopy
TIM	Traditional Indian Medicine
TKDL	Traditional Knowledge Digital Library
TLC	Thin layer chromatography
TPA	12-O-Tetradecanoylphorbol-13-acetate
TRIPs	Trade-Related Aspects of Intellectual Property Rights
UDHR	Universal Declaration of Human Rights
UNCED	United Nations Conference on Environment and Development
VR	Virtual reality
WGS	Whole-genome shotgun
WHE	World Herbal Encyclopedia
WHO	World Health Organization
WIPO	World Intellectual Property Organization
WWF	World Wide Fund
XRD	X-ray powder diffraction
ZFN	Zinc finger nuclease

Factors Affecting Biodiversity and Conservation of Medicinal Plants

1

Acharya Balkrishna, Priya Yadav, Priyanka Yadav, Aakanksha Saini, Brijesh Kumar, Netrapal Singh, Dushyant Sharma, and Ved Priya Arya

Abstract

Biodiversity, a vital pillar of ecological equilibrium and human well-being, underscores the complex link between ecosystems, species diversity, and human health. This complex relationship gains importance due to the substantial contribution of natural sources and remarkably diverse plant species within varied ecosystems to a significant portion of traditional and contemporary medicinal treatments. Preserving biodiversity is paramount to ensuring an abundant medicinal plant repository that significantly contributes to global healthcare. The fusion of time-honoured wisdom and modern scientific progress offers a potential avenue for innovative drug discovery and the creation of safe and efficacious therapeutic agents. In light of the challenges posed by factors such as habitat

A. Balkrishna
Patanjali Herbal Research Department, Patanjali Research Foundation, Haridwar, Uttarakhand, India

University of Patanjali, Haridwar, Uttarakhand, India
e-mail: acharya.balkrishnapri@patanjali.res.in

P. Yadav · P. Yadav · A. Saini · B. Kumar · N. Singh · D. Sharma (✉)
Patanjali Herbal Research Department, Patanjali Research Foundation, Haridwar, Uttarakhand, India
e-mail: priya.yadav1@patanjali.res.in; priyanka.yadav@patanjali.res.in; aakanksha.saini@patanjali.res.in; brijesh.kumar@patanjali.res.in; netrapal.singh@patanjali.res.in; dushyant.sharma@patanjali.res.in; phrd.publications@patanjali.res.in

V. P. Arya
Patanjali Herbal Research Department, Patanjali Research Foundation, Haridwar, Uttarakhand, India

University of Patanjali, Haridwar, Uttarakhand, India

Centre of Excellence, Patanjali Ayurved Hospital, Haridwar, Uttarakhand, India
e-mail: vedpriya.arya@patanjali.res.in

© The Author(s), under exclusive license to Springer Nature Singapore Pte Ltd. 2024

M. Nandave et al. (eds.), *Ethnopharmacology and OMICS Advances in Medicinal Plants Volume 1*, https://doi.org/10.1007/978-981-97-2367-6_1

degradation, overharvesting, and shifting climates, the chapter addresses preserving medicinal plant diversity. It delves into the synergy between conservation endeavours and the accessibility of medicinal plants, accentuating the significance of establishing sustainable botanical reserves. However, realising these potential mandates a comprehensive approach that precedes conservation tactics and the sustainable exploitation of the Earth's botanical heritage for medicinal purposes.

Keywords
Biodiversity · Medicinal plants · Healthcare · Diversity · Conservation

Abbreviations

IPCC	Intergovernmental Panel on Climate Change
IUCN	International Union for Conservation of Nature
MAPs	Medicinal and aromatic plants
MDR	Multi-drug resistance
NABARD	National Bank for Agriculture and Rural Development

1.1 Biodiversity of Medicinal Plants

1.1.1 Introduction

The term "biodiversity" emerged approximately in 1985, but its underlying conceptual and political roots were established in the preceding decades. The connection between the diversity of life forms and human welfare is evident in the period before the term's inception. This era, often called the "prehistory" of biodiversity, witnessed a considerable emphasis on individual species' significance to human society in discussions about the ongoing species extinction crisis. This emphasis was in addition to recognising the inherent value of these species (Farnham 2007; Mazur and Lee 1993). Biodiversity encompasses a wide-ranging concept that pertains to the diverse variations found in the natural world, containing differences in both frequency and quantity. Different scientists have defined biodiversity in other ways with similar meanings. It involves the array of plants, animals, and microorganisms, the genetic information they carry, and the intricate ecosystems they establish. This tapestry of biodiversity has evolved over billions of years through natural processes, but human activities have increasingly influenced it (Silvestro et al. 2022). Scientific exploration places significant emphasis on biodiversity. It encapsulates the richness of all organisms inhabiting Earth, spanning all organisational levels, which includes life forms from terrestrial, aquatic, and freshwater realms, spanning every corner of the globe. It extends to have all living entities, not limited to visible or consumable ones—encompassing bacteria, viruses, plants, fungi, invertebrates, and vertebrates alike (Morton and Hill 2014).

Further, as a facet of an environment, biodiversity explicitly denotes the variety inherent within and among organisms, living entities, biotic communities, and their interactions, which will apply to both natural occurrences and human-modified environments. To gauge biodiversity, metrics such as genetic diversity, species identification and abundance, species combinations, biological communities, biotic processes, and their attributes (such as abundance, biomass, coverage, rate, and structure) can all be employed (DeLong 1996). This measurement can be carried out across various scales, ranging from microhabitats to the entire biosphere. While ecological processes are often highlighted for their role in preserving biodiversity, including these processes in the concept of biodiversity has yet to be universally agreed upon. The term "biodiversity" emerged for convenience, and, as such, its definition is inherently challenging. However, its inception has spurred research, primarily in ecology and systematics, with scientists compelled to contribute due to the recognition that biodiversity is beneficial to humans and vital for ecosystem functioning. Establishing a workable definition remains necessary, driven by the interconnectedness of these issues with biodiversity conservation (Fisher et al. 1943).

1.1.2 Classification of the Biodiversity

This diversity can be classified into three primary types: (1) species diversity, (2) genetic diversity, and (3) ecosystem diversity. Species diversity refers to the variety of species coexisting within a particular habitat or region, highlighting the richness and variety of life forms. Genetic diversity delves into the genetic variation present within individual species, allowing them to adapt to changing environments and ensuring the resilience of populations. Ecosystem diversity captures the range of distinct habitats and ecosystems across the Earth, each with its unique assemblage of species and ecological processes. Together, these types of biodiversity foster the stability of ecosystems, provide resources for human well-being, and contribute to the overall health and balance of our planet's biosphere.

1.1.2.1 Genetic Diversity

Genetic diversity within a species pertains to the variation among its genetic components (genes) inherited from generation to generation. Diversity, at its core, stems from genetic variation, with the level of gene diversity influencing speciation. Genetic variety is crucial for natural selection, enabling populations to adapt to changing environments (Verma 2016). While environmental differences often enhance diversity within a species, not all animal groups exhibit uniform genetic diversity. Preserving genetic diversity necessitates safeguarding different populations of a species (Teixeira and Huber 2021), which encompasses the genetic information found in various populations of plants, animals, and microorganisms. Genetic variation arises from gene mutations and recombination, introducing novel genetic information. Genetic diversity contributes to evolutionary change and artificial selective breeding, influencing the form and function of organisms. Some genes governing fundamental processes remain conserved, while others exhibit considerable variation, as seen in the mammalian immune system (Fabian et al. 2021). Studying genetic diversity involves comparing individuals or populations and assessing variation within a population—a pivotal aspect of biodiversity. Genetic variability shapes individuals, populations, species, and the Earth's life kingdoms. Genotypes reflect genetic variety among individuals and populations (Pauls et al. 2013). Genetic variability, distinct from diversity, defines the propensity for genetic traits to fluctuate among populations (Laikre et al. 2010; Uniyal et al. 2006).

1.1.2.2 Species Diversity

While, historically, species have served as the fundamental descriptive units in the realm of living organisms, the term "biodiversity" is often and inaccurately employed interchangeably with "species diversity", specifically denoting "species richness", which signifies the count of species in a given location (Chiarucci et al. 2011). The customary approach to discussing global biodiversity involves enumerating the total species count across various taxonomic groups (Heydari et al. 2020). The projection estimates approximately 8.7 million eukaryotic species worldwide, with a standard error of ±1.3 million, of which roughly 2.2 million are marine species, with a standard error of ±0.18 million. Even though taxonomic classification has been ongoing for 250 years and more than 1.2 million species are presently documented in a central database, previous findings indicated that around 86% of Earth's existing species and 91% of oceanic species remain undiscovered or undescribed (Mora et al. 2011). Species diversity characterises the variety of living organisms, quantified through species abundance and richness measures. The count of species within a specific area is termed its species richness, while the proportional representation of each species constitutes species abundance. Equal abundance across all species yields significant diversity due to low variance.

Conversely, diversity diminishes if one species account for 96 individuals while others each have one. Species in a population do not inherently differ to the same degree. Categorising species based on their ecological roles adds further dimension (Rawat and Agarwal 2015). When considering species alone, it becomes evident that insects and microbes dominate Earth's biota. Although species play a central role in evolutionary mechanisms and the overall diversity hinges on species birth and extinction, accurately discerning and tallying all species proves challenging for systematists due to differing perspectives on what defines a species among different organism groups. This divergence contributes to the insufficiency of relying solely on species diversity to determine biodiversity. The ecological significance of a species within a community structure can directly influence overall biological diversity. For example, a tropical rainforest tree supporting a diverse native invertebrate fauna contributes more to global biological diversity conservation than a European alpine species (Swingland 2001; Hillebrand et al. 2018).

1.1.2.3 Ecosystem Diversity

An ecosystem represents a community of living organisms interacting among themselves and with non-living elements of their environment. An ecosystem consists of a group of species engaging with their physical surroundings. These ecosystems, ranging from vast entities like the Great Barrier Reef to minute spaces like the back of a spider crab's shell, serve as habitats for various creatures such as plants, sponges, algae, and worms. Ecosystem diversity, in turn, pertains to the assortment of habitats where organisms naturally exist, coupled with the diversity of life forms within these habitats. At the community and ecosystem levels, three layers of diversity are recognised. First is alpha diversity, which focuses on diversity within a community, followed by beta diversity (Gaston 2000), often linked to warm climates and high primary productivity (Field et al. 2009). Notably, marine biodiversity tends to peak along coasts in the Western Pacific due to elevated sea surface temperatures, and a latitudinal gradient in species diversity exists (Tittensor et al. 2010). Biodiversity typically clusters in hot spots (Myers et al. 2000) and has shown an increasing trend over time (McPeek and Brown 2007), although this growth is expected to slow down (Rabosky 2009). Due to their incorporation of abiotic components and influences from factors like soil and climate, ecosystems are distinct from genes and species. This challenge is overcome by gauging the diversity of individual species constituting the ecosystem, considering parameters such as relative abundance and quantity across various size categories, trophic levels, or taxonomic groups.

Consequently, an ecosystem with numerous plant species might be less diverse than one featuring a range of species, predators, and herbivores. The measurement of ecosystem diversity needs a singular standardised metric, as varying aspects might carry differing weights in different assessments, affecting the conservation ranking of other areas. When examining beta diversity, which concerns the variation in species composition between sites, the most reliable approach is to directly compare species composition between sites (Swingland 2001).

1.1.3 History of Medicinal Plant Usage

The utilisation of medicinal plants for healing dates to the earliest times of human existence. The link between humans and their exploration of natural sources for remedies traces its origins to ancient history, supported by abundant proof from diverse sources such as written records, well-preserved landmarks, and even authentic botanical treatments (Jamshidi-Kia et al. 2017). The most ancient, recorded evidence of the utilisation of medicinal plants in creating remedies has been uncovered on a Sumerian clay tablet discovered in Nagpur, dating back around 5000 years. This tablet contained 12 formulations for drug production, mentioning more than 250 different plants, including certain alkaloid-rich ones like poppy, henbane, and mandrake (Petrovska 2012). Around 2500 BC, Emperor Shen Nung authored the Chinese compendium *Pen T'Sao*, focusing on roots and grasses. This ancient book confers 365 drugs derived from dried medicinal plant parts.

Interestingly, several of these substances continue to be utilised today. Examples include *Rhei rhizoma*, camphor, *Theae folium*, *Podophyllum*, the sizeable yellow gentian, ginseng, jimson weed, cinnamon bark, and ephedra (Wiart 2006). The Vedas, sacred texts of India, also reference using plants in healing, including many spice plants that continue to be widely used today, like nutmeg, pepper, and clove, which have their origins in India (Tucakov 1971). Dating back to around 1550 BC, the Ebers Papyrus is a compilation of 800 prescriptions involving 700 plant species and medicinal substances employed for treatments. These therapies include the use of plants like pomegranate, castor oil plant, aloe, Senna, garlic, onion, fig, willow, coriander, juniper, and common centaury (Glesinger 1954; Tucakov 1964). Some essential medicinal systems, including Ayurveda, Siddha, Unani, Chinese, Persian, Japanese, Korean, Aboriginal, African, etc., have also been reported in the literature and the uses of plants (Yuan et al. 2016). These studies have revealed that individuals have endeavoured to discover remedies for easing pain and treating various ailments throughout the ages. Across every era and successive century in the evolution of humanity and advanced civilisations, the therapeutic qualities of specific medicinal plants have been recognised, documented, and transmitted to succeeding generations. The advantages of one community have been handed down to another, leading to the enhancement of established properties and the revelation of novel ones, persisting until the present era. The unceasing and everlasting fascination of people with medicinal plants has led to the contemporary and refined methods of their preparation and utilisation that we witness today.

1.1.4 Relevance of Medicinal Plants with Modern Medicines

The World Health Organization has determined that 80% of individuals in developing nations rely on botanical remedies as their primary source of healthcare (World Health Organization 2011). A survey conducted by the World Health Organization revealed that traditional medical practitioners attend to roughly 8% of patients in India, 85% in Burma, and 90% in Bangladesh. Even though India accounts for just 2.4% of the world's total land area, it possesses about 8% of the planet's biodiversity, encompassing around 49,000 plant species, with 4900 being unique to the region. Despite approximately 265,000 species of seed plants globally, less than 50% of them have comprehensively explored their chemical composition and medicinal benefits (Bhat 2021). Over time, as medical advancements have progressed, synthetic medications have increasingly supplanted natural remedies, even though they may carry specific adverse effects. While numerous synthetic drugs effectively contribute to the well-being of humanity and are widely utilised, there remain a significant number of medications associated with severe side effects (Nisar et al. 2018). Roughly 8% of hospital admissions in the United States of America occur as a result of unfavourable or unintended reactions caused by synthetic medications (George 2011). Each year, over 100,000 people are killed as a result of these toxicities. It indicated that the number of persons murdered in the United States by pharmaceutical medications is at least three times higher than the number of people killed by drunk drivers (Karimi et al. 2015).

Further, over the past 30 years, antimicrobial resistance has given rise to multidrug resistance (MDR), which is a direct result of the frequent and consistent utilisation of single drugs targeting the same therapeutic purpose. As a result, it serves as a cautionary message for pharmaceutical researchers, urging them to develop innovative strategies against MDR. Moreover, it presents an opening to explore a fresh range of antibiotics to combat antimicrobial resistance (Anand et al. 2019). In this regard, plants might be a potent source as they produce and utilise diverse metabolic substances to discourage pathogens and predators. Thus, exploring these metabolites from plants holds excellent potential for discovering innovative bioactive compounds, which could be leveraged to create potent antimicrobial medications to tackle multi-drug-resistant pathogens (Kar et al. 2022). In a recent review, 15 plants with their used parts (*Allexis cauliflora* (leaves), *Persea americana* (Stones), *Entada africana* (bark), *Pentaclethra macrophylla* (bark), *Entada abyssinica* (leaves and roots), *Alkanna tinctoria* (leaves), *Artemisia absinthium* (bark), *Martynia annua* (bulk), *Adiantum capillus-veneris* (bark), *Zanthoxylum armatum* (bark), *Swertia chirata* (bark)) have been found to have potential to kill resistant bacterial strains (*Escherichia coli*, *Klebsiella* spp., *Pseudomonas aeruginosa*, *Proteus mirabilis*, *Staphylococcus aureus*, *S. epidermidis*, *Acinetobacter baumannii*, *K. pneumoniae*, *Providencia stuartii*, *Enterococcus faecium*) (Keita et al. 2022). The advantages draw scientists towards exploring novel compounds sourced from plants, characterised by minimal side effects and potent disease-fighting capabilities. In addition, the ongoing expansion in the utilisation of medicinal plants is leading to a significant reduction in both the diversity and quantity of these valuable botanical resources. Hence, there is an urgent need to investigate the present circumstances and focus on conserving and safeguarding medicinal plants.

1.1.5 Medicinal Plant Diversity

The world stands at a critical juncture, prompted by the imminent threat stemming from a steep increase in the global population, swift and occasionally unregulated industrial growth, as well as unchecked deforestation, the excessive use of natural resources, pollution, and the overarching spectre of worldwide climate change. These factors underscore the urgency for rapid and resolute action to protect the Earth's biodiversity and ensure the sustainable advancement of human civilisation on a global scale. Till now, approximately 350,000 plant species (WCVP 2020) have been recorded worldwide, out of which 35,407 have been found to have medicinal values (MPNS 2023). As outlined in the 2019 report on the worldwide utilisation of herbal remedies, encompassing herbal pharmaceuticals, dietary supplements, and functional foods, the demand for these products is experiencing remarkable growth. This trend is marked by an annual expansion rate of approximately 6%, signifying a substantial increase in global sales. By 2023, the cumulative revenue is anticipated to soar to an impressive USD 130 billion. Within this dynamic landscape, the segment that contributes significantly to this surge is herbal pharmaceuticals, which recorded an impressive USD 51 billion in sales during the year 2017, which underscores the prominence of herbal-based medical treatments in contemporary healthcare practices and consumer preferences. This surge in demand and revenue demonstrates the enduring popularity of herbal medicines and their acceptance in diverse health-related industries worldwide (Howes et al. 2020). The above factors contribute to the unanticipated harvesting of medicinal plants from the wild. Additionally, there is a lack of awareness among people about cultivating and preserving these valuable plants with medicinal properties.

As of the latest assessment conducted by the International Union for Conservation of Nature (IUCN) regarding threatened species, a substantial number of plant species, approximately 24,914 to be precise, have been meticulously classified as facing various degrees of endangerment. Among these species, 5336 are presently categorised under the critically endangered classification, signifying an exceptionally high risk of extinction if urgent conservation actions are not undertaken. Furthermore, the IUCN assessment reveals that 10,202 plant species fall within the endangered category, implying a significant vulnerability to extinction soon unless effective preservation measures are executed promptly. Additionally, a noteworthy figure of 9376 plant species is classified as vulnerable, highlighting their susceptibility to threats that could push them towards the endangered or critically endangered status if protective efforts are not implemented (IUCN 2022).

1.1.6 Diversity and Conservation Status of Medicinal Plants in India

India has been regarded as a storehouse or botanical sanctuary of plant genetic variety due to its diverse agro-climatic conditions and regional landscapes. It is one of the world's 12 most biodiverse countries, harbouring 2 unique biodiversity hot spots out of 18 hot spots in the world. India possesses approximately 8% of the global biodiversity, encompassing around 49,000 plant species, of which 4900 are native to the region. In India, medicinal plants thrive across 15 distinct agro-climatic zones, containing an impressive array of 17,000–18,000 flowering plants. Approximately 6000–7000 have been documented for their therapeutic applications within various medical systems such as Ayurveda, Siddha, Unani, and homoeopathy. This compilation constitutes half of India's higher flowering plant species (Subrat 2002; Dhar et al. 2000). Approximately 70% of India's medicinal plants are found in the tropical forests of the Western Ghats. According to an estimate, 95% of the medicinal plants utilised in the Indian herbal industry are sourced from the wild. Despite these medicinal plant species used by various Indian communities in different ecosystems, only about 10% of these species are actively traded (Kala et al. 2006; Sharma and Kumari 2021). The Himalayan and Northeastern regions are well renowned for their high medicinal and aromatic plant biodiversity, which includes 2500 (213%) medicinal species (Hegde 2003; Maiti 2004; Kala et al. 2006; GoI 2011; Uniyal et al. 2006).

More than 90% of medicinal plants found in India face a critically endangered status due to practices marked by unsustainable gathering, utilisation, overexploitation, or unskilled harvesting methods (Kumari et al. 2011). Additionally, data need to be present concerning the utilisation of medicinal plants within non-codified and non-commercial indigenous healthcare traditions, which heavily rely on plant species specific to particular regions (Pareek et al. 2005; Sharma and Pandey 2013). These traditions are adhered to by approximately 1 million folk healers and 140 million rural households. Notably, out of the 242 species in high commercial demand (>100 MT per year), a significant 72% are primarily sourced from wild plants, encompassing elements like roots, bark, wood stem, or the entire plant. This scenario sparks apprehensions regarding the preservation of these species. The conjunction of unsustainable harvesting practices, insubstantial replenishment endeavours, extended plant life cycles, climate shifts, and development initiatives have collectively contributed to a swift decline in the accessibility of medicinal plants across India in the preceding three decades. This deterioration has led to the vulnerability of many species at both national and global levels. In response, the Government of India (GoI) has put forth a range of regulatory structures and policies to address this issue (Gowthami et al. 2021).

The Indian Government, in collaboration with the National Bank for Agriculture and Rural Development (NABARD), has introduced a range of policies to promote the cultivation of medicinal and aromatic plants (MAPs) among the population. These initiatives are designed to incentivise individuals by providing them with loans and financial support. The primary objective is to encourage increased cultivation of these valuable plant species, which have significant medicinal and aromatic properties, thus contributing to rural development and the sustainable utilisation of natural resources. Through these concerted efforts, individuals are empowered to cultivate MAP, leading to potential economic growth and enhanced conservation of plant biodiversity (Gulati 2018).

In addition to the attempts above, scientific procedures like in situ and ex situ conservation and social initiatives can also be used to conserve wild medicinal plants or other vulnerable species.

1.1.7 Factors Influencing Biodiversity of Medicinal Plants

The biodiversity of medicinal plants is influenced by a multitude of interconnected factors like variations in temperature, the intensity of light, increase in the level of carbon dioxide, UV radiations, ozone depletion, salinity, flood, drought, etc. that shape the distribution, abundance, and diversity of these valuable plant species.

1.1.7.1 Climatic Factors

Environmental conditions are crucial since each plant species needs unique ecological conditions for proper growth. According to the Intergovernmental Panel on Climate Change (IPCC), mean annual temperature rises at a pace of 0.06–0.1 °C every year, with a CO_2 rise of 407.4 ppm. Global warming is anticipated to alter vital land ecosystems by shifting habitat ranges and distributions. This shift dramatically impacts the spatial arrangement of plants and could heighten the risk of extinction for endangered flora. Recent research has concentrated on the potential consequences of climate change on vegetation and protected areas, causing significant worries about the fate of various animal and plant species (Guo et al. 2016).

The climatic changes affect not only the usual behaviour of the plants but also their physiology, which in turn affects secondary metabolites and may lead to the death of the plant (Pachauri et al. 2014). In a study conducted in the Himalayan region focused on the impact of climate change on medicinal plants, notable findings included the upward displacement of vegetation by 200–500 m, earlier flowering, prolonged seed dormancy, diminished seed germination rates, the expansion of invasive species to sub-temperate Himalayan zones, shifts from upright to prostrate plant growth, migration of alpine species to cold desert conditions, and the concerning trends of reduced fertility, heightened mortality rates due to climate change, and instances of habitual abortion among Himalayan musk deer (Ratha et al. 2012). In 2021, a list of 20 Indonesian medicinal plant species that were uncommon or endemic and vulnerable to extinction due to climate change was provided (Cahyaningsih et al. 2021).

Additionally, it has been noted that climatic factors can induce changes at the genetic level. An examination in Mexico on *Brickellia veronicifolia* collected from different locations and climates unveiled significant genetic diversity across 360 distinct loci. This investigation showcased a broad spectrum of patterns among the six studied groups. The count of obtained bands varied from 12 to 31, contingent on the inter-primer Binding Site primer employed, with molecular weights spanning from 250 to 5000 bp (Pacheco-Hernández et al. 2021). Furthermore, the impact of climate change was evident in three key plants essential to the local community in Morocco, *Thymus vulgaris*, *Mentha pulégium*, and *Rosmarinus officinalis*, and an increase in secondary metabolites occurred with a 50% rise in temperature and water pressure (Laftouhi et al. 2023). Thus, the preceding research clearly shows that climate change can substantially influence the variety and diversity of medicinal plants, and there is an urgent need for their conservation.

1.1.7.2 Geographical Distribution

The correlation between geographical distribution and medicinal plants is a significant aspect of biodiversity and traditional medicine. The natural habitats and ecosystems where medicinal plants grow are crucial in determining their availability, potency, and composition of bioactive compounds. For example, plants adaptable to arid climates might be more common in desert areas, while those requiring cooler temperatures might be concentrated in higher altitudes. Studying the elevation-based patterns of medicinal plant distribution and their various uses can aid in identifying resource-rich areas within the region. This knowledge can then be applied to enhance the protection of these valuable resources (Rokaya et al. 2012). Local understanding of these geographical distributions has been passed down through generations and is a crucial component of traditional medicine in many cultures (Kunwar et al. 2022). Indigenous communities often possess intricate knowledge about where to find specific plants for treating various ailments based on their understanding of the plants' habitats.

However, the impact of environmental changes due to factors like climate change, deforestation, and urbanisation can disrupt the geographical distribution of medicinal plants. This disruption can lead to decreased availability of specific plants in their natural habitats, potentially affecting traditional medicine practices and the availability of raw materials for pharmaceutical industries. In a study conducted at different altitudes, the peak species diversity for vascular plants occurred between 1500 and 2500 m in altitude, while it was observed at 1000 m for medicinal vascular plants (Rokaya et al. 2012; Gryntes and Vetaas 2002). Studying the correlation between geographical distribution and medicinal plants helps researchers, conservationists, and policymakers understand the vulnerability of these plants to changing environments. It also underscores the importance of sustainable harvesting practices and habitat preservation to ensure the continued availability of valuable medicinal resources (Tshabalala et al. 2022).

1.1.7.3 Soil and Microbiota

The various types of soil present significantly influence plant species diversity within a given ecosystem. Soil types play a pivotal role in shaping the composition and abundance of plant species due to their distinct physical, chemical, and nutrient characteristics. Different plant species have evolved to thrive in specific soil conditions, resulting in a dynamic interplay between soil and vegetation (ur Rahman et al. 2021). For instance, well-drained soils with good aeration often support a broader range of plant species compared to poorly drained soils, which might restrict the types of plants that can flourish. As an example, a strong correlation between soil types, texture, and the growth as well as the maturation of medicinal plants has been observed and highlighted that loam soil stood out as a relatively optimal soil type for cultivating medicinal plants of the root/stem variety (Liu et al. 2007).

Additionally, variations in soil pH, texture, and nutrient availability directly impact the establishment and growth of different plant species. Toxic substances in soil are assimilated and accumulated in medicinal plants, eventually entering the food chain that jeopardise medical security and presents potential hazards to human well-being (Ara et al. 2021; Fu et al. 2020).

Besides this, the correlation between soil microbes and medicinal plants is a fascinating and intricate relationship that underscores the complex balance within ecosystems. Soil microbes, including bacteria, fungi, and archaea, play a pivotal role in maintaining the health and vitality of medicinal plants (Kushwaha et al. 2020). These microorganisms contribute to nutrient cycling, aiding in the decomposition of organic matter and releasing essential nutrients crucial for medicinal plants' growth and development. Moreover, certain soil microbes engage in symbiotic relationships with these plants, forming mycorrhizal associations that enhance nutrient uptake and water absorption, thereby bolstering the overall resilience of medicinal plants (Chen et al. 2023). Beyond nutrient provision, soil microbes also contribute to the synthesis of bioactive compounds within medicinal plants (Wang et al. 2022). Some microbes participate in intricate biochemical processes that trigger the production of secondary metabolites, such as alkaloids, flavonoids, and terpenoids, which often possess therapeutic properties (Haruna and Yahaya 2021). These compounds are vital for the plant's survival and hold immense value for human health, forming the basis of various traditional and modern medicines. As such, the collaboration between soil microbes and medicinal plants highlights the profound interdependence between the microbial world and the plant kingdom, offering insights into sustainable cultivation practices and potential avenues for discovering novel therapeutic compounds. Hence, understanding the intricate relationship between soil types, soil microbes, and plant species diversity is essential for effective ecosystem management and conservation strategies, allowing for preserving a rich array of plant life.

1.1.7.4 Human Activities, Habitat Destruction, and Biodiversity Loss

Human actions have notably contributed to the destruction of habitats and the loss of biodiversity, resulting in a troubling reduction in the variety of life forms on our planet (Singh et al. 2021). Activities such as urbanisation, deforestation, industrialisation, and agriculture have changed natural habitats, often leading to the disappearance and extinction of many species. This alteration and fragmentation of ecosystems create challenges for species in finding appropriate places to live and resources to sustain themselves, ultimately leading to a decline in biodiversity (Upadhyay 2020; Borges et al. 2020).

The ongoing deterioration of forests and modifications in the characteristics of plants of medicinal significance is a significant issue (Gilliam 2007; Wyatt and Silman 2010). Within the Indian subcontinent, approximately 350 million people residing in rural areas, along with 84% of ethnic groups in or near forested regions (Mehta and Shah 2003), continue to rely on the resources found within forests for their sustenance. These communities traditionally adhere to indigenous medical methods and depend significantly on medicinal plants that flourish near these forests. The vanishing of medicinal plant species can also disrupt ecosystems, causing a ripple effect on various other organisms (Martin et al. 2020). Further, the disruption frequently results in the disappearance of distinct microenvironments and ecological niches crucial for the growth of different medicinal plant species (Singh et al. 2022). These plants often share intricate relationships with pollinators, herbivores, and other species, and their absence can disturb these vital interactions, causing imbalances in ecosystems.

Furthermore, the extinction of specific medicinal plants could signify the permanent loss of potential sources for innovative medical discoveries (Miththapala 2006). Tackling these challenges requires a focused approach to conservation. The safeguarding of medicinal plant habitats can be facilitated by establishing protected areas, adopting sustainable land-use practices, and initiatives for habitat restoration. Collaborations between local communities, scientists, and policymakers are essential for developing strategies that ensure these valuable plants' sustainable harvesting, cultivation, and propagation (Hamilton 2004). Recognising the interconnection between preserving habitats, conserving biodiversity, and maintaining human health underscores the necessity for a comprehensive strategy to secure medicinal plant availability for generations to follow.

1.2 Conclusion and the Way Forwards

Biodiversity stands as a foundational element contributing to the health and resilience of our planet, assuming a pivotal role in upholding ecosystem equilibrium and furnishing vital services that bolster human well-being. Across cultures and civilisations, medicinal plants have been an inseparable component of human health and welfare for ages, providing a natural reservoir of curative compounds. Regrettably, the assortment of medicinal plants is currently imperilled by the consequences of habitat degradation, excessive extraction, and shifts in climate patterns. As species dwindle and age-old wisdom fades, the potential loss of reservoirs for future medical innovations and the cultural heritage enshrined in these plants becomes palpable. In response, endeavours aimed at conserving the diversity of medicinal plants entail ventures to shield habitats, advocate sustainable collection techniques, and chronicle traditional insights. Furthermore, there is a burgeoning enthusiasm for scientific exploration that corroborates the effectiveness of conventional applications and unearths novel compounds for medical purposes.

References

Anand U, Jacobo-Herrera N, Altemimi A, Lakhssassi N (2019) A comprehensive review on medicinal plants as antimicrobial therapeutics: potential avenues of biocompatible drug discovery. Metabolites 9(11):258

Ara T, Nisa WU, Aziz R, Rafiq MT, Gill RA, Hayat MT, Afridi U (2021) Health risk assessment of hexachlorocyclohexane in soil, water and plants in the agricultural area of Potohar region, Punjab, Pakistan. Environ Geochem Health 43:1–17

Bhat SG (2021) Medicinal plants and its pharmacological values. In: Natural medicinal plants. IntechOpen, London

Borges PA, Gabriel R, Fattorini S (2020) Biodiversity erosion: causes and consequences. In: Life on land. Springer International Publishing, Cham, pp 81–90

Cahyaningsih R, Phillips J, Brehm JM, Gaisberger H, Maxted N (2021) Climate change impact on medicinal plants in Indonesia. Glob Ecol Conserv 30:e01752

Chen W, Modi D, Picot A (2023) Soil and phytomicrobiome for plant disease suppression and management under climate change: a review. Plants 12(14):2736

Chiarucci A, Bacaro G, Scheiner SM (2011) Old and new challenges in using species diversity for assessing biodiversity. Philos Trans R Soc B Biol Sci 366(1576):2426–2437

DeLong DC (1996) Defining biodiversity. Wildl Soc Bull 24:738–749

Dhar U, Rawal RS, Upreti J (2000) Setting priorities for conservation of medicinal plants—a case study in the Indian Himalaya. Biol Conserv 95(1):57–65

Fabian DK, Fuentealba M, Dönertaş HM, Partridge L, Thornton JM (2021) Functional conservation in genes and pathways linking ageing and immunity. Immun Ageing 18(1):1–18

Farnham TJ (2007) Saving nature's legacy: origins of the idea of biological diversity. Yale University Press, New Haven

Field R, Hawkins BA, Cornell HV, Currie DJ, Diniz-Filho J, Alexandre F, Guégan J-F, Kaufman DM, Kerr JT, Mittelbach GG, Oberdorff T, O'Brien EM, Turner JRG (2009) Spatial species-richness gradients across scales: a meta-analysis. J Biogeogr 36(1):132–147. https://doi.org/10.1111/j.1365-2699.2008.01963.x

Fisher RA, Corbet AS, Williams CB (1943) The relation between the number of species and the number of individuals in a random sample of an animal population. J Anim Ecol 12:42–58

Fu Y, Dou X, Lu Q, Qin J, Luo J, Yang M (2020) Comprehensive assessment for the residual characteristics and degradation kinetics of pesticides in Panax notoginseng and planting soil. Sci Total Environ 714:136718

Gaston KJ (2000) Global patterns in biodiversity. Nature 405(6783):220–227. https://doi.org/10.1038/35012228. PMID 10821282

George P (2011) Concerns regarding the safety and toxicity of medicinal plants—an overview. J Appl Pharm Sci 1(6):40–44

Gilliam FS (2007) The ecological significance of the herbaceous layer in temperate forest ecosystems. Bioscience 57(10):845–858

Glesinger L (1954) Medicine through centuries. Zora, Zagreb, pp 21–38

GoI (2011) Report of steering committee on AYUSH for 12th five year plan-ST 2 AYUSH: report of the steering committee on AYUSH for 12th five year plan (2012–17). Government of India, New Delhi

Gowthami R, Sharma N, Pandey R, Agrawal A (2021) Status and consolidated list of threatened medicinal plants of India. Genet Resour Crop Evol 68(6):2235–2263

Gryntes JA, Vetaas OR (2002) Species richness and altitude a comparison between stimulation models and interpolated plant species richness along the Himalayan gradient Nepal. Am Nat 159:294–304

Gulati R (2018) Medicinal and aromatic plants biodiversity of India—a review. Int Arch Appl Sci Technol 9(4):130–137

Guo Y, Wei H, Lu C, Gao B, Gu W (2016) Predictions of potential geographical distribution and quality of Schisandra sphenanthera under climate change. PeerJ 4:e2554

Hamilton AC (2004) Medicinal plants, conservation and livelihoods. Biodivers Conserv 13:1477–1517

Haruna A, Yahaya SM (2021) Recent advances in the chemistry of bioactive compounds from plants and soil microbes: a review. Chem Afr 4(2):231–248

Hegde GR (2003) Medicinal plants in India: status and prospects. In: Hiremath KG (ed) Recent advances in environmental sciences. Discovery Publishing House, New Delhi, pp 36–44

Heydari M, Omidipour R, Greenlee J (2020) Biodiversity, a review of the concept, measurement, opportunities, and challenges. J Wildlife Biodivers 4(4):26–39

Hillebrand H, Blasius B, Borer ET, Chase JM, Downing JA, Eriksson BK et al (2018) Biodiversity change is uncoupled from species richness trends: consequences for conservation and monitoring. J Appl Ecol 55(1):169–184

Howes MJR, Quave CL, Collemare J, Tatsis EC, Twilley D, Lulekal E et al (2020) Molecules from nature: reconciling biodiversity conservation and global healthcare imperatives for sustainable use of medicinal plants and fungi. Plants People Planet 2(5):463–481

IUCN (2022) The IUCN red list of threatened species. Version 2022-2. https://www.iucnredlist.org. Accessed 28 Aug 2023

Jamshidi-Kia F, Lorigooini Z, Amini-Khoei H (2017) Medicinal plants: past history and future perspective. J Herbmed Pharmacol 7(1):1–7

Kala CP, Dhyani PP, Sajwan BS (2006) Developing the medicinal plants sector in northern India: challenges and opportunities. J Ethnobiol Ethnomed 2:1–15

Kar A, Banerjee R, Devi I, Mukherjee PK, Rahman M (2022) Medicinal plants in antimicrobial resistance—perspectives towards drug development. In: Evidence-based validation of herbal medicine. Elsevier, Amsterdam, pp 299–335

Karimi A, Majlesi M, Rafieian-Kopaei M (2015) Herbal versus synthetic drugs; beliefs and facts. J Nephropharmacol 4(1):27

Keita K, Darkoh C, Okafor F (2022) Secondary plant metabolites as potent drug candidates against antimicrobial-resistant pathogens. SN Appl Sci 4(8):209

Kumari P, Joshi GC, Tewari LM (2011) Diversity and status of ethno-medicinal plants of Almora district in Uttarakhand, India. Int J Biodivers Conserv 3(7):298–326

Kunwar RM, Baral B, Luintel S, Uprety Y, Poudel RC, Adhikari B et al (2022) Ethnomedicinal landscape: distribution of used medicinal plant species in Nepal. J Ethnobiol Ethnomed 18(1):34

Kushwaha RK, Rodrigues V, Kumar V, Patel H, Raina M, Kumar D (2020) Soil microbes-medicinal plants interactions: ecological diversity and future prospect. In: Plant microbe symbiosis. Springer, Berlin, pp 263–286

Laftouhi A, Eloutassi N, Ech-Chihbi E, Rais Z, Abdellaoui A, Taleb A et al (2023) The impact of environmental stress on the secondary metabolites and the chemical compositions of the essential oils from some medicinal plants used as food supplements. Sustainability 15(10):7842

Laikre L, Schwartz MK, Waples RS Ryman N (2010) Compromising genetic diversity in the wild: unmonitored large-scale release of plants and animals. Trends Ecol Evol 25(9):520–529.

Liu Y, Zhang ZS, He YL, Zhang BG, Li XE (2007) Quality of crude traditional Chinese drugs and ecological environment. In: Modernisation of traditional Chinese medicine and materia materia-world science and technology, vol 9. Elsevier, Amsterdam, pp 65–69

Maiti S (2004) Inventory, documentation and status of medicinal plants research in India. In: Batugal PA, Kanniah J, Young LS, Oliver JT (eds) Medicinal plants research in Asia, volume 1: the framework and project workplans. International Plant Genetic Resource Institute-Regional Office for Asia, the Pacific and Oceania, Serdang

Martin C, Doyle J, LaFrance J, Lefthand MJ, Young SL, Three Irons E, Eggers MJ (2020) Change rippling through our waters and culture. J Contemp Water Res Educ 169(1):61–78

Mazur A, Lee J (1993) Sounding the global alarm: environmental issues in the US National News. Soc Stud Sci 23(4):681–720. https://doi.org/10.1177/030631293023004003

McPeek MA, Brown JM (2007) Clade age and not diversification rate explains species richness among animal taxa. Am Nat 169(4):E97–E106. https://doi.org/10.1086/512135. PMID 17427118

Mehta AK, Shah A (2003) Chronic poverty in India: incidence, causes and policies. World Dev 31(3):491–511

Miththapala S (2006) Conserving medicinal species: securing a healthy future. International Union for Conservation of Nature and Natural Resources, Gland

Mora C, Tittensor DP, Adl S, Simpson AG, Worm B (2011) How many species are there on earth and in the ocean? PLoS Biol 9(8):e1001127

Morton S, Hill R (2014) What is biodiversity, and why is it important. In: Biodiversity: science and solutions for Australia. CSIRO Publishing, Clayton, pp 55–68

MPNS Version 12 (2023) Medicinal plant names services. Royal Botanic Gardens, Kew. http://www.kew.org/mpns

Myers N, Mittermeier RA, Mittermeier CG, da Fonseca GAB, Kent J, Mittermeier CG, Da Fonseca GAB, Kent J (2000) Biodiversity hotspots for conservation priorities. Nature 403(6772):853–858. https://doi.org/10.1038/35002501. Bibcode:2000 Natur. 403..853M. PMID 10706275

Nisar B, Sultan A, Rubab SL (2018) Comparison of medicinally important natural products versus synthetic drugs—a short commentary. Nat Prod Chem Res 6(2):308

Pachauri RK, Allen MR, Barros VR, Broome J, Cramer W, Christ R et al (2014) Climate change 2014: synthesis report. In: Contribution of working groups I, II and III to the fifth assessment report of the intergovernmental panel on climate change. IPCC, Geneva, p 151

Pacheco-Hernández Y, Villa-Ruano N, Lozoya-Gloria E, Barrales-Cortés CA, Jiménez-Montejo FE, Cruz-López MDC (2021) Influence of environmental factors on the genetic and chemical diversity of Brickellia veronicifolia populations growing in fragmented shrublands from Mexico. Plants 10(2):325

Pareek SK, Gupta V, Bhatt KC, Negi KS, Sharma N (2005) Plant genetic resources: horticultural crops. In: Medicinal and aromatic plants. Narosa Publishing House Pvt. Ltd, New Delhi, pp 279–289

Pauls SU, Nowak C, Bálint M, Pfenninger M (2013) The impact of global climate change on genetic diversity within populations and species. Mol Ecol 22(4):925–946

Petrovska BB (2012) Historical review of medicinal plants' usage. Pharmacogn Rev 6(11):1

Rabosky DL (2009) Ecological limits and diversification rate: alternative paradigms to explain the variation in species richness among clades and regions. Ecol Lett 12(8):735–743. https://doi.org/10.1111/j.1461-0248.2009.01333.x. PMID 19558515

Ratha KK, Sruajani MS, Arya JC, Joshi GC (2012) Impact of climate change on diversity of Himalayan medicinal plant: a threat to Ayurvedic system of medicine. Int J Res Ayurveda Pharm 3(3):327–331
Rawat US, Agarwal NK (2015) Biodiversity: concept, threats and conservation. Environ Conserv J 16(3):19–28
Rokaya MB, Münzbergová Z, Shrestha MR, Timsina B (2012) Distribution patterns of medicinal plants along an elevational gradient in central Himalaya, Nepal. J Mt Sci 9:201–213
Sharma S, Kumari K (2021) An overview on calendula officinalis Linn.: (pot marigold). J Adv Sci Res 12(03 Suppl 2):13–18
Sharma N, Pandey R (2013) Conservation of medicinal plants in tropics. In: Normah MN, Chin HF, Reed BM (eds) Conservation of tropical plant species. Springer, New York, pp 437–487. https://doi.org/10.1007/978-1-4614-3776-5
Silvestro D, Goria S, Sterner T, Antonelli A (2022) Improving biodiversity protection through artificial intelligence. Nat Sustain 5(5):415–424
Singh V, Shukla S, Singh A (2021) The principal factors responsible for biodiversity loss. Open J Plant Sci 6(1):11–14
Singh PK, Prajapati SK, Sunita K, Chaturvedi RK (2022) Disturbance induced changes in diversity of medicinal plants in a dry tropical forest of India. Front For Glob Change 4:718930
Subrat N (2002) Ayurvedic and herbal products industry: an overview. In: Proceeding of wise practices and experiential learning in the conservation and management of Himalayan medicinal plants, Kathumandu, Nepal
Swingland IR (2001) Definition of biodiversity. In: Encyclopedia of biodiversity, vol 1. Academic, New York, pp 377–391
Teixeira JC, Huber CD (2021) The inflated significance of neutral genetic diversity in conservation genetics. Proc Natl Acad Sci 118(10):e2015096118
Tittensor DP, Mora C, Jetz W, Lotze HK, Ricard D, Berghe E, Vanden Worm B, Jetz W, Lotze HK, Ricard D, Berghe EV, Worm B (2010) Global patterns and predictors of marine biodiversity across taxa. Nature 466(7310):1098–1101. https://doi.org/10.1038/nature09329. Bibcode:2010 Natur.466.1098T. PMID 20668450
Tshabalala T, Mutanga O, Abdel-Rahman EM (2022) Predicting the geographical distribution shift of medicinal plants in South Africa due to climate change. Conservation 2(4):694–708
Tucakov J (1964) Pharmacognosy. Institute for Textbook Issuing in SR, Beograd, pp 11–30
Tucakov J (1971) Healing with plants–phytotherapy, vol 180. Culture, Beograd, p 90
Uniyal SK, Singh KN, Jamwal P, Lal B (2006) Traditional use of medicinal plants among the tribal communities of Chhota Bhangal, Western Himalaya. J Ethnobiol Ethnomed 2(14):1–8
Upadhyay RK (2020) Markers for global climate change and its impact on social, biological and ecological systems: a review. Am J Clim Chang 9(3):159
ur Rahman A, Khan SM, Ahmad Z, Alamri S, Hashem M, Ilyas M et al (2021) Impact of multiple environmental factors on species abundance in various forest layers using an integrative modeling approach. Glob Ecol Conserv 29:e01712
Verma AK (2016) Biodiversity: its different levels and values. Int J Environ Sci 7(2):143–145
Wang G, Ren Y, Bai X, Su Y, Han J (2022) Contributions of beneficial microorganisms in soil remediation and quality improvement of medicinal plants. Plants 11(23):3200
WCVP (2020) World checklist of vascular plants, version 2.0. Royal Botanic Gardens, Kew. Published online, http://wcvp.science.kew.org/
WHO (World Health Organization) (2011) The world traditional medicines situation, in traditional medicines: global situation, issues and challenges, vol 3. World Health Organization, Geneva, pp 1–14
Wiart C (2006) Ethnopharmacology of medicinal plants: Asia and the Pacific (pp. 1–50), vol 10. Humana Press, New Jersey, pp 978–971
Wyatt JL, Silman MR (2010) Centuries-old logging legacy on spatial and temporal patterns in understory herb communities. For Ecol Manag 260(1):116–124
Yuan H, Ma Q, Ye L, Piao G (2016) The traditional medicine and modern medicine from natural products. Molecules 21(5):559

Assessment of the Medicinal Flora for Treating Skin Disorders in Indonesia

2

Inggit Puji Astuti, Ria Cahyaningsih, R. Syamsul Hidayat, Titi Juhaeti, Iin P. Husaini, Endah Suhaendah, R. Vitri Garvita, Dewi Maharani, Hary Wawangningrum, Saniyatun Mar'atus Solihah, and Rusmanto

Abstract

The immense biodiversity of Indonesia has provided a home to a broad range of medicinal plants, which have been used for years to cure various illnesses, including skin disorders. This chapter investigates the astonishing diversity of medicinal plants found in Indonesia and their potential for treating skin diseases, with a few examples of essential medicinal plants in Indonesia and the related information in treating skin disorders such as *Centella asiatica* (L.) Urb., *Tinospora crispa* (L.) Hook. f. & Thomson, and *Andrographis paniculata* (Burm. f.) Wall. ex Nees. Using these medicinal plants in modern healthcare systems could provide sustainable and alternative approaches to addressing skin health concerns, benefiting Indonesians and people worldwide looking for natural and effective remedies for skin disorders. As interest in traditional medicine and natural cures grows, it becomes increasingly important to support conservation

Inggit Puji Astuti, Ria Cahyaningsih, R. Syamsul Hidayat, Titi Juhaeti, Iin P. Husaini, Endah Suhaendah, R. Vitri Garvita, Dewi Maharani, Hary Wawangningrum, Saniyatun Mar'atus Solihah and Rusmanto contributed equally with all other contributors.

I. P. Astuti · R. Cahyaningsih (✉) · R. S. Hidayat · T. Juhaeti · I. P. Husaini · E. Suhaendah · R. V. Garvita · D. Maharani · H. Wawangningrum · Rusmanto
Research Center for Plant Conservation, Botanical Gardens, and Forestry, National Research and Innovation Agency (BRIN), Kawasan Sains dan Teknologi (KST) Soekarno, Bogor, West Java, Indonesia
e-mail: ingg001@brin.go.id; ria.cahyaningsih@brin.go.id; rsya001@brin.go.id; titi002@brin.go.id; iinp003@brin.go.id; enda051@brin.go.id; rade021@brin.go.id; dewi047@brin.go.id; hary008@brin.go.id; rusm004@brin.go.id

S. M. Solihah
Conservation Department, Mitra Natura Raya, Bogor Botanic Gardens, Bogor, West Java, Indonesia
e-mail: sani.solihah@kebunraya.id

© The Author(s), under exclusive license to Springer Nature Singapore Pte Ltd. 2024
M. Nandave et al. (eds.), *Ethnopharmacology and OMICS Advances in Medicinal Plants Volume 1*, https://doi.org/10.1007/978-981-97-2367-6_2

efforts that conserve Indonesia's unique biodiversity for sustainable use, particularly for future generations.

Keywords

Biodiversity · Medicinal plants · Indonesia · Skin disorders · Conservation and sustainable use

Abbreviations

ADHD	Attention-deficit hyperactivity disorder
GCMS	Gas chromatography-mass spectrometry
HDI	Human development index
PASS	Prediction of activity spectra for substances
POWO	Plants of the World
SC	Stratum corneum
WHO	World Health Organization

2.1 Introduction

Indonesia is an archipelago country situated between two continents, Asia and Australia. As an archipelagic country with a tropical climate and high rainfall, Indonesia has a high biodiversity (Retnowati et al. 2019). Of those data, Indonesian medicinal plants may vary, from at least 1845 species (Widjaja 2014) to 5490 species (Cahyaningsih et al. 2021a, b, c). Moreover, Indonesia has various local cultures of indigenous people. Among the many cultures owned by traditional tribes in Indonesia, indigenous knowledge and practices related to natural resource management are the most notable, including those related to the utilization of plants for medicinal purposes. Natural drugs from plants play an essential role in treating skin disorders. The therapeutic use of medicinal plants is still popular due to several advantages, such as fewer side effects, better patient tolerance, and relatively less expensive and acceptable due to a long history of use (Tabassum and Hamdani 2014; Eraslan et al. 2020). For these reasons, several plants have been researched as potential treatments for skin disorders.

Treatment using medicinal plants that grow in the environment around the community or are cultivated is a treatment for minor diseases based on beliefs and experiences experienced by the Indonesian community, which is then developed under the culture of the community. Generally, traditional medicines that utilize medicinal plants meet appropriate criteria such as high prevalence, high incidence, spread over large areas, and low health service facilities in the community but are easily recognized by the community (Siregar et al. 2020) The study results (Rahayu and Andini 2019) stated that several factors that affect the use of medicinal plants are significantly related to the area of residence, age, level of education, and employment. Besides, using medicinal plants for health maintenance and overcoming disease disorders is still very much needed due to soaring medical costs, environmental changes, lifestyles, and disease changes (Vera and Yanti 2020). On the other hand, the diversity and population of medicinal plants are degraded due to new population growth, land use, forest clearing, and climate change. Therefore, for the sustainable use of medicinal plants, efforts are needed to conserve these medicinal plants (Cahyaningsih et al. 2021a, b, c)

Diseases that commonly occur in Indonesian society include skin diseases or disorders. According to the 2015 Indonesia Health Profile data, skin disorders are ranked third out of the ten most common disorders in outpatient hospitals across Indonesia. Data from the Ministry of Health of the Republic of Indonesia for 2012 showed that the prevalence of skin disease in Indonesia was 8.46%; then, it increased in 2013 to 9% (Agustina et al. 2022). Then, according to the 2020 Indonesia Health Profile, Indonesia has a skin disease prevalence rate of 0.49 (49%), while the prevalence of infectious skin diseases worldwide is around 300 million cases per year (Desmawati 2015). Various plants are usually used to treat skin diseases, especially those containing essential oils, such as the research results by Rubiah et al. (2015) in Aceh, which produced 57 species to treat skin diseases. This data is similar to the results of Hidayat's (2022) recapitulation of 28 ethnic groups with 60 species for skin disorders (Table 2.1). Several skin diseases include leprosy, dermatitis, scabies, and tinea versicolor. Common skin problems include dry skin; rough, scaly texture on the hands, feet, or face; acne, skin rashes; contact dermatitis or skin inflammation; and abrasion or loss of the epidermal layer (Isro'in and Andarmoyo 2012). One that is common is scabies, which is an infectious skin condition resulting from a *Sarcoptes scabiei* mite infection. Clinical symptoms include itching to cause lesions. Scabies sufferers worldwide are reported around 300 million cases per

Table 2.1 Ten types of diseases that are more commonly treated by medicinal plants in Indonesian society (28 ethnic samplings)

No	The type of disease	Number of treatment cases	Number of ethnic users of medicinal plants	Number of species used
1.	Stomachaches/diarrhea	93	20	50
2.	Skin disease	79	21	60
3.	Cough	76	22	62
4.	Malaria	58	21	32
5.	Toothache	53	24	21
6.	Canker sore	42	15	34
7.	Back pain	40	14	39
8.	Liver	38	13	36
9.	Rheumatism	36	14	28
10.	Stomach ulcer	30	16	25

year. In Indonesia, this disease ranks third out of the 12 most common skin diseases and is still a problem of infectious diseases (Wulandari 2018). Other types of skin diseases experienced by the community are generally caused by the pathogen *Staphylococcus aureus* bacteria, such as acne, boils, scabies, tinea versicolor, and wound infections.

Natural medications using plants play an essential role in the treatment of skin disorders. The therapeutic use of medicinal plants is still widespread because of several advantages, such as often having fewer side effects, better patient tolerance, and being relatively less expensive and acceptable due to a long history of use (Tabassum and Hamdani 2014; Eraslan et al. 2020). For these reasons, several plants have been researched as potential treatments for skin disorders. Given the rich plant diversity, high cases of skin disorders, and the benefit of natural medication, in this chapter, we try to explore the incredible diversity of medicinal plants, whether native or introduced plants, which may be found in Indonesia and their potential to cure skin disorders.

2.2 About Indonesia

2.2.1 Indonesia Country and Its Flora

Indonesia's unique geographical position with a constellation of large and small islands from Asia, Wallacea, and Australia simultaneously makes Indonesia have high biodiversity and endemicity (Widyatmoko 2019). There are at least 17,000 islands spread from Sabang to Merauke, and they are included in 37 provincial governments. Indonesia's territory covers about 9 million km^2 (with the ratio of land to sea 2:7). Forests make up 88.4 million hectares, or 47.1% of the total land area. Indonesia is only about 1.3% of the earth's area, but it has a very high level of biodiversity (Kusmana and Hikmat 2015a, b).

Based on the Indonesian Biodiversity Strategy and Action Plan 2015–2020 document, the diversity of Indonesia flora species is classified into spore plants and seed plants, with 91,251 species (6% of the world's spore plants) and 19,232 species (8% of the world's seed plants), respectively (Bappenas 2016). These plants are divided into two major groups based on the area of distribution, namely, Indo-Malayan and Indo-Australian, separated by the Wallace line and the Lydekker line. The first group, Indonesia-Malayan, is the West Indonesia region. The islands included in this area are Sumatra, Kalimantan, Java, and Bali, while the rest include Indo-Australians in eastern Indonesia (Batubara and Prastya 2020; Kusmana and Hikmat 2015a, b; Santosa et al. 2021). Apart from having ecological value, the flora (plants) also have high economic value in terms of their function as medicinal and ornamental plants (Kuswanto et al. 2021). As a country of biodiversity, most Indonesians depend on natural resources (Rahayu et al. 2020). Agriculture and forestry are two strategic fields in national development. Community interaction with biodiversity is reflected in daily life, especially in traditional medicine, which has been embedded in the nation's lifestyle for centuries (Arozal et al. 2020; Nugroho et al. 2020).

2.2.2 Indonesian People and Medicinal Plant Use

Indonesia is a highly populated country with a population reaching 278.69 million by mid-2023 (Statistics 2023). With over 400 distinct ethnic groups and sub-ethnic groups dispersed around the country, Indonesia is also rich in cultural diversity. Java, Sunda, Manado, Kalimantan, and various other regions still use plants as traditional medicine, which is a hereditary inheritance (Adiyasa and Meiyanti 2021). Even a statistics census in 2010 reported 1340 ethnic groups in the country.

More than 30 million of them live in forest areas (2.48%) and around forest areas (23.42%), according to survey data from the forestry sector in 2014 (Yuwono et al. 2015). These communities consist of hundreds of ethnic groups that depend on forest resources. This high ethnic diversity in Indonesia results in diverse cultures, traditions, and local wisdom that differ from one ethnicity or region to another. They generally have traditional knowledge of using plants, especially plants that are efficacious for treating certain diseases. This conventional knowledge is fundamental local wisdom to maintain and preserve its existence in Indonesia because the loss of local knowledge can be assumed to be the burning of the great Indonesian library.

Indonesia has various local cultures of indigenous people. Among the many cultures owned by traditional tribes in Indonesia, indigenous knowledge and practices related to natural resource management are the most notable, including those associated with the utilization of plants for medicinal purposes. Knowledge of people of various ethnicities in Indonesia is essential for preserving medicinal plants as part of flora (Hidayat 2021)

Such ethnic diversity contributes to the spread of this traditional knowledge. Traditional knowledge is usually passed down orally and often to a specific person. Therefore, knowledge about the use of medicinal plants is often possessed by tribal leaders, village heads, elders, village heads, shamans, or tribal communities (Bhagawan et al. 2021; Jadid et al. 2020) This local wisdom is vital for the sustainability of diversity and utilization of medicinal plants in Indonesia (Az-Zahra et al. 2021)

2.2.3 Data on Medicinal Plants and Diseases to Cure in Indonesia

Data on the potential of medicinal plants in Indonesia is still very diverse; some say at least 1845 species (Widjaja 2014), 3500 species (Suryanto and Setiawan 2013), 6000 species (Nugraha and Keller 2011), 7000 species (Handayani 2017), 7500 species (Sutrisna 2016), 9000 species (Salim and Munadi 2017), and even 9600 species (Nugroho 2017). Recent data showed that 5490 species identified have medicinal records (Cahyaningsih et al. 2021a, b, c).

Meanwhile, Heyne's data (1987) showed that there are 5000 officially recorded species of valuable plants, consisting of 1259 wood-producing species; 1050 medicinal plant species (21%); 984 food plant species (vegetables, fruits, seeds, and tubers); 520 species producing oil, resins, dyes, and other natural chemical compounds (10%); 328 species of animal feed; and 885 species of plants used for various purposes. Species as medicinal ingredients (21%) and chemicals (10%) dominate their utilization (Kartawinata 2010). Meanwhile, according to Kusmana and Hikmat (2015a, b), about 42% of medicinal plants are found in lowland tropical rainforests, 18% in seasonal forests, 4% in coastal, and 3% in mangrove forests.

People in Indonesia have carried out the use of plants as medicinal ingredients for a long time. With the existing ethnic diversity, using plants as medicine is also increasingly diverse. Of the 28 ethnicities sampled in Hidayat's study (2022), in total, 608 species of medicinal plants were obtained from 431 genera and 125 families. This number is slightly above the results of a study by Permanasari (2001) on the use of medicinal plants in 48 ethnic groups with 574 species from 117 families but still below the total collection of medicinal plants owned by the Bogor Botanical Gardens, 760 species (Hidayat et al. 2011). Most medicinal plants recorded were wild (460 species or 76%), and only 148 (24%) were domesticated or cultivated. This number seems still tiny compared to the information stating that there are thousands of species of medicinal plants in Indonesia, as mentioned in the previous paragraph.

The number of medicinal plant species in each bioregion is as follows: 126 species recorded in Sumatra, 223 recorded in Java, 119 recorded in Kalimantan, 126 recorded in Sulawesi, 242 recorded in Bali-Nusa Tenggara, 100 recorded in Moluccas, and 105 recorded in Papua. The recapitulation results are 608 species, with the most significant family being Leguminosae with 41 species, and Compositae and Lamiaceae with 29 and 26 species, respectively. There are 20 contributing families with more than or equal to 10 species, as shown in Fig. 2.1.

Various types of diseases can be treated using these medicinal plants. According to Purnamasari (2018), the 10 most common diseases in Indonesia are stroke (21.1%), heart disease (12.9%), diabetes mellitus (6.7%), tuberculosis (5.7%), blood pressure complications high (5.3%), chronic lung disease (4.9%), liver

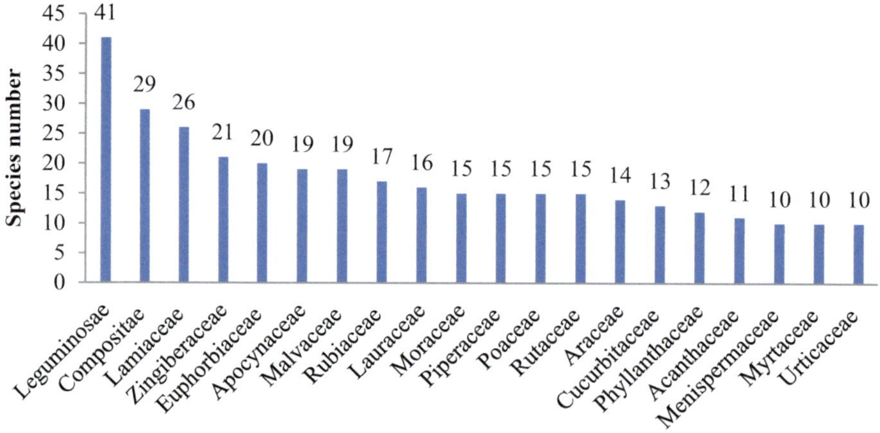

Fig. 2.1 Top 20 families of medicinal plants, based on 28 ethnic sampling

disease (2.7%), traffic accidents (2.6%), pneumonia (2.1%), and diarrhea (1.9%). However, based on the results of Hidayat's research (2022), based on a sampling of 28 ethnicities, the 10 most common diseases treated by medicinal plants are as follows (Table 2.1).

Diarrhea or stomachaches, skin disease, and cough are health problems with alternative medicinal plant species that can be used the most. Each is 62 species for coughs, 60 species for skin diseases, and 50 species for stomachaches. Cough is a health disorder common in every region and is affected at various age levels, so many alternative medicinal plants are used. Besides that, the character of cough that requires the warm nature of plants is found in many plants such as *Piper betle* L., *Zingiber officinale* Roscoe, and *Mentha arvensis* L.

2.3 Medicinal Plants for Treating Skin Disorders in Indonesia

2.3.1 Skin and Its Disorders

2.3.1.1 Skin

Human skin, the body's outer covering, is the largest organ that contains many specialized cells and structures (Reddy et al. 2022). The skin effectively prevents or neutralizes most physical, chemical, or biological attacks, expanding the senses (touch, taste, temperature, pain) as a physical barrier between organs and the environment, and through sweat, the skin plays an essential role in the control of thermoregulation. The skin also functions as an immune organ. Its role in regulating lymphocytes appears to be another crucial protective function (Walker et al. 1990; Gianfaldoni et al. 2020). The skin's barrier function resides in the stratum corneum (SC), consisting of a layer of corneocytes enriched with intercellular membrane proteins and lipids, such as ceramides, cholesterol, and free fatty acids. Corneocytes are rapidly and continuously replaced to maintain skin hydration, flexibility, and structural integrity and repair damaged skin (Goldsmith et al. 2012).

Skin that is not maintained well can cause various skin disorders. The skin, being the largest organ in the body, can become irritated, clogged, or inflamed, leading to various symptoms like redness, swelling, burning, and itching. Skin problems such as dermatitis and hives can be brought on by allergies, irritants, and certain infections. Skin diseases, including skin, hair, nails, and associated glands, affect the body's surfaces. They can involve all skin layers and result in infections, tumors, or inflammation occurrences.

2.3.1.2 Skin Disorders

Skin disorder is a vital disease group often found in the community, whether infectious or noninfectious. Skin diseases affect people of all ages, yet children are one of the most common victims of skin diseases since their skin is thin and delicate (Vora et al. 2012). Skin diseases are a common problem among young adults. Skin conditions can negatively impact every element of life, including relationships, employment choices, and academic performance. Skin conditions can have a terrible effect on a person's physical and emotional health in children and adolescents. In terms of infectious, the disease can be caused by bacteria, parasites, viruses, and fungi. In contrast, noninfectious skin diseases can be caused by accidents, allergic reactions, autoimmune, and drug reactions (Sahala et al. 2016).

Skin disorders occur worldwide at a significant level and pose a significant public health burden in developing and developed countries (Karimkhani et al. 2017). The skin disease often shows a physiological, psychological, and social impact (Zhang et al. 2019). According to a survey in 2020 by the All Parliamentary Group on Skin, 98% of patients with a skin disorder report that their condition affects their emotional and psychological well-being. Nevertheless, only 18% have received psychological support (Medical Research Foundation 2023). These skin diseases impact health-related quality of life due to the high rate of diseases in the rural community and the associated impact (Wootton et al. 2018). For example, 20% of children and 8% of adults suffer from eczema, an inflammatory skin condition. It is strongly linked to severe impacts on quality of life and psychological and psychiatric illnesses such as anxiety, depression, and attention-deficit hyperactivity disorder (ADHD) (Medical Research Foundation 2023).

The pattern and prevalence of skin disorders in young people are influenced by a number of factors, including food, environmental factors, skin care practices, personal hygiene, and race. Various skin diseases can be caused by several factors, such as lousy environment and daily habits, climate change, viruses, bacteria, allergies, immune system, and others (Pardiansyah 2015). With a diverse pattern driven by numerous factors, the incidence of skin disorders in children ranges from 34% to 87.7%. Research in 2016–2018 at the Skin and Venereology Polyclinic division of Dr. M. Djamil Padang Hospital in Indonesia, conducted on 1–18-year-old children, showed 277 new patients, including 123 women (44.4%) and 154 men (55.6%). According to the results, dermatitis (29.9%) was the most prevalent disease group, with fungal infections (16.9%), parasite infestations (13.4%), viral infections (13.3%), and bacterial infections (6.9%) following in order of prevalence. The majority of skin conditions affect boys. The most prevalent skin conditions are fungus infections, which are followed by parasite infestations. Among adolescents, a common skin disease is acne (acne vulgaris), a skin disease caused by an accumulation of oil that causes clogged facial skin pores, triggering bacterial activity and inflammation of the skin (Nurjanah et al. 2018).

The most common skin infection due to parasites with a high prevalence is scabies (Gustia et al. 2020), which is caused by the mite *Sarcoptes scabiei* var. *hominis* (Heukelbach and Feldmeier 2006). The World Health Organization (WHO) classified scabies as a "neglected tropical disease" (Casulli 2021). The systematic review highlights the prevalence of scabies worldwide, showing that scabies are still a severe, increasing disease that occurs globally and is clustered in developing countries (Schneider et al. 2023). The countries with a low human development index (HDI) are more commonly affected by scabies outbreaks (Romani et al. 2015). The prevalence of scabies is high in certain countries, ranging from 32.1% to 74%, especially in densely populated places like prisons, boarding schools, and orphanages. Indonesia is one of the most heavily affected countries worldwide (Widaty et al. 2022).

Skin diseases caused by viral, bacterial, fungal, or parasitic infections can cause both primary infections and can continue to become secondary infections. Primary infection is a primary skin infection with typical clinical symptoms caused by a single pathogen, for example, impetigo, folliculitis, and boils. *Staphylococcus aureus*, β-hemolytic *Streptococcus*, and coryneform bacteria are the most common primary skin pathogens. These organisms usually enter through a skin wound as an effect of an insect bite. Meanwhile, secondary infection is an infection that occurs in diseased skin (Baron 1996).

2.3.2 General Medicinal Plant Use for Treating Skin Disorders

Several studies have been conducted in Indonesia for treating skin disorders. A study in some subdistricts of Ternate Island describes that 15 species of medicinal plants are used to treat skin disorders, where the most often used species are *Cocos nucifera* L., *Psidium guajava* L., and *Carica papaya* L. (Umagap 2018). Another study by Kartina et al. (2022) in North Kalimantan showed there are 19 medicinal species for treating skin disorders, where *Selaginella intermedia* (Blume) Spring and *Jatropha curcas* L. have inhibitory potential in melanin formation on B16 melanoma cells. A study in Probolinggo, East Java, showed that *Piper betle* L. is used to treat hives, *Aloe vera* (L.) Burm. f. is used to treat burns, and *Senna alata* (L.) Roxb. is used to treat ringworm (Jadid et al. 2020).

One of the types of skin disorders that often occurs is a wound. The quest for natural product-based medications in treating wounds, including cuts and burns, is highly pertinent, given that wound curing is a complex process of tissue restoration. The management and treatment of wounds can be significantly improved by using plant extracts. Kanedi et al. (2022) showed 20 potential plants in Indonesia that can be developed into wound-healing medications. These plants are *Anredera cordifolia* (Ten.) Steenis, *Ageratum conyzoides* L., *Allium sativum* L., *Aloe vera* L., *Senna alata* L., *Coffea canephora* Pierre ex A. Froehner, *Colocasia esculenta* L., *Cocos nucifera* L., *Chromolaena odorata* L., *Euphorbia tirucalli* L., *Ficus deltoidea* Jack, *Impatiens balsamina* L., *Melaleuca cajuputi* Powell, *Melastoma polyanthum* L., *Musa acuminata* Colla, *Poikilospermum suaveolens* (Blume) Merr., *Piper nigrum* L., *Plantago major* L., *Saurauia vulcani* Korth., and *Zea mays* L. Using medicinal plants on the skin is mostly still crude extracts presented as a paste, gel, or liquid rubbed on the wound.

Another use of medicinal plants on the skin is as a cosmetic. Plants, as one of the natural sources of cosmetic ingredients, can be used to synthesize some functional inorganic materials, usually called green synthesis (Raajshree and Durairaj 2017). The development of herbal cosmetics in Indonesia is multiplying. Data from the Ministry of the Trade Republic of Indonesia (2009) showed that the global import markets for cosmetics are expanding fast, averaging between 17% and 15% yearly expansion. Cosmetics are designed to deal with oil production, acne, and wrinkles. The herbs should possess many qualities such as antibacterial, antiseptic, emollient, anti-inflammatory, and antioxidant. Some plants in Indonesia for commonly used for herbal cosmetics include *Curcuma longa* L. (as antioxidants), *Centella asiatica* L. urb. (as antiaging), and *Piper betle* L. (as liquid bath soap) (Wathoni et al. 2018). *Curcuma longa* L. has an activity to inhibit melanogenesis as a skin whitening and as an anti-acne agent (Batubara and Prastya 2020).

2.3.3 Recent Data on Medicinal Plants in Indonesia for Treating Skin Disorders

Data and information about medicinal plants' potential for treating skin disorders in Indonesia were collected from several references, such as books, journals, and ethnobotanical reports. About 700 species from 136 families used for cuts (wounds), acne, burns, smallpox, boils, scabies, and ulcers can be treated. The medicinal plants are processed into a decoction to drink or to bathe; otherwise, they are applied topically like a paste. Aerial parts (including roots, rhizomes, and tubers), bark or stems, leaves, flowers, fruit, seeds, other parts (sap and water), and whole parts might be used in medication (Fig. 2.2).

2.3.4 A Few Examples of Medicinal Plants Used for Treating Skin Disorders by Indonesia People

Here, we explore ten species of medicinal plants that Indonesian people usually use to treat skin disorders.

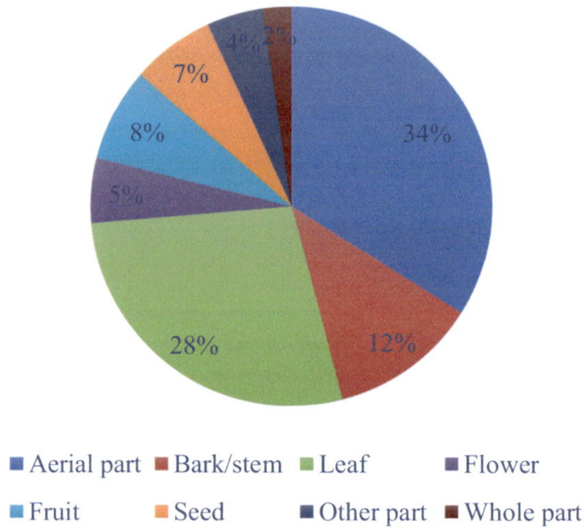

Fig. 2.2 Plant parts for treating skin disorders in Indonesia

2.3.4.1 *Centella asiatica* (L.) Urban

Centella asiatica (L.) Urban is classified in the Apiaceae family and *Centella* genus (Hargono et al. 1999; POWO 2023). It has 38 synonyms, including the homotypic synonym *Hydrocotyle asiatica* L. and heterotypic synonyms *Centella asiatica* f. *crispata* (Maxim.) H. Hara, *Centella asiatica* var. *boninensis* (Nakai ex Tuyama) H. Hara, *Centella asiatica* var. *crispata* (Maxim.) Y. Yabe, *Centella asiatica* var. *cristata* Makino, *Centella boninensis* Nakai ex Tuyama, *Centella hirtella* Nannf., *Centella ulugurensis* (Engl.) Domin, *Chondrocarpus asiaticus* Nutt., *Chondrocarpus triflorus* Nutt., *Glyceria asiatica* Nutt., *Glyceria triflora* Nutt., *Hydrocotyle abyssinica* Gand., *Hydrocotyle artensis* Montrouz., *Hydrocotyle asiatica* var. *crispata* Maxim., *Hydrocotyle asiatica* var. *monantha* F. Muell., *Hydrocotyle brasiliensis* Scheidw., *Hydrocotyle brevipedata* St.-Lag., *Hydrocotyle brevipes* DC., *Hydrocotyle dentata* A. Rich., *Hydrocotyle ficarifolia* Stokes, *Hydrocotyle ficarioides* Lam., *Hydrocotyle hebecarpa* DC., *Hydrocotyle inaequipes* DC., *Hydrocotyle indivisa* Banks & Sol. ex Hook. f., *Hydrocotyle leptostachys* Spreng., *Hydrocotyle lunata* Lam., *Hydrocotyle lurida* Hance, *Hydrocotyle natans* Thunb., *Hydrocotyle nummularioides* A. Rich., *Hydrocotyle pallida* DC., *Hydrocotyle reniformis* Walter, *Hydrocotyle sarmentosa* Salisb., *Hydrocotyle sylvicola* Cordem., *Hydrocotyle thunbergiana* Spreng., *Hydrocotyle ulugurensis* Engl., *Hydrocotyle wightiana* Wall, and *Trisanthus cochinchinensis* Lour. (POWO 2023).

C. asiatica is native and distributed in the Caucus, tropical Southeast Asia, subtropics, and eastern Australia, with more diverse species found in South Africa (Hargono et al. 1999; POWO 2023). It has many local names such as Asiatic pennywort, Indian pennywort, spadeleaf, pennywort India, and Gotu kola (English). Indonesian people call it pegagan or daun kaki kuda, and antanan, while the tribes called it peugaga (Aceh); jalukap (Banjar); Kajalukap (dayak); daun kaki kuda (Melayu); ampagaga (Batak); antanan gede, antanan, and dulang sontak (Sundanese); tikusan (Madura); piduh and taiduh (Bali); kori-kori (Halmahera); panigowang, gagan-gagan, rendeng, cowek-cowekan, and pane gowang (Java); pegago (Minangkabau); dogauke or sandanan or gogauke (Papua); kalotidi manora (Maluku), bebele and bebile (Lombok); semanggen (Indramayu, Cirebon); pagaga (Makassar); daun tungke (Bugis); Pigago (Minang); daun tapak kudo (solok); and jelukap/jalukap (Kutai/Borneo) (Hargono et al. 1999; Sutardi 2016).

C. asiatica is an evergreen plant that spreads with stolons up to 2.5 cm long and is rooted in nodes. It has a single kidney-shaped green to yellow-green leaf with a 1–7 cm diameter on old leaves. The flowers grow in the armpits of the green and yellow-green leaves. The flower stalks are purple, purple-red, green, or yellow-green, with purple-red being the most dominant color (Hargono et al. 1999; Sutardi 2016; Darmalaksana et al. 2018), as illustrated in Fig. 2.3. This plant can grow in moist soil and shade with sufficient sunlight, ranging from lowlands to an altitude of 2500 m asl (Hargono et al. 1999).

Sundanese tribes consume *C. asiatica* as a vegetable (Silalahi 2014), while the Mee tribe in Papua uses it for their traditional wound medication (Ramandey and Bunei 2021). *C. asiatica* contains antinociceptive, anti-inflammatory, antioxidant (Mareta 2020), and antibacterial properties (Azzahra and Hayati 2018) due to its

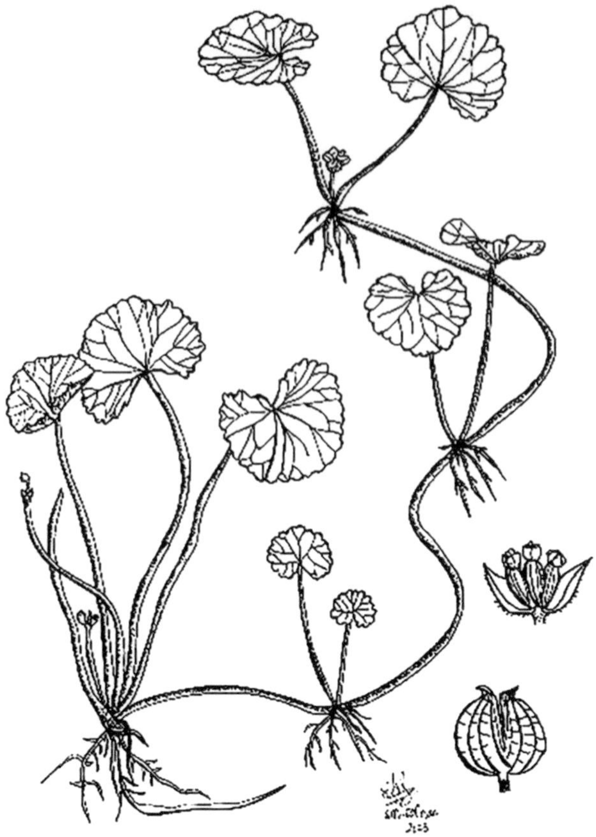

Fig. 2.3 Illustration of *Centella asiatica* (L.) Urb

alkaloids, saponins, flavonoids, steroids, and triterpenoids (Sutardi 2016; Irham et al. 2019; Djoko et al. 2020). The leaf extract effectively healed grade II burns in rats and accelerated regeneration without side effects or death in rabbits (Rismana et al. 2013; Darmalaksana et al. 2018). It also effectively improves skin inflammation and reduces psoriasis or chronic skin inflammation (Lin et al. 2023). Additionally, *C. asiatica* extract cream reduces pain in CTR surgery scars (Balevi and Balevi 2023). The effectiveness of *C. asiatica* extract can be enhanced when formulated in a gel form. It cools on the skin and dries quickly (Fujiastuti and Sugihartini 2015).

Its active compounds, such as asiaticoside and Asiatic acid, madecassoside, and madecassic acid, promote collagen production, which aids in skin cell regeneration and healing (Sutardi 2016; Dewi et al. 2018; Irham et al. 2019). Asiaticoside, which asiaticoside contains one esterified carboxylate with a sugar group and one main alcohol group, glycol., is a triterpenic glycoside/asiaticoside derived from alphamarin with two glucose molecules and one rhamnose. Asiaticosides are part of triterpenoids that act as natural antibiotics and are able to repair and stimulate blood cells, strengthen skin cells, and boost the immune system. Madecassoside also belongs to the triterpenoid group that repairs cell damage, synthesizes collagen, and reduces keloid formation in primary fibroblasts derived from human keloids. Asiatic acid is an anti-inflammatory and antiseptic used to heal wounds (Sutardi 2016; Irham et al. 2019). According to Suprihatin et al. (2023), the active content of asiaticoside can be increased by tillage, such as improving moisture and applying fertilizer. We could find some herbal products online (Fig. 2.4).

2.3.4.2 *Tinospora crispa* (L.) Hook. f. & Thomson

Tinospora crispa (L.) Hook. f. & Thomson belong to the Menispermaceae family and *Tinospora* genus (Ahmad et al. 2016). It has 17 homotypic synonyms (*Chasmanthera crispa* (L.) Baill., *Cocculus crispus* (L.) DC., *Menispermum crispum* L.) and heterotypic synonyms (*Cocculus bantamensis* Blume, *Cocculus rimosus* Blume, *Cocculus verrucosus* (Roxb.) Wall., *Menispermum bantamense*

Fig. 2.4 Some herbal products from *Centella asiatica* from the online market

(Blume) Spreng., *Menispermum rimosum* (Blume) Spreng., *Menispermum tuberculatum* Lam., *Menispermum verrucosum* Roxb., *Nephroia bantamensis* (Blume) Miers, *Tinospora gibbericaulis* Hand.-Mazz., *Tinospora masters ii* Diels, *Tinospora rumphii* Boerl., *Tinospora thorelii* Gagnep., *and Tinospora tuberculata* (Lam.) Beumée ex K. Heyne) (POWO 2023).

T. crispa grows naturally in wet tropical biomes from Arunachal Pradesh to South Yunnan and West and Central Malesia, including Indonesia (Yusuf et al. 1999; POWO 2023). It has several local names, for example, in Indonesia, it is called andawali (Sundanese), brotowali, or putrowali (Javanese) (Yusuf et al. 1999; Ahmad et al. 2016).

T. crispa is a shrub that can climb up to 15 m. Its old stems are fleshy with nodules, while young stems are brown and slightly fleshy with thin epidermal skin. The stem contains milky white sap. The heart-shaped leaves are single leaves with pointed tips, a width of 7–12 cm, a length of 6–12 cm, and a petiole length of 5–15 cm. The yellow or greenish flowers usually have 3 petals, and the ellipsoidal fruits are orange with a length of about 2 cm (Fig. 2.5). *T. crispa* can be found in primary and mixed deciduous forests and is widely found in secondary forests. It can grow up to an altitude of 1000 m above sea level (Yusuf et al. 1999; Ahmad et al. 2016).

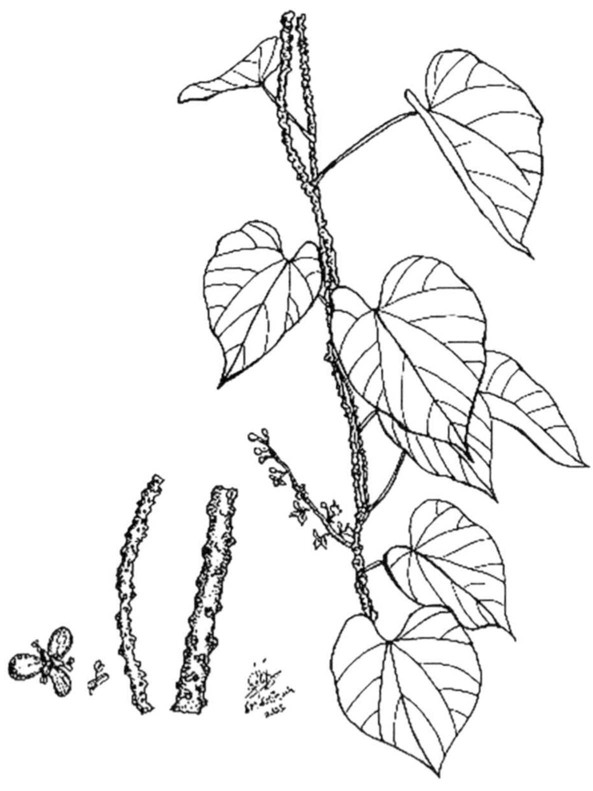

Fig. 2.5 Illustration of *Tinospora crispa* (L.) Hook. f. & Thomson

T. crispa is used by the Sasak tribe in West Lombok as a skin medicine to treat wounds and itching (Jannah and Ridwan 2013; Yamin et al. 2018). It contains active chemical compounds like flavonoids, terpenoids, alkaloids, lignans, nucleosides, and sterols, along with picoretin, berberine, and palmitin, which have bactericidal effects (Ahmad et al. 2016; Roni et al. 2022; Yusriani et al. 2018). The stem extract cream can inhibit the growth of *Propionibacterium acnes* bacteria and can be used as an acne drug (Purwoko et al. 2020).

Phytochemical analysis results of *T. crispa* revealed the presence of alkaloids, flavonoids and flavone glycosides, triterpenes, diterpenes, and diterpene glycosides, cis-type furanoditerpenoids, clerodane, lactones, sterols, lignans, and nucleosides. The study by Ahmad et al. (2016) showed that the crude extract and isolated compounds of *T. crispa* have various pharmacological activities such as anti-inflammatory, antioxidant, immunomodulatory, cytotoxic, antimalarial, cardioprotective, and antidiabetic activities. The active chemical compounds in *T. crispa* include flavonoids, terpenoids, alkaloids, lignans, nucleosides, and sterols (Ahmad et al. 2016; Roni et al. 2022). Warsinah et al. (2020) explained that *T. crispa* acts as an antioxidant because ethanol extract and water fraction contain flavonoid glycoside compounds from the flavonol group, where the hydrolysis process of water fractions using chloric acid increases antioxidant activity. Yusriani et al. (2018) explained that *T. crispa* also contains picoretin, berberine, and palmatine compounds, which include alkaloid group compounds, saponins and tannins that are many in the stem and have bactericidal effects. Giving *T. crispa* stem extract cream can inhibit the growth of *Propionibacterium acnes* bacteria, which are acne-causing bacteria. According to Purwoko et al. (2020), based on the GCMS (gas chromatography-mass spectrometry) test results, there is a leading chemical compound hexamethylcyclotrisiloxane, and PASS (prediction of activity spectra for substances) analysis contains antineoplastic, antiseborrheic, oxygen scavenger which can be used as an acne drug. The herbal products can be found online (Fig. 2.6).

2.3.4.3 *Andrographis paniculata* (Burm. f.) Wall. ex Nees

This species belongs to the Acanthaceae family and *Andrographis* genus (Mishra et al. 2007). It has synonyms: *Andrographis paniculata* var. *glandulosa* Trimen, *Andrographis subspathulata* C.B. Clarke, *Justicia latebrosa* Russell ex Wall., *Justicia paniculata* Burm. f., and *Justicia stricta* Lam. ex Steud. (POWO 2023).

Fig. 2.6 Some herbal products from *Tinospora crispa* from the online market

A. paniculata is native to Assam, Bangladesh, India, and Sri Lanka and was introduced to Malaysia, Indonesia, the West Indies, and the Americas (POWO 2023). Generally, it is known as "King of Bitters" for its extremely bitter taste. Its local names are Sambiloto, Pepaitan, Ampadu, Ki oray, Ki peurat, Takilo, Bidara, Sadilata, Sambilata, and Takila (Indonesian); Quasabhuva (Arabic); La-Sa-Bee (Lao); Chiorta, Hempedu Bumi (Malay); Nelaberu (Kannada); Nain-e Havandi (Persian); Vubati (Manipuri); Oli-kiryata and Hnakhapui (Mizo); Bhuinimba (Oriya); Chooraita (Panjabi); Senshinren (Japanese); Kirayat and Kalpanath (Hindi); Kalmegh (Assamese); Vhadlem Kiratyem (Konkani); Sambiloto, Nelavepu, and Kalpa (Marathi); Kiriyattu (Malayalam); Chirette verte and Roi des amers (French); The Creat and King of Bitters (English); Kariyatu (Gujarati); Chuan Xin Lian (Chinese); Se-ga-gyi (Burmese); Kalmegh (Bengali); Acılar S¸ahı, Acılar Kralı, Acı Pas¸a, and Acı Bey (Turkish); Acılar Xanı (khanı) (Azerbaijani); Aluy, Lekha, and Sinta (Philippines); Andrografis (Russian); Kalmegha, Bhunimba, and Yavatikta (Sanskrit); Green Chiratta (Scandinavian); H¯ın Kohomba or Heen Kohomba (Sinhalese); Andrografis (Spanish); Nilavembu (Tamil); Nilavembu (Telugu); Fa-Talai-Jorn and Fah-talai-jon (jone) (Thai); Kalmegh, Kariyat, and Mahatita (Urdu); and Xuyen T ˆ am Li ˆ en (Vietnamese) (de Padua et al. 1999; Jarukamjorn and Nemoto 2008; Mishra et al. 2007; Valdiani et al. 2012; Sharma and Sharma 2013; Hossain et al. 2014).

A. paniculata is a herbaceous, high, branching annual plant. Height of plant: 30–110 cm. Dark green stem that is 30–100 cm long and 2–6 mm in diameter. The leaves are glabrous and have a length of 2–12 cm and a width of 1–3 cm; the leaf structure is lanceolate with a pinnate shape. The flowers are white with a characteristic purple pattern on the petals; the flowers have small petals and are scattered in panicles or bunches. Seed capsules are linear-oblong, acute at both ends, size 1.9 cm × 0.3 cm, color yellowish brown, shape subquadrate, numerous (Mishra et al. 2007; Jarukamjorn and Nemoto 2008; Anju et al. 2012; Boopathi 2000) (Fig. 2.7).

Systematic studies on the chemistry of *A. paniculata* that have been carried out by Chao and Lin (2010) and Hossain et al. (2014) resulted in andrographolide, neoandrographolide, quinic acids, deoxyandrographolide, chloroform extract, bitter diterpene, lactones, diterpene glucosides, and flavonoids. Andrographolide is the main component isolated from the plant leaves, and it is a bicyclic diterpenoid lactone. Gorter isolated this bitter component in its pure form in 1911.

The infusion, sap from the crushed leaves, or leaf paste of *A. paniculata* has been used for treating skin eruptions or abscesses (de Padua et al. 1999; Siew et al. 2014). However, this plant is used traditionally globally to treat many illnesses besides skin disorders, such as fever, liver disorder, diarrhea, snake bites, and anthelmintics (Boopathi 2000; Jarukamjorn et al. 2008; Borhanuddin et al. 1994; Kabir et al. 2014; Akbar 2011). Today, users can benefit from online products without hassle in preparing the herbal (Adrian 2021), as shown in Fig. 2.8.

Fig. 2.7 Illustration of *Andrographis paniculata* (Burm. f.) Wall. ex Nees

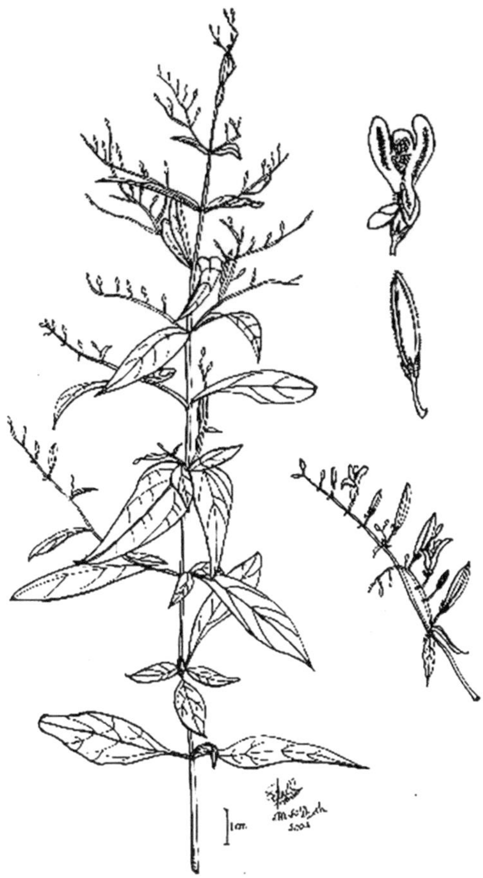

2.3.4.4 *Cymbopogon nardus* (L.) Rendle

Cymbopogon nardus (L.) Rendle is a species of Poaceae family and *Cymbopogon* genus. This species has several synonyms: *Andropogon nardus* L., *Sorghum nardus* (L.) Kuntze, *Andropogon hamulatus* Nees *ex* Steud., *Andropogon nardus* var. *prolixus* Stapf, *Andropogon nardus* var. *zeyheri* Hack., *Andropogon nardus* var. *validus* Stapf, *Andropogon pseudohirtus* Steud., *Andropogon grandis* Nees *ex* Steud., *Andropogon confertiflorus* Steud., *Andropogon thwaitesii* Hook. f., *Andropogon citrosus* Steud., *Andropogon nardus* subsp. *grandis* (Nees ex Steud.) Hack., *Andropogon nardus* var. *luridus* Hook. f., *Andropogon nardus* subsp. *nilagiricus* Hack., *Cymbopogon nardus* subsp. *nilgiricus* Hack., *Andropogon nilagiricus* Hochst., *Cymbopogon prolixus* (Stapf) E. Phillips, *Cymbopogon virgatus* Stapf ex Bor, *Lagurus paniculatus* Burm. f. *Cymbopogon afronardus* Stapf, *Cymbopogon nardus* var. *confertiflorus* (Steud.) Bor, *Cymbopogon thwaitesii* (Hook. f.) Willis, *Cymbopogon validus* (Stapf) Stapf ex Burtt Davy, *Cymbopogon validus* var. *lysocladus* Stapf, *Cymbopogon nardus* var. *luridus* (Hook. f.) N. Rama Rao, *Cymbopogon confertiflorus* (Steud.) Stapf, and *Cymbopogon claessensii* Robyns (POWO 2023).

Fig. 2.8 Some herbal products from *Andrographis paniculata* from the online market

Cymbopogon nardus is native to the country of Angola, Assam, Bangladesh, Botswana, Burundi, Cambodia, Cape Provinces, East Himalaya, Free State, India, Kenya, KwaZulu-Natal, Laos, Lesotho, Madagascar, Mozambique, Myanmar, Northern Provinces, Rwanda, Seychelles, Sri Lanka, Sudan, Swaziland, Tanzania, Uganda, Vietnam, West Himalaya, Zaïre, and Zimbabwe and introduced to Indonesia (POWO 2023). There are many local names for *C. nardus*, such as Ceylon oil, citronella, and citronella oil. It is known as ganja (Hindi), kamakher (Bengali), usadhana (Marathi), kamachipillu (Tamil), kamkshi-kasuvu (Telgu), kava (Punjabi), Sereh wangi (Indonesia), Citronelle (French), Java Citronellgras (German), Ya Xiang mao (Chinese), Ta krai hom (Thai), Lenabatu citronella (Sri Lanka), Cu sa (Vietnam), and Naid grass (India) (Kaur et al. 2021).

C. nardus is a fragrant, evergreen grass. This grass clamps onto itself and grows to a height of 2.5 m from rootstock. It is a grass with small leaf blades that lives a long time. The panicles are 8–10 mm long with long, silky hairs around them, and they are 15–30 cm long and narrow with unbranched inflorescence racemes. The spikelets have flat, concave winged keels on the back and are stalkless. Above it branches, forming a huge decompound nodding panicle. Culms have a diameter of up to 10 mm at the base. They are polished, are solid, and have nodes that are finely glabrescent and black. The leaves have a ribbonlike blade at the greenish-green end, a white midrib, and an upper cauline that is over 9 mm wide before narrowing to the base. The leaves are long and narrow, measuring 0.5–1.6 cm broad and 1 m long. The leaf's apex is filiform and glaucous beneath with scabrous margins and is glabrous except at the top of the sheath (Fig. 2.9) (Nakahara et al. 2003; Kaur et al. 2021).

Several studies state that *C. nardus* treats redness, irritation, toothaches, and skin inflammation. Additionally, it may be able to treat intestinal parasites, fever,

Fig. 2.9 Illustration of *Cymbopogon nardus* (L.) Rendle

rheumatism, headaches, menstruation and digestive issues, infectious diseases, exhaustion, and postpartum wash. The essential oil of *C. nardus* has long been used as a blood tonic and to treat gastritis, indigestion, flatulence, stomachaches, intestinal cramps, and irritable bowel syndrome. It has been used as a diuretic, vermifuge, antipyretic, aromatic tea, and mental disorder treatment since the Vedic era. Furthermore, it treats colds, headaches, and the flu when used in aromatherapy. Nowadays, the essential oil obtained is found to have values in soaps, toiletry, perfumery, tobacco, cosmetics, insect repellent packaging, body care products, and the pharmaceutical industries (Kaur et al. 2021). *C. nardus* is also used as pediculosis. The oil is mixed with coconut oil, and the mixture is applied on the head (Saikia et al. 2006).

The essential oil from *C. nardus* is obtained by distilling part of the plant, roots, stems, or leaves, yet the yield and the best quality originated from the leaves (Saputra et al. 2020), and some herbal products can be found online (Fig. 2.10). The geographic origin, ecological and climatic factors, developmental phases, harvest period, and genetic factors all affect the chemical composition of citronella essential oil. The chemical composition also depends upon the method of extraction and drying techniques (Kaur et al. 2021). The chemical composition of *C. nardus* essential oil was studied. Its significant constituents included camphene, β-caryophyllene, limonene, myrcene, terpinolene, borneol, citronellol, geraniol, linalool, piperitol, citral (cis and trans), citronellal, methyl heptane, citronellic acid, piperine, citronellyl acetate, caryophyllene oxide, geranyl acetate, geranyl butyrate, methyl eugenol, chavicol, eugenol, methyl isoeugenol, nerol, ocimene, lemon, η-propyl alcohol,

Fig. 2.10 Some herbal products from *Cymbopogon nardus* from the online market

4-terpineol, menthane, α-terpinene, α-thujiene, α-terpineol, αpinene, and β-pinene (Kaur et al. 2021).

2.3.4.5 *Graptophyllum pictum* (L.) Griff.

Graptophyllum pictum is a species belonging to the Acanthaceae family and *Graptophyllum* genus. Synonyms of *G. pictum* are *Graptophyllum hortense* Nees, *Justicia picta* L., and *Marama picta* (L.) Raf., *Graptophyllum hortense* var. *luridosanguineum* (Sims) Chitt, *Graptophyllum hortense* var. *rubrum* Hassk, *Graptophyllum hortense* var. *viride* Hassk, *Graptophyllum medioauratum* Linden, *Graptophyllum medioauratum* Linden ex K. Koch, *Graptophyllum nortonii* Van Geert, *Graptophyllum pictum* var. Blume *albums*, *Graptophyllum pictum* var. *luridosanguineum* (Sims) Bremek & Backer, *Graptophyllum picturatum* (W.) Bull, *Graptophyllum versicolor* Veitch ex J. Dix, *Justicia carayana* F. Newman ex Nees, *Justicia ecbolium* Blanco, *Justicia klotzschiana* Hoffm ex Nees, *Justicia picta* var. *luridosanguineum* Sims, and *Justicia violacea* Noronha (POWO 2023).

G. pictum is native to New Guinea (POWO 2023). It has some local names such as caricature plant (English), Indonesian Handeuleum (Sundanese), Daun ungu (Central Java), Pudingsimalur (Sumatra), Temen (Bali), Kadi-kadi (Maluku: Ternate), Dongo-dongo (Maluku: Tidore), Pudding black (Bukit Tiga puluh, Riau, and Jambi), and High (Babang Buyangge, Central Sulawesi) (Dharma 1985; Sangat et al. 2000).

G. pictum is a shrub plant, tall between 1.5 and 8 m, with a woody stem, erect, purple-greenish, segmented, and slippery surface. This plant has a single leaf, with pointed end, the base leaf is tapered, the edge leaf is flat, reinforcement the leaves pinnate, surface sheets of leaf part are shiny, with 15–25 cm length and 5–11 cm breadth and purple in color. This species has an compound inflorescence, flower supports several florets, flower trumpet shaped, petals attached at the base of flower, fruit box oblong, purple brownish, seeds whole, round and white (Dharma 1985) (Fig. 2.11).

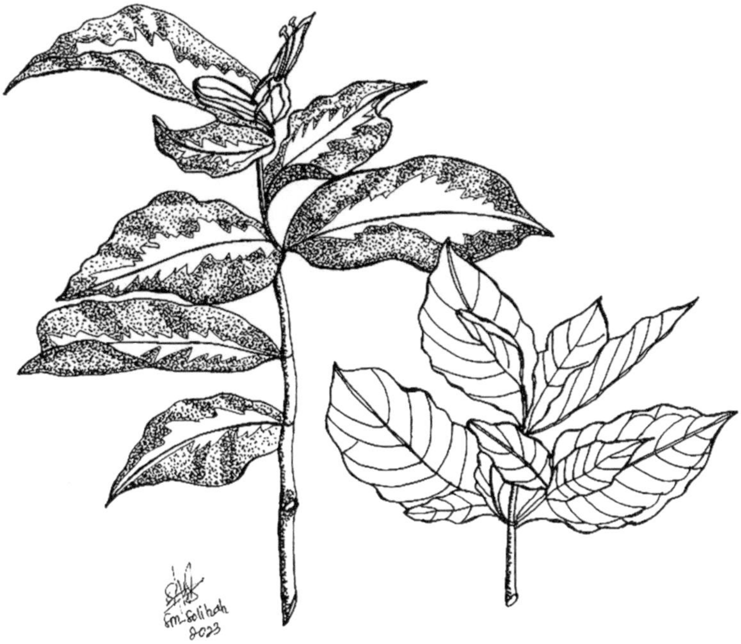

Fig. 2.11 Illustration of *Graptophyllum pictum* (L.) Griff

Several tribes in Indonesia use this plant: Suluan (Banggai, Central Sulawesi), Sundanese (Bandung, Subang, Sukabumi, Bogor), and Ternate (North Maluku) (Dharma 1985; Sangat et al. 2000). The standard part used as medicine is leaves. Among the existing compounds in *G. pictum* are flavonoids, alkaloids, tannins, saponins, and glycosides. Flavonoids, tannins, and saponins in plants work as an anti-inflammatory. It can cure boils, swelling, ulcers, and other skin diseases (Wijaya and Utami 2018; Sartika and Indradi 2021). This plant's herbal products could be found online (Fig. 2.12).

2.3.4.6 *Alpinia galanga* (L.) Willd.

Alpinia galanga (L.) Willd. is a member of the Zingiberaceae family and *Alpinia* genus. Synonyms of this species are *Alpinia bifida* Warb., *Alpina alba* (Retz.) Roscoe, *Alpinia galanga* subsp. *pyramidata* (Blume) K. Schum, *Alpinia carnea* grif, *Alpinia pyramidata* Blume, *Alpinia rheedei* Wight, *Alpine viridiflora* Griff, *Amomum galanga* (L.) Lour, *Amomum medium* Lour, *Galanga major* Garsault, *Galanga officinalis* Rompis, *Hellenia alba* (Retz) Willd., *Hellenia alba* Retz, *Languas galanga* (L.) Stuntz, *Langkuas pyramida* (Blume) Merr, *Langkuas vulgare* J. Koenig, *Maranta galanga* L., *Zingiber galanga* (L.) Stokes, Stoke *medium Zingiber*, and *Zingiber sylvestre* Gaertn (POWO 2023).

A. galanga is native to Indonesia and neighboring countries (POWO 2023). Local names of this plants are laos (general, Java, Madurese, Sasak); laja (Sundanese); isem, kalawasan, laja, and lahwas (Bali); langkuwas (Kalimantan);

Fig. 2.12 Some herbal products from *Graptophyllum pictum* from the online market

lingkuwas (Sulawesi, Manado); ringkuwas (Sulawesi, Minahasa); Lingkoboto (Sulawesi, Gorontalo); laja and langkuwasa (Sulawesi, Makasar); Aliku (Sulawesi, Bugis); lengkueneh (Sumatra, Aceh); lengkueus (Sumatra, Gayo); Kelawas (Batak Karo); halawas (Nias); lengkuweh (Sumatra, Minangkabau); lawas (Lampung); galangal (Malay); lawase (Maluku, Seram); laawasi and glawasi (Maluku, Ambon); lauwasel (Maluku, Saparua); lagoase (Maluku, Buru); and galiasa (Maluku, Halmahera) (Sangat et al. 2000).

A. galanga is a perennial shrub, a height reaching 2 m, with a stem pseudo consisting of fronds fused, intersected, green vaginal discharge. Leaves are single, oval, edge leaves flat, tip leaf taper, base leaf blunt, 25–50 cm long, wide 7–15 cm, spine-leaf pinnate, stalk short. The inflorescence compound appears at the end rod and supports lots of floret flowers. Flower with leaf petals is green, crown-shaped tube, white. Fruit is round, complex, whole seeds. Seeds are rounded and colored black-brownish (Fig. 2.13).

A. galanga is used by the people of Indonesia: Sakai tribe, village Sebangau, District Mandau, Regency Bengkalis Riau, Village community Cibugel, Subdistrict Darmaraja, Regency Sumedang (Sundanese), Village community Gambirmanis, district Pracimantoro, Regency Wonogiri (Java), Punan Lisum, Muara Tuboq and Muara Kebaq villages, District Tabang, Regency Kutai, East Kalimantan. Bolaangmongondow in Kotamobagu, North Sulawesi, Indigenous people Seram Island living in the village Poka, Suli, Salamada and Kotabaru, Ambon (Sangat et al. 2000).

The central part used in the *A. galanga* is the rhizome. However, all parts have been in use for a long time as traditional treatments for various types of disease due to the plant's antifungal, anti-tumor, antimicrobial, anti-inflammatory, antidiabetic, antioxidant, and antiulcer characteristics. Many active substances have been extracted from the plant, including 1,8-cineol, 1-S-1-acetoxychavicol acetate, 1-S-1-acetoxyeuginol acetate, 1-bisabolene, 1-bergamotene, 1-pinene, 1-sitsterylArabinoside (AG-7), 1-sitsterylDiglucoside (AG-8), and 1-sitsterylArabinoside (AG-8) (Chouni and Paul 2018).

Rhizome of *A. galanga* contains eugenol, which functions as an antifungal so that it can be used to treat and cure skin disorders, especially those caused by fungi, such as *Tinea versicolor*, ringworm, eczema, pimples, scabs, ulcers, and other

Fig. 2.13 Illustration of *Alpinia galanga* (L.) Willd

diseases (Silvina 2006). Traditionally, rhizomes galangal are pounded fine or shredded, and then smeared on the part affected skin by *T. versicolor* (Dewi 2020). Herbal products of *A. galanga* found online are shown in Fig. 2.14.

2.3.4.7 *Senna alata* (Linnaeus) Roxburgh

Senna alata (Linnaeus) Roxburgh is classified as part of the Fabaceae family and *Senna* genus. Synonym names of *S. alata* are *Cassia alata* L., *Cassia herpetica* Jacq., *Cassia alata* var. *rumphiana* DC., *Cassia alata* var. *perennis* Pamp., *Cassia bracteate* L.f., *Herpetica rumphiana* (DC.) J. Presl., *Herpetica alata* (L.) Raf., and *Cassia arayatensis* Náves ex Fern.-Vill. (POWO 2023).

Senna alata is native to the tropical America continent and was introduced in Indonesia (POWO 2023). This species has various names in different regions, such as seven golden candlesticks, candlestick cassia, Christmas candle, ringworm bush, gelenggang, and empress candle plant. Another name for *Senna alata* is ringworm cassia, reflecting its historical use in traditional medicine. In local contexts, it is also referred to as ketepeng, ketepeng cina, ketepeng kebo, ketepeng badak, haya mara, tabankun, and kupang-kupang highlighting its significance across cultures (Garvita et al. 2023; Oladeji et al. 2020; POWO 2023).

Fig. 2.14 Some herbal products from *Alpinia galanga* from the online market

This unique shrub, whether annual or biannual, exudes an unpleasant scent and reaches a modest height of 1–4 m (illustrated in Fig. 2.15). The leaves, boasting a yellowish-green hue, are broad and consist of 5–14 pairs of leaflets, with the distal leaf blade usually larger with a curved apex. The flowers are bright yellow with a simple shape, zygomorphous, gathered together so that their shape resembles a dense spike. It has seven stamens and two stamens that stand out due to their considerable length, accompanied by a pubescent ovary. As the shrub matures, it produces a tetragonal pod measuring 10–16 × 1.5 cm, featuring winged angles and a mature brown coloration. Within this pod, an abundance of diamond-shaped blackish-brown seeds, numbering as many as 60, can be found. The seed type is orthodox with hard skin (Garvita et al. 2023; Hennebelle et al. 2009; Oladeji et al. 2020).

Various parts of this plant (flowers, leaves, stems, or roots) are used as traditional medicine. *Senna alata* remedies have a history of being employed as mixtures and infusions to address various ailments, including asthma, hemorrhoids, dysentery, typhoid, ringworms, diabetes, malaria, hepatitis, yellow fever, decoction, and skin conditions (itching, skin rash, blotch, scabies, eczema, and herpes) (Esan et al. 2021; Hennebelle et al. 2009; Oladeji et al. 2020). It is because those various parts have arrays of bioactive chemical compounds that have been recognized for their potential therapeutic efficacy. For instance, these natural compounds encompass flavonoids (such as rhein, chrysophanol, kaempferol, luteolin, rhamnetin, santal, aloe-emodin, and glycosides), anthraquinones (including aloe-emodin, alatinone, and alatonal), carbohydrates, fatty acids (like oleic, palmitic, and linoleic acids), as well as steroids and terpenoids (such as β-sitosterol, stigmasterol, and campesterol) (Esan et al. 2021; Hennebelle et al. 2009; Oladeji et al. 2020). The essential oil of *Senna alata* leaves was obtained by hydrodistillation and analyzed by GCMS. The major constituents are linalool (23.0%), borneol (8.6%), pentadecanal (9.3%), and α-terpineol (5.9%) (Hennebelle et al. 2009).

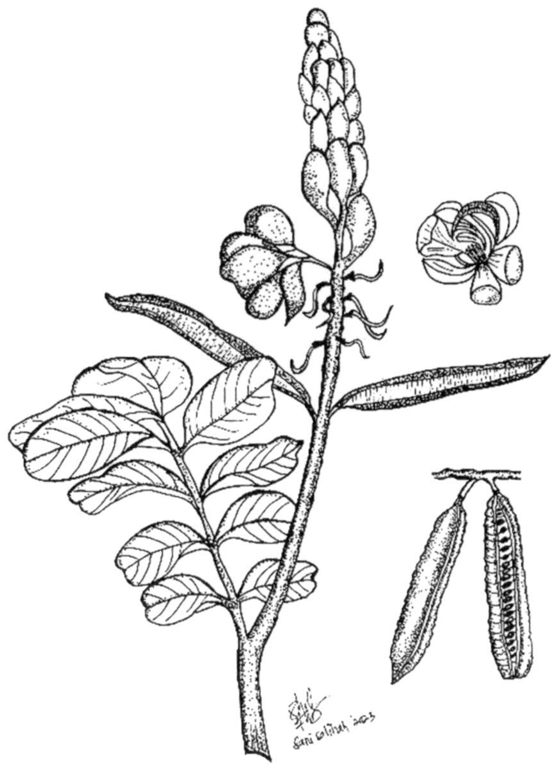

Fig. 2.15 Illustration of *Senna alata* (Linnaeus) Roxburgh

The antidiabetic impact of infectious diseases is using leaves and fruits. The leaves had been decoctioned and macerated before given orally; in the meantime, fruit had been decoctioned before given orally. Consuming leaves and flower buds orally by boiling them first can be an anti-inflammatory. The leaves, when utilized in infusion or decoction and taken orally as an extract, can address digestive problems such as constipation, stomachache, liver ailments, and prehepatic jaundice. The application of leaves, both topically and orally, using techniques like maceration, powdering, decoction, crushing, and infusion, can be utilized as a remedy for a range of skin ailments such as dermatitis, skin rashes, athlete's foot, shingles, eczema, and mycosis (Hennebelle et al. 2009).

Currently, *Senna alata* leaves, flowers, and bark are utilized to treat diverse skin infections and ailments. This plant has been identified in Thailand for managing conditions like itching, fungal skin infections, scabies, ringworm, urticaria, and shingles. This effectiveness is attributed to its bioactive compounds, including anthranols, anthrones, flavonoids, phenols, tannins, and anthracene derivatives. In Indonesia, grinding or rubbing leaves directly was observed to be used on the surface of the skin infected with fungus so that the fungus disappears. In the Kedung Peluk village, Sidoarjo community, *Senna alata* have a value as dermatological, and leaves were the only part used. The leaves are mashed until blended and then affixed to the itchy body parts (Efida 2022; Garvita et al. 2023; Oladeji et al. 2020). A concoction made from the leaves showcases potent inhibition of *Escherichia coli*, *Streptococcus pyogenes*, *Pseudomonas cepacia*, *Staphylococcus aureus*, *Pseudomonas aeruginosa*, and *Klebsiella pneumoniae*. Moreover, soap from *S. alata* seeds which contains cannabinoids and apigenin has been proven to treat boils, eczema, ringworm, breast abscesses, and impetigo. Furthermore, evaluations of the leaf, stem-bark, flower exudates, and ethanolic leaf extract against clinical isolates of *T. verrucosum*, *M. canis*, *T. mentagrophytes*, *E. floccosum*, *B. dermatitidis*, *A. flavus*, and *C. albicans* reveal robust inhibitory effects against the causative pathogens (Oladeji et al. 2020). Some herbal products from this plant could be found online (Fig. 2.16).

2.3.4.8 *Curcuma longa* Linn.

C. longa have synonym names, namely, *Curcuma longa* L. (*Kua domestica* Medik., *Stissera curcuma* Giseke., *Amomum curcuma* Jacq., *Amomum latifolium* Lam., *Curcuma tinctoria* Guibourt, *Curcuma brog* Valeton., *Curcuma ochrorhiza* Valeton., *Curcuma longa* var. *vanaharidra* Velay., Pandrav., J.K.George & Varapr., *Curcuma domestica* Valeton., and *Curcuma soloensis* Valeton.) (POWO 2023).

C. longa is native to India and spread worldwide, including Indonesia (POWO 2023). This plant is known by different names, not only turmeric or turmeric. Commonly recognized as yellow ginger and sun root, *C. longa* belongs to the Zingiberaceae family, including ginger. In India it is called haldi and manjal, in

Fig. 2.16 Some herbal products from *Senna alata* from the online market

many cultures known as yellow root or curcuma; in Indonesia, *C. longa* has many different names, including Kunyir (Komering), Konéng (Sunda), Cahang (Dayak Panyamabung), Unin (Ambon), Alawaha (Gorontalo), Huni (Bima), Janar (Banjar), Gorachi (Ternate), Kunir (Jawa), and Konyè' (Madura) (da Mata et al. 2020; Prasad and Anggarwal 2011).

Curcuma longa has a cylindrical rhizome, is yellow to orange in color, has many branches, and emits an aroma. The leaves are oval to elliptical with pointed tips, alternately arranged in two rows. The petiole length is from 50 to 115 cm, leaf blade length is from 76 to 115 cm, and leaf width is from 38 to 45 cm. Flowers are bracts without flowers, are tapered at the top, and have a white, green, or reddish-purple color (Fig. 2.17). *C. longa* flowers are bisexual, with sepal sizes ranging from 0.8 to 1.2 cm. Terminally on the pseudo-stem is an inflorescence stem, 12–20 cm long, containing many flowers. The bracts are pale green and oblong-ovate with a 3–5 cm blunt tip (Kocaadam and Şanlier 2017; Prasad and Anggarwal 2011).

Rhizomes are the most commonly used plant part as a culinary spice, food dye, and traditional medicine. Essential oil, which contains turmerone and diarylheptanoids, is the main component of the rhizome. Apart from that, there are also coloring substances in the form of curcuminoids which consist of mostly curcumin

Fig. 2.17 Illustration of *Curcuma longa* Linn

(70–80%), demethoxycurcumin (18–20%), bisdemethoxycurcumin (2–10%), 5′-methoxycurcumin, and dihydrocurcumin, along with secondary metabolites which are found to be natural antioxidants (da Mata et al. 2020; Prasad and Anggarwal 2011; Nelson et al. 2017; Sharifi-Rad et al. 2020). Curcumin, an orange-yellow dye practically insoluble in water, is the best characteristic of the compounds responsible for its biological actions (da Mata et al. 2020).

Over centuries, curcumin, an extracted chemical compound from *Curcuma longa*, has found traditional use as a remedy for infectious skin infections, weight management, digestion issues, infectious diseases, inflammatory conditions, cancer prevention, antioxidative effects, microbial defense, antineoplastic agent, and cutaneous and gastrointestinal inflammation (da Mata et al. 2020; Prasad and Anggarwal 2011; Sharifi-Rad et al. 2020; Vollono et al. 2019). Curcumin exhibited favorable availability and efficacy when applied externally, particularly when integrated into innovative compositions like hydrogels, collagen membranes, β-cyclodextrin-curcumin chitosan-alginate sponges, alginate foams, polymeric dressings, nanoparticle complexes, and nano-emulsions. This shows that curcumin is suitable for use as a therapeutic ingredient to treat skin diseases. Examining the potential interactions between curcumin and other chemicals commonly employed in topical skin treatments could offer valuable knowledge for creating novel combination preparations targeted toward specific conditions (Vollono et al. 2019).

Throughout history, curcumin has been employed for treating skin ailments, such as insect bites, burns, and acne, with origins tracing back to ancient times. We could find some herbal products from *C. longa* online (Fig. 2.18).

2.3.4.9 *Piper betle* L.

Piper betle L. is classified in the Piperaceae family and *Piper* genus. It has some synonyms, including homotypic synonym names, *Chavica betle* (L.) Miq. and *Piperi betlum* (L.) St.-Lag., and heterotypic synonym names: *Artanthe hexagyna* Miq., *Betela mastica* Raf., *Chavica auriculata* Miq., *Chavica blumei* Miq., *Chavica canaliculata* (Opiz) C.Presl, *Chavica densa* Miq., *Chavica siriboa* (L.) Miq., *Cubeba melamiri* Miq., *Cubeba siriboa* (L.) Miq., *Macropiper potamogetonifolium* (Opiz) Miq., *Piper anisodorum* Náves ex Fern.-Vill., *Piper anisodorum* Blanco, *Piper bathicarpum* C.DC., *Piper betel* Blanco, *Piper betle* var. *amplifolium* C.DC., *Piper pinguispicum* C.DC. & Koord., *Piper malamiri* Blume, *Piper marianum* Opiz, *Piper philippinense* C.DC., *Piper betle* f. *densum* (Blume) Fosberg, *Piper siriboa* L., *Piper betle* var. *marianum* (Opiz) C.DC., *Piper malarayatense* C.DC., *Piper malamiris* L., *Piper betle* f. *marianum* (Opiz) Fosberg, *Piper puberulinodum* C.DC., *Piper fenixii* C.DC., *Piper betle* var. *siriboa* (L.) C.DC., *Piper bidentatum* Stokes, *Piper blancoi* Merr., *Piper blumei* (Miq.) Backer, *Piper canaliculatum* Opiz, *Piper carnistilum* C.DC., *Piper chavya* Buch.-Ham. ex Wall., *Piper densum* Blume, *Piper macgregorii* C.DC., *Piper potamogetonifolium* Opiz, *Piper rubroglandulosum* Chaveer. & Mokkamul, *Piper saururus* Burm., *Piper silletianum* P.K.Mukh., and *Piper betle* var. *densum* (Blume) C.DC. (POWO 2023).

P. betle is native to Cambodia, Laos, Lesser Sunda Is., Malaya, Myanmar, the Philippines, Thailand, and Vietnam, but also distributed in Indonesia, especially in

Fig. 2.18 Some herbal products from *Curcuma longa* from the online market

Lesser Sunda Island (POWO 2023). This plant has several local names such as betel, betel vine, betel pepper, betle pepper, pan, sire (English), sirih (Indonesia), sireh (Melayu), amo (Ambon), suruh (Jawa), bido (Ternate), and base (Bali). This plant is not native to Indonesia but has spread across the archipelago, starting with Java, which is located in the western part of the island, and then starting at Mount Sindoro to the east, with an altitude of 125–1400 m above sea level. *P. betle* requires a place to grow that has shade and is always moist. It thrives in ravines and small rivers, forests in villages, and moist thickets (Backer and Bakhuizen Van Den Brink 1965).

P. betle is a semishrub or climbing vine with glossy, heart-shaped leaves (Fig. 2.19). This vine has a height of up to 1.5 m; has greenish-brown stem, round, and segmented; and is the place where the roots emerge. The single leaf is heart-shaped, pointed at the tip, grows alternately, is stalked, and gives off a pleasant smell when squeezed, with dimensions of 5–8 cm in length and 2–5 cm in width. It has compound flowers in granular form, and there are protective leaves of ±1 mm in elliptical shape. The fruit is round and grayish-green in color. The roots are taper, round, and yellowish brown (Munawaroh et al. 2021).

P. betle is a volatile plant (veranderlijke plant) that adapts to the environmental conditions where it grows, so it has long been cultivated by tribes in the archipelago, used as a garden plant or around the house. According to Moeljanto and Mulyono (2003), a betel leaf is a canal or vine plant, one type of plant widely used for treatment. Leaves are mainly used as medicine for skin disorders such as wounds, burns,

Fig. 2.19 Illustration of *Piper betle* L

itchy rash, swelling, itching, and body odor, especially by the Kanayatan Dayak tribe and Nias tribe. They mashed or boiled the leaves, then affixed them to the skin, drank them, or used them as a bath.

P. betle is rich in chemical constituents such as essential oils, hydroxychavicol, chavicol, chavibetol, allypykatekol, carvacrol, eugenol, eugenol methyl ether, *p*-terpenenna, sesquiterpenes, phenylpropane, tannins, diastase, sugar, and starch. Within the scope of indigenous peoples, *P. betle* is used as a hereditary culture, as a medicinal plant (phytopharmaca), and in various traditional ceremonies (Moeljanto and Mulyono 2003). Furthermore, Munawaroh et al. (2021) explained that, apart from being used as a conventional medicine because it contains essential oils, betel nut is also used as a ritual plant and in traditional ceremonies. As herbal products, we could find them online (Fig. 2.20).

2.3.4.10 *Areca catechu* L.

Areca catechu is classified as the Arecaceae family and *Areca* genus. It has several synonyms, including homotypic synonyms, *Areca faufel* Gaertn. and *Areca hortensis* Lour., and heterotypic synonyms: *Sublimia areca* Comm. Ex Mart, *Areca catechu* var. *batanensis* Becc., *Areca catechu* var. *longicarpa* Becc., *Areca catechu* var. *silvatica* Becc., *Areca catechu* Burm. f., *Areca macrocarpa* Becc., *Areca himalayana* Griff. ex H. Wendl., *Areca nigra* Giseke ex H. Wendl., *Areca catechu* f.

Fig. 2.20 Some herbal products from *Piper betle* from the online market

Fig. 2.21 Illustration of *Areca catechu* L

communis Becc., *Areca catechu* var. *nigra* Giseke, and *Areca catechu* var. *alba* Blume (POWO 2023) (Fig. 2.21).

A. catechu is native to the Philippines but distributed to many countries, including Indonesia (POWO 2023). It has several local names: *Betel palm*, *Areca*, *Betelnut Palm* (English), Pinang (Indonesia), Pineung (Aceh), Pining (Batak Toba), Jambe (Jawa), Penang (Medan), Batang mayang (Karo), and Batang pinang (Minangkabau) (Socfindo Conservation 2023).

The leaf, flower, seed, and fruit of *A. catechu* are traditionally used for medicine and aromatherapy. It can cure skin disorders, intestinal worms, and back pain and

Fig. 2.22 Some herbal products from *Areca catechu* from the online market

strengthens teeth and gums. The young seeds are pounded and applied directly to the burnt skin or boiled and used the water to clean wounds and other skin illnesses (Sudarsono et al. 1996). The Dayak Benuaq tribe in East Kalimantan traditionally uses its fruit skin to cure acne (Pribady et al. 2019).

More than 59 compounds were isolated from *A. catechu*, including steroids, alkaloids, fatty acids, flavones, tannins, and triterpenes. The extracts and compounds isolated from *A. catechu* have many pharmacological activities (Peng et al. 2015; Socfindo Conservation 2023). Modern research has shown that *A. catechu* has a wide range of pharmacological activities, including antiparasitic effects on the nervous, digestive, and cardiovascular systems; antioxidant, antibacterial, antifungal, anti-inflammatory, and analgesic, anti-allergic, and regulatory effects on blood glucose and lipids; as well as effects on other diseases (Gilani et al. 2004; Liu et al. 2011; Peng et al. 2015). It has the potential to be antiaging (Lee and Choi 1999) and cure burns grade IIA (Sandhiutami et al. 2023). We could find some herbal products from this plant online (Fig. 2.22).

2.4 Conclusion

Due to Indonesia's incredible biodiversity, many medicinal plants can potentially treat a wide range of skin disorders. The Indonesian community has been able to effectively utilize the therapeutic benefits of these plants, thanks to the traditional knowledge and practices passed down through generations.

In order to validate the safety and efficacy of medicinal plants for treating skin disorders, it is crucial to find a balance between retaining traditional knowledge and carrying out scientific studies. By incorporating these various medicinal plants into contemporary healthcare systems, we may be able to manage skin issues in alternative and sustainable ways that assist Indonesians and people worldwide looking for safe and efficient alternative skincare solutions. It is critical to support conservation efforts to save Indonesia's distinctive biodiversity for future generations as we continue exploring the potential of these medicinal plants.

References

Adiyasa MR, Meiyanti M (2021) Pemanfaatan Obat Tradisional di Indonesia: Distribusi dan Faktor Demografis yang Berpengaruh. Jurnal Biomedika Dan Kesehatan 4(3):130–138. https://doi.org/10.18051/jbiomedkes.2021.v4.130-138

Adrian K (2021) 4 Manfaat Daun Sambiloto untuk Kulit yang Sayang Dilewatkan. https://www.alodokter.com/4-manfaat-daun-sambiloto-untuk-kulit-yang-sayang-dilewatkan. Accessed 15 Aug 2023

Agustina F, Zakaria R, Santi TD (2022) Hubungan Personal Hygiene Dengan Keluhan Penyakit Kulit Masyarakat Desa Tuwi Kayee Kecamatan Panga Kabupaten Aceh Jaya Tahun 2022. J Health Med Sci 1(4):142–149

Ahmad W, Jantan I, Bukhari SNA (2016) *Tinospora crispa* (L.) Hook. f. & Thomson: a review of its ethnobotanical, phytochemical, and pharmacological aspects. Front Pharmacol 7:59. https://doi.org/10.3389/fphar.2016.00059

Akbar S (2011) *Andrographis paniculata*: a review of pharmacological activities and clinical effects. Altern Med Rev 16(1):66–77

Anju D, Jugnu G, Kavitha S, Arjun N, Sandeep D (2012) A review on medicinal prospective of *Andrographis paniculata* Nees. J Pharm Sci Innov 1(1):1–4

Arozal W, Louisa M, Soetikno V (2020) Selected Indonesian medicinal plants for the management of metabolic syndrome: molecular basis and recent studies. Front Cardiovasc Med 7:82. https://doi.org/10.3389/fcvm.2020.00082

Azzahra F, Hayati M (2018) Uji Aktivitas Ekstrak Daun Pegagan (*Centella asiatica* (L). Urb) Terhadap Pertumbuhan *Streptococcus mutans*. Jurnal B-Dent 5(1):9–19

Az-Zahra FR, Sari NLW, Saputry R, Nugroho GD, Sunarto, Pribadi T, Setyawan AD (2021) Traditional knowledge of the Dayak Tribe (Borneo) in the use of medicinal plants: a review. Biodiversitas 22(10):4633–4647. https://doi.org/10.13057/biodiv/d221057

Backer A, Bakhuizen Van Den Brink RC (1965) Flora of Java (Spermatophytes only), vol 2. Angiospermae, families. pp 111–160

Balevi MMD, Balevi A (2023) The effect of *Centella asiatica* cream on scar development in patients who underwent open carpal tunnel release surgery. J Hand Ther 36(4):962–966

Bappenas (2016) Indonesian biodiversity strategy and action plan (IBSAP) 2015-2020. In: Kementerian Perencanaan Pembangunan Nasional/BAPPENAS. Kementerian Lingkungan Hidup dan Kehutanan (KLHK) dan Lembaga Ilmu Pengetahuan Indonesia (LIPI), Jakarta

Baron S (ed) (1996) Medical microbiology, 4th edn. University of Texas Medical Branch, Galveston

Batubara I, Prastya ME (2020) Potential use of Indonesian medicinal plants for cosmetic and oral health: a review. Jurnal Kimia Valensi 6(1):118–132

Bhagawan WS, Barsyaif UA, Hidayat MA (2021) Pendekatan Etnobotani Tumbuhan Obat Untuk Permasalahan Seksual Suku Tengger Di Desa Argosari, Lumajang, Indonesia. Jurnal Tumbuhan Obat Indonesia 14(2):99–110. https://doi.org/10.22435/jtoi.v14i2.4169

Boopathi C (2000) *Andrographis spp.*: a source of bitter compounds for medicinal use. Anc Sci Life 19(3–4):164–168

Borhanuddin M, Shamsuzzoha M, Hussain AH (1994) Hypoglycaemic effects of *Andrographis paniculata* Nees on nondiabetic rabbits. Bangladesh Med Res Counc Bull 20(1):24–26

Cahyaningsih R, Phillips J, Magos Brehm J, Gaisberger H, Maxted N (2021a) Climate change impact on medicinal plants in Indonesia. Global Ecol Conserv 30:e01752. https://doi.org/10.1016/j.gecco.2021.e01752

Cahyaningsih R, Magos Brehm J, Maxted N (2021b) Gap analysis of Indonesian priority medicinal plant species as part of their conservation planning. Glob Ecol Conserv 26:e01459. https://doi.org/10.1016/j.gecco.2021.e01459

Cahyaningsih R, Magos Brehm J, Maxted N (2021c) Setting the priority medicinal plants for conservation in Indonesia. Genet Resour Crop Evol 68(5):2019–2050. https://doi.org/10.1007/s10722-021-01115-6

Casulli A (2021) New global targets for NTDs in the WHO roadmap 2021-2030. PLoS Negl Trop Dis 15(5):e0009373. https://doi.org/10.1371/journal.pntd.0009373

Chao W-W, Lin B-F (2010) Isolation and identification of bioactive compounds in *Andrographis paniculata* (Chuanxinlian). Chin Med 5:17. https://doi.org/10.1186/1749-8546-5-17

Chouni A, Paul S (2018) A review on phytochemical and pharmacological potential of Alpinia galanga. Pharm J 10(1):9–15

da Mata IR, da Mata SR, Menezes RRC, Faccioli LS, Bandeira KK, Bosco SMD (2020) Benefits of turmeric supplementation for skin health in chronic diseases: a systematic review. Crit Rev Food Sci Nutr 61(20):3421–3435

Darmalaksana IGN, Jayawardhita AAG, Dada IKA, Sudimartini LM (2018) Gerusan Daun Pegagan Mempercepat Kesembuhan Luka Bakar pada Tikus Putih. Buletin Veteriner Udayana 10(2):137–146

de Padua LS, Bunyapraphatsara N, Lemmens RHMJ (eds) (1999) Plant resources of South-East Asia no. 12: (1) medicinal and poisonous plants 1. Backhuys Publishers, Leiden, p 711

Desmawati (2015) Hubungan personal hygiene Dan Sanitasi Lingkungan Dengan Kejadian Skabies Di Pondok Pesantren AlKautsar Pekanbaru. Jurnal Online Mahasiswa 2(1):628–637

Dewi IK (2020) Sifat Fisik Sediaan Ekstrak Rimpang Lengkuas (*Alpinia galanga*) dengan Variasi kadar CMC-Na. Jurnal Penelitian dan Kajian Ilmiah Kesehatan 6(2):215–220

Dewi NLA, Adnyani LPS, Pratama RBR, Yanti NND, Manibuy JI, Warditiani NK (2018) Pemisahan, Isolasi, dan Identifikasi Senyawa Saponin dari Herba Pegagan (*Centella asiatica* L. Urban). Jurnal Farmasi Udayana 7(2):68–76. https://doi.org/10.33394/bioscientist.v1i2.790

Dharma AP (1985) Tanaman obat tradisional Indonesia. PN Balai Pustaka, Jakarta. 291p

Djoko W, Taurhesia S, Djamil R, Simanjuntak P (2020) Standardisasi Ekstrak Etanol Herba Pegagan (*Centella asiatica*). Sainstech Farma 13(2):118–123

Efida M (2022) Inovasi produk berbasis ekstrak daun ketepeng cina sebagai gel antipruritic. Berita Artikel segar dan orosinal. https://bekelsego.com/inovasi-produk-berbasis-ekstrak-daun-ketepeng-cina-sebagai-gel-antipruritik/

Eraslan ZB, Genc GE, Kultur S (2020) Review article: medicinal plants traditionally used to treat skin diseases in Turkey – Eczema, Psoriasis, Vitiligo. J Fac Pharm Ankara 44(1):137–166. https://doi.org/10.33483/jfpau.586114

Esan VI, Ayanbamiji TA, Abodunrin AD (2021) Physiology of breaking seed dormancy and increasing seed germination in *Senna alata* (L) Roxb seeds in Nigeria. Asian J Agric Res 15(1):1–6. https://doi.org/10.3923/ajar.2021.1.6

Fujiastuti T, Sugihartini N (2015) Sifat Fisik dan Daya Iritasi Gel Ekstrak Etanol Herba Pegagan (*Centella asiatica* L.) Dengan Variasi Jenis Gelling Agen. Pharmacy 12(1):11–20. ISSN 1693-3591

Garvita RV, Ismaini L, Damayanti F, Lailaty IQ, Surya MI et al (2023) A study on seed germination of *Senna alata* (L.) (ketepeng). IOP Conf Earth Environ Sci 1192:012016. https://doi.org/10.1088/1755-1315/1192/1/012016

Gianfaldoni S et al (2020) Skin and skin disease throughout life. In: Smoller B, Bagherani N (eds) Atlas of dermatology, dermatopathology and venereology. Springer, Cham. https://doi.org/10.1007/978-3-319-45134-3_15-1

Gilani AH, Ghayur MN, Saify ZF, Ahmed SP, Choudhary MI, Khalid A (2004) Presence of cholinomimetic and acetylcholinesterase inhibitory constituents in betel nut. Life Sci 75(20):2377–2389. https://doi.org/10.1016/j.lfs.2004.03.035

Goldsmith LA, Katz SI, Gilchrest BA, Pallrer AS, Leffel DJ, Wolf KF (2012) Fitzpatrick's dermatology in general medicine, 8th edn. McGraw-Hill, New York

Gustia R, Yenny SW, Octari S (2020) Karakteristik penyakit kulit pada anak di poliklinik kulit dan kelamin RSUP. Dr. M. Djamil Padang periode 2016-2018. J Kedokt Syiah Kuala 20(3):143–146

Handayani (2017) Traditional to rational and modern phytopharmaca. In: Proceeding of Surabaya international health conference 2017, pp 23–34

Hargono D, Lastari P, Astuti Y, van den Bergh MH (1999) *Centella asiatica* (L.) Urban. In: de Padua LS, Bunyapraphatsara N, Lemmens RHMJ (eds) Plant resources of South-East Asia no

12(1): medical and poisonous plants 1. PROSEA Foundation, Bogor. Database record: prota4u.org/prosea

Hennebelle T, Weniger B, Joseph H, Sevser S, Bailleul F (2009) Senna alata. Fitoterapia 80:385–393. https://doi.org/10.1016/j.fitote.2009.05.008

Heukelbach J, Feldmeier H (2006) Scabies. Lancet 367(9524):1767–1774. https://doi.org/10.1016/S0140-6736(06)68772-2

Heyne K (1987) Tumbuhan Berguna Indonesia Jilid lll. Terjemahan. Badan Litbang Kehutanan. Yayasan Sarana Wana Jaya, Jakarta

Hidayat S (2021) Pemanfaatan Tumbuhan Obat oleh Beberapa Etnis di Indonesia. Prosiding seminar nasional PMEI ke V, pp 177–185

Hidayat S (2022) Nilai Konservasi Tumbuhan Obat di Kebun Raya Bogor. Dissertation. IPB University, Bogor

Hidayat S, Zuhud EAM, Widyatmoko D (2011) Konservasi Ex Situ Tumbuhan Obat di Kebun Raya Bogor. Forum Pascasarjana 34(4):257–265

Hossain MS, Urbi Z, Sule A, Hafizur Rahman KM (2014) *Andrographis paniculata* (Burm. f.) wall. ex Nees: a review of ethnobotany, phytochemistry, and pharmacology. Sci World J 2014:274905. https://doi.org/10.1155/2014/274905. 28p

Irham WH, Tamrin, Lamek M, Marpongahtun (2019) Bioactive compounds in pegagan leaf (*Centella asiatica* L. urban) for wound healing, 012019. J Phys Conf Ser 1232. https://doi.org/10.1088/1742-6596/1232/1/012019

Isro'in L, Andarmoyo S (2012) Personal Hygiene Konsep Proses Dan Aplikasi Dalam Praktik Keperawatan. Graha Ilmu, Yogyakarta

Jadid N, Kurniawan E, Himayani CES, Andriyani, Prasetyowati I, Purwani KI, Muslihatin W, Hidayati D, Tjahjaningrum ITD (2020) An ethnobotanical study of medicinal plants used by Tengger Tribe in Ngadisari Village, Indonesia. PLos One 15(7):e0235886. https://doi.org/10.1371/journal.pone.0235886

Jannah H, Ridwan R (2013) Pengobatan Tradisional Suku Sasak Berbasis Ilmiah di Kabupaten Lombok Barat. Bioscientist 1(2):116–125. https://doi.org/10.33394/bioscientist.v1i2.790

Jarukamjorn K, Nemoto N (2008) Pharmacological aspects of *Andrographis paniculata* on health and its major diterpenoid constituent andrographolide. J Health Sci 54(4):370–381. https://doi.org/10.1248/jhs.54.370

Kabir MH, Hasan N, Rahman MM, Rahman MA, Khan JA, Hoque NT, Bhuiyan MRQ, Mou SM, Jahan R, Rahmatullah M (2014) A survey of medicinal plants used by the Deb barma clan of the Tripura tribe of Moulvibazar district, Bangladesh. J Ethnobiol Ethnomed 10:19. https://doi.org/10.1186/1746-4269-10-19

Kanedi M, Handayani K, Setiawan WA (2022) Plants potential to be developed for wound healing medicine in Indonesia. Int J Front Life Sci Res 3(1):39–44. https://doi.org/10.53294/ijflsr.2022.3.1.0062

Karimkhani C, Dellavalle RP, Coffeng LE et al (2017) Global skin disease morbidity and mortality: an update from the global burden of disease study 2013. JAMA Dermatol 153(5):406–412

Kartawinata K (2010) Dua abad mengungkap kekayaan flora dan ekosistem Indonesia. Sarwono Prawirohardjo Memorial Lecture X. LIPI Press, Jakarta

Kartina K, Batubara I, Kuspradini H, Kusuma IW, Wahyuni WT, Egra S, Mitsunaga T (2022) Diversity of medicinal plants used for skincare by Bulungan tribe in North Kalimantan, Indonesia and its melanin biosynthesis inhibition. Biodiversitas 23(3):1246–1253. https://doi.org/10.13057/biodiv/d230308

Kaur H, Bhardwaj U, Kaur R (2021) *Cymbopogon na*rdus essential oil: a comprehensive review on its chemistry and bioactivity. J Essent Oil Res 33(3):205–220

Kocaadam B, Şanlier N (2017) Curcumin, an active component of turmeric (*Curcuma longa*), and its effects on health. Crit Rev Food Sci Nutr 57(13):2889–2895. https://doi.org/10.1080/10408398.2015.1077195

Kusmana C, Hikmat A (2015a) Keanekaragaman hayati flora di Indonesia (The biodiversity of flora in Indonesia). Jurnal Pengelolaan Sumberdaya Alam dan Lingkungan 5(2):187–198

Kusmana C, Hikmat A (2015b) The biodiversity of flora in Indonesia. J Nat Resour Environ Manag 5(2):187–198. https://doi.org/10.19081/jpsl.5.2.187

Kuswanto L, Chusna NA, Purnomo E, Krisantini, Ahmad MU (2021) Identification and documentation of wild species with ornamental potential at Mount Prau, Central Java, Indonesia. Ornam Hortic 28(1):110–119

Lee KK, Choi J-D (1999) The effects of *Areca Catechu* L extract on antiaging. Int J Cosmet Sci 21:285–295

Lin P, Shi HY, Lu YY, Lin J (2023) *Centella asiatica* alleviates psoriasis through JAK/STAT3-mediated inflammation: an in vitro and in vivo study. J Ethnopharmacol 317:116746

Liu ST, Young GC, Lee YC, Chang YF (2011) A preliminary report on the toxicity of arecoline on early pregnancy in mice. Food Chem Toxicol 49(1):144–148

Mareta CA (2020) Efektifitas Pegagan (*Centella asiatica*) sebagai Antioksidan. Jurnal Medika Hutama 2(1):390–394

Medical Research Foundation (2023) Addressing the impact of skin disorders in children and young people. Medical Research Foundation, London. https://www.medicalresearchfoundation.org.uk/news/addressing-the-impact-of-skin-disorders-in-children-and-young-people

Ministry of the Trade Republic of Indonesia (2009) Study of Foreign Development Ministry of Trade Republic of Indonesia

Mishra SK, Sangwan NS, Sangwan RS (2007) *Andrographis paniculata* (Kalmegh): a review. Pharmacogn Rev 1(2):283–298

Moeljanto DR, Mulyono M (2003) Khasiat dan Manfaat Daun Sirih. Agromedia Pustaka, Bandung

Munawaroh E, Solihah SM, Yuzammi Y (2021) Mengenal Tumbuhan Sirih (*Piper spp.*): Manfaat, Budidaya, dan Budaya. Pusat Penelitian Konservasi Tumbuhan dan Kebun Raya-LIPI, Bogor

Nakahara K, Alzoreky SN, Yoshihashi T, Nguyen TTH, Trakoontivakorn G (2003) Chemical composition and antifungal activity of essential oil from from *Cymbopogon nardus* (Citronella grass). Jpn Agric Res Q 37(4):249–252

Nelson KM, Dahlin JL, Bisson J, Graham J, Pauli GF, Walters MA (2017) The essential medicinal chemistry of curcumin. J Med Chem 60(5):1620–1637. https://doi.org/10.1021/acs.jmedchem.6b00975

Nugraha AS, Keller PA (2011) Revealing indigenous Indonesian traditional medicine: anti-infective agents. Nat Prod Commun 6(12):1953–1966

Nugroho AW (2017) Review: Konservasi Keanekaragaman Hayati Melalui Tanaman Obat Dalam Hutan di Indonesia dengan Teknologi Farmasi: Potensi dan Tantangan. Jurnal Sains dan Kesehatan 1(7):377–383

Nugroho Y, Soendjoto MA, Suyanto, Matatula J, Alam S, Wirabuana PYAP (2020) Traditional medicinal plants and their uitilization by local communities around Lambung Mangkurat Education Forests, South Kalimantan, Indonesia. Biodiversitas 23(1):306–314

Nurjanah N, Aprilia BE, Fransiskayana A, Rahmawati M, Nurhayati T (2018) Senyawa bioaktif rumput laut dan ampas teh sebagai antibakteri dalam formula masker wajah. Jurnal Pengolahan Hasil Perikanan Indonesia 21(2):304–316

Oladeji OS, Adelowo FE, Oluyori AP, Bankole DT (2020) Ethnobotanical description and biological activities of *Senna alata*. Evid Based Complement Alternat Med 2020:2580259. https://doi.org/10.1155/2020/2580259

Pardiansyah R (2015) Association between personal protective equipment with contact dermatitis in scavengers. Jurnal Majority 4(4):80–87

Peng W, Liu YJ, Wu N, Sun T, He XY, Gao YX, Wu CJ (2015) *Areca catechu* L. (Arecaceae): a review of its traditional uses, botany, phytochemistry, pharmacology and toxicology. J Ethnopharmacol 164:340–356

Permanasari T (2001) Kajian Pemanfaatan Tumbuhan Obat oleh Berbagai Etnis di Indonesia. PhD thesis. IPB (Institut Pertanian Bogor)

Plants of the World Online (POWO) (2023) Kew science. Royal Botanic Gardens, Kew. https://powo.science.kew.org/. Diakses tanggal 17 Agustus 2023

Prasad S, Anggarwal BB (2011) Chapter 13. Turmeric, the golden spice. In: Herbal medicine: biomolecular and clinical aspects, 2nd edn. CRC Press/Taylor & Francis, Boca Raton

Pribady HK, Ardana M, Rusli R (2019) Potensi Ekstrak Kulit Buah Pinang sebagai Antibakteri *Propionibacterium acne* Penyebab Jerawat. Proc Mulawarman Pharm Conf 10(1):100–103. https://doi.org/10.25026/mpc.v10i1.370
Purnamasari D (2018) The emergence of non-communicable diseases in Indonesia. Acta Med Indones 50(4):273–274
Purwoko Y, Kusumaningrum HP, Sugiarti L, Hapsari HA (2020) Aplikasi Konsorsium Tanaman Herbal Untuk Mengatasi Jerawat Akibat Autoimun Suatu Upaya Pengembangan Traditional Biomedicine. Cendekia J Pharm 4(1):10–25
Raajshree KR, Durairaj B (2017) Evaluation of the antityrosinase and antioxidant potential of zinc oxide nanoparticles synthesized from the brown seaweed-turbinaria canoides. Int J Appl Pharm 9(5):116–120
Rahayu SM, Andini AS (2019) Ethnobotanical study on medicinal plants in Sesaot Forest, Narmada,West Lombok, Indonesia. Biosaintifika 11(2):234–242. https://doi.org/10.15294/biosaintifika.v11i2.19314
Rahayu M, Ibo LK, Arimukti SD, Susiarti S (2020) Diversity of medicinal plants in yard in several villages in Indonesia. IOP Conf Ser Earth Environ Sci 572:012004. https://doi.org/10.1088/1755-1315/572/1/012004
Ramandey J, Bunei P (2021) Indentifikasi Tanaman Pegagan (*Centela asiatika* L.) sebagai Tanaman Obat bagi Masyarakat suku Mee di Distrik Tigi Timur Kabupaten Deiya. Jurnal Pertanian dan Peternakan 6(1):23–31
Reddy KV, Yachawad AV, Pathan RR, Syed IJ (2022) Herbal medicinal plants used in treatment of skin disorders: overview. GIS Sci J 9(5):2009–2024
Retnowati A, Rugayah, Rahajoe JS, Arifiani D (2019) Status Keanekaragaman Hayati Indonesia: Kekayaan Jenis Tumbuhan dan Jamur Indonesia. LIPI Press, Jakarta
Rismana E, Idah R, Prasetyawan Y, Olivia B, Erna Y (2013) Efektivitas Khasiat Pengobatan Luka Bakar Sediaan Gel Mengandung Fraksi Ekstrak Pegagan Berdasarkan Analisis Hidroksiprolin dan Histopatologi pada Kulit Kelinci. Buletin Penelitian Kesehatan 41(1):45–60
Romani L, Steer AC, Whitfeld MJ, Kaldor JM (2015) Prevalence of scabies and impetigo worldwide: a systematic review. Lancet Infect Dis 15(8):960–967
Roni A, Kurnia D, Hafsyah N (2022) Penetapan Kadar Flavonoid Dan Aktivitas Antioksidan Pada Ekstrak Batang Brotowali (*Tinospora Crispa* L.) Dengan Metode Cuprac. Jurnal Ilmiah Ibnu Sina 7(1):165–173
Rubiah R, Djufri D, Muhibbuddin M (2015) Ethnobotanical srudy for skin disease drug by society in Pidie District. Biologi Edukasi: Jurnal Ilmiah Pendidikan Biologi 7(1):34–41
Sahala MA, Soedarman S, Rizky LA, Natanegara AP, Advani MS, Sungkar S (2016) The prevalence of skin diseases and its association with hygiene behavior and level of education in a Pesantren, Jakarta Selatan 2013. eJournal Kedokteran Indonesia 4(2):119–124. https://doi.org/10.23886/ejki.4.6288.119-24
Saikia AP, Ryakala VK, Sharma P, Goswami P, Bora U (2006) Ethnobotany of medicinal plants used by Assamese people for various skin ailments and cosmetics. J Ethnopharmacol 106:149–157
Salim Z, Munadi E (2017) Info Komoditi Tanaman Obat. Badan Pengkajian dan Pengembangan Perdagangan Kementerian Perdagangan Republik Indonesia, Jakarta
Sandhiutami NMD, Fahleni F, Miftahurrohmah N, Widhiyasari NKA, Azalia A, Amalia I (2023) Enhanced wound healing effect of Areca catechu L. ointment via antibacterial activity and anti-inflammatory process at grade IIA burns in rats. J Herbmed Pharmacol 12(3):388–398. https://doi.org/10.34172/jhp.2023.42
Sangat HM, Zuhud EAM, Damayanti EK (2000) Kamus penyakit dan tumbuhan obat Indonesia (etnofitomedika). Yayasan Obor Indonesia, Jakarta
Santosa E, Susila AD, Widodo WD, Nasrullah N, Ruwaida IP, Sari R (2021) Exploring fruit tree species as multifunctional greenery: a case of its distribution in Indonesian cities. Sustainability 13:7835. https://doi.org/10.3390/su13147835
Saputra NA, Wibisono HS, Darmawan S, Pari G (2020) Chemical composition of *Cymbopogon nardus* essential oil and its broad-spectrum benefit. IOP Conf Ser Earth Environ Sci 415:012017

Sartika S, Indradi RB (2021) Pharmacological activities of Daun Ungu Plants (*Graptophyllum pictum* L. Griff). Indones J Biol Pharm 1(2):88–96

Schneider S, Wu J, Tizek L, Ziehfreund S, Zink A (2023) Prevalence of scabies worldwide—an updated systematic literature review in 2022. J Eur Acad Dermatol Venereol 2023:1–9

Sharifi-Rad J, Rayess YE, Rizk AA, Sadaka C, Zgheib R, Zam W, Martins N (2020) Turmeric and its major compound curcumin on health: bioactive effects and safety profiles for food, pharmaceutical, biotechnological and medicinal applications. Front Pharmacol 11:01021. https://doi.org/10.3389/fphar.2020.01021

Sharma M, Sharma R (2013) Identification, purification and quantification of andrographolide from *Andrographis paniculata* (burm. F.) Nees by HPTLC at different stages of life cycle of crop. J Curr Chem Pharm Sci 3(1):23–32

Siew YY, Zareisedehizadeh S, Seetoh WG, Neo SY, Tan CH, Koh HL (2014) Ethnobotanical survey of usage of fresh medicinal plants in Singapore. J Ethnopharmacol 155(3):1450–1466

Silalahi M (2014) Etnobotani Dan Valuasi Ekonomi Pegagan (*Centella asiatica* L. Urban) Sebagai Upaya Pendidikan Konservasi. Jurnal Dinamika Pendidikan 7(1):1–8. ISSN 1410 4695 (uki.ac.id)

Silvina. (2006) Uji Banding Efektivitas Ekstrak Rimpang Lengkuas (*Alpinia Galanga*) 10% Dengan Ketokonazol 2% Secara In Vitro Terhadap Pertumbuhan Candida albicans Pada Kandidiasis Vaginalis. Fakultas Kedokteran Universitas Diponegoro Semarang, Skripsi

Siregar RS, Tanjung AF, Siregar AF, Salsabila S, Bangun IH, Mulya MO (2020) Studi literatur tentang pemanfaatan tanaman obat tradisional. In: Scenario (Seminar of social sciences engineering and humaniora), pp 385–391

Socfindo Conservation (2023) Index tanaman. https://www.socfindoconservation.co.id/index. Accessed 26 Aug 2023

Statistics (2023) Statistical year book of Indonesia 2023. Badan Pusat Statistik

Sudarsono, Pudjoarinto A, Gunawan D, Wahyuonos, Donatus IA, Drajad M (1996) Tumbuhan Obat. Pusat Penelitian Obat Tradisional Universitas Gajah Mada, Yogyakarta

Suprihatin A, Eliza Y, Ratmini NPS, Raharjo B, Purwanto BH, Syukur A (2023) *Centella asiatica* in Inceptisol Ngipiksari, Yogyakarta. IOP Conf Ser Earth Environ 1172(1):012018. https://doi.org/10.1088/1755-1315/1172/1/012018

Suryanto R, Setiawan D (2013) Struktur Data Datawarehouse Tanaman Obat Indonesia dan Hasil Penelitian Obat Tradisional. SESINDO 2013

Sutardi S (2016) Kandungan Bahan Aktif Tanaman Pegagan dan Khasiatnya untuk Meningkatkan Sistem Imun Tubuh. Jurnal Litbang Pertanian 3(3):121–130. https://doi.org/10.21082/jp3.v35n3.2016

Sutrisna EM (2016) Herbal Medicine: Suatu Tujuan Farmakologis. Muhammadiyah University Press, Surakarta

Tabassum N, Hamdani M (2014) Plants used to treat skin diseases. Pharmacogn Rev 8(15):52–60. https://doi.org/10.4103/0973-7847.125531

Umagap WA (2018) Inventarisasi Jenis Tanaman Obat Tradisional yang Berkhasiat untuk Pengobatan Kulit di Beberapa Kelurahan Pulau Ternate dan Ternate Selatan. Techno: Jurnal Penelitian 7(1):66–75

Valdiani A, Kadir MA, Tan SG, Talei D, Abdullah MP, Nikzad S (2012) Nain-e havandi andrographis paniculata present yesterday, absent today: a plenary review on underutilized herb of Iran's pharmaceutical plants. Mol Biol Rep 39(5):5409–5424

Vera Y, Yanti S (2020) Penyuluhan pemanfaatan tanaman obat dan obat tradisional Indonesia untuk pencegahan dan penanggulangan penyakit hipertensi di Desa Salam Bue. Jurnal Education and Development 8(1):11–14

Vollono L, Falconi M, Gaziano R, Lacovelli F, Dika E, Terracciano C, Campione E (2019) Potential of curcumin in skin disorders. Nutrients 11(9):2169. https://doi.org/10.3390/nu11092169

Vora R, Bodiwala N, Patel S, Krishna S (2012) Prevalence of various dermatoses in school children of Anand district. Natl J Commun Med 3(1):100–103

Walker HK, Hall WD, Hurst JW (1990) Clinical methods: the history, physical and laboratory examinations, 3rd edn. Butterworths, Boston

Warsinah W, Baroroh HN, Harwoko H (2020) Phytochemical analysis and antioxidant activity of Brotowali (*Tinospora crispa* L. Mier). Stem Molekul 15(2):73–78. https://doi.org/10.20884/1.jm.2020.15.2.533

Wathoni N, Haerani A, Yuniarsih N, Haryanti R (2018) A review on herbal cosmetics in Indonesia. Int J App Pharm 10(5):13–16. https://doi.org/10.22159/ijap.2018v10i5.28102

Widaty SE, Miranda E, Cornain EF, Rizky LA (2022) Scabies: update on treatment and efforts for prevention and control in highly endemic settings. J Infect Dev Ctries 16(2):244–251

Widjaja (2014) Kekinian Keanekaragaman Hayati Indonesia 2014. In: Igarss 2014 (issue 1)

Widyatmoko D (2019) Strategi dan Inovasi Konservasi Tumbuhan Indonesia untuk Pemanfaatan secara Berkelanjutan. In: Proceedings SNPBS (Seminar Nasional Pendidikan Biologi Dan Saintek Ke-IV 2019), pp 1–22

Wijaya A, Utami LW (2018) Uji fisik sediaan gel dengan ekstrak daun wungu (Graptophyllum pictum (L) Griff) dengan kombinasi humektan propilen glikol dan gliserin. J Kefarmasian Akfarindo. 31:16–22. https://doi.org/10.37089/jofar.v0i0.27

Wootton CI, Bell S, Philavanh A, Phommachack K, Soukavong M, Kidoikhammouan S, Mayxay M (2018) Assessing skin disease and associated health-related quality of life in a rural Lao community. BMC Dermatol 18(1):11. https://doi.org/10.1186/s12895-018-0079-8

Wulandari A (2018) Hubungan personal hygiene dan sanitasi lingkungan dengan kejadian skabies pada santri di pesantren Ulumul Qur'an Kecamatan Bebesen Kabupaten Aceh Tengah. Glob Health Sci 3(4):322–328

Yamin M, Burhanudin, Jamaluddin, Nasruddin (2018) Pengobatan Dan Obat Tradisional Suku Sasak Di Lombok. Jurnal Biologi Tropis 18(1):1411–9587

Yusriani Y, Ermawati E, Dewi R (2018) Uji Daya Hambat Krim Ekstrak Batang Brotowali (*Tinospora crispa* L.) terhadap Propionibacterium acnes. Jurnal Kesehatan yamasi Makassar 2(2):1–6

Yusuf UK, Horsten SFAJ, Lemmens RHMJ (1999) *Tinospora crispa* (L.) Hook.f. & Thomson. In: de Padua LS, Bunyapraphatsara N, Lemmens RHMJ (eds) Plant resources of South-East Asia No 12(1): medical and poisonous plants 1. PROSEA Foundation, Bogor

Yuwono Y, Marhaeni H, Tusianti E, Larasaty P, Nugroho A (2015) Analisis Rumah Tangga Sekitar Kawasan Hutan di Indonesia. Badan Pusat Statistik, Jakarta

Zhang XJ, Wang AP, Shi TY, Zhang J, Xu H, Wang DQ, Feng L (2019) The psychosocial adaptation of patients with skin disease: a scoping review. BMC Public Health 19:1404. https://doi.org/10.1186/s12889-019-7775-0

Endophytic Fungal Diversity of Medicinal Plants and Their Role During Stress Response

3

Ankita Chandel, Harleen Kaur, Ravneet Kaur, Swapnil Singh, Neetu Jagota, and Ashish Sharma

Abstract

In the plant kingdom, endophytes and, in particular, endophytic fungi have been given immense importance simply due to the diversity of roles they take up for plants and based on the simple fact that almost 80% of perennial plants are associated with one or the other endophytic fungal partner. Furthermore, it has also been shown that up to six different endophytes can be present in a single plant species, and one endophytic partner can form relations with different plant species. On the other hand, medicinal plants have also been an essential part of the human health management system from times immemorial. As many medicinal plant species are also perennial, it becomes necessary to believe that they may also enter associations with endophytic fungi among other partners. In the present manuscript, we try to present a comprehensive picture of various associations formed by the endophytic fungal partners with their medicinal host plants. We shall explore the diversity of fungal species associated with plants, the diversity of medicinal plants playing host to endophytes, and the types of relationships existing between them.

Keywords

Endophytes · Symbiosis · Medicinal compounds · Biotic stress management · Abiotic stress management

A. Chandel · H. Kaur · R. Kaur · S. Singh · N. Jagota · A. Sharma (✉)
Department of Botany and Environment Science, DAV University, Jalandhar, Punjab, India
e-mail: ashish10210@davuniversity.org

© The Author(s), under exclusive license to Springer Nature Singapore Pte Ltd. 2024

M. Nandave et al. (eds.), *Ethnopharmacology and OMICS Advances in Medicinal Plants Volume 1*, https://doi.org/10.1007/978-981-97-2367-6_3

3.1 Introduction

Fungi are heterotrophic eukaryotes that divide through the processes of meiosis and mitosis to create elongated polar cells (hyphae) or budding cells known as yeast (Naranjo-Ortiz and Gabaldón 2019). In the Tree of Life, these microbial lifestyles developed separately (Hawksworth and Lücking 2018). Eukaryotes in this region are among the most biologically diverse in the world distribution, and diversity of fungi, particularly endophytes, are poorly understood because of their complicated morphological structure and lifestyle, which frequently depend on other organisms or fungi, endophyte tissues intracellularly or extracellularly throughout all, or a portion of the vegetative life cycle (Li et al. 2016a, b). It is also challenging to distinguish isolates by morphological characteristics because many fungal species share similar structural types, like mycelium made up of more than a single cell covered by a tubular cell wall (Lücking et al. 2020). Furthermore, some of these techniques involve enriching the environment with specific substrates to boost the variety of fungal varieties (Mourad et al. 2018), as well as the use of molecular techniques to identify isolates (Noriler et al. 2018).

3.2 Relationship Between Plants and Endophytic Fungus

With specialized adaptations on the side of the plant and the fungi, the interaction between an endophyte and a host plant evolved (Verma et al. 2012). Depending on the fungus, the plant, or both, the interaction is complex and varied. The endophyte's origin is not yet thoroughly understood. It can fall under two categories, viz., exogenous and endogenous. Endophytes are grown from plant chloroplast and mitochondria, as implied by the endogenous premise, and, as a result, share the host's genetic heritage. The exogenous hypothesis, however, contends that endophytes enter their host plant by surface contact, induced channels, or root wounds (Hondelmann et al. 2020). It is known as neutralism that fungal endophytics are typically dormant once they have entered the tissues of their host. According to mutualism or antagonism, other fungal endophytes may be dormant for a long time before environmental factors favor their activation (Verma et al. 2016).

According to broad consensus (Sherameti et al. 2005), elements, including the genetic makeup of the plants in question, their ecological habitats, and the nutrients in the soil, have a significant impact on the quantity and quality of raw medicines obtained from medicinal plants. However, over the past few years, it has become increasingly clear that endophytic fungi or endophytes have a significant impact on the quantity and quality of crude medicines through particular fungal-host interactions, indicating the need for a more profound knowledge of endophytic connections. Increasing the manufacturing of crude medicines requires mushrooms and herbs. Despite the fact that endophytic fungi are one of the most prominent components of plant microecosystems and have a considerable impact on the rise and development of host plants, there is still minimal understanding of the precise interactions between endophytic fungus and their host plants. By modifying the environment under which medicinal plants develop, such as by adding a specific group of endophytic fungi to plants to improve the quality and quantity of the medicine, it is possible to produce better medicines in the optimal manner (Firáková et al. 2007). Suppose the fungus-host relationships and their metabolic mechanisms under cultivation conditions are well understood. In that case, an alternative method for the direct production of desired drugs may be developed (Kumaran et al. 2009), which is possible by using biotechnology of specific medicinal plants and endophytic fungi. This industrial method of manufacturing can take the role of the conventional approach to producing drugs, which is based mainly on natural medicinal plants. The term "endophytic fungi" refers to diasporic and mesospheric ascomycetes that "symptomatically inhabit plant endosperms beneath the epidermal cell layer, where they colonize healthy and living tissue through dormant infections." In temperate areas and tropical rainforests, endophytic fungi naturally occur and have a wide variety of biological characteristics. Terrestrial host species number 300,000. There are one or more endophytic fungus species for every plant species. Endophytic fungi are a wide-ranging, polyphyletic class of microorganisms that can grow asymptomatically in a number of healthy tissues above and below ground, including the stems, leaves, and roots of living plants. Endophytic fungi are thought to exist in nature in over a million different species (Faeth and Fagan 2002). A mycorrhizal endophyte, a pasture endophyte, and a non-pasture endophyte are the three primary ecological groupings. Schultz divided endophytic fungi into the host plants, and endophytic fungal bioactive substances are crucial for increasing the adaptability of both the endophytic fungi and their host plants, for example, by boosting tolerance to biotic and abiotic stress.

Additionally, these substances can promote the synthesis of a variety of well-known and novel physiologically active secondary metabolites (Rodriguez et al. 2009), which are significant sources of medication that can be used and used by people. Due to chemotaxis, which is the movement of endophytic fungus by specific chemicals produced by the host plants, it is understood that this process is not random. While this is happening, several kinds of secondary metabolites, such as saponin and essential oils from medicinal plants, are developing as a defense mechanism against pathogens, including endophytic fungi. As a result, endophytic fungi encountered a barrier when trying to colonize secondary metabolites. In order to overcome this challenge, endophytic fungi need to release specific nontoxic enzymes, including cellulases, lactases, xylanases, and proteases. These enzymes serve to break down the secondary compounds before engaging with the defensive mechanisms of the host plant. After entering the host plant's tissues, endophytic fungi go into a condition of dormancy (latency), either for the duration of the host plant's whole life cycle (neutralism) or for a shorter period until the situation is favorable, for a more extended period (mutualism or hostility). According to Sieber (2007), endophytic fungi benefit from the host's ontogenetic condition. Endophytic fungi and their host plants have developed a variety of connections throughout lengthy cohabitation and evolutionary processes through interactions that have been classified as follows: (a) mutualism, (b) antagonism, and (c) commensalism. In the population structure of an endophytic fungus, the genetic background, nutrient availability, and ecological niches of the therapeutic host plants are thought to serve as selection pressure factors, which in turn provide benefits like stimulated growth and improved resistance. Various bioactive components have been collected over time due to diseases (Rodriguez et al. 2009) and herbivores, some of which can be employed as human medications.

Consequently, the interactions between endophytic fungus and their host plants can affect the makeup of specific bioactive chemicals that are accessible to humans (Firáková et al. 2007). We examined studies on endophytic fungi and medicinal plants from the past 30 years in this paper, paying particular attention to variables that may affect the population structure and distribution of endophytic fungus as well as the advantages endophytic fungi have for their host plant (Firáková et al. 2007). For the practical synthesis of desired pharmaceutical chemicals, biotechnology should be able to design a system that mimics the mutualistic or antagonistic symbiotic circumstances between endophytic fungi and their host plants.

The advantages that some endophytic fungi offer to their host plants following colonization were further disclosed by our chosen reference-based investigation. These positive interactions can be explained from three different angles. To promote the growth of host plants, numerous endophytic fungi can produce a range of plant hormones (Waqas et al. 2012). Alkaloids, diterpenes, flavonoids, and isoflavonoids are only a few examples of the bioactive substances that some endophytic fungi create to help their host plants fight biotic and abiotic challenges (Firáková et al. 2007; Rodriguez et al. 2009). Third, certain endophytic fungi may encourage the buildup of secondary metabolites, including significant medicines or therapeutic components, which plants first created. Both host plants are capable of producing these metabolites, according to the sources consulted (Shwab and Keller 2008). We provide three potentially helpful types of endophytic fungus-host relationships in response to the significance of the three factors.

3.3 Modes of Host-Endophyte Transfer

Endophytic fungi can reproduce in three different ways:

(a) Vertical or intercellular transmission: Infected plants use seeds or hyphae to pass the virus to their progeny in this way, which is how *Neotyphodium* spp. spreads (Gebretsadik et al. 2022).
(b) Horizontal (intracellular) transmission: This type is carried out by *Epichloë* spp. and involves the dissemination of gametes from infected plants (Chen et al. 2022). Gebretsadik et al. (2022) define mixed mediation as a situation in which mediation is both vertical and horizontal. In order to start an infection or germinate contaminated seeds, vertical transmission is an asexual reproduction method that includes the transfer of intercellular hyphae from an infected plant to a non-plant (Daley et al. 2017). The formation of an infectious gamete that passes through the fungus' sexual life cycle is necessary for horizontal transmission. The stroma surrounding the grass flag leaf is the sole place for these

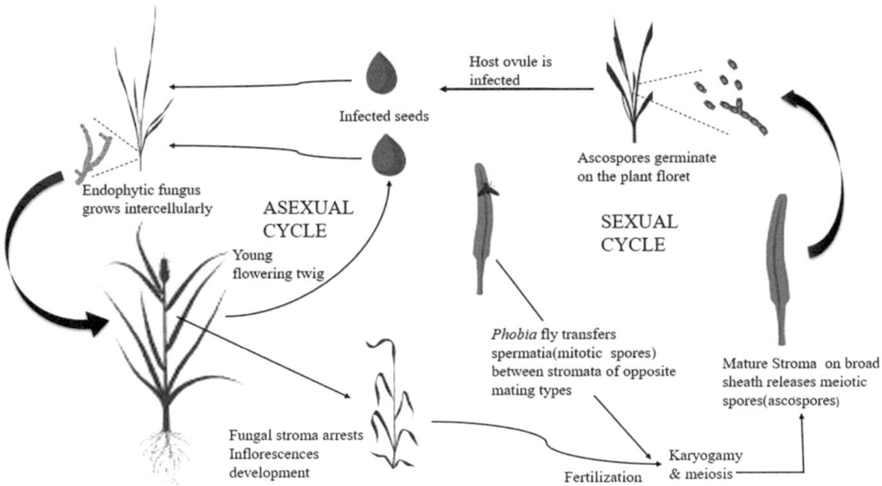

Fig. 3.1 Schematic representation of life cycle (asexual and sexual) of *Epichloë festucae* symbiotic with *Festuca* sp. (in the book: *Sustainable crop production*. License CC BY 3.0, redrawn on Biorender)

spores (like some *Epichloë* spp.) to live. The propagation of the fungus affects the leaves as soon as it emerges. The leaf cuticle then extends intercellularly after penetrating via the interior stomata of healthy tissues (Verma et al. 2016). However, a third reproductive characteristic is used by the majority of *Epichloë* spp. According to Fig. 3.1 (Verma et al. 2016), endophytes are transferred both vertically (systemic) and horizontally (intracellularly).

3.4 Medicinal Plants and Fungal Associations

Endophytic fungi live inside plant tissue like leaves, flowers, bark, scales, twigs, reproductive organs, and roots without manifesting any symptoms of a disease. Endophytes have been demonstrated to have a number of positive impacts on plants, including improved growth and disease resistance (Palem et al. 2016). They are commonly found in different plant groupings, such as medicinal plants, grasses, trees, and vegetables. Although it has been shown that endophytic interactions with plants can shield them from infections, the link between endophytes and host plants and pathogens is complex and not yet entirely understood (Li et al. 2016a, b). Endophytes may multiply in the intercellular spaces of plants because of the abundance of nutrients there, including potassium, sulfur, calcium, carbohydrates, organic acids, and different amino acids. Endophytic fungi stimulate growth factors in their host plants, which directly benefits them and indirectly prevents illnesses (Yashavantha Rao 2020). By providing resistance to several diseases and dry conditions, endophytic bacteria enhance seedling growth and survival. Host endophytes may be able to improve host plant immunity against many diseases. However, certain herbs have an insufficient immune response to some invading pathogens that impact their growth and function. These endophytes either directly or indirectly support systems for plant growth.

A very important step in the identification of endophytic fungi by mycologists is morphological identification (Yashavantha Rao 2020). On standard media, fungus is developed, sexual reproduction is examined, and growth properties are determined. According to taxonomical categories, a few tests, including subculturing, a rapid test for purity, an evaluation of sexual reproduction, conidiogenesis, and an estimation of radial growth and media temperature, have been added recently (Maehara et al. 2013). These endophytes can target different medicinal plants through a variety of biocontrol mechanisms. Plants can defend themselves against diseases infiltrating the endosphere by secreting signaling chemicals. Through various signaling molecules, these plants engage with symbiotic, pathogenic, neutral, or related microorganisms in multiple ways. Better communication between plants and other species is facilitated by host-endophyte associations, as shown in Table 3.1 (Yashavantha Rao 2020).

3.5 Assortment of Endophytic Fungi with Medicinal Plants

Plant tissues from the leaves, stems, and roots included 328 different fungus isolates in total (Hafeez et al. 2023). Molecular characterization was used to identify 12 species of endophytic fungus. Due to its numerous natural products and antioxidant action, *Crocus sativus* L. (family Iridaceae) has been widely utilized as an antibacterial, antidepressant, digestive, anticancer, and anticonvulsant medicine. Endophytic fungi were isolated from 5 different *Crocus sativus* tissue regions (bark, wing, leaf, petal, and stigma) and discovered a total of 32 endophytic fungal groups, which were split into 7 classes (Lu et al. 2023). Thirty-four endophytic fungi from 10 genera and 16 species of the traditional Chinese medicine *Salvia miltiorrhiza* were isolated by Wang et al. (2021), and *Epicoccus* sp. and *Colletotrichum gloeosporioides* endophytic fungi from 10 genera and 16 species of the traditional Chinese medicine *Salvia miltiorrhiza* were isolated by Wang et al. (2021), and *Epicoccus* sp. and *Colletotrichum gloeosporioides* showed a potent inhibitory impact on 5 pathogens. Sororicides showed a full inhibitory impact on five pathogens. The drought-resistant mint species *Endostemon obtusifolius*' roots and leaves were used to isolate a total of 11 endophytic fungi (Ogbe et al. 2023). The five endophytic fungi *Colletotrichum tropicicola*, *Fusarium keratoplasticum*, *F. oxysporum*, *F. solani*, and *Trichoderma longibrachiatum* were also isolated from wild *Dendrobium nobile* leaf segments. In traditional Chinese medicine, the plant *Codonopsis pilosula* has properties that include spleen strengthening, lung strengthening, and fluid creation (Luan et al. 2021). *Fusarium*, *Aspergillus*, *Alternaria*, *Penicillium*, *and Plectosphaerella* were the dominant families among 205 strains of endophytic fungus that were collected from 6 Gansu provinces in China. More than 30 different species of endophytic fungi have been isolated from *Vernonia anthelmintica* by researchers, which holds a long history of use to treat various disorders of the skin, CNS, kidneys, and

Table 3.1 An assortment of different fungal endophytes found associated with medicinal plants and metabolites obtained from them

Host plant	Tissues	Endophytic fungi	Major medicinal components	References
Aconitum heterophyllum	Root, stem, and leaf	*Arthrinium, Chaetomium, Purpureocillium, Alternaria, Penicillium, Aspergillus, Cladosporium, and Bjerkandera* sp.	Alkaloids, diterpenes, flavonoids, saponins, and tannins	Hafeez et al. (2023)
Crocus sativus	Corm, scape, leaf, petal, and stigma	*Sistotrema, Bjerkandera, and Penicillium* sp.	Alkaloids, phenolic compounds, saponins, terpenoids, carotenoids, flavonoids, and anthocyanins	Lu et al. (2023)
Salvia miltiorrhiza	Root	*Fusarium, Epicoccum, Aspergillus, Coprinellus, Dictyosporium, Colletotrichum, Rhizoctonia, Phomopsis, and Pithomyces*	Salvianolic acid A, salvianolic acid B, *Litho* sp., *Ovatospora* sp., *Chaetomium* sp., *Thielavia* sp., and *Aspergillus* sp., tanshinone I, tanshinone IIA, tanshinone IIB	Wang et al. (2021)
Mint (Endostemon obtusifolius)	Root and leaf	*Fusarium* sp.	Menthol, menthone, neomenthol, and iso-menthone	Ogbe et al. (2023)
Dendrobium officinale	Leaf	*Fusarium* sp., *Trichoderma* sp., *and Colletotrichum* sp.	Polysaccharides	Sarsaiya et al. (2020)
Codonopsis pilosula	Root	*Alternaria* sp., *Aspergillus* sp., *Penicillium* sp., *and Plectosphaerella* sp.	Polysaccharides	Fan et al. (2023)
Vernonia anthelmintica	Flower	*Aspergillus* sp., *Ovatospora* sp., *Chaetomium* sp., *Thielavia* sp., *and others*	Fatty acids, steroids, chalcones, flavonoids, terpenes, phenolic acids, and other substances	Niu et al. (2022)
Taxus brevifolia	Root, stem, and leaf	*Taxomyces and Renae*	Paclitaxel	Adeleke and Babalola (2021)
Taxodium distichum	Leaf, petal	*Alternaria alternata*	Taxol	Adeleke and Babalola (2021)

(continued)

Table 3.1 (continued)

Host plant	Tissues	Endophytic fungi	Major medicinal components	References
Diphylleia sinensis	Root	*Penicillium implicatum*	Podophyllotoxin and chaetominine	Ejaz et al. (2020)
Citrus limon	Root and leaf	*Penicillium digitatum*	Tryptoquialanine A, tryptoquialanine C, fumiquinazoline A, deoxytryptoquialanone, citrinadin A, deoxycitrinadin A, chrysogenamide A	Costa et al. (2019)
Sabina recurse	Leaf	*Fusarium oxysporum*	Cryptocandin, cyclosporin xanthoviridicatins	Jia et al. (2016)
Senecio kleinii	Root	*Phoma* sp.	Atrovenetinone, sclerodine	Hussain et al. (2015)
Solanum xanthocarpum	Flower	*Aspergillus terreus*	Lovastatin	Parthasarathy and Sathiyabama (2015)
Streptomyces spp.	Root and stem	*Streptomyces* spp.	Glacial acetic acid, linoleic acid, 4-methylvaleric acid, 4-hexenoic acid, dehydroacetic acid, heptanedioic acid, 2-methyl butyric acid, and 1-*p*-methen-8-ol	Zhou et al. (2018)

metabolic health among others. *Fusarium* sp., *Aspergillus* sp., *and Penicillium* sp. may be separated from the majority of medicinal plants. Because of their diverse biological roles, they are recognized as primary candidates for the development of new treatment investigation of medicinal plant endophytic fungus, as shown in Table 3.2 (Niu et al. 2022).

Additionally, it has been suggested that endophytic fungi are metabolically more active than their free-living counterparts, which accounts for their capacity to produce several secondary metabolites. It's also crucial to remember that endophytic fungi are the source of several essential herbal medicines, like the cancer medication paclitaxel (El-Hawary et al. 2016). Endophytic fungi are a little-understood class of microbes despite their potential as a source of novel natural compounds and their widespread distribution (Seetharaman et al. 2017). Endophytic fungi from the genera *Colletotrichum*, *Fusarium*, *Alternaria*, and *Aspergillus* that have been isolated from medicinally significant plants exhibit a variety of biological activities, including anticancer, antimicrobial, antifungal, immunomodulatory, antituberculosis, and antioxidant activities. These activities are widely utilized in the agrochemical and pharmaceutical industries (Pan et al. 2015). Although several fungi are frequently referred to as endophytes, some fungi are occasionally seen colonizing hosts and are separated just once or twice from multiple samples. In the literature, endophytes of diverse plant tissues have been named *Alternaria*, *Cladosporium*,

Table 3.2 Association of various types of endophytic fungi with medicinal plants along with their medicinal properties

Medicinal plants	Endophytic fungi	Medicinal properties	References
Capsicum annuum	*Alternaria alternata*	Cardiovascular protective and antitumor	Cui et al. (2012)
Rheum palmatum	*Fusarium solani*	Antibacterial, anti-inflammatory, and hepato-protective	Zhang et al. (2012)
Cajanus cajan	*Fusarium solani, Fusarium oxysporum, and Fusarium proliferatum, Hypocrea lixii*	Antioxidant, antitumor (lung cell carcinoma)	Zhao et al. (2013)
Cinchona ledgeriana	*Phomopsis, Diaporthe, Schizophyllum, Penicillium, Fomitopsis, and Arthrinium*	Antiparasitic (malari)	Maehara et al. (2013)
Forsythia suspensa	*Colletotrichum gloeosporioides*	Antioxidant, anti-inflammatory, antipyretic, and hypolipidemic	Zhao et al. (2013)
Nerium indicum	*Geomyces* sp.	Cardiovascular and cerebrovascular protective and acetylcholinesterase inhibitor	You et al. (2013)
Piper nigrum	*Colletotrichum gloeosporioides*	Antibacterial, antifungal, hepato-protective, antipyretic, anti-inflammatory, anticonvulsant, insecticidal, and antioxidant	Kaul et al. (2013)
Digitalis lanata	*Alternaria* spp., *Penicillium* spp., *and Aspergillus* spp.	Cardiotonic	Devari et al. (2014)
Macleaya cordata	*Fusarium proliferatum*	Antioxidant	Wang et al. (2014)
Coleus forskohlii	*Rhizoctonia bataticola*	Glaucoma, anticancer, anti-HIV, and heart-healthy	Mir et al. (2015)
Fritillaria cirrhosa	*Fusarium redolens*	Antitussive and expectorant	Pan et al. (2015)
Catharanthus roseus	*Fusarium oxysporum, Talaromyces radicus, and Eutypella* spp.	Antitumor	Palem et al. (2016)
Cephalotaxus hainanensis	*Alternaria tenuissima*	Antitumor (chronic myeloid leukemia)	Hu et al. (2016)
Rhodiola rosea	*Phialocephala fortinii*	Antioxidant	Cui et al. (2016)
Ginkgo biloba	*Fusarium oxysporum*	Renal, cardiovascular, and respiratory protective	Na et al. (2016)
Sabina recurse	*Fusarium oxysporum*	Antiviral, antimalaria, antiparasitic activity	Jia et al. (2016)

(continued)

Table 3.2 (continued)

Medicinal plants	Endophytic fungi	Medicinal properties	References
Salvia miltiorrhiza	*Phoma glomerata*	Cardiovascular and cerebral protection	Li et al. (2016a, b)
Solanum nigrum	*Aspergillus flavus*	Antitumor	El-Hawary et al. (2016)
Torreya taxifolia	*Penicillium chrysogenum*	Antiviral, antimalaria, antiparasitic activity	Jia et al. (2016)
Passiflora incarnata	*Alternaria alternata, Colletotrichum capsici, and Chryseobacterium taiwanense*	Antitumor (hepatic carcinoma)	Seetharaman et al. (2017)
Citrus limon	*Penicillium digitatum*	Antiviral, antimalaria, antiparasitic	Costa et al. (2019)
Dendrobium officinale	*Fusarium* sp., *Trichoderma* sp.	Antioxidant	Sarsaiya et al. (2020)
Diphylleia sinensis	*Penicillium implicatum*	Antiparasitic, antiviral	Ejaz et al. (2020)
Sinopodophyllum hexandrum	*Fusarium solani*	Anticancer/antitumor	Ejaz et al. (2020)
Taxus brevifolia	*Taxomyces and Renae*	Anticancer/antitumor	Adeleke and Babalola (2021)
Taxodium distichum	*Alternaria alternata*	Cardiovascular protective	Adeleke and Babalola (2021)
Vernonia anthelmintica	*Aspergillus* sp., *Ovatospora* sp., *Chaetomium* sp., *Thielavia* sp., *and others*	Antitumor	Niu et al. (2022)
Codonopsis pilosula	*Alternaria* sp., *Aspergillus* sp., *Penicillium* sp., *and Plectosphaerella* sp.	Cardiovascular protective	Fan et al. (2023)
Mint (*Endostemon obtusifolius*)	*Fusarium* sp.	Antioxidant, antiparasitic	Ogbe et al. (2023)

Colletotrichum, *Guignardia*, *Phoma*, *Phomopsis*, *Phyllosticta*, and *Xylaria* (Li et al. 2016a, b). It has been questioned if our isolated species of *Aspergillus*, *Gibberella*, *and Penicillium are actually endophytic fungi* (Maehara et al. 2013).

3.6 Stress Management Through Endophytic Fungi

Because they can withstand different pressures that a plant would experience in nature, endophytes have become more and more popular among industrial and agricultural scientists, which covers both biotic stress from things like pests, nematodes, and other omnivores and abiotic stress from things like both chemical and drought stress (Seleiman et al. 2021).

3.7 Biotic Stress Management: *Biopesticides and Endophytic Fungi*

Fungal endophytes can produce biopesticides for their host because they give the plant insecticides and antimicrobials (Gong et al. 2020). Defense mechanisms such as antixenosis and antibiosis stop insects from feeding or lessen their effectiveness (Hondelmann et al. 2020). Endophytic fungi create and expel host-safe, tissue-specific bioactive compounds (Liarzi et al. 2016). The majority of plant phytochemicals and a number of additional substances fall under the category of defensive compounds, which can be divided into various functional classes (Verma et al. 2016).

3.7.1 Endophytic Fungus as Nematicides

Fungal endophytes serve as nematode-toxicating agents since they contain these substances. The tomato plant-parasitic nematode *Meloidogyne incognita*'s nematicidal activity was reportedly reduced by the endophytic fungus *Fusarium oxysporum* (Dhayanithy et al. 2019).

3.7.2 Endophytic Fungus and Phytohormone Production

By solubilizing phosphate and boosting the uptake of phosphorus, plant hormones, and other crucial metabolites, endophytes have a positive impact on plant development (Firáková et al. 2007). Endophytes, including those from the genus *Colletotrichum*, the genus *Piriformospora indica*, the genus *Sebacina vermifera*, and the genus *Penicillium*, have drawn particular attention because they frequently create enzymes and metabolites that aid plant growth in adverse environments (Redman et al. 2011). Gibberellic acid, which promotes the growth of rice and soybean, has been reported to be released by *Cladosporium sphaerospermum*, an endophyte of *Glycine max* (Hamayun et al. 2010). A pestalotin analog, a metabolite with gibberellin activity that boosts the germination rate, has also been found to be produced by *Pestalotiopsis microspores* (Li et al. 2018).

Endophytes like *Fusarium tricinctum* and *Alternaria alternata* created indole-acetic acid derivatives that helped plants develop. *P. indica* root endophytes modify phytohormones that are important for the growth and development of the host plant. Additionally, they improved nutrient translocation and absorption. It is crucial for phosphorus and nitrogen absorption from the soil (Johnson et al. 2014).

3.7.3 Endophytic Fungus and Enzymes

Endophytic fungi's degradation of soil biomass is a crucial step in recycling spent nutrients and improving soil quality. Extracellular hydrolases, including pectinase, cellulase, xylanase, and proteinase, which protect the pathogen's released molecules and those obtaining nourishment from the host, have been discovered to be synthesized and produced by fungal endophytes (Leo et al. 2016). By converting macromolecules like polysaccharides, proteins, and organophosphates into micromolecules that are carried through cells for metabolic functions, these enzymes help the host during the symbiotic phase. More than 50 endophytic fungal strains, including hydrolases and proteases, were identified and described by Sunitha et al. (2013). Endophytic fungi isolated from the plant *Cardiospermum halicacabum* enhance plant growth by creating a variety of extracellular enzymes, enabling it to withstand challenging circumstances. Chitin molecules, including carbon and nitrogen, are broken down and recycled in large part by the fungal enzymes known as chitinases (Chathurdevi and Gowrie 2016).

3.8 Abiotic Stress Management

Endophytic biocompounds produced by endophytes, which operate as stress relievers, can either (a) activate response mechanisms immediately after being exposed to stress or (b) reduce stress (Gong et al. 2020).

3.8.1 Endophytic Fungi and Drought Stress

Crop yield is seriously threatened by drought due to the shortage of groundwater, which can include plant death before germination, damage to membrane integrity, and alterations in the productivity of photosynthesis and reactive oxygen species (Seleiman et al. 2021). The impacts of the drought, which are still very much present, are less severe on mycelium-infected plants than they are on uninfected plants. Because of this, plants with fungal endophytes are more tolerant of drought. They accomplish this by altering the amount of solute that accumulates in the tissues of the leaf, which in turn slows evaporation or results in a thicker cuticle. *Neotyphodium* sp. was endophytic and has been demonstrated through stomatal opening, shutting, and osmoregulation to increase grasses' ability to drought (Chhipa and Deshmukh 2019). The infected drought-stressed rice plants with inoculating *Fusarium culmorum* and *Curvularia protuberata* results to reveal an increase in biomass relative to uninoculated plants (Redman et al. 2011).

3.8.2 Salinity Stress and Endophytic Fungi

In recent decades, high salinity has limited worldwide agricultural productivity due to extreme weather circumstances and the indiscriminate use of agricultural land (Chhipa and Deshmukh 2019). The term "soil salinity" describes the accumulation of soluble salts in the soil that change its physicochemical characteristics and consequently reduce agricultural production (Hu et al. 2011). Superoxide dismutase (SOD) is one kind of reactive oxygen species (ROS). Plants produce it during salt or osmotic stress. By secreting antioxidant compounds in the plant tissues, the fungal endophyte *Piriformospora indica* enhances the resistance to salt stress (Baltruschat et al. 2008).

3.8.3 Endophytic Fungus and Temperature Stress

Because high temperatures can promote protein aggregation and protein denature, they hinder crop growth (Hussain et al. 2018). On the other side, low temperatures can impair metabolism by preventing intermolecular and enzyme interactions that alter the shape of the protein cell membrane (Andreas et al. 2016). It has been demonstrated that endophyte *Curvularia* spp. benefit plants which include tomatoes, watermelon, and wheat (Kaur 2020). Contrary to endophyte-free plants in regions where the soil temperature is over 57 °C, *Dichanthelium lanuginosum* tissues contain *Curvularia* sp., which protects the plant from stress brought on by temperature variations (Kaur 2020). It has been demonstrated that the grass *D. lanuginosum* can survive in temperatures ranging from 38 to 65 °C, which is believed to be brought on by an endophytic relationship with Mycoviruses called "(CThTV) *Curvularia* thermal tolerance virus," and *C. protuberate* fungi are also present. In geothermal soils and with rice seeds that have grown in a chilly environment in a lab setting, *C. protuberance* has been reported to be successful (Redman et al. 2011).

3.8.4 Endophytic Fungi Control Heavy Metal Stress

As we know, heavy metals cannot be degraded, and heavy metal pollution brought on by growing industrialization has recently gained attention, which loses and reduces the root growth of plants (Kaur 2020). Many domesticated as well as medicinal plants are vulnerable to the harmful effects of heavy metals, which inhibit root growth (Singh et al. 2011). They have mechanisms for metal chelation in fungus endophytes. By enhancing the antioxidant system, which leads to detoxification and enables the plant to flourish in contaminated soil, they increase the resistance of plants to heavy metals, as shown in Fig. 3.2 (Wang et al. 2017). Siderophores are "metal-chelating" substances that plant roots create and release into the immediate rhizosphere (Kaur 2020), which is accomplished by combining heavy metals to form complexes that are difficult for plant roots to absorb. Siderophores can influence how well heavy metals are removed from the rhizosphere, and *P. fortinii* and

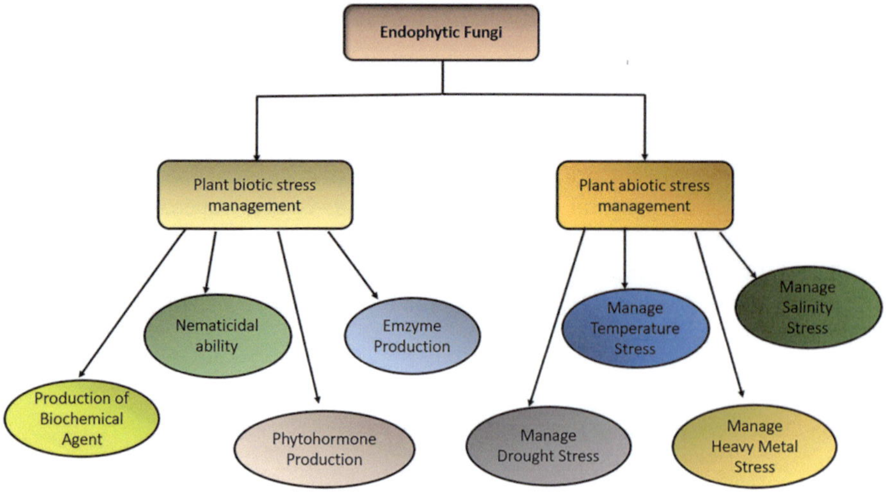

Fig. 3.2 Important role of endophytic fungi in medicinal plants for stress management

Rhizodermea veluwensis have been found to produce them to aid in preventing plants from absorbing heavy metals (Yamaji et al. 2016).

3.9 Conclusion

The habitat and ecological niche of endophytic fungi are developed within plant cells. The plants are unaffected by them since they dwell inside plant cells. The interactions between these mushrooms and their surroundings can be both advantageous and detrimental. Compared to most fungal strains, they have more incredible promise in the medical and pharmaceutical industries.

They have antimicrobial, antibacterial, and anticancer properties. According to endophytic research, they can boost plant growth, decrease insect feeding, and increase resilience to disease and drought. Also regarded as a significant source of secondary metabolites and novel bioactive chemicals for biological control are endophytic fungi. Endophytes have positive effects on people, the environment, and plants in various ways, according to what research has shown. In the domains of pharmacology, agronomy, and drug development, they can be utilized to manage hazardous pathogenic endoparasites.

References

Adeleke BS, Babalola OO (2021) Pharmacological potential of fungal endophytes associated with medicinal plants: a review. J Fungi 7(2):147. https://doi.org/10.3390/jof7020147

Andreas LB, Jaudzems K, Stanek J, Lalli D, Bertarello A, Le Marchand T, Pintacuda G (2016) Structure of fully protonated proteins by proton-detected magic-angle spinning NMR. Proc Natl Acad Sci U S A 113(33):9187–9192. https://doi.org/10.1073/pnas.1602248113

Baltruschat H, Fodor J, Harrach BD, Niemczyk E, Barna B, Gullner G, Skoczowski A (2008) Salt tolerance of barley induced by the root endophyte Piriformospora indica is associated with a strong increase in antioxidants. New Phytol 180(2):501–510. https://doi.org/10.1111/j.1469-8137.2008.02583.x

Chathurdevi G, Gowrie SU (2016) Endophytic fungi isolated from medicinal plant—a source of potential bioactive metabolites. Int J Curr Pharm Res 8:50–56

Chen KH, Marcón F, Duringer J, Blount A, Mackowiak C, Liao HL (2022) Leaf mycobiome and mycotoxin profile of warm-season grasses structured by plant species, geography, and apparent black-stroma fungal structure. Appl Environ Microbiol 88(21):e00942–e00922. https://doi.org/10.1128/aem.00942-22

Chhipa H, Deshmukh SK (2019) Fungal endophytes: rising tools in sustainable agriculture production. In: Endophytes and secondary metabolites. Springer International Publishing AG, Cham, pp 1–24. https://doi.org/10.1007/978-3-319-76900-4_26-1

Costa JH, Wassano CI, Angolini CFF, Scherlach K, Hertweck C, Pacheco Fill T (2019) Antifungal potential of secondary metabolites involved in the interaction between citrus pathogens. Sci Rep 9(1):18647

Cui Y, Yi D, Bai X, Sun B, Zhao Y, Zhang Y (2012) Ginkgolide B produced endophytic fungus (Fusarium oxysporum) isolated from Ginkgo biloba. Fitoterapia 83(5):913–920. https://doi.org/10.1016/j.fitote.2012.04.009

Cui J, Guo T, Chao J, Wang M, Wang J (2016) Potential of the endophytic fungus Phialocephala fortinii Rac56 found in Rhodiola plants to produce salidroside and p-tyrosol. Molecules 21(4):502. https://doi.org/10.3390/molecules21040502

Daley DK, Brown KJ, Badal S (2017) Pharmacognosy fundamentals, applications and strategies. Elsevier, Amsterdam, pp 413–421

Devari S, Jaglan S, Kumar M, Deshidi R, Guru S, Bhushan S, Shah BA (2014) Capsaicin production by Alternaria alternata, an endophytic fungus from Capsicum annum; LC-ESI-MS/MS analysis. Phytochemistry 98:183–189. https://doi.org/10.1016/j.phytochem.2013.12.001

Dhayanithy G, Subban K, Chelliah J (2019) Diversity and biological activities of endophytic fungi associated with Catharanthus roseus. BMC Microbiol 19. https://doi.org/10.1186/s12866-019-1386-x

Ejaz M, Javed S, Hamza M, Tabassum S, Abubakar M, Ullah I (2020) Fungal endophytes are effective alternatives and novel sources of anticancer drugs. Punjab Univ J Zool 35:13–24

El-Hawary SS, Mohammed R, Abouzid SF, Bakeer W, Ebel R, Sayed AM, Rateb ME (2016) Solamargine production by a fungal endophyte of Solanum nigrum. J Appl Microbiol 120(4):900–911. https://doi.org/10.1111/jam.13077

Faeth SH, Fagan WF (2002) Fungal endophytes: common host plant symbionts but uncommon mutualists. Integr Comp Biol 42(2):360–368. https://doi.org/10.1093/icb/42.2.360

Fan L, Li Y, Wang X, Leng F, Li S, Zhu N, Wang Y (2023) Culturable endophytic fungi community structure isolated from Codonopsis pilosula roots and effect of season and geographic location on their structures. BMC Microbiol 23(1):132. https://doi.org/10.1186/s12866-023-02848-3

Firáková S, Šturdíková M, Múčková M (2007) Bioactive secondary metabolites produced by microorganisms associated with plants. Biologia 62(3):251–257. https://doi.org/10.2478/s11756-007-0044-1

Gebretsadik KG, Zhang Y, Chen J (2022) Screening and evaluation for antixenosis resistance in wheat accessions and varieties to grain aphid, Sitobion miscanthi (Takahashi) (Hemiptera: Aphididae) (Hemiptera: Aphididae). Plants 11(8):1094. https://doi.org/10.3390/plants11081094

Gong Z, Xiong L, Shi H, Yang S, Herrera-Estrella LR, Xu G et al (2020) Plant abiotic stress response and nutrient use efficiency. Sci China Life Sci 63(5):635–674. https://doi.org/10.1007/s11427-020-1683-x

Hafeez S, Yaqoob S, Magray AR, Kamili AN, Ganai BA (2023) Molecular characterization of fungal endophyte diversity isolated from Aconitum heterophyllum: a critically endangered medicinal plant of Kashmir Himalaya. Int Microbiol 26(3):651–662. https://doi.org/10.1007/s10123-023-00331-7

Hamayun M, Khan SA, Khan AL, Rehman G, Kim YH, Iqbal I, Lee IJ (2010) Gibberellin production and plant growth promotion from pure cultures of Cladosporium sp. MH-6 isolated from cucumber (Cucumis sativus L.). Mycologia 102(5):989–995. https://doi.org/10.3852/09-261

Hawksworth DL, Lücking R (2018) Fungal diversity revisited: 2.2 to 3.8 million species. In: Heitman J et al (eds) The fungal kingdom. ASM Press, Washington, DC, pp DC79–DC95

Hondelmann P, Paul C, Schreiner M, Meyhöfer R (2020) Importance of antixenosis and antibiosis resistance to the cabbage whitefly (Aleyrodes proletella) in brussels sprout cultivars. Insects 11(1):56

Hu H, Tian F, Hu H (2011) Soil particle size distribution and its relationship with soil water and salt under mulched drip irrigation in Xinjiang of China. Sci China Technol Sci 54(6):1568–1574. https://doi.org/10.1007/s11431-010-4276-x

Hu X, Li W, Yuan M, Li C, Liu S, Jiang C, Liu Y (2016) Homoharringtonine production by endophytic fungus isolated from Cephalotaxus hainanensis Li. World J Microbiol Biotechnol 32(7):110. https://doi.org/10.1007/s11274-016-2073-9

Hussain H, John M, Al-Harrasi A, Shah A, Hassan Z, Abbas G et al (2015) Phytochemical investigation and antimicrobial activity of an endophytic fungus Phoma sp. J King Saud Univ Sci 27(1):92–95. https://doi.org/10.1016/j.jksus.2014.08.001

Hussain HA, Hussain S, Khaliq A, Ashraf U, Anjum SA, Men S, Wang L (2018) Chilling and drought stresses in crop plants: implications, cross talk, and potential management opportunities. Front Plant Sci 9:393. https://doi.org/10.3389/fpls.2018.00393

Jia M, Chen L, Xin HL, Zheng CJ, Rahman K, Han T, Qin LP (2016) A friendly relationship between endophytic fungi and medicinal plants: a systematic review. Front Microbiol 7:906. https://doi.org/10.3389/fmicb.2016.00906

Johnson JM, Alex T, Oelmüller R (2014) Piriformospora indica: the versatile and multifunctional root endophytic fungus for enhanced yield and tolerance to biotic and abiotic stress in crop plants. Trop Agric 52(2):103–122

Kaul S, Ahmed M, Zargar K, Sharma P, Dhar MK (2013) Prospecting endophytic fungal assemblage of Digitalis lanata Ehrh. (foxglove) as a novel source of digoxin: a cardiac glycoside. 3 Biotech 3(4):335–340. https://doi.org/10.1007/s13205-012-0106-0

Kaur T (2020) Fungal endophyte-host plant interactions: role in sustainable agriculture. In: Sustainable crop production. IntechOpen, London, p 211

Kumaran RS, Muthumary J, Kim EK, Hur BK (2009) Production of taxol from Phyllosticta dioscoreae, a leaf spot fungus isolated from Hibiscus rosa-sinensis. Biotechnol Bioprocess Eng 14:76–83. https://doi.org/10.1007/s12257-008-0041-4

Leo VV, Passari AK, Joshi JB, Mishra VK, Uthandi S, Ramesh N et al (2016) A novel triculture system (CC3) for simultaneous enzyme production and hydrolysis of common grasses through submerged fermentation. Front Microbiol 7:447. https://doi.org/10.3389/fmicb.2016.00447

Li P, Wu Z, Liu T, Wang Y (2016a) Biodiversity, phylogeny, and antifungal functions of endophytic fungi associated with zanthoxylum bungeanum. Int J Mol Sci 17(9):1541. https://doi.org/10.3390/ijms17091541

Li X, Zhai X, Shu Z, Dong R, Ming Q, Qin L, Zheng C (2016b) Phoma glomerata D14: an endophytic fungus from salvia miltiorrhiza that produces salvianolic acid. Curr Microbiol 73(1):31–37. https://doi.org/10.1007/s00284-016-1023-y

Li SJ, Zhang X, Wang XH, Zhao C (2018) Novel natural compounds from endophytic fungi with anticancer activity. Eur J Med Chem 156:316–343. https://doi.org/10.1016/j.ejmech.2018.07.015

Liarzi O, Bucki P, Braun Miyara S, Ezra D (2016) Bioactive volatiles from an endophytic Daldinia cf. concentrica isolate affect the viability of the plant parasitic nematode meloidogyne javanica. PLoS One 11(12):e0168437. https://doi.org/10.1371/journal.pone.0168437

Lu J, Wang J, Zhang J, Zhu Y, Qin L, Zhu B (2023) Diversity of culturable endophytic fungi in Crocus sativus and their correlation with crocin content. Curr Microbiol 80(2):73. https://doi.org/10.1007/s00284-023-03177-4

Luan F, Ji Y, Peng L, Liu Q, Cao H, Yang Y et al (2021) Extraction, purification, structural characteristics and biological properties of the polysaccharides from Codonopsis pilosula: a review. Carbohydr Polym 261:117863. https://doi.org/10.1016/j.carbpol.2021.117863

Lücking R, Aime MC, Robbertse B, Miller AN, Ariyawansa HA, Aoki T et al (2020) Unambiguous identification of fungi: where do we stand and how accurate and precise is fungal DNA barcoding? IMA Fungus 11:14. https://doi.org/10.1186/s43008-020-00033-z

Maehara S, Simanjuntak P, Maetani Y, Kitamura C, Ohashi K, Shibuya H (2013) Ability of endophytic filamentous fungi associated with Cinchona ledgeriana to produce cinchona alkaloids. J Nat Med 67(2):421–423. https://doi.org/10.1007/s11418-012-0701-8

Mir RA, Kaushik PS, Chowdery RA, Anuradha M (2015) Elicitation of forskolin in cultures of Rhizactonia bataticola—a phytochemical synthesizing endophytic fungi. Int J Pharm Pharm Sci 7(10):185–189. ISSN: 0975-1491

Mourad EF, Sarhan MS, Daanaa HA, Abdou M, Morsi AT, Abdelfadeel MR, Hegazi NA (2018) Plant materials are sustainable substrates supporting new technologies of plant-only-based culture media for in vitro culturing of the plant microbiota. Microbes Environ 33(1):40–49. https://doi.org/10.1264/jsme2.ME17135

Na R, Jiajia L, Dongliang Y, Yingzi P, Juan H, Xiong L et al (2016) Indentification of vincamine indole alkaloids producing endophytic fungi isolated from Nerium indicum, Apocynaceae. Microbiol Res 192:114–121. https://doi.org/10.1016/j.micres.2016.06.008

Naranjo-Ortiz MA, Gabaldón T (2019) Fungal evolution: major ecological adaptations and evolutionary transitions. Biol Rev Camb Philos Soc 94(4):1443–1476. https://doi.org/10.1111/brv.12510

Niu L, Rustamova N, Ning H, Paerhati P, Lu C, Yili A (2022) Diversity and biological activities of endophytic fungi from the flowers of the medicinal plant Vernonia anthelmintica. Int J Mol Sci 23(19):1935. https://doi.org/10.3390/ijms231911935

Noriler SA, Savi DC, Aluizio R, Palácio-Cortes AM, Possiede YM, Glienke C (2018) Bioprospecting and structure of fungal endophyte communities found in the Brazilian biomes, pantanal, and cerrado. Front Microbiol 9:1526. https://doi.org/10.3389/fmicb.2018.01526

Ogbe AA, Gupta S, Stirk WA, Finnie JF, Van Staden J (2023) Growth-promoting characteristics of fungal and bacterial endophytes isolated from a drought-tolerant mint species Endostemon obtusifolius (E. Mey. ex Benth.) N. E. Br. Plants 12(3):638. https://doi.org/10.3390/plants12030638

Palem PPC, Kuriakose GC, Jayabaskaran C (2016) Correction: an endophytic fungus, Talaromyces radicus, isolated from Catharanthus roseus, produces vincristine and vinblastine, which induce apoptotic cell death. PLoS One 11(4):1–6. https://doi.org/10.1371/journal.pone.0153111

Pan BF, Su X, Hu B, Yang N, Chen Q, Wu W (2015) Fusarium redolens 6WBY3, an endophytic fungus isolated from Fritillaria unibracteata var. wabuensis, produces peimisine and imperialine-3β-d-glucoside. Fitoterapia 103:213–221. https://doi.org/10.1016/j.fitote.2015.04.006

Parthasarathy R, Sathiyabama M (2015) Lovastatin-producing endophytic fungus isolated from a medicinal plant Solanum xanthocarpum. Nat Prod Res 29(24):2282–2286. https://doi.org/10.1080/14786419.2015.1016938

Redman RS, Kim YO, Woodward CJ, Greer C, Espino L, Doty SL, Rodriguez RJ (2011) Increased fitness of rice plants to abiotic stress via habitat adapted symbiosis: a strategy for mitigating impacts of climate change. PLoS One 6(7):e14823. https://doi.org/10.1371/journal.pone.0014823

Rodriguez RJ, White JF Jr, Arnold AE, Redman RS (2009) Fungal endophytes: diversity and functional roles. New Phytol 182(2):314–330. https://doi.org/10.1111/j.1469-8137.2009.02773.x

Sarsaiya S, Jain A, Jia Q, Fan X, Shu F, Chen Z, Chen J (2020) Molecular identification of endophytic fungi and their pathogenicity evaluation against Dendrobium nobile and Dendrobium officinale. Int J Mol Sci 21(1):316. https://doi.org/10.3390/ijms21010316

Seetharaman P, Gnanasekar S, Chandrasekaran R, Chandrakasan G, Kadarkarai M, Sivaperumal S (2017) Isolation and characterization of anticancer flavone chrysin (5,7-dihydroxy flavone)-

producing endophytic fungi from Passiflora incarnata L. leaves. Ann Microbiol 67(4):321–331. https://doi.org/10.1007/s13213-017-1263-5

Seleiman MF, Al-Suhaibani N, Ali N, Akmal M, Alotaibi M, Refay Y, Battaglia ML (2021) Drought stress impacts on plants and different approaches to alleviate its adverse effects. Plants 10(2):259. https://doi.org/10.3390/plants10020259

Sherameti I, Shahollari B, Venus Y, Altschmied L, Varma A, Oelmüller R (2005) The endophytic fungus Piriformospora indica stimulates the expression of nitrate reductase and the starch-degrading enzyme glucan-water dikinase in tobacco and Arabidopsis roots through a home-odomain transcription factor that binds to a conserved motif in their promoters. J Biol Chem 280(28):26241–26247. https://doi.org/10.1074/jbc.M500447200

Shwab EK, Keller NP (2008) Regulation of secondary metabolite production in filamentous ascomycetes. Mycol Res 112(2):225–230. https://doi.org/10.1016/j.mycres.2007.08.021

Sieber TN (2007) Endophytic fungi in forest trees: are they mutualists? Fungal Biol Rev 21(2–3):75–89. https://doi.org/10.1016/j.fbr.2007.05.004

Singh HP, Kaur G, Batish DR, Kohli RK (2011) Lead (Pb)-inhibited radicle emergence in Brassica campestris involves alterations in starch-metabolizing enzymes. Biol Trace Elem Res 144(1–3):1295–1301. https://doi.org/10.1007/s12011-011-9129-3

Sunitha VH, Nirmala Devi D, Srinivas C (2013) Extracellular enzymatic activity of endophytic fungal strains isolated from medicinal plants. World J Agric Sci 9(1):1–9

Verma VC, Singh SK, Kharwar RN (2012) Histological investigation of fungal endophytes in healthy tissues of Azadirachta indica A. Juss. Kasetsart J (Nat Sci) 46:229–237

Verma V, Ravindran P, Kumar PP (2016) Plant hormone-mediated regulation of stress responses. BMC Plant Biol 16:86. https://doi.org/10.1186/s12870-016-0771-y

Wang XJ, Min CL, Ge M, Zuo RH (2014) An endophytic sanguinarine producing fungus from Macleaya cordata, Fusarium proliferatum BLH51. Curr Microbiol 68(3):336–341. https://doi.org/10.1007/s00284-013-0482-7

Wang LW, Wang JL, Chen J, Chen JJ, Shen JW, Feng XX, Chen FY (2017) A novel derivative of (-) mycousnine produced by the endophytic fungus Mycosphaerella nawae, exhibits high and selective immunosuppressive activity on T cells. Front Microbiol 8:1251. https://doi.org/10.3389/fmicb.2017.01251

Wang GK, Yang JS, Huang YF, Liu JS, Tsai CW, Bau DT, Chang WS (2021) Culture separation, identification and unique anti-pathogenic fungi capacity of endophytic fungi from Gucheng salvia miltiorrhiza. In Vivo 35(1):325–332. https://doi.org/10.21873/invivo.12263

Waqas M, Khan AL, Kamran M, Hamayun M, Kang SM, Kim YH, Lee IJ (2012) Endophytic fungi produce gibberellins and indoleacetic acid and promotes host-plant growth during stress. Molecules 17(9):10754–10773. https://doi.org/10.3390/molecules170910754

Yamaji K, Watanabe Y, Masuya H, Shigeto A, Yui H, Haruma T (2016) Root fungal endophytes enhance heavy-metal stress tolerance of Clethra barbinervis growing naturally at mining sites via growth enhancement, promotion of nutrient uptake and decrease of heavy-metal concentration. PLoS One 11(12):e0169089. https://doi.org/10.1371/journal.pone.0169089

Yashavantha Rao HC (2020) Fascinating fungal endophytes associated with medicinal plants: recent advances and beneficial applications. In: Microbial endophytes. Elsevier, Amsterdam, pp 263–289. https://doi.org/10.1016/B978-0-12-818734-0.00011-5

You X, Feng S, Luo S, Cong D, Yu Z, Yang Z, Zhang J (2013) Studies on a rhein-producing endophytic fungus isolated from Rheum palmatum L. Fitoterapia 85(1):161–168. https://doi.org/10.1016/j.fitote.2012.12.010

Zhang Q, Wei X, Wang J (2012) Phillyrin produced by Colletotrichum gloeosporioides, an endophytic fungus isolated from Forsythia suspensa. Fitoterapia 83(8):1500–1505. https://doi.org/10.1016/j.fitote.2012.08.017

Zhao J, Li C, Wang W, Zhao C, Luo M, Mu F et al (2013) Hypocrea lixii, novel endophytic fungi producing anticancer agent cajanol, isolated from pigeon pea (Cajanus cajan [L.] Millsp.). J Appl Microbiol 115(1):102–113. https://doi.org/10.1111/jam.12195

Zhou J, Diao X, Wang T, Chen G, Lin Q, Yang X, Xu J (2018) Phylogenetic diversity and antioxidant activities of culturable fungal endophytes associated with the mangrove species Rhizophora stylosa and R. mucronata in the South China Sea. PLoS One 13(6):e0197359. https://doi.org/10.1371/journal.pone.0197359

Traditional Knowledge Databases of Medicinal Plants

4

Acharya Balkrishna, Aakanksha Saini, Priya Yadav, Priyanka Yadav, Brijesh Kumar, Dushyant Sharma, Netrapal Singh, and Ved Priya Arya

Abstract

The traditional system of medicine harnesses the potency of medicinal plants for remedying a variety of ailments, and this medicinal knowledge is deeply embedded in a multitude of global cultures. The repositories of traditional knowledge provide invaluable information and play a pivotal role in conserving, disseminating, and enriching our understanding of the therapeutic potential of flora across the globe. Medicinal plant databases have now become indispensable in organizing new scientific literature emerging in the realm of herbal medicine, as well as novel drug discoveries in this age of information. The information of existing databases available throughout the world in digital form was retrieved through Google Search. Among the assorted databases, World Herbal Encyclopedia

A. Balkrishna
Patanjali Herbal Research Department, Patanjali Research Institute, Haridwar, Uttarakhand, India

University of Patanjali, Haridwar, Uttarakhand, India
e-mail: acharya.balkrishnapri@patanjali.res.in

A. Saini · P. Yadav · P. Yadav · B. Kumar · D. Sharma · N. Singh (✉)
Patanjali Herbal Research Department, Patanjali Research Institute, Haridwar, Uttarakhand, India
e-mail: aakanksha.saini@patanjali.res.in; priya.yadav1@patanjali.res.in; priyanka.yadav@patanjali.res.in; brijesh.kumar@patanjali.res.in; dushyant.sharma@patanjali.res.in; netrapal.singh@patanjali.res.in

V. P. Arya
Patanjali Herbal Research Department, Patanjali Research Institute, Haridwar, Uttarakhand, India

University of Patanjali, Haridwar, Uttarakhand, India

Centre of Excellence, Patanjali Ayurved Hospital, Haridwar, Uttarakhand, India
e-mail: vedpriya.arya@patanjali.res.in

© The Author(s), under exclusive license to Springer Nature Singapore Pte Ltd. 2024

M. Nandave et al. (eds.), *Ethnopharmacology and OMICS Advances in Medicinal Plants Volume 1*, https://doi.org/10.1007/978-981-97-2367-6_4

(WHE) was found to be the most extensive compendium, collating traditional knowledge of approximately 60,000 medicinal plants that are used in different conventional therapeutic systems across the globe. This chapter is focused on comprehensive details of other existing databases of medicinal plants from various countries and their unique features.

Keywords

Traditional knowledge database · Medicinal plants · Traditional medicinal systems · World Herbal Encyclopedia · Medicinal plant database

Abbreviations

AYUSH	Ayurveda, Yoga, Naturopathy, Unani, Siddha, Sowa-Rigpa, Homeopathy
SCM	Sasang constitutional medicine
TCM	Traditional Chinese medicine
TIM	Traditional Indian medicine
TM	Traditional medicine
WHE	World Herbal Encyclopedia

4.1 Introduction

The practice of utilizing medicinal plants for healing is as ancient as human history. The relationship between humans and their exploration of natural substances for therapeutic purposes can be traced back to ancient times, ~60,000 years ago, supported by abundant evidence found in diverse sources such as written records, preserved historical sites, and even the use of original herbal remedies (Petrovska 2012; Yuan et al. 2016; Jamal 2023). The global adoption of herbal remedies and phytonutrients, as well as nutraceuticals, is experiencing swift growth. These products are now being turned to by numerous individuals worldwide to address a range of health issues within diverse national healthcare frameworks. The utilization of herbal medicinal items and supplementary nutrients has undergone a significant surge in the last 30 years, with no fewer than 80% of the global population depending on them as a component of their primary healthcare approach (Ekor 2014; Yusupova et al. 2023). In numerous parts of the globe, "traditional medicine" holds a vital role within healthcare systems and is a part of the healthcare systems, which is evident in various Asian nations such as Thailand, India, Malaysia, Indonesia, Japan, China, and Taiwan, as well as several European countries, particularly those in Central Europe, including Poland, Hungary, Austria, Germany, Italy, and Switzerland use herbal medicinal products and herbal medicines (Zeng et al. 2019; Heinrich et al. 2023). Within the healthcare system of China, traditional herbal medicine has been incorporated, with traditional Chinese medicine (TCM) being employed for both outpatient and inpatient treatment in hospitals; around 95% of general medical facilities now feature departments dedicated to traditional medicine (Li et al. 2022; Fathifar et al. 2023; Nie et al. 2023). In Germany, herbal medicine constitutes a significant component of traditional naturopathy, alongside five other core elements: phytotherapy, hydrotherapy, exercise therapy, dietetic therapy, and "lifestyle regulation" therapy. Moreover, a wide array of healthcare practitioners, including physicians, nurses, pharmacists, and nonmedical practitioners specializing in complementary and alternative medicine, actively participate in the realm of herbal medicine (Joos et al. 2012). The utilization of biopharmaceutical products holds a worldwide competitive edge, with citizens from various nations benefiting from them. For instance, China leads with a utilization rate of 90%, followed by Chile at 71%, Colombia at 40%, France at 49%, Canada at 78%, the United Kingdom at 60%, the United States at 42%, and Germany at 73% (Sumarya et al. 2020). Furthermore, within India, the acronym AYUSH stands for Ayurveda, Yoga, Naturopathy, Unani, Siddha, Sowa-Rigpa, and Homeopathy. This acronym represents a well-structured sector that offers healthcare services across both the public and private domains. Recognizing the efficacy of AYUSH systems in alleviating the burden of diseases, endeavors have been ongoing for several decades to foster the growth of these systems and integrate them with conventional medicine. The pursuit of effective integration strategies aims to facilitate communication and mutual comprehension among diverse medical approaches. This approach also involves the comprehensive assessment of medical care, ensuring the fair allocation of resources, establishing training and educational initiatives for practitioners of both traditional and conventional medicines, and ultimately cultivating a comprehensive healthcare framework that encompasses all aspects of well-being (Rudra et al. 2017; Mohan et al. 2023; Patel et al. 2023; Sattigeri 2023). Even in the present day, two prominent enduring practices are traditional Indian medicine (TIM), specifically Ayurveda, and traditional Chinese medicine (TCM). These traditions have intertwined to enrich the wealth of knowledge concerning medicinal plants. Their combined efforts have led to a substantial collection of plants and phytomedicines that are currently employed globally (Elahee et al. 2019). Furthermore, they have played a crucial role in establishing guidelines for herbal drugs within the pharmaceutical sector (Pandey et al. 2013). Nevertheless, to effectively bring it into the mainstream, it is crucial to establish operational integration involving communication, sharing of information, and cross-referrals between conventional medical systems and AYUSH. Until a short while ago, knowledge concerning medicinal herbs was confined to journals, manuals, and textbooks. However, a novel avenue for disseminating information about medicinal herbs has emerged with the proliferation of scientific databases (Hidayat et al. 2023). Numerous databases that hold all the data needed to expedite the development of new medications have been created by researchers. Database technologies unquestionably made it easier to access more complicated data through a variety of new and innovative ways. These databases amass and furnish information about medicinal herbs, encompassing details about their constituents, the structures of compounds in both 2D and 3D, associated target proteins, pertinent diseases, and metabolic toxicity. This wealth of information is indispensable for scientific research on medicinal herbs, catering to the needs of researchers, medical professionals, and pharmacists alike. The databases play a pivotal role in various facets of biological research, offering insights into genes, proteins, and intricate tools for data analysis (Britan et al. 2018; Anand et al. 2019). The utility of these databases hinges on three key factors: the breadth of information they contain, accurate interpretation of that information, and the consistent updating of their content. Present-day biomedical databases exhibit variations in terms of data volume, specialization, functionality, and access methods. Nonetheless, most of these databases are extensive in size and incorporate intricate data from multiple sources. The category of medicinal herbs databases falls under the canopy of biomedical databases and shares similar characteristics. These databases compile an extensive array of information, including taxonomy, common names, geographical distribution, medicinal applications, plant parts used, and up-to-date scientific insights like physicochemical properties, constituents, genomic details, mechanisms of action, and other specific parameters related to medicinal plants. The information stored in these databases is predominantly presented in the form of textual descriptions or images (Dankar et al. 2018; Fathifar et al. 2023).

4.2 Traditional Medicine Systems

Historical and philosophical prospects of traditional medicinal systems of the world were unleashed to determine the era to which they belong and to depict their fundamental concepts regarding constituents, nature, and characteristics of man. Worldwide, 9 medicinal systems and more than 900 healing practices were observed belonging to ancient eras of more than 15,000 years. Transmission of therapeutic information via silk and sea routes was also decoded along with the history of travelers in ancient and classical times (Balkrishna 2022a, b, c, d, e, f, 2023). Traditional medicine encompasses culture-bound, experiential medical practices that leverage natural elements to diagnose, prevent, and treat diseases and promote overall health. These practices are not contingent upon their explicability, as pointed out by Che et al. (2017). Each community, in its civilian capacity, possesses a comprehensive healthcare system deeply rooted in its cultural and societal heritage. The oldest medical system in the world is the Ayurveda. Various tribes used therapeutic plants in prehistoric times. There is evidence that the Indus Valley civilization's inhabitants used this system of medicine. With the help of people like Charaka, Sushruta, Vagbhata, and others, Ayurveda developed during the Vedic and post-Vedic eras. Monks who practiced Buddhism were crucial in the spread of Ayurveda. But after the tenth century, when the Muslims invaded India, Ayurveda was decimated, and the Unani medical system took over. Ayurveda once again rose to prominence during the pre-independence era. It reached new heights following Independence when it was included in the Drug and Cosmetic Act. The creation of an *Ayurvedic Pharmacopoeia* aids in the standardization of Ayurvedic medications so they can compete with the allopathic medical system on a global scale. Core to traditional medicine is the utilization of natural resources such as plants, animals, and minerals. The treatment methodologies are guided by indigenous beliefs, practical expertise, and knowledge passed down through generations. Renowned traditional medical systems, like traditional Chinese medicine (TCM), Ayurveda from India, and Unani medicine from the Arab world, exemplify this approach. Despite variations in therapeutic principles, theoretical foundations, and associated technologies among these traditional medical systems, they share specific defining characteristics, as highlighted by Majeed et al. (2023).

In the realm of traditional medicine, the practice of "clinical trials" dates back to ancient times. Many clinical studies have been carried out in China for traditional Chinese medicine (TCM) and have garnered an extensive wealth of experience and progress over countless years (Chen et al. 2021; Luo et al. 2021; Lu et al. 2019; Sun et al. 2021). This progress pertains to various aspects, including preparation methods, herb selection, identification of medicinal substances, and optimal timing for gathering different plants. It is imperative to focus on refining the processing techniques and regulating dosages within TCM to enhance the effectiveness of medications and mitigate their potential toxicity. Substantial volumes of data have been garnered through clinical experiments, effectively contributing to the advancement of modern pharmaceuticals. Leveraging natural products, TCM holds several advantages over alternative medical approaches in areas such as identifying primary compounds and potential drug candidates, assessing drug-like effects, and exploring the physical, chemical, pharmacokinetic, and toxicological properties of substances. The successful application of any facet of TCM could unexpectedly play a pivotal role in the development of novel drugs, resulting in numerous advantages, notably significant cost savings. Presently, traditional Chinese medicine stands as an integral component of the Chinese public health system, inseparable from its framework (Yuan et al. 2016).

Kampo, originating in Japan, has evolved from traditional Chinese medicine (TCM) introduced during the fifth and sixth centuries. Japanese practitioners adapted TCM to their context, resulting in Kampo. A recent study reveals that Japanese physicians often incorporate Kampo medicines, sometimes even as their preferred choice, alongside treatments like radiotherapy or chemotherapy for cancer patients, which showcases the integration of Western medicine with traditional medicine (T.M.). As Kampo's popularity grows alongside Western medicine, there's a recognized need to study how these two medical approaches interact (Yuan et al. 2016; Kuchta and Cameron 2022; Ohbe et al. 2023; Okayasu et al. 2023).

Unani, a holistic system with 2500 years of history tracing back to ancient Greece, gained global attention since the WHO's increased focus on T.M. in the mid-1970s. India notably integrated Unani into its national healthcare system (Khan et al. 2022). Conversely, traditional medicine in Australia faces a decline due to conventional medicine, while Israel's diverse population leans towards modern medicine.

Despite skepticism from some Western medical practitioners, T.M. systems are widely embraced. The potential for synergizing modern scientific techniques with T.M.'s ethnobotanical and ethnomedical wisdom holds promise for developing new medicines. Table 4.1 summarizes the characteristics of various T.M. systems.

Table 4.1 Characteristics of various traditional medicine systems

Name	Country	Characteristics	Status	References
Ayurveda	India	Ayurveda employs natural elements to address the root cause of illnesses by restoring equilibrium. The core principle of Ayurveda is to follow a healthy lifestyle for preventing the emergence of imbalances and unnecessary suffering. Ayurveda includes a combination of multiple herbs which is carefully blended in specific proportions to achieve an optimal therapeutic outcome while minimizing toxicity	There are over 4 lakh registered Ayurveda practitioners, and an official govt.-recognized organization oversees the quality of Ayurveda education and practice	Mukherjee et al. (2017), Parasuraman et al. (2014), Patwardhan (2014)
Traditional Chinese medicine	China	Traditional Chinese medicine (TCM) draws its foundation from the principles of Yinyang and Wuxing. A TCM prescription consists of a combination of different medicinal components that work harmoniously to produce a cooperative and enhanced effect. A standard prescription typically comprises four essential elements, each assigned a specific role within the formula: the principal component, secondary component, supportive component, and auxiliary component	Traditional Chinese medicine (TCM) and conventional medical practices are integrated throughout various levels of the healthcare system, and they are included in both public and private insurance coverage. Most general hospitals have dedicated TCM departments that offer services to both inpatients and outpatients. TCM is gaining growing global recognition, interest, and acceptance	Shi et al. (2022), Yuan et al. (2016), Wang et al. (2016)

Table 4.1 (continued)

Name	Country	Characteristics	Status	References
Unani medicine	Unani medicine, which originated from the Greco-Arabic medical tradition and was refined during the era of Arab civilization and also followed in India	It considers the human body, mind, and spirit as interconnected components, viewing the human body as a unified entity composed of four fundamental elements, each associated with distinct temperaments. These temperaments are a reflection of an individual's physical traits and inherent disposition. Any imbalance in these temperaments can render the body vulnerable to various health conditions	India recognizes Unani as a legitimate healthcare system to address the population's health needs and has granted it official recognition. The World Health Organization (WHO) has also recognized Unani as a viable alternative healthcare approach. Unani stands as a significant traditional medical system	Parveen et al. (2020, 2022), Poulakou-Rebelakou et al. (2015)
Kampo	Japan	Over 1400 years, Kampo has evolved and seamlessly integrated with traditional Japanese healing practices. It regards each individual as a holistic entity where the interplay between the mind and body is crucial. In Kampo's philosophy, diseases are believed to stem from imbalances in both mental and physical aspects and herbal remedies are considered to have an equally profound impact on the body and the soul. Kampo therapy prioritizes the patient as a complete being, focusing on their overall well-being rather than just the specific ailment	Kampo has been integrated into Japan's healthcare system, allowing all citizens to access government-approved Kampo herbal formulas	Yakubo et al. (2014), Okamoto et al. (2014)

(continued)

Table 4.1 (continued)

Name	Country	Characteristics	Status	References
Sasang constitutional medicine (SCM)	Korea	SCM categorizes individuals into four Sasang types, namely, Tae-Yang, So-Yang, Tae-Eum, and So-Eum, based on their inherent characteristics. SCM is a holistic approach that shares theoretical similarities with personalized medicine, providing tailored treatments specific to each person's constitution for a wide range of issues	Korean SCM services are available from traditional medicine doctors in both private and public hospitals, with coverage provided by both national medical insurance and private insurance	Yoo et al. (2012), So et al. (2016)
Traditional Aboriginal medicine	Australia	The indigenous populations of Australia hold the belief that health issues can stem from three distinct sources: natural bodily factors, evil spirits, or practices associated with witchcraft	At present, there exists a sole national folk organization in operation. Between 2010 and 2011, approximately 32.1% of the principal indigenous healthcare organizations in Australia offered traditional medicine services to some extent. The preservation of traditional Aboriginal medicine is threatened due to the impacts of colonization, putting it at risk of fading away	Kim and Noble (2014), Oliver (2013)

Table 4.1 (continued)

Name	Country	Characteristics	Status	References
Traditional medicine in Africa	Africa	Practitioners of traditional medicine adopt a holistic approach to patient care, aiming to restore the mental and social balance of individuals by societal norms and connections. The widespread availability of traditional medicine stands as a critical factor contributing to its popularity in Africa, demonstrating a deep appreciation for cultural heritage	Around 80% of the African population incorporates traditional medicine into their healthcare, either alongside or independently of conventional medicine. Traditional medicine serves as the primary healthcare choice for up to 80% of the population in Ghana and Ethiopia. In Ghana, the traditional medical system is integrated into the national healthcare system, resulting in a relatively well-structured approach to healthcare delivery	Ozioma and Chinwe (2019), Kasilo et al. (2019), Cock et al. (2023)
Russian herbal medicine	Russia	Because of Russia's unique geographical setting, Russian herbal therapy has incorporated and embraced traditional medicinal practices originating from both Asia and Europe. The Russian Federation adheres to the State Pharmacopoeia of the USSR, which exclusively includes 32 out of the total 83 individual plant monographs	Russia recognizes herbal therapy as an established and distinct field within the realm of medicine, legitimizing herbal medicinal products as officially recognized treatments. According to a recent survey, 14% of the Russian population regularly incorporate herbal remedies into their healthcare routines, while 44% occasionally turn to them for medicinal purposes	Shikov et al. (2014, 2021), Conroy (2021)

4.3 Traditional Knowledge Databases

Medicinal plant databases serve as crucial tools for documenting and sharing valuable insights into the characteristics and applications of various plant species in the realm of therapeutic use. Over recent years, a multitude of such databases have emerged, each with its distinct focus and contributions to the fields of herbal medicine and pharmaceutical discovery. Examples of these databases include WHE (World Herbal Encyclopedia), GreenMolBD, HERB, TCMPG, Phytochemdb, SuperTCM, UNaProd (Universal Natural Product Resource), MedPServer, AromaDb, CMAUP, ETM-DB (Ethiopian Traditional Medicine Database), IMPPAT, KampoDB, SymMap, YaTCM, PHCD, MPOD, MPD3 (Medicinal Plants Database for Drug Designing), VIETHERB, MPDB 2.0, PharmDB-K, TM-MC, AfroDb, TCM-ID, TCM Systems Pharmacology Database and Analysis Platform (TCMSP), and TCM Database@Taiwan. These databases stock extensive collections of information regarding plant species, encompassing their chemical constituents, pharmacological properties, and traditional applications. Researchers, herbalists, and healthcare practitioners heavily rely on these repositories to delve into and harness the therapeutic potential of medicinal plants, thereby facilitating groundbreaking drug development endeavors and the preservation of traditional healing wisdom. These digital archives stand as indispensable resources in the ongoing quest to unlock the curative qualities found within nature's botanical reserves. The various databases from different countries, along with their URLs, are shown in Table 4.2.

In 2012, Ningthoujam and their team conducted a review that explored various approaches to storing ethnopharmacological data related to medicinal herbs within databases to establish minimum standards for the development of medicinal plant databases. They identified several challenges associated with sharing information in the development of herbal databases. These challenges encompassed ethnobotanical concerns, such as the absence of a benchmark model, issues related to intellectual property rights, multiple taxonomies, conservation strategies, and biopiracy, as well as technical challenges, including irregular updates, undisclosed publication years, website inaccessibility or relocation, evolving hardware and software, system obsolescence, and high maintenance costs (Ningthoujam et al. 2012).

Another review provided an analytical overview of databases of different natural products, highlighting their strengths, weaknesses, limitations, and future database-building trends. Xie et al.'s study introduced a substantial number of natural product databases, such as TCM Database@Taiwan, TCM-ID, CEMTDD, SuperToxic, and SuperNatural. Additionally, this study presented recommendations for enhancing the role of medicinal herbs databases as crucial contributors to future drug discovery endeavors (Xie et al. 2015). In recent times, increase of focus on medicinal plants for drug discovery has spurred the creation of numerous new databases, often featuring advanced functionalities. In contrast, some older databases have become obsolete. Contemporary medicinal herbs databases place increased emphasis on phytochemical and physicochemical properties, reflecting the renewed interest in herbal sources for drug discovery. Evaluating the current status and contents of medicinal herbs databases and offering recommendations based on this assessment

Table 4.2 Various medicinal plant databases

Name	Country	URL	References
WHE (World Herbal Encyclopedia)	India	https://patanjali.res.in/world-herbal-encyclopedia.php and https://wheportal.com/	Balkrishna (2023)
GreenMolBD	Bangladesh	https://www.greenmolbd.gov.bd/	Hosen et al. (2021)
HERB	China	http://herb.ac.cn	Fang et al. (2021)
TCMPG	China	http://cbcb.cdutcm.edu.cn/TCMPG/	Meng et al. (2022)
Phytochemdb	Bangladesh	https://phytochemdb.com/	Mahmud et al. (2022)
SuperTCM	Germany	http://tcm.charite.de/supertcm	Chen et al. (2021)
UNaProd (Universal Natural Product Resource)	Iran	https://unaprod.com	Naghizadeh et al. (2020)
MedPServer	India	http://bif.uohyd.ac.in/medserver	Potshangbam et al. (2019)
AromaDb	India	http://bioinfo.cimap.res.in/aromadb/	Kumar et al. (2018)
CMAUP	Singapore	http://bidd.group/CMAUP	Zeng et al. (2019)
ETM-DB: Ethiopian Traditional Medicine Database	Ethiopia	http://biosoft.kaist.ac.kr/etm	Bultum et al. (2019)
IMPPAT	India	https://cb.imsc.res.in/imppat	Mohanraj et al. (2018)
KampoDB	Japan	http://wakanmoview.inm.u-toyama.ac.jp/kampo/	Sawada et al. (2018)
SymMap	China	http://www.symmap.org	Wu et al. (2019)
YaTCM	China	http://cadd.pharmacy.nankai.edu.cn/yatcm/home	Li et al. (2018)
PHCD	Iran	http://persianherb.com	Karimi-Jafari et al. (2022)
MPOD	China	http://medicinalplants.ynau.edu.cn	He et al. (2022)
MPD3: Medicinal Plants Database for Drug Designing	Pakistan	https://www.bioinformation.info	Mumtaz et al. (2017)
VIETHERB	Vietnam	http://vietherb.com.vn	Nguyen-Vo et al. (2018)
MPDB 2.0	Bangladesh	https://www.medicinalplantbd.com	Hussain et al. (2021)
PharmDB-K	South Korea	http://pharmdb-k.org http://biomart.i-pharm.or	Lee et al. (2015)
TM-MC	Northeast Asia	http://informatics.kiom.re.kr/compound	Kim et al. (2015)

(continued)

Table 4.2 (continued)

Name	Country	URL	References
AfroDb	Africa	http://african-compounds.org/about/afrodb/	Ntie-Kang et al. (2013)
TCM-ID	China	http://www.megabionet.org/tcmid	Xue et al. (2013)
TCM Systems Pharmacology Database and Analysis Platform (TCMSP)	China	https://old.tcmsp-e.com/tcmsp.php	Ru et al. (2014)
TCM Database@Taiwan	China	http://tcm.cmu.edu.tw/	Chen and Hofmann (2011)

can be instrumental in shaping more recent and effective database designs. This review's primary objective is to investigate existing medicinal herbs databases, with a particular focus on their unique capabilities and information sources.

Furthermore, we will explore the relationships, specifically those between herbs and their constituent ingredients/compounds, and highlight emerging trends in the development of future databases (Nguyen-Vo et al. 2020; Sorokina and Steinbeck 2020). Users can easily explore, search for, retrieve, and visualize pertinent data within these databases, which encompass connections between various database elements. Improving the capabilities of conventional knowledge databases can be accomplished by creating multilingual databases, offering dosage details, comprehensively listing side effects and interactions, documenting user experiments, establishing an extensive ontology for medicinal plants, assigning distinct identifiers or names via certified entities or databases, and facilitating data submissions (Fathifar et al. 2023).

4.4 World Herbal Encyclopedia (WHE)

The inception of the World Herbal Encyclopedia aimed to meet the current demand for a centralized repository containing information on medicinal plants utilized worldwide for therapeutic purposes, as outlined by Balkrishna in 2023 (Balkrishna 2023). It stands as the most extensive database of medicinal plants available to date. The World Herbal Encyclopedia comprises a total of 111 volumes, with the initial 5 volumes delving into the historical context of plant-based therapeutic systems practiced across various global communities and ethnicities.

This comprehensive resource encompasses approximately 60,000 medicinal plants carefully chosen from diverse plant-based medicinal traditions worldwide, selected based on their significant therapeutic value and geographical distribution. The volumes are meticulously organized in alphabetical order, commencing with the genus "Algae" and concluding with the last genus "Angiosperms," thus encompassing a wide array of plant species globally recognized for their medicinal properties.

The documentation encompasses over 7500 genera of plants, featuring a staggering 2.5 lakh vernacular names in more than 2000 languages spoken around the world. It constitutes an authentic record of more than 3 lakh folk formulations and

962 healing practices from different cultures worldwide. For the purpose of upholding the precision and dependability of scientific information, it is cross-referenced with an extensive repository of over 600,000 references. Additionally, the Encyclopedia boasts an impressive compilation of 35,000 meticulously crafted paintings and 30,000 line drawings illustrating various medicinal plants. Additionally, the Encyclopedia houses the Patanjali Research Foundation Herbarium, containing approximately 30,000 medicinal plants. This herbarium has earned accreditation from the Index Herbarium at the New York Botanical Garden, adding to its credibility. Notably, the Encyclopedia incorporates a novel "Vedic nomenclature" system, developed with the collaboration of expert botanists, Sanskrit translators, and Ayurveda authorities. This nomenclature system combines the ancient and scientific aspects of the Sanskrit language with the principles of Latin-based binomial nomenclature, and it is applied throughout the Encyclopedia. The main features of this system are highlighted as follows.

In the Vedic system of plant naming, the primary focus is on the external and physical characteristics of plants, such as their roots, stems, leaves, flowers, fruits, and other organs. This approach facilitates easy identification by linking plant names to these external features (Balkrishna 2022a, b, c, d, e, f). In order to maintain the brevity of plant names, the study of etymology has led to the condensation of these characteristics into shorter forms. This approach results in concise yet meaningful names for plants, facilitating their easy identification in their natural environment without requiring laboratory equipment. This nomenclature system also incorporates gender, similar to the Latin-based binomial method. In the names of genera, a single word indicates the gender, and in compound names, the gender is determined by the final word. Species names follow the same gender rules as genera. As a result, the same species name can be used in different genders depending on the genera. In Vedic plant nomenclature, suffix words are used in the names of genera to indicate the six classes of the plant kingdom, such as Angiosperms. Suppose the name already includes this indicative mark. In that case, it is not repeated, as seen in terms like Nalika, Andadhrik, and tvak, which already contain the expressive mark "ka & k" (का & क्) for the Angiosperm group (Balkrishna 2022a, b, c, d, e, f). Currently, the upper plant kingdom, according to popular plant nomenclature, comprises approximately 17,000 genera and around 360,000 species, resulting in over 1.3 million names or words. When considering unique names, this count exceeds 210,000. "Vedic nomenclature" utilizes a systematic terminology database to create approximately 17,000 genera names and 5000 species names, thereby establishing a scientific method for plant nomenclature. In the event of the discovery of new plant genera in the future, new terminologies can be developed based on their physical features. However, for new species, existing terminologies can be applied to name plants (Balkrishna 2022a, b, c, d, e, f). Vedic Taxonomy represents a well-designed scientific approach to plant nomenclature, utilizing the rules above and principles. It offers easy access to various plant groups through systematically assigned "Vedic Plant Names." This methodology simplifies the process of memorizing genera and species names by providing concise and meaningful terminologies under usage rules. It is anticipated that this nomenclature pattern will greatly assist botanists, researchers, and plant enthusiasts by simplifying plant identification, nomenclature, and memorization, all without the need for laboratory equipment (Balkrishna 2022a, b, c, d, e, f, 2023). The World Herbal Encyclopedia, described in this context, will enrich the knowledge of students, researchers, and scholars interested in medicinal plants and their applications. It stands as one of the most comprehensive and reliable references globally and will prove valuable to various communities, including forest dwellers, rural and urban populations, and medical professionals.

4.5 Future Perspective and Conclusion

Traditional medicinal plants hold the potential for discovering novel pharmaceutical compounds from the natural world. The role of conventional knowledge online databases, with their evolving approaches, content, and structures, is crucial in achieving this objective. Nevertheless, studies have revealed that these databases exhibit diversity in content, quality, and data representation. Establishing commonly accepted standards for constructing these databases, encompassing data elements, minimum dataset requirements, quality, and functionality, can prove instrumental in facilitating the integration of information and showcasing the intricate field of ethnopharmacology. This chapter gives insights into the current landscape of traditional medicine and knowledge databases, emphasizing the need for gathering more valuable information from diverse sources such as scientific literature and specialized databases, employing various computational approaches, and utilizing software tools. Through these databases, users can readily navigate, search, download, and visualize relevant data, including the interrelationships among database components.

Furthermore, it is proposed that enhancing the functionality of traditional knowledge databases could be achieved through the creation of multilingual databases, providing dosage information, detailing side effects and interactions, documenting user experiments, establishing a comprehensive ontology for medicinal plants, assigning unique identifiers or names through certified organizations or databases, and enabling data submissions. Medicinal plants have historically served as remedies for various human ailments, and contemporary interest in plant-based products for modern drug discovery has surged. However, the limited understanding of the mechanisms of action in therapeutic contexts has hindered research into harnessing herbal medicinal systems for modern therapeutics. Much of the existing knowledge regarding the application of medicinal plants in treating diseases is based on anecdotal human experiences rather than scientific experimentation, such as randomized control trials. The creation of a comprehensive medicinal plant database incorporating modern biomedical database features could open new avenues for discovering drug candidates based on herbal sources.

References

Anand U, Jacobo-Herrera N, Altemimi A, Lakhssassi N (2019) A comprehensive review on medicinal plants as antimicrobial therapeutics: potential avenues of biocompatible drug discovery. Metabolites 9(11):258. https://doi.org/10.3390/metabo9110258

Balkrishna A (2022a) Saumitreya-Mahodadhiḥ Nighaṇṭuḥ. Divya Prakashan, Haridwar

Balkrishna A (2022b) Saumitreya-Mahodadhiḥ. Divya Prakashan, Haridwar

Balkrishna A (2022c) Saumitreya-Nirvacana-Kośaḥ. Divya Prakashan, Haridwar

Balkrishna A (2022d) Saumitreya-Saṃhitā. Divya Prakashan, Haridwar

Balkrishna A (2022e) Vaidika-Pādapa-Vargikī. Divya Prakashan, Haridwar

Balkrishna A (2022f) World herbal encyclopedia. Divya Prakashan, Haridwar

Balkrishna A (2023) World herbal encyclopedia. J Res Ayurvedic Sci 7(Suppl S1):92–94

Britan A, Cusin I, Hinard V et al (2018) Accelerating annotation of articles via automated approaches: evaluation of the neXtA5 curation-support tool by neXtProt. Database 2018:1–16

Bultum LE, Woyessa AM, Lee D (2019) ETM-DB: integrated Ethiopian traditional herbal medicine and phytochemicals database. BMC Complement Altern Med 19:212–223

Che CT, George V, Ijinu TP, Pushpangadan P, Andrae-Marobela K (2017) Traditional medicine, pharmacognosy. Academic Press, New York, pp 15–30. https://doi.org/10.1016/B978-0-12-802104-0.00002-0

Chen CY, Hofmann A (2011) TCM database@Taiwan: the world's largest traditional Chinese medicine database for drug screening in silico. PLoS One 6:e15939–e15944

Chen Q, Springer L, Gohlke BO et al (2021) SuperTCM: a bio-cultural database combining biological pathways and historical linguistic data of Chinese materia medica for drug development. Biomed Pharmacother 144:1–3

Cock IE, Ndlovu NA, Van Vuuren SF (2023) The traditional use of southern African medicinal plants to treat oral pathogens, and studies into their relevant antimicrobial properties. S Afr J Bot 153:258–279

Conroy MS (2021) Herbal medicine in Russia's history: the use of herbal medicine for infectious diseases in Russia's history. In: Herbal medicine: back to the future: volume 4, infectious diseases. Bentham Science, Sharjah, p 90

Dankar FK, Ptitsyn A, Dankar SK (2018) The development of large-scale de-identified biomedical databases in the age of genomics—principles and challenges. Hum Genomics 12:1–15

Ekor M (2014) The growing use of herbal medicines: issues relating to adverse reactions and challenges in monitoring safety. Front Pharmacol 4:177. https://doi.org/10.3389/fphar.2013.00177. PMID: 24454289; PMCID: PMC3887317

Elahee SF, Mao H, Shen X (2019) Traditional Indian medicine and traditional Chinese medicine: a comparative overview. Chin Med Cult 2(3):105–113

Fang S, Dong L, Liu L et al (2021) HERB: a high-throughput experiment- and reference-guided database of traditional Chinese medicine. Nucleic Acids Res 49:D1197–D1206

Fathifar Z, Kalankesh LR, Ostadrahimi A, Ferdousi R (2023) New approaches in developing medicinal herbs databases. Database (Oxford) 2023:baac110. https://doi.org/10.1093/database/baac110. PMID: 36625159; PMCID: PMC9830469

He S, Yang L, Ye S et al (2022) MPOD: applications of integrated multi-omics database for medicinal plants. Plant Biotechnol J 20:1–3

Heinrich M, Sharma SK, Suetterle U, Bhamra SK (2023) Herbal medicine use in the U.K. and Germany and pharmacy practice—a commentary. Res Soc Adm Pharm 19(3):535–540

Hidayat S, Subositi D, Batubara I, Munawaroh E, Afandi S, Cahyaningsih R (2023) Scientific databases for conservation of medicinal plants. In: Medicinal plants: biodiversity, biotechnology and conservation. Springer Nature Singapore, Singapore, pp 265–287

Hosen SZ, Junaid M, Alam MS et al (2021) GreenMolBD: nature derived bioactive molecules' database. Med Chem 18:724–733

Hussain N, Chanda R, Abir RA et al (2021) MPDB 2.0: a large scale and integrated medicinal plant database of Bangladesh. BMC Res Notes 14:1–6

Jamal A (2023) Embracing nature's therapeutic potential: herbal medicine. Int J Multidiscip Sci Arts 2(1):117–126

Joos S, Glassen K, Musselmann B (2012) Herbal medicine in primary healthcare in Germany: the patient's perspective. Evid Based Complement Alternat Med 2012:294638. https://doi.org/10.1155/2012/294638. Epub 2012 Dec 31. PMID: 23346197; PMCID: PMC3549419

Karimi-Jafari MH, Firouzi R, Ashouri M, Poursoleiman A (2022) A database of chemical compositions of Persian medicinal herbs. ChemRxiv 2022. https://doi.org/10.26434/chemrxiv-2022-8rrwp

Kasilo OMJ, Wambebe C, Nikiema JB, Nabyonga-Orem J (2019) Towards universal health coverage: advancing the development and use of traditional medicines in Africa. BMJ Glob Health 4(Suppl 9):e001517

Khan AA, Janardhanan R, Kishore J, Karwasra R, Saeed S, Das BC et al (2022) Unani system of medicine and it's development in contemporary healthcare: a comprehensive review. J Pharm Negat Results 13:634–645

Kim JY, Noble D (2014) Recent progress and prospects in Sasang constitutional medicine: a traditional type of physiome-based treatment. Prog Biophys Mol Biol 116:76–80

Kim SK, Nam S, Jang H et al (2015) TM-MC: a database of medicinal materials and chemical compounds in northeast Asian traditional medicine. BMC Complement Altern Med 15:1–8

Kuchta K, Cameron S (2022) Kampo medicine in a modern context: ethnopharmacological perspectives. Front Pharmacol 13:971254

Kumar Y, Prakash O, Tripathi H et al (2018) AromaDb: a database of medicinal and aromatic plant's aroma molecules with phytochemistry and therapeutic potentials. Front Plant Sci 9:1081

Lee JH, Park KM, Han DJ et al (2015) PharmDB-K: integrated bio-pharmacological network database for traditional Korean medicine. PLoS One 10:1–14

Li B, Ma C, Zhao X et al (2018) YaTCM: yet another traditional Chinese medicine database for drug discovery. Comput Struct Biotechnol J 16:600–610

Li Y, Zhao Y, Niu X, Zhou W, Tian J (2022) The efficiency evaluation of municipal-level traditional Chinese medicine hospitals using data envelopment analysis after the implementation of hierarchical medical treatment policy in Gansu Province, China. Inquiry 59:469580221095799

Lu CL, Li X, Zhou HM, Zhang C, Yang YY, Feng RL et al (2019) Randomised controlled trials of traditional Chinese medicine in cancer care published in Chinese: an overview. Lancet 394:S26

Luo H, Yang M, Tang QL, Hu XY, Willcox ML, Liu JP (2021) Characteristics of registered clinical trials on traditional Chinese medicine for coronavirus disease 2019 (COVID-19): a scoping review. Eur J Integr Med 41:101251. https://doi.org/10.1016/j.eujim.2020.101251

Mahmud S, Paul GK, Biswas S et al (2022) Phytochemdb: a platform for virtual screening and computer-aided drug designing. Database 2022:1–14

Majeed M, Nagabhushanam K, Prakasan P, Mundkur L (2023) The pursuit of natural medicine—a current perspective, nutrition science, marketing nutrition, health claims, and public policy. Academic Press, New York, pp 173–192

Meng F, Tang Q, Chu T et al (2022) TCMPG: an integrative database for traditional Chinese medicine plant genomes. Hortic Res 9:1–8

Mohan R, Pandey V, Kurien N (2023) India as Global Center for Traditional Medicine: needs, initiatives, and challenges. J Ayurveda 17(2):150–153

Mohanraj K, Karthikeyan BS, Vivek-Ananth RP et al (2018) IMPPAT: a curated database of Indian medicinal plants, phyto-chemistry and therapeutics. Sci Rep 8:1–17

Mukherjee PK, Harwansh RK, Bahadur S, Banerjee S, Kar A, Chanda J et al (2017) Development of Ayurveda—tradition to trend. J Ethnopharmacol 197:10–24

Mumtaz A, Ashfaq UA, Ul Qamar MT et al (2017) MPD3: a useful medicinal plants database for drug designing. Nat Prod Res 31:1228–1236

Naghizadeh A, Hamzeheian D, Akbari S et al (2020) UNaProd: a universal natural product database for materia medica of Iranian traditional medicine. Evid Based Complement Alternat Med 2020:1–14

Nguyen-Vo T-H, Le T, Pham D et al (2018) VIETHERB: a database for Vietnamese herbal species. J Chem Inf Model 59:1–9

Nguyen-Vo TH, Nguyen L, Do N et al (2020) Plant metabolite databases: from herbal medicines to modern drug discovery. J Chem Inf Model 60:1101–1110
Nie H, Han Z, Nicholas S, Maitland E, Huang Z, Chen S, Shi X (2023) Costs of traditional Chinese medicine treatment for inpatients with lung cancer in China: a national study. BMC Complement Med Ther 23(1):5
Ningthoujam SS, Talukdar AD, Potsangbam KS et al (2012) Challenges in developing medicinal plant databases for sharing ethnopharmacological knowledge. J Ethnopharmacol 141:9–32
Ntie-Kang F, Zofou D, Babiaka SB et al (2013) AfroDb: a select highly potent and diverse natural product library from African medicinal plants. PLoS One 8:1–15
Ohbe H, Sasabuchi Y, Jo T et al (2023) Kampo medicine in ICUs in Japan between 2010 and 2020. J Anesth 37:654–656. https://doi.org/10.1007/s00540-023-03190-8
Okamoto H, Iyo M, Ueda K, Han C, Hirasaki Y, Namiki T (2014) Yokukan-san: a review of the evidence for use of this Kampo herbal formula in dementia and psychiatric conditions. Neuropsychiatr Dis Treat 10:1727–1742
Okayasu T, Mitani K, Kitahara T (2023) Reviewing Kampo medicine (traditional Japanese Herbal Medicine) for otology/neurotology diseases. Auris Nasus Larynx 51:25
Oliver SJ (2013) The role of traditional medicine practice in primary health care within Aboriginal Australia: a review of the literature. J Ethnobiol Ethnomed 9:46
Ozioma EOJ, Chinwe OAN (2019) Herbal medicines in African traditional medicine. Herb Med 10:191–214
Pandey MM, Rastogi S, Rawat AKS (2013) Indian traditional ayurvedic system of medicine and nutritional supplementation. Evid Based Complement Alternat Med 2013:376327
Parasuraman S, Thing GS, Dhanaraj SA (2014) Polyherbal formation: concept of Ayurveda. Pharmacogn Rev 8:73–80
Parveen A, Parveen R, Akhatar A, Parveen B, Siddiqui KM, Iqbal M (2020) Concepts and quality considerations in Unani system of medicine. J AOAC Int 103(3):609–633
Parveen A, Ahmad M, Parveen B, Parveen R, Iqbal M (2022) The traditional system of Unani medicine, its origin, evolution and Indianisation: a critical appraisal. Indian J Tradit Knowl 21(3):511–521
Patel G, Brosnan C, Taylor A (2023) Understanding the role of context in health policy implementation: a qualitative study of factors influencing traditional medicine integration in the Indian public healthcare system. Health Sociol Rev 32:1–17
Patwardhan B (2014) Bridging Ayurveda with evidence-based scientific approaches in medicine. EPMA J 5:1–7
Petrovska BB (2012) Historical review of medicinal plants' usage. Pharmacogn Rev 6(11):1–5. https://doi.org/10.4103/0973-7847.95849. PMID: 22654398; PMCID: PMC3358962
Potshangbam AM, Polavarapu R, Rathore RS et al (2019) MedPServer: a database for identification of therapeutic targets and novel leads pertaining to natural products. Chem Biol Drug Des 93:438–446
Poulakou-Rebelakou E, Karamanou M, George A (2015) The impact of ancient Greek medicine in India: the birth of Unani medicine. Acta Med Hist Adriat 13(2):323–328
Ru J, Li P, Wang J et al (2014) TCMSP: a database of systems pharmacology for drug discovery from herbal medicines. J Cheminform 6:1–6
Rudra S, Kalra A, Kumar A, Joe W (2017) Utilization of alternative systems of medicine as health care services in India: evidence on AYUSH care from NSS 2014. PLoS One 12(5):e0176916
Sattigeri VJ (2023) Role of government and research organizations in rejuvenating and re-affirming the value of traditional knowledge systems—AYUSH, as a case in point. J Res Ayurved Sci 7(5):55
Sawada R, Iwata M, Umezaki M et al (2018) KampoDB, database of predicted targets and functional annotations of natural medicines. Sci Rep 8:11216
Shi J, Yang Y, Zhou X, Zhao L, Li X, Yusuf A et al (2022) The current status of old traditional medicine introduced from Persia to China. Front Pharmacol 13:953352

Shikov AN, Pozharitskaya ON, Makarov VG, Wagner H, Verpoorte R, Heinrich M (2014) Medicinal plants of the Russian Pharmacopoeia; their history and applications. J Ethnopharmacol 154:481–536
Shikov AN, Narkevich IA, Flisyuk EV, Luzhanin VG, Pozharitskaya ON (2021) Medicinal plants from the 14th edition of the Russian Pharmacopoeia, recent updates. J Ethnopharmacol 268:113685
So JH, Kim JW, Nam JH, Lee BJ, Kim YS, Kim JY, Do JH (2016) The web application of constitution analysis system-SCAT (Sasang Constitution Analysis Tool). J Sasang Constitut Med 28(1):1–10
Sorokina M, Steinbeck C (2020) Review on natural products databases: where to find data in 2020. J Cheminform 12:1–51
Sumarya IM, Suarda IW, Sudaryati NLG, Sitepu I (2020) Benefits of biopharmaca products towards healthy Indonesia. J Phys Conf Ser 1469:012133. https://doi.org/10.1088/1742-6596/1469/1/012133
Sun X, Li L, Liu Y, Wang W, Yao M, Tan J et al (2021) Assessing clinical effects of traditional Chinese medicine interventions: moving beyond randomized controlled trials. Front Pharmacol 12:713071
Wang J, Guo Y, Li GL (2016) Current status of standardization of traditional Chinese medicine in China. Evid Based Complement Alternat Med 2016:9123103
Wu Y, Zhang F, Yang K et al (2019) SymMap: an integrative database of traditional Chinese medicine enhanced by symptom mapping. Nucleic Acids Res 47:D1110–D1117
Xie T, Song S, Li S et al (2015) Review of natural product databases. Cell Prolif 48:398–404
Xue R, Fang Z, Zhang M et al (2013) TCMID: traditional Chinese medicine integrative database for herb molecular mechanism analysis. Nucleic Acids Res 41:D1089–D1095
Yakubo S, Ito M, Ueda Y, Okamoto H, Kimura Y, Amano Y, Togo T, Adachi H, Mitsuma T, Watanabe K (2014) Pattern classification in kampo medicine. Evid Based Complement Alternat Med 2014:535146
Yoo J, Lee E, Kim C, Lee J, Lixing L (2012) Sasang constitutional medicine and traditional Chinese medicine: a comparative overview. Evid Based Complement Alternat Med 2012:980807
Yuan H, Ma Q, Ye L, Piao G (2016) The traditional medicine and modern medicine from natural products. Molecules 21(5):559. https://doi.org/10.3390/molecules21050559. PMID: 27136524; PMCID: PMC6273146
Yusupova ZA, Baratjon Ogli SF, Abduqunduzovna MZ (2023) Medicinal plants growing in our republic medicinal properties. Period J Mod Philos Soc Sci Hum 15:5–7
Zeng X, Zhang P, Wang Y, Qin C, Chen S, He W, Tao L, Tan Y, Gao D, Wang B, Chen Z, Chen W, Jiang YY, Chen YZ (2019) CMAUP: a database of collective molecular activities of useful plants. Nucleic Acids Res 47(D1):D1118–D1127. https://doi.org/10.1093/nar/gky965. PMID: 30357356; PMCID: PMC6324012

Ethnopharmacognosy of Medicinal Plants: Challenges and Biotechnological Advancements

5

Gaurav Upadhyay and Priya Thakur

Abstract

Ethnopharmacognosy represents a multidisciplinary research field, drawing insights from various disciplines and scientific methods. It shares a close association with integrative medicine, exploring the significance of ethnomedicine in healthcare and complementary therapies across diverse domains. The collaborative approaches employed have the potential to enhance healthcare safety, fostering sustainable well-being in marginalized regions. This chapter delves into ethnopharmacognosy, emphasizing anthropological perspectives on plant utilization and the process of drug development from ethnomedicine. The exploration begins with the validation of ethnomedicines, tracing the research development in ancestral medicine. The chapter also discusses traditional medicine-inspired approaches to drug development, intellectual property rights, biodiversity conservation, and transformative research and development paradigms such as Ayurgenomics, ayurvedic practices, ethnopharmacognosy, network pharmacology, reverse pharmacology, etc. These strategies prove effective in discovering novel drugs from ethnopharmacognosy, showcasing interdisciplinary and transdisciplinary connections leading to nature-based and ancestral drug discoveries.

Keywords

Ethnomedicines · Ayurveda · Pharmacognosy · Biodiversity

G. Upadhyay (✉)
Ram-Eesh Institute of Vocational and Technical Education, Greater Noida, Uttar Pradesh, India

P. Thakur
MET Faculty of Pharmacy, MIT Campus, Moradabad, Uttar Pradesh, India

© The Author(s), under exclusive license to Springer Nature Singapore Pte Ltd. 2024
M. Nandave et al. (eds.), *Ethnopharmacology and OMICS Advances in Medicinal Plants Volume 1*, https://doi.org/10.1007/978-981-97-2367-6_5

5.1 Introduction

Ethnopharmacognosy is an interdisciplinary field aimed at examining the pharmacological underpinnings and anthropological justifications of how human cultures employ plants, animals, fungi, microbes, and minerals as medicines. A realistic assessment of an indigenous drug is an interdisciplinary approach to enhance its medical potential. Richard Evans Schultes and Tony Swain stated that "*people whom we have chosen to consider members of less-advanced societies have consistently looked to the Plant Kingdom*." As chemists, pharmacologists, and botanists, shouldn't we make use of the numerous and diverse resources available to us in order to learn? The global economy is driving an acceleration of the monetization of indigenous and local knowledge, despite the fact that the search for new treatments has traditionally been driven by financial gains (Patwardhan 2005). But as society has become more interconnected, concerns have grown that "freezing people in time should not be our goal in the interim." Instead, we must work out ways to ensure that, in a pluralistic, interconnected world, everyone may benefit from modernity without having to renounce their ethnic identity in order to do so. The words "complementary medicine" and "alternative medicine" describe a wide range of medical practices that are not fully included in the country's dominant healthcare system and are not a part of that country's own customs or conventional medicine (Li and Xing 2016). They are utilized in some countries in place of traditional medicine. Therefore, it is important to distinguish between "traditional medicine," which refers to a complex therapeutic environment, and "traditional medicines," which are understood to be the pharmaceuticals utilized inside a particular traditional medicine (Buenz et al. 2004).

Traditional medicine and medicinal plants are commonly employed in urban contexts as alternatives to modern medical care and as means of self-medication for minor and chronic illnesses, but they are especially favored in less affluent rural communities or during economic downturns (Chakravarty 2010). The adoption of a holistic viewpoint on human health is typical of the majority of "traditional" or regional medical systems. Hypothesis-driven methodologies are advantageous for the assessment of descriptive field data as well as the recognition of anthropological components in traditional medicines. Medical anthropology research questions are well-defined and testable help to strengthen and promote the relationship between the social and natural sciences, as well as prevent the drawing arbitrary for conclusions (Butler et al. 2014). The bio-cultural examination and description of "traditional," indigenous, or local pharmacopoeias, as well as the experimental evaluation of materia medica through biological test systems, are the main focuses of ethnopharmacognosy (Buenz et al. 2007).

The bioassays used to support the observed health claims vary in their depth and breadth of representation of the molecular features and complexity of the condition or symptoms that a medicinal drug is used to treat in "traditional" medicine. Contributing to evidence-based medicine is ethnopharmacognosy's ultimate goal, and to do this, clear negative findings may be just as significant as positive ones (Buenz et al. 2006).

5.2 The State of Ethnopharmacognosy at the Moment in Drug Discovery

An increasing number of new physiologically active chemicals derived from natural sources are being produced. Natural product libraries beat combinatorial chemistry libraries in terms of hit rate, despite a historical tendency in screening efforts toward combinatorial chemistry. Part of the reason for this is that natural products don't work with traditional high-throughput screening paradigms. It is likely due to the potential physiological roles of the primary and secondary metabolites under investigation that natural product libraries exhibit such a high hit rate (Adams et al. 2009). In particular, natural product libraries include a higher percentage of biologically relevant compounds but fewer chemicals overall than combinatorial chemistry libraries. Most people worldwide rely on traditional medicine, and those who practice these conventional treatments employ natural items as therapeutic instruments. According to recent studies, the traditional ethnopharmacognosy practice of using plant-based remedies to identify bioactive chemicals is helpful. The historical usage of botanicals and other biomaterials in a traditional medical system, which was based on trial and error, might theoretically yield significant information on the biological activity of the chemicals in standard conventional medicines (Scott and Hewett 2008).

Recent evolutionary research supports these findings, demonstrating that people using traditional medical systems in various geographic locations independently integrated related plants into their conventional medical systems (Buenz et al. 2006).

Although there is a great deal of potential for using ethnopharmacognostic data in drug research, the path forward is not totally obvious. A continual decline in biodiversity has resulted from the extinction of source species, and traditional medical knowledge is being lost generation after generation. Documentation of the application of natural materials in ethnopharmacognosy is therefore crucial. Since the source species and ethnopharmacognostic knowledge are vanishing, it is not viable to find novel biologically active compounds using ethnopharmacognostic instruments; thus, a concentrated effort is needed (Soelberg et al. 2016).

5.3 Ethnopharmacognosis: Difficulties

"Traditional" medicine systems sometimes recommend complex herbal mixtures and extracts with many components. In polypharmacology, the principle of "one disease, one target, and one drug" is more appropriate for treating diseases with numerous causes, including degenerative and chronic illnesses, than it is for biomedicine (Barrett et al. 2015). This means that the principle does not apply in polypharmacology. A range of moderately active metabolites may be present in an extract, and these metabolites may disrupt and inhibit the function of different proteins within a particular signaling network, leading to additive pharmacological effects. In this regard, it was also clear how crucial it is to adopt a more holistic in vivo strategy, whether through clinical trials or animal investigations, and how reductionist in vitro methodologies are ill-suited for discovering synergistic characteristics or prodrugs. It has also been shown how systems biology and omics techniques are conceptually closer to the holistic perspective of ethnopharmacognosy, given that in vivo research is not always practical (Ningthoujam et al. 2012).

For example, in metabonomics, the goal is to quantify as many metabolic parameters as possible to assess and better understand how complex biological systems are altered, perturbed, and operate when exposed to specific external stimuli, such as plant extracts with therapeutic properties. More and more emphasis is being placed on omic approaches, particularly with regard to traditional Chinese medicine (TCM) (Barrett et al. 2015). The holistic standard of Chinese and Ayurvedic medicine is best adhered to by using a systems approach for examining quality, safety, and rationalizing polypharmacological mechanisms or network pharmacognosy exerted by complex mixes of chemicals and extracts. The vast arrays of therapeutic applications for traditional medicines that have been documented in the literature starkly contrast with their activity as determined by one-dimensional bioassays (Ningthoujam et al. 2012).

It is common practice to apply extracts or natural compounds in therapeutically dubious amounts to test systems until the desired impact is seen. The story typically stops at this point since only a small number of in vitro observations can be confirmed in vivo. Many plant species and crude medications are recommended and treated simultaneously for multiple diseases and symptoms within one culture and even to a greater extent between distinct cultures, making it challenging to link ethnomedical indications with biological test techniques. Additionally, ethnopharmacopoeias are cultural constructs that are always changing and exchanging (Bajorath 2017).

The local commercialization and use of pharmaceuticals can coexist with and be complemented by the local use of medicinal plants and related cultural healing beliefs. Syncretism developments are frequently the result of such coexistence, as rural indigenous groups in Mexico and Peru have demonstrated. Intensified globalization is accelerating the exchange of local and global pharmacopoeias through print media, television, the Internet, and, in addition to international corporate interests, creating a feedback loop. In discussions about medical anthropology and drug discovery, the influx of global phytotherapeutic and biological information into local pharmacopoeias could result in circular reasoning (Weckerle et al. 2011).

The perception in Western societies that the use of such products would be free of health risks, especially when compared to pharmaceuticals, has contributed to the global commercialization of herbal medicine and the rise of what is known as complementary and alternative medicine as a social phenomenon over the past few decades (Sharma and Sarkar 2013). In contrast, the philosophy and standards of Chinese medicine, which experimentally exploit the distinctions between the therapeutic uses of toxic and nontoxic herbal formulations and medications, do not support such a perception (Barrett et al. 2015). As an illustration, intentional or unintentional adulterations and incorrectly labeled herbal medications pose a persistent and latent health concern. In this overview of perspectives, we emphasize the development and historical heritage of ethnopharmacognosy, the significance of allelochemicals, and the relevance of human food. An examination of bioprospecting ancient literature, phylogenetic methods in ethnopharmacognosy, and evolutionary ecology illustrates the relationship between phytotherapy and drug discovery (Prinz et al. 2011). This is demonstrated through an analysis of bioprospecting ancient literature, phylogenetic techniques in ethnopharmacognosy, and evolutionary ecology (Kindscher et al. 2013).

5.4 Recent Advances

Traditional medical knowledge is still being lost, which can be due to the extinction of some species or the replacement of natural product-based treatments with contemporary pharmaceuticals. There is no chance of determining the therapeutic benefits of an extinct organism. Recent interesting initiatives, however, have exploited historical texts to identify traditional medical applications that have been lost or replaced by contemporary medicine (Natural Product Libraries 2017). It is feasible to revive lost traditional medicine knowledge through the knowledge found in ancient manuscripts, and these efforts have been successful, leading to at least one favorable clinical evaluation of the natural substance extracted from historical herbal literature.

Drug discovery will still draw on ancient medical books, but efforts now extend beyond medicine. In the United States Virgin Islands, for instance, records from the slave trade have been exploited to locate forgotten traditional medical knowledge (Harvey et al. 2015).

The proper documentation of the historical use of natural products is a prerequisite for the effort focused on recovering lost historical information. However, the diversity of methods used to record the uses of therapeutic raw materials presents a unique problem, and the variable quality of the ethnopharmacognostic data is one of the ongoing difficulties. Thankfully, criteria for strict standards have been created, enabling statistical approaches to be used as the foundation for a bioinformatics analysis evaluating how well these existing ethnopharmacognostic data, which are of wildly varied quality, meet anticipated models (Li and Vederas 2009).

Although it is crucial to remember that even most peer-reviewed reports of drug function are irreproducible, this restriction on ethnopharmacognosy is explicit; these difficulties in ethnopharmacognosy are to be expected (Harvey et al. 2015).

Natural product libraries are accessible and practical tools for online natural product screening procedures. Until recently, most ethnopharmacognosy resources have been extensive medical books like the *Ayurvedic Pharmacopoeia of India*. These tools are not appropriate for in silico screening projects. Fortunately, several databases have been created to screen ethnopharmacognostic data in silico, and some of them have been applied with outstanding results. Although these databases try to describe how an organism is used in conventional medicine, the relationship between traditional uses and these resources for natural products is not always well documented, and as was mentioned above, the variety of methods used to record these historical uses can be problematic (Li and Vederas 2009).

5.5 Imminent Risks to the Bases of Ethnopharmacognosy

The loss of knowledge about traditional medicine across generations, the depletion of natural product materials used in traditional medicines, and the historical absence of a legal framework to conduct bioprospecting are the three biggest risks associated with integrating ethnopharmacognosy in the drug discovery process. It is simple to stop the transmission of customs from one generation to the next. While the extinction of traditional medicine animals may also appear straightforward, the reality is more intricate because it is hard to pass down generational knowledge about the application of a natural product if the creature is extinct (Sukuru et al. 2009).

The tragedy of the commons fits the conventional economic concept of traditional medicine. As the species grows rarer, the remaining species increase in value for use in conventional medications and hence increase in popularity, ultimately causing the species to go extinct. In the wild, people gather more than half of the plants used as medicines. It has been shown that both plants and animals utilized in traditional medicine preparations are on the verge of extinction due to this tendency toward extinction. Because of this, rhinoceros horn has a value equal to that of gold (Vining 1990).

5.6 Omics: Using Ethnopharmacognosy to Benefit from the Polypharmacy of Natural Products

Traditional medical systems integrate a variety of chemicals in accordance with a predetermined methodology; some compounds are thought to have a therapeutic influence, while others help to augment that effect and maybe mitigate side effects. Thus, synergy is important in conventional therapy. This might be brought on by different substances present in a single creature or by substances produced from different organisms (Akhtar et al. 2016). This synergy may take place at the level of pharmacodynamics, altering blood pressure or other systems through antagonist activities or allosteric effects on the same receptor, or at the level of pharmacokinetics, affecting bioavailability. Synergy is notoriously challenging for bioassay-guided fractionation, which is unfortunate. Thanks to omics techniques, opportunities for ethnopharmacognosy are growing (Balandrin et al. 1985).

Many of the synergistic drugs utilized in conventional medicine, for instance, comprise numerous receptors or proteins that are not used in typical bioassays. Therefore, clinical studies of traditional medicines are the most effective means of overcoming the inherent limits of in vivo assays, which may overlook the content of the therapeutic advantage of a traditional medicine therapy (Akhtar et al. 2016). The safety barriers for some of these medications with a long history of use are substantially lower when compared to a novel molecular entity. Additionally, by accumulating metabolomics data for these investigations, it is possible to analyze the entire reaction accurately enough to reduce the number of patients (Irwin et al. 2012).

5.7 Health and Ethnopharmacognosy

Experts questioned whether or not we should "set a place for diet in ethnopharmacognosy" in light of the "extra-nutritive" nature of food, phytochemistry, and associated pharmacognosy. Vegetables that are both domesticated and wild-collected have coevolved as food and medicine, according to the evolutionary perspective on health and nutrition. Many edible wild plants have been reported to have pharmacological activity and to have positive effects on the kidneys, stomach, and heart in addition to having diuretic qualities in regional and popular folklore (Leuenroth et al. 2007). Foods that are regularly consumed may yet retain significant levels of pharmacologically active plant metabolites, even though domestication has reduced the overall chemical content of staple crops. Thus, the focus moves from the physiologic requirements for vitamins, pigments (carotenoids, lycopene), and fatty acids that nutrition sciences take into account to the pharmacological characteristics of allele chemicals found in food, i.e., the bioactivity of secondary metabolites that evolved under ecological constraints such as herbivory (Fabricant and Farnsworth 2001).

Moreover, flavonoids and other polyphenolic substances with high concentrations of polyphenols do not fit into the more restrictive concept of ethnopharmacognosy because their antioxidant qualities in vitro are rarely transferred to in vivo settings. There was evidence that flavonoids did not evolve to scavenge free radicals, as suggested by evolutionary theory (Leuenroth et al. 2007).

5.8 Traditional Medicines: Globalization Challenges

Traditional remedies and medicinal plants in particular have been a part of human history for millennia, with knowledge being passed down from one generation to the next. In many parts of the world, including Asia and Europe, written records stretch back thousands of years (Cox 2000).

For example, the so-called *Shennong Bencao Jing*, also known as Shennong's Classic of Materia Medica, is a noteworthy compendium of the traditional use of 365 Chinese remedies, and Dioscurides documented the therapeutic application of approximately 600 medicinal plants in ancient Greece. Worldwide commerce, in particular medicinal plants, began as travel increased along major trade routes connecting Europe, Asia, and other continents (Van der Kooy et al. 2009).

Many countries throughout the world today use herbal and traditional medicine to aid with healthcare. Herbal medicines are a common therapeutic option used in conjunction with other therapies such as physical therapy, diet, and therapeutic movement. This trend of mixing herbal and conventional treatments is becoming more and more popular (O'Reilly et al. 2014).

5.9 Conclusion

The foundation of ethnopharmacognosy is the understanding that humans have historically employed natural products as medicines. It was evident that using natural product chemical libraries to create new medications was a successful strategy, and knowledge about a natural product's therapeutic characteristics can be gleaned via ethnopharmacognostic medicinal understanding. The discovery of artemisinin and other therapies demonstrates the benefit of utilizing an earlier understanding of how natural compounds might be used medicinally. It is also obvious that prodrugs and synergy affect how traditional medicines work. In the current pharmaceutical repertoire around the world, these topics are not well explored.

Traditional knowledge is a resource that has historically been undervalued, and ethnopharmacognosy's current contribution to drug development has often been patchy. The future goal is to standardize the adoption of ethnopharmacognosy to optimize the drug discovery process and expedite the discovery of new treatments, particularly given the resources currently available. It might be advantageous to start the drug development process using natural ingredients. We concur with that recommendation and add this: The potential to increase the hit rate and hasten the identification of new treatments lies in frontloading the screening procedure with ethnopharmacognostic data.

References

Adams M, Berset C, Kessler M, Hamburger M (2009) Medicinal herbs for the treatment of rheumatic disorders—a survey of European herbals from the 16th and 17th century. J Ethnopharmacol 121(3):343–359

Akhtar MT, Mushtaq MY, Verpoorte R, Richardson MK, Choi YH (2016) Zebrafish as a model for systems medicine R&D: rethinking the metabolic effects of carrier solvents and culture buffers determined by 1H NMR metabolomics. OMICS J Integr Biol 20(1):4252

Bajorath J (2017) Compound data mining for drug discovery. Methods Mol Biol 1526:247–256

Balandrin MF, Klocke J, Wurtele ES, Bollinger WH (1985) Natural plant chemicals: sources of industrial and medicinal materials. Science 228(4704):1154–1159

Barrett P, Flower A, Lo V (2015) What's past is prologue: Chinese medicine and the treatment of recurrent urinary tract infections. J Ethnopharmacol 167:86–96

Buenz EJ, Schnepple DJ, Bauer BA, Elkin PL, Riddle JM, Motley TJ (2004) Techniques: bioprospecting historical herbal texts by hunting for new leads in old tomes. Trends Pharmacol Sci 25(9):494–498

Buenz EJ, Johnson HE, Beekman EM, Motley TJ, Bauer BA (2005) Bioprospecting Rumphius's *Ambonese herbal*: volume I. J Ethnopharmacol 96(1–2):57–70

Buenz EJ, Bauer BA, Johnson HE, Tavana G, Beekman EM et al (2006) Searching historical herbal texts for potential new drugs. BMJ 333(7582):1314–1315

Buenz EJ, Bauer BA, Schnepple DJ, Wahner-Roedler DL, Vandell AG, Howe CL (2007) A randomized phase I study of *Atuna racemosa*: a potential new anti-MRSA natural product extract. J Ethnopharmacol 114(3):371–376

Butler MS, Robertson AA, Cooper MA (2014) Natural product and natural product derived drugs in clinical trials. Nat Prod Rep 31(11):1612–1661

Chakravarty R (2010) Preserving traditional knowledge: initiatives in India. IFLA J 36(4):294–299

Cox PA (2000) Will tribal knowledge survive the millennium? Science 287(5450):44–45

Fabricant DS, Farnsworth NR (2001) The value of plants used in traditional medicine for drug discovery. Environ Health Perspect 109(Suppl. 1):69–75
Harvey AL, Edrada-Ebel R, Quinn RJ (2015) The re-emergence of natural products for drug discovery in the genomics era. Nat Rev Drug Discov 14(2):111–129
Irwin JJ, Sterling T, Mysinger MM, Bolstad ES, Coleman RG (2012) ZINC: a free tool to discover chemistry for biology. J Chem Inf Model 52(7):1757–1768
Kindscher K, Corbett S, McClure K (2013) A statistical analysis of medicinal plants: a case study of plant families in Kansas and the Great Plains. Trans Kans Acad Sci 116(3–4):149–155
Leuenroth SJ, Okuhara D, Shotwell JD, Markowitz GS, Yu Z et al (2007) Triptolide is a traditional Chinese medicine-derived inhibitor of polycystic kidney disease. PNAS 104(11):4389–4394
Li JW, Vederas JC (2009) Drug discovery and natural products: end of an era or an endless frontier? Science 325(5937):161–165
Li DL, Xing FW (2016) Ethnobotanical study on medicinal plants used by local Hoklos people onHainan Island, China. J Ethnopharmacol 194:358–368
Natural Product Libraries (2017) Natl. Cent. Complement. Integr. Health. Natural Product Libraries, Bethesda, MD. Natl. Inst. Health. https://nccih.nih.gov/grants/naturalproducts/libraries
Ningthoujam SS, Talukdar AD, Potsangbam KS, Choudhury MD (2012) Challenges in developing medicinal plant databases for sharing ethnopharmacological knowledge. J Ethnopharmacol 141(1):9–32
O'Reilly LP, Luke CJ, Perlmutter DH, Silverman GA, Pak SC (2014) *C. elegans* in high-throughput drug discovery. Adv Drug Deliv Rev 69:247–253
Patwardhan B (2005) Ethnopharmacology and drug discovery. J Ethnopharmacol 100(1–2):50–52
Prinz F, Schlange T, Asadullah K (2011) Believe it or not: how much can we rely on published data on potential drug targets? Nat Rev Drug Discov 10(9):712
Scott G, Hewett ML (2008) Pioneers in ethnopharmacology: the Dutch East India company (VOC) at the cape from 1650 to 1800. J Ethnopharmacol 115(3):339–360
Sharma V, Sarkar IN (2013) Bioinformatics opportunities for identification and study of medicinal plants. Brief Bioinform 14(2):238–250
Soelberg J, Davis O, Jager AK (2016) Historical versus contemporary medicinal plant uses in the US Virgin Islands. J Ethnopharmacol 192:74–89
Sukuru SCK, Jenkins JL, Beckwith RE, Scheiber J, Bender A et al (2009) Plate-based diversity selection based on empirical HTS data to enhance the number of hits and their chemical diversity. J Biomol Screen 14(6):690–699
Van der Kooy F, Maltese F, Choi YH, Kim HK, Verpoorte R (2009) Quality control of herbal material and phytopharmaceuticals with MS and NMR based metabolic fingerprinting. Planta Med 75(07):763–775
Vining LC (1990) Functions of secondary metabolites. Ann Rev Microbiol 44(1):395–427
Weckerle CS, Cabras S, Castellanos ME, Leonti M (2011) Quantitative methods in ethnobotany and ethnopharmacology: considering the overall flora—hypothesis testing for over-and underused plant families with the Bayesian approach. J Ethnopharmacol 137(1):837–843

Ethnobotanical and Conservational Studies of Medicinal Plants: A Case Study of *Dolichandrone falcata* (Wall. Ex DC.) Seem. in Western Ghats, India

6

Suhas Daswadikar, Mandar Muley, Satyendra Prasad, and Prakash Itankar

Abstract

The traditional use of plants as medicine by tribal people led to the discovery of several natural products and the development of modern medicine. Natural products, either as standardized plant extracts or pure compounds and their structural analog, have historically provided extensive opportunities for new drug leads. The present chapter discusses the potential of ethnobotanical studies with particular reference to conservation of important medicinal plants in the Western Ghats of India. Several tribes of Western Ghats still depend majorly on the indigenous systems of medicine, and this region has a valuable heritage of herbal remedies. The present case study on *Dolichandrone falcata* (Wall. ex DC.) Seem. of Bignoniaceae family discusses its ethnobotanical and ethnomedicinal properties. The given plant is considered an endemic rare plant of India, commonly called a trumpet vine. The unexplored plant has several medicinal values, such as antidiabetic, anti-inflammatory, analgesic, antiestrogenic, antimicrobial, antinociceptive, antibacterial, anti-hemorrhoidal, and immunomodulatory. It has been used for centuries in India to treat a wide range of illnesses. Phytochemically, it contains steroids, amino acids, alkaloids, flavonoids, saponins, polyphenols, terpenoids, and cardiac glycosides. This chapter attempts to provide thorough and current information on the traditional uses, phytochemistry, taxonomy, ethnopharmacology, pharmacology and preclinical studies of the plant in order to understand its pharmacognostic profile better. The chapter covers pertinent publications, online databases, and testimonials of traditional healers. This chapter outlines the most recent scientific research on this plant. It aims to raise awareness of the issue of its decreasing availability for the extension of its benefits to humans, particularly in terms of its health benefits. It also makes an appeal to all

S. Daswadikar · M. Muley · S. Prasad · P. Itankar (✉)
Department of Pharmaceutical Sciences, Rashtrasant Tukadoji Maharaj Nagpur University, Nagpur, Maharashtra, India

© The Author(s), under exclusive license to Springer Nature Singapore Pte Ltd. 2024

M. Nandave et al. (eds.), *Ethnopharmacology and OMICS Advances in Medicinal Plants Volume 1*, https://doi.org/10.1007/978-981-97-2367-6_6

stakeholders for the implementation of various conservational and sustainable strategies for rare plant species.

Keywords

Ethnobotany · Ethnobotanical appraisal · Conservation of medicinal plants · *Dolichandrone falcata*

6.1 Introduction

Humans have relied on plants for survival and growth since the dawn of humanity. Earlier, they were food gatherers and hunters, but they gradually shifted their focus to plants that were useful for other purposes, such as artifacts, shelter, and healthcare. Over many centuries, knowledge about how to use plants for food, medicine, agriculture, and other purposes became available as traditional knowledge. Indigenous people across the globe have created their own methods for employing plants for health treatment while adhering to their traditions, culture, folklore, and dietary patterns. This information has been passed down from one generation to the next.

Traditional plant-based healing practices are still employed by people all over the world due to their economic cost and accessibility. The majority of rural populations and tribal communities who live in forested areas rely on locally available medicinal plants as a primary health intervention, and this has become an essential component of their culture.

6.2 Historical Perspectives of Ethnobotanical Studies

Eventually, a new science known as ethnobotany has emerged as a result of the gathered, diverse traditional knowledge. Ethnobotany is the study of the interactions between people and plants for medicine, food, shelter, clothing, fuel, fodder, and other household needs. It is concerned with the relationship between indigenous plants and the residents (Ahmed et al. 2013; Pei et al. 2020).

In order to preserve and utilize the knowledge of indigenous people, it is essential to document traditional knowledge through ethnobotanical investigations. An initial study in the last few decades has focused mainly on developing inventories of plants from a particular locality or ethnic group. Outside of the ethnobotanists, numerous scientists, naturalists, and intellectuals began to emphasize the importance of ethnobotanical research. Ethnobotanical studies can give insights into how communities interact with their natural resources on a regional basis. Ethnobotanical research has the potential to bring together local and scientific knowledge in order to promote the cause of biocultural conservation (Gaoue et al. 2017; Prance 2007). This information was not effectively documented because of illiteracy and widespread myths and beliefs in the communities. Ethnobotanical recording is becoming important for overall human well-being. The vast expanses of the Indian subcontinent, as well as its diverse flora and wildlife, remain mostly unexplored. People who live near forests and in those areas, as well as tribal cultures, use a variety of plants as medicine. The World Health Organization (WHO) has acknowledged that traditional medicines derived from plants are used for primary healthcare by 80% of the world's population, particularly in underdeveloped nations (Sukumaran et al. 2021; World Health Organization 2016).

India is one of the most biodiverse countries. The political boundaries of India cover a diverse range of ecozones, including the desert, high mountains, highlands, tropical and temperate forests, swamplands, plains, grasslands, and islands. Western Ghats, also known as "Sahyadri," constitute around 1600 km mountain range, and this originates from south of the Tapti River and extends up to Kanyakumari in Southern India. It traverses the states of Gujrat, Maharashtra, Goa, Karnataka, Kerala, and Tamil Nadu. The Western Ghats are known for their rich biodiversity and natural heritage (Sivaperuman and Venkataraman 2018).

6.3 A Case Study of *Dolichandrone falcata* (Wall. Ex DC.) Seem

The present reports discuss one of the endemic rare plants of India, *Dolichandrone falcata* (Wall. ex DC.) Seem. which is a member of the Bignoniaceae family, commonly called as trumpet vine. This unexplored plant has several ethnomedicinal values, such as antidiabetic, anti-inflammatory, analgesic, antiestrogenic, antimicrobial, antinociceptive, antibacterial, anti-hemorrhoidal, and immunomodulatory (Bagwan and Kishore 2013). Indigenous and tribal peoples have long used *Dolichandrone falcata* for medical purposes and a variety of pharmacological effects around the world, including India (Khilare and Saindanshiv 2005; Dhaswadikar et al. 2022).

6.3.1 Distribution and Taxonomy

The *Dolichandrone falcata* species are endemic and found in tropical climatic zones of India. The plant is majorly distributed in forest regions and outskirts of the Western Ghats of the Indian peninsula (Sprague 1919). The Plant List (n.d.), a part of the World Flora Online Consortium, enlisted synonyms of *Dolichandrone falcata* (Wall. ex DC.) Seem., viz., *Bignonia falcata* K.D. Koenig ex DC, *Bignonia spathacea* Roxb., *Dolichandrone lawii* Seem., and *Spathodea falcata* Wall. Ex DC.

6.3.2 Ethnobotanical Appraisal of *Dolichandrone falcata* (Wall. Ex DC.) Seem

Ethnobotanical knowledge supplements the modern evidence-based medical system. Traditionally, the native plant species are utilized for primary healthcare interventions in nearly every region of its occurrence (Muley et al. 2023). However, not much work has been documented on further investigating the claims of ethnomedicinal usage in light of present therapeutic investigations or the phytoconstituents isolated from *Dolichandrone falcata.*

6.3.2.1 Ethnobotanical Uses of Leaves

Traditionally, the leaves have been used by the tribes of Maharashtra as a folk treatment in situations of food poisoning and abortifacient (Kirtikar and Basu 1999). Patients suffering from piles are treated with 100 ml of leaf juice and 100 ml of curd twice a day for a week. In people with diabetes, leaf powder is administered with water (Dhaswadikar et al. 2022; Korpenwar and Borkar 2011). The leaf decoction is used in the treatment of leprosy patients (Dushing and Patil 2010).

6.3.2.2 Ethnobotanical Uses of Stem Bark

The decoction of stem bark is given internally for 3 days to alleviate menorrhagia and leukorrhea (Prashantkumar and Vidyasagar 2006). Internal administration of root bark powder and infusion is used to treat acute rheumatism (Mali and Bhadane 2011). Corneal opacity is treated with *D. falcata* leaves, along with several other herbs which are crushed with goat's milk. The resulting extract is administered as an eye drop twice daily until the condition is resolved (Sreeramulu et al. 2013).

6.3.2.3 Ethnobotanical Uses of Fruit

Dried powder of fruits is administered internally as a stomachic (Kamble et al. 2008).

6.3.3 Phytochemical Constituents of *Dolichandrone falcata* (Wall. Ex DC.) Seem

Different forms of flavonoids, including flavones, flavanones, and flavonol, are present in the leaves of *D. falcata*. Chrysin, a flavone, chrysin-7-rutinoside, and chrysin-7-glucoside were isolated from the leaves of *D. falcata* (Subramanian et al. 1972). Another flavonoid that has been identified is pectolinarigenin, which is a dihydroxyflavone (Salve and Bhuktar 2017). Several quinone derivatives have been isolated from the heartwood of *D. falcata* like α-lapachone, lapachol, aloe-saponarin II, 8-hydroxydehydroiso α-lapachone, and 3,8-dihydroxydehydroisoα-lapachone (Aparna et al. 2009). α-Lapachone is a naphthoquinone derivative (Burnett and Thomson 1967). There are two glycosides present in the plant, namely, dolichandroside A and verbascoside. It also contains several secondary metabolites like sterol, coumarin derivatives, monoterpenes, phenols, volatile oils, etc. (Aparna et al. 2009) (Fig. 6.1).

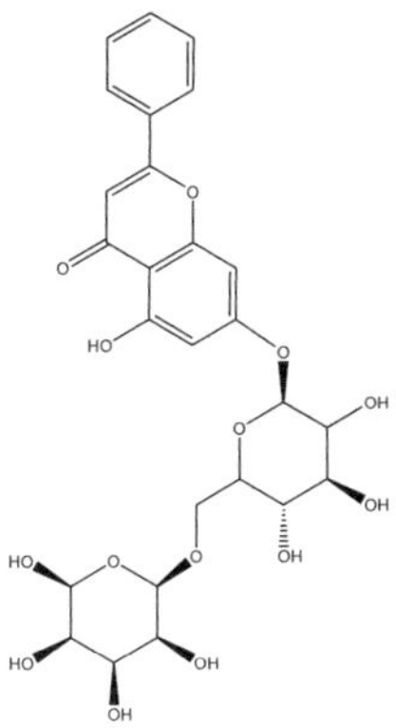

Chrysin 7-rutinoside

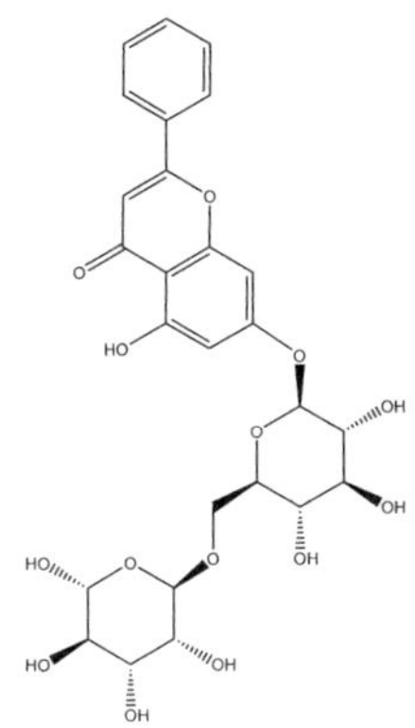

Chrysin 7-glucoside

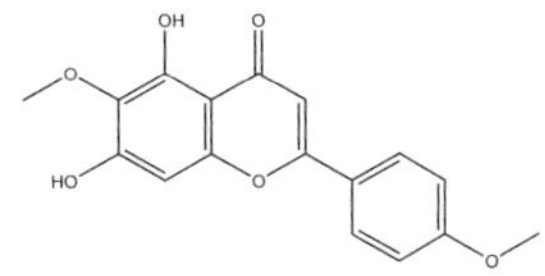

Pectolinarigenin

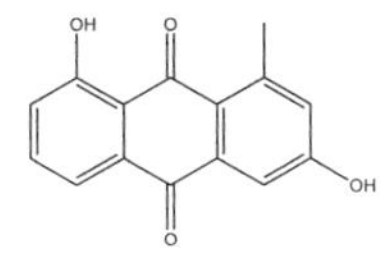

Aloe Saponarin II

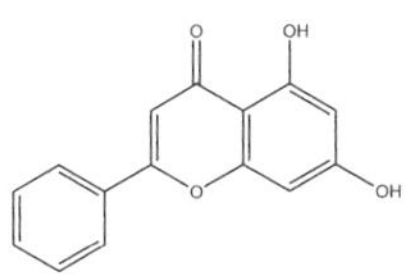

Chrysin

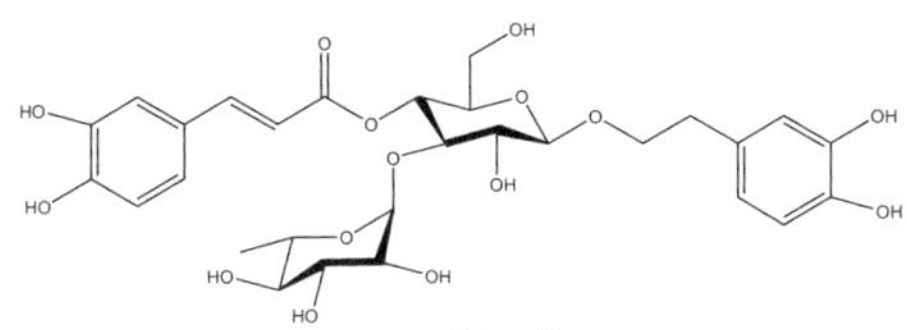

Verbascoside

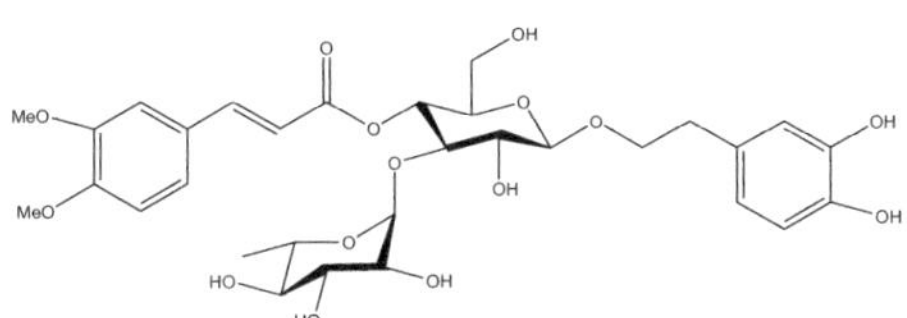

Dolochandroside A

Fig. 6.1 Major chemical constituents isolated from *Dolichandrone falcata*

6.3.4 Pharmacological Activity and Preclinical Study of *Dolichandrone falcata*

6.3.4.1 Anxiolytic Effects

Dolichandrone falcata has been reported to exhibit antinociceptive effects. Methanolic extract (DFBM) and ethyl acetate extract (DFBEA) of *D. falcata* stem bark were administered orally to mice at 100, 200, and 400 mg/kg and evaluated against anxiolytic effects using elevated plus maze (EPM) and marble-burying test (MBT) assay. The anxiolytic effects were found to be significant in a dose-dependent manner, similar to that of reference standards, diazepam and fluoxetine (Badgujar and Surana 2010).

6.3.4.2 Anti-Hemorrhoidal Activity

Our group reported that ethanolic extract of leaves of *Dolichandrone falcata* (DFE) has a significant anti-hemorrhoidal effect compared to standard Pilex. The croton oil-induced hemorrhoidal rat model was utilized for the study. It was found that oral administration of DFE at 200 mg/kg, along with carboxymethyl cellulose, exhibited a more substantial anti-hemorrhoidal effect. The Inflammation score was calculated based on clinical hemorrhoidal area and severity index. The hemorrhoidal tissues were also subjected to biochemical profiling, which revealed a decline in the levels of IL-1β, IL-6, and TNF-α expression with restoration of altered antioxidants and liver enzymes. In addition, histopathological studies confirmed the tissue recovery (Dhaswadikar et al. 2022).

6.3.5 Acute Oral Toxicity Study

Acute oral toxicity studies of DFE were performed by our group according to OECD (Organisation for Economic Co-operation and Development) guideline 425. DFE was administered orally to female Wistar rats fasted overnight, followed by individual monitoring of around 48 h for any behavioral or neurological abnormalities, such as diarrhea, feeding behavior, convulsions, tremors, lacrimation, sleep, and salivation as a marker for toxicity. For confirmation, the rats were kept under observation for 2 weeks to determine the mortality rate, if any (OECD 2008; Dhaswadikar et al. 2022).

6.4 Conservational Needs of Medicinal Plants

Even now, the majority of therapeutic plants are still gathered from the wild reserves. Numerous plant species populations have declined from their natural habitat due to the ongoing commercial exploitation of these plants. The supply of raw plant materials, which are extensively required by both the pharmaceutical sector and Agrofood sectors, is creating an unavoidable depletion of deciduous trees. In order to ensure their availability to the industry as well as to those involved with traditional systems of healing, it is necessary to cultivate these endemic plant species. Prompt action is required for their preservation, cultivation, and widespread reproduction. The herbal industry's increasing demand cannot be fulfilled by in-place conservation of these resources alone (Singh and Gautam 1997).

6.4.1 Conservational Constraints

- Inadequate assessment of resource availability and availability of specific species.
- A lack of scientific farming for reliable raw material sources and readily available quantities.

- People don't seem to care very much about conservation.
- Problems and the extent of sustainable development due to ignorance.
- A lack of infrastructure and marketing channels. Additionally, there are no manufacturing or processing facilities at all.
- Poor documentation of local medical traditions and practices and inadequate survey of medicinal plants.
- Inadequate financial backing for business endeavors.
- Lack of dedicated nongovernmental organizations in this subject and a successful extension service.
- Lack of a directory of village/traditional healers, farmers, traders, and business owners engaged in the state's medicinal plant industry for networking and coordination.
- Inadequately trained and competent labor for tasks involving medicinal plants (Kumar et al. 2014).

6.4.2 Conservation Strategies

A growing number of medicinal plants are being procured, mostly from wild resources. In fact, during the past few decades, the demand for forest resources has grown by 8–15% annually throughout Europe, North America, and Asia (Simmonds and Ross 2006). There is a threshold point at which a species' capacity to reproduce is irreversibly compromised. There are several sets of recommendations for the preservation of deciduous medicinal plants, including both in situ and ex situ conservation. Botanical gardens and seed banks are essential models for ex situ conservation and future replanting; natural reserves and wild nurseries are typical examples of preserving plant medicinal effectiveness in their natural habitat (Chen et al. 2016).

6.4.2.1 In Situ Conservation

The majority of medicinal plants are endemic species, and their therapeutic benefits mainly result from the presence of secondary metabolites that react to stimuli in their natural habitats but may not manifest themselves in cultured environments (Figueiredo and Grelle 2009). We can preserve natural communities and their rich web of relationships by conserving entire communities in their natural habitat (Gepts 2006). Furthermore, in situ conservation reinforces the connection between resource protection and sustainable use by increasing the amount of diversity that may be preserved. Across the globe, in situ conservation efforts have concentrated on creating protected areas and using an ecosystem-oriented approach rather than a species-oriented one (Forest et al. 2007). Implementing effective regulation of medicinal plants within growth habitats is required for thriving in situ conservation (Volis and Blecher 2010).

Natural Reserves

The decline of medicinal plant resources is primarily due to habitat deterioration and destruction. Significant wild resources are preserved in natural reserves, which were established to maintain and replenish biodiversity (Rodríguez et al. 2007). More than 12,700 protected areas have been created globally, covering 13.2 million km^2, or 8.81% of the land area of the planet. More than 12,700 protected areas have been developed globally, covering 13.2 million km^2, or 8.81% of the land area of the planet. Analyzing the contributions and ecosystem functions of specific habitats is necessary for conserving medicinal plants through safeguarding important natural habitats (Liu et al. 2001).

Wild Nurseries

Because of costs and competing land uses, it is not practical to designate every natural wild plant habitat as a protected area. For species-focused domestication and cultivation of threatened medicinal plants, a forest nursery is set up in a protected area, a natural habitat, or a location that is just a short distance from the plant's native environment (Strandby and Olsen 2008). Despite the fact that many wild species' populations are under stress owing to overexploitation, habitat degradation, and invasive species, fantastic nurseries provide an excellent technique for in situ conservation of medicinal plants that are rare, endangered, and in demand (Chen et al. 2016).

6.4.2.2 Ex Situ Conservation

Ex situ conservation is an excellent supplement to in situ conservation, particularly for overused and endangered medicinal plants with sluggish growth, limited quantity, and high susceptibility to replanting diseases (Yu et al. 2010). Ex situ conservation, which is commonly used as an immediate strategy to preserve medicinal plant resources, seeks to cultivate and naturalize vulnerable species in order to assure their longevity and, ultimately, to generate considerable volumes of raw plant materials, which is required in herbal industries (Swarts and Dixon 2009). Several species of once-wild medicinal plants may have their reproductive components gathered and conserved in seed banks for future replanting, in addition to retaining therapeutic potency when cultivated in gardens distant from their original habitat (Hamilton 2004).

Botanical Gardens

Botanical gardens are crucial to ex situ preservation. They can protect ecosystems to help rare and endangered plant species survive. Despite the fact that living collections often only contain a few representatives of each species, they are not very useful for genetic conservation. Botanical gardens have many distinctive characteristics. They frequently feature a taxonomically and ecologically diverse flora and involve an extensive range of plant species that are grown together under similar conditions. Botanic gardens can contribute to the conservation of medicinal plants by creating guidelines for their propagation and culture, as well as initiatives for domestication and variety breeding (Havens et al. 2006).

Seed Banks

It is advised to use seed banks to help conserve the biological and genetic diversity of wild plant species since they provide a better means of preserving the genetic diversity of many medicinal plants ex situ than through botanic gardens. The Millennium Seed Bank Project at the Royal Botanic Gardens in Britain is the most notable seed bank. Plant samples may be accessed from seed banks reasonably quickly for the evaluation of their characteristics, which is helpful information for protecting the surviving natural populations. How to restore plant species to the wild and how to actively support the restoration of wild populations are the challenging issues of seed banking (Li and Pritchard 2009).

6.5 Conclusion

By combining ethnobotanical knowledge with conservation efforts, it is feasible to comprehend and protect the vast richness of therapeutic plants. This chapter delves into the case study of *Dolichandrone falcata*, an iconic species in India's Western Ghats, to evaluate the applicability of its traditional uses and conservation status. Recent scientific research has supported several conventional applications of the *Dolichandrone falcata*. There are bioactive chemicals present in plants, according to phytochemical investigations, which are consistent with their traditional medical uses. This plant species lacks authentic phytochemical studies and pharmacological evaluations.

The current ecosystems are under tremendous pressure from rapid urbanization, habitat loss, and overexploitation. Considering the traditional significance of *Dolichandrone falcata*, its conservation status in this context is of the utmost importance. Integrated conservation methods are essential to protect the future of *Dolichandrone falcata* and other medicinally important plants. These might include community involvement, sustainable harvesting methods, and habitat restoration. Additionally, raising awareness of the plant's ecological and cultural significance can inspire regional protection efforts (Chen et al. 2016).

Participating local communities in conservation efforts can have a significant positive impact. A holistic approach can be achieved by merging traditional knowledge into conservation measures. *Dolichandrone falcata's* survival is supported by empowering indigenous groups to take part in its preservation, which also protects traditional knowledge for future generations.

Dolichandrone falcata is a case study that shows how ethnobotanical insights, therapeutic potential, and conservation goals are all intertwined. We can provide the foundation for comprehensive conservation efforts that go beyond plant species to include cultural legacy and ecological balance by acknowledging indigenous wisdom and incorporating it with scientific study.

Cultivation and conservation of such deciduous medicinal plants can be carried out in collaboration with forest department through different projects like therapeutic plant gardens, plant reservoirs, etc. Lack of scientific evidence and ignorance about the potential of botanicals often lead to their depletion from natural reservoirs. The scientific and social communities need to take responsibility by organizing awareness programs with the help of Gram panchayats and the Department of Forest. The future perspectives of this study can be supplemented with tissue culture for the conservation of important medicinal plants. In addition, pharmacodynamic and clinical studies for potential management of the condition of various diseases can be further proposed.

References

Ahmed E, Arshad M, Saboor A, Qureshi R, Mustafa G, Sadiq S, Chaudhari SK (2013) Ethnobotanical appraisal and medicinal use of plants in Patriata, New Murree, evidence from Pakistan. J Ethnobiol Ethnomed 9(1):1

Aparna P, Tiwari AK, Srinivas PV, Ali AZ, Anuradha V, Rao JM (2009) Dolichandroside A, a new α-glucosidase inhibitor and DPPH free-radical Scavenger from Dolichandrone falcata seem. Phytother Res 23(4):591–596

Badgujar VB, Surana SJ (2010) Anxiolytic effects of Dolichandrone falcata Seem., Bignoniaceae, stem-bark in elevated plus maze and marble burying test on mice. Revista Brasileira de Farmaco 20:773–780

Bagwan AS, Kishore ME (2013) Phytochemical and antimicrobial activity of *Dolichandrone falcata* stem on some drug resistance bacterial species. Asian J Microbiol Biotechnol Environ Sci 15:379–384

Burnett AR, Thomson RH (1967) Quinones. Part VIII. Dehydro-α-and-β-lapachone. J Chem Soc C Organic:1261–1264

Chen SL, Yu H, Luo HM, Wu Q, Li CF, Steinmetz A (2016) Conservation and sustainable use of medicinal plants: problems, progress, and prospects. Chin Med 11:1

Dhaswadikar SR, Parmar KM, Kamble SK, Kathuria I, Dhobi M, Birajdar A, Prasad SK, Itankar PR (2022) Anti-hemorrhoidal potential of standardized leaf extract of Dolichandrone falcata. Phytomedicine Plus 2(1):100172

Dushing YA, Patil DA (2010) Studies on ethnomedicine in Buldhana district of Maharashtra (India). J Phytology 2(12):35–41

Figueiredo MS, Grelle CE (2009) Predicting global abundance of a threatened species from its occurrence: implications for conservation planning. Divers Distrib 15(1):117–121

Forest F, Grenyer R, Rouget M, Davies TJ, Cowling RM, Faith DP, Balmford A, Manning JC, Procheş Ş, van der Bank M, Reeves G (2007) Preserving the evolutionary potential of floras in biodiversity hotspots. Nature 445(7129):757–760

Gaoue OG, Coe MA, Bond M, Hart G, Seyler BC, McMillen H (2017) Theories and major hypotheses in ethnobotany. Econ Bot 71:269–287

Gepts P (2006) Plant genetic resources conservation and utilization: the accomplishments and future of a societal insurance policy. Crop Sci 46(5):2278–2292

Hamilton AC (2004) Medicinal plants, conservation and livelihoods. Biodivers Conserv 13:1477–1517

Havens K, Vitt P, Maunder M, Guerrant EO, Dixon K (2006) Ex situ plant conservation and beyond. Bioscience 56(6):525–531

Kamble SY, More TN, Patil SR, Pawar SG, Bindurani R, Bodhankar SL (2008) Plants used by the tribes of Northwest Maharashtra for the treatment of gastrointestinal disorders. Indian J Tradit Knowl 7(2):321–325

Khilare CJ, Saindanshiv SE (2005) Survey of medicinal plants and their conservation for sustainable health hygine and environment. Biodivers Conserv 65

Kirtikar KR, Basu BD (1999) Indian medicinal plants, vol 3. International Book Distributors

Korpenwar AN, Borkar SU (2011) Indigenous medicinal plants used by traditional healers of buldhana tahsil, Dist. Buldhana (MS). Int J Recent Trends Sci Technol 1(1):14–19

Kumar V, Babu S, Revale AK, Meena RK, Ranjan MK, Desai BS (2014) Cultivation of medicinal plants in natural ecosystem in Gujarat (India): constraints and conservation need. Journal of Plant Development Sciences 6(3):425–435
Li DZ, Pritchard HW (2009) The science and economics of ex situ plant conservation. Trends Plant Sci 14(11):614–621
Liu J, Linderman M, Ouyang Z, An L, Yang J, Zhang H (2001) Ecological degradation in protected areas: the case of Wolong Nature Reserve for giant pandas. Science 292(5514):98–101
Mali PY, Bhadane VV (2011) Ethno-medicinal wisdom of tribals of Aurangabad district (MS). Indian J Nat Prod Resour 2(1):102–109
Muley MM, Doshi SM, Goyal A, Jachak SM (2023) Ethnopharmacology and phytochemistry of selected species of Boerhavia occurring in India: a review. Curr Tradit Med 9(2):93–113
OECD (2008) Guidelines for testing of chemicals. Guidance no. 425. Acute oral toxicity: up and down procedure (UDP). / http://iccvam.niehs.nih.gov/SuppDocs/FedDocs/OECD/OECDtg425.pdf S (Accessed 03.10.08)
Pei S, Alan H, Wang Y (2020) Vital roles for ethnobotany in conservation and sustainable development. Plant Divers 42(6):399
Prance GT (2007) Ethnobotany, the science of survival: a declaration from Kaua'i. Econ Bot 61(1):1–2
Prashantkumar P, Vidyasagar GM (2006) Documentation of traditional knowledge on medicinal plants of Bidar district, Karnataka. Indian J Tradit Knowl 05(3):295
Rodríguez JP, Brotons L, Bustamante J, Seoane J (2007) The application of predictive modelling of species distribution to biodiversity conservation. Divers Distrib 1:243–251
Salve SD, Bhuktar AS (2017) Pharmacognosy and phytochemistry of dolichandrone falcata (Wall Ex Dc) Seem. BIOINFOLET Q J Life Sci 14(4b):444–447
Simmonds M. Ross IA (2006) Medicinal plants of the world: volume 3 Chemical constituents, traditional and modern medicinal uses, Humana Press Inc., New Jersey (2005) ISBN 1–58829–129-4 pp. 623. Phytochemistry;67(7):754
Singh BM, Gautam PL (1997) Bioresources of medicinal and aromatic plants in India, their conservation and related issues. Kurukshetra XLVI(3):9–13
Sivaperuman C, Venkataraman K (eds) (2018) Indian hotspots: vertebrate faunal diversity, conservation and management, vol 2. Springer Singapore
Sprague TA (1919) Dolichandrone and Markhamia. Bulletin of miscellaneous information (Royal Botanic Gardens, Kew). 1919(8):302–14
Sreeramulu N, Suthari S, Ragan A, Raju VS (2013) Ethno-botanico-medicine for common human ailments in Nalgonda and Warangal districts of Telangana, Andhra Pradesh, India. Ann Plant Sci 02(07):220–229
Strandby U, Olsen CS (2008) The importance of understanding trade when designing effective conservation policy–the case of the vulnerable Abies guatemalensis Rehder. Biol Conserv 141(12):2959–2968
Subramanian SS, Nagarajan S, Sulochana N (1972) Chrysin-7-rutinoside from the leaves of Dolichandrone falcata. Phytochemistry 11(1):438–439
Sukumaran S, Sujin RM, Geetha VS, Jeeva S (2021) Ethnobotanical study of medicinal plants used by the Kani tribes of Pechiparai Hills, Western Ghats, India. Acta Ecol Sin 41(5):365–376
Swarts ND, Dixon KW (2009) Terrestrial orchid conservation in the age of extinction. Ann Bot 104(3):543–556
The Plant List (n.d.). http://www.theplantlist.org/tpl1.1/record/kew-320745. Accessed 08.09.2023
Volis S, Blecher M (2010) Quasi in situ: a bridge between ex situ and in situ conservation of plants. Biodivers Conserv 19:2441–2454
World Health Organization (2016) Traditional medicine fact sheet no. 134. 2008. URL: http://www.who.int/mediacentre/factsheets/fs134/en/. Accessed 10.09.09
Yu H, Xie C, Song J, Zhou Y, Chen S (2010) TCMGIS-II based prediction of medicinal plant distribution for conservation planning: a case study of rheum tanguticum. Chin Med 5:1–9

Bridging Cultures and Medicine: Quantitative Insights in Ethnopharmacology

7

Varsha Srivastava, Areeba Insaf, and Sayeed Ahmad

Abstract

For the discovery and creation of novel medicines, ethnopharmacology—the study of conventional medical practices and plant-based treatments utilized by various cultures—has become a crucial resource. This chapter examines the critical role that quantitative and comparative methodologies play in advancing our understanding of ethnopharmacology. By using strict quantitative methods and making cultural comparisons, researchers have been able to give important information on the efficacy, safety, and mechanisms of conventional therapies. The key findings from recent studies that assessed the therapeutic potential of ethnomedicinal plants using a range of quantitative approaches, such as pharmacological testing, bioinformatics analysis, and clinical trials, are summarized in this chapter. Many quantitative and comparative methodologies used in ethnopharmacology, their benefits, drawbacks, and contributions to our knowledge of conventional medical practices have all been covered in this chapter. The chapter also emphasizes the need of doing comparative study in diverse cultural contexts to identify variations in plant use, food preparation methods, and cultural beliefs. The need for multidisciplinary approach in the context of ethnopharmacological research is emphasized in this chapter, bridging the gap between conventional wisdom and cutting-edge scientific knowledge.

Keywords

Ethnopharmacology · Pharmacological assays · Bioinformatics · Clinical trials · Traditional knowledge · Interdisciplinary research

V. Srivastava · A. Insaf · S. Ahmad (✉)
Bioactive Natural Product Laboratory, Department of Pharmacognosy and Phytochemistry, School of Pharmaceutical Education and Research, Centre of Excellence in Unani Medicine (Pharmacognosy and Phytochemistry), Jamia Hamdard, New Delhi, India

© The Author(s), under exclusive license to Springer Nature Singapore Pte Ltd. 2024

M. Nandave et al. (eds.), *Ethnopharmacology and OMICS Advances in Medicinal Plants Volume 1*, https://doi.org/10.1007/978-981-97-2367-6_7

7.1 Introduction

7.1.1 Overview of Ethnopharmacology and its Significance in Modern Drug Discovery

An interdisciplinary area termed as ethnopharmacology investigates how different civilizations have used plants, animals, and other natural substances for medical purposes throughout history. It aims to find and develop novel pharmaceuticals based on the traditional knowledge of indigenous people and bridges the gap between traditional and contemporary medicine (Ghorbani and Farzaneh Naghibi 2005). The traditional knowledge of indigenous and local cultures about medicinal plants and the treatment offered is preserved and documented with the aid of ethnopharmacology. This wisdom has been passed down through the years and frequently includes useful knowledge on the medicinal characteristics of natural substances (Kumar et al. 2019). Traditional medicine served as the basis for many pharmacological medications. By discovering bioactive chemicals from conventional treatments, ethnopharmacological research can provide new pathways for medication development. Novel medications have benefited greatly from the use of natural materials derived from plants, fungi, and animals (Harvey 2008). Ethnopharmacology advocates the conservation of medicinal plants and animals to enable the sustainable use of biodiversity. It is possible to take action to safeguard these species from overharvesting and habitat loss by recognizing the significance of these species in conventional medicine. Research on ethnopharmacology frequently entails cooperation with regional populations and indigenous peoples. By utilizing their traditional knowledge sustainably, it may also create economic prospects for these people while respecting their cultural legacy. The use of traditional medicine may provide information on the efficacy and safety of using natural medicines. Ethnopharmacologists carry out research to support conventional claims and identify the best doses and formulations for the creation of contemporary drugs (Chakrabarty and Kaur 2021; Heywood 2011). Researchers prioritize which natural products should be tested for bioactivity in the laboratory using ethnopharmacological data. Identifying substances with potential medicinal characteristics is made easier by high-throughput screening and other contemporary approaches. Novel drug targets or mechanisms of action may be discovered as a result of ethnopharmacological investigations (Periwal et al. 2020; Harvey 2008). This information may prove useful in the creation of new pharmaceuticals with novel mechanisms of action that might combat adverse drug reactions or resistance. Traditional treatments could shed light on the possible therapeutic applications of well-known substances. Research in ethnopharmacology may result in the repurposing of already approved medications for use in novel conditions. Ethnopharmacology can address health inequities in impoverished groups by integrating traditional medicine into contemporary drug development. It may result in the creation of accessible, inexpensive medications that meet the demands of many social groups (Harvey 2008).

7.1.2 Importance of Quantitative and Comparative Approaches in Advancing Ethnopharmacological Research

In order to advance ethnopharmacological research, quantitative and comparative methods are essential since they bring scientific rigor and objectivity to the investigation of conventional medicine (Leonti and Weckerle 2015). Researchers may systematically assess the efficacy and safety of conventional treatments using quantitative approaches like bioassays and pharmacological tests, which provides vital information for drug development. Additionally, comparative studies make it possible to spot trends, parallels, and contrasts among various cultural practices, which helps researchers choose potential leads and comprehend the bigger picture of traditional medicine. Ethnopharmacological study incorporates quantitative and comparative methodologies, which not only confirms traditional knowledge but also improves its use in contemporary healthcare, thereby aiding in the development of new treatments and the preservation of cultural legacy (Pirintsos et al. 2022).

7.1.3 Quantitative Approaches in Ethnopharmacology

In ethnopharmacology, quantitative methods are essential for converting conventional wisdom into science based on facts. To assess the therapeutic potential of medicinal plants and other naturally occurring compounds used in traditional medicine, these procedures systematically gather and analyze numerical data (Dapar et al. 2020). The term "quantitative methods" refers to a broad variety of procedures, including pharmacokinetic and pharmacodynamic investigations as well as bioassays and phytochemical tests. Researchers can find active substances, evaluate their potency, and establish the best doses by assessing the bioactivity, chemical makeup, and pharmacological effects of conventional treatments. This quantitative rigor helps ethnopharmacological research become more credible and repeatable while also making it easier to incorporate traditional medicine into contemporary healthcare systems, opening the door to the development of new medications and evidence-based therapeutic interventions (Hadi and José Closs 2016).

7.1.4 Future Directions in Quantitative Insights in Ethnopharmacology

Future directions in quantitative insights in ethnopharmacology are likely to be marked by increased interdisciplinary collaboration, advanced analytical techniques, and a focus on personalized medicine. Integrating ethnopharmacological data with genomics, metabolomics, and bioinformatics will allow for a deeper understanding of the molecular mechanisms underlying traditional remedies. Machine learning and data mining will help identify novel drug candidates from the vast repository of ethnobotanical knowledge. Moreover, the ethical and legal aspects of ethnopharmacological research will gain prominence, with efforts to ensure equitable benefits for indigenous communities (Rodriguez et al. 2021). This evolving field will continue to bridge the gap between traditional wisdom and modern science, paving the way for innovative drug discovery, conservation of medicinal plants, and culturally respectful healthcare practices (Saxena and Rao 2023).

Future developments in quantitative insights in ethnopharmacology are probably going to be characterized by an emphasis on personalized therapy, more multidisciplinary collaboration, and sophisticated analytical approaches (Dapar et al. 2020). A greater comprehension of the molecular mechanisms behind conventional treatments will be possible by fusing ethnopharmacological data with genomes, metabolomics, and bioinformatics. The large database of ethnobotanical knowledge will be used to find new medication candidates with the use of machine learning and data mining. Additionally, with initiatives to provide equal benefits for indigenous populations, the ethical and legal issues of ethnopharmacological research will take front stage. This developing discipline will minimize the knowledge gap between conventional wisdom and contemporary research, opening the door to novel medication development, the preservation of medicinal plants, and cross-culturally sensitive medical procedures (Subramanian et al. 2006).

7.2 Ethnobotanical Surveys and Data Collection Methods

7.2.1 Quantitative Analysis of Usage of Medicinal Plants, Preparation Methods, and Dosage

In order to get useful information and insights from conventional knowledge, ethnopharmacology relies heavily on quantitative investigation of medicinal plant use, preparation techniques, and dosage (Sailaja et al. 2016). Quantitative analysis can be applied to each aspect as follows.

7.2.1.1 Medicinal Plant Use

Frequency of Mention: Researchers can measure the frequency with which a certain plant or cure is brought up in community interviews or surveys. This makes it easier to recognize the most frequently utilized medicinal herbs (Tugume et al., 2016).

Use-Value Index: Based on the reported applications of various plants within a community, a quantitative indicator, such as the Use-Value Index, may be used to prioritize and rank the significance of each plant.

7.2.1.2 Preparation Methods

Quantifying Methods The processes used to manufacture medical remedies can be documented, and their prevalence can be measured, by researchers. This entails keeping track of information on additives, plant parts utilized, and extraction methods.

Comparative Analysis Insights into the diversity of traditional practices can be gained through comparative analysis, which reveals regional or cultural variances in preparation techniques.

7.2.1.3 Dosage

Dosage Frequency The frequency at which a certain dosage or administration schedule is advised for a given therapy can be quantified. The perceived efficacy and safety of the therapy are determined using this information (Kruse et al. 1991).

Dosage Quantification Traditional healers or members of the community could occasionally offer precise dosage quantities or ratios. Understanding the actual use of conventional medicine might require an understanding of this information (Kruse et al. 1991).

Researchers often use structured questionnaires or interviews to get information from knowledgeable informants inside the community in order to undertake quantitative analysis in ethnopharmacology (Holmes 2023). The quantitative data may be analyzed and understood using statistical approaches such as frequency distribution, correlation analysis, and descriptive statistics. Researchers can more accurately evaluate the therapeutic potential of traditional remedies, spot trends, and direct further research into the bioactivity and safety of these treatments by quantifying plant use, preparation techniques, and dosage information. This advances evidence-based ethnopharmacological research (Adams and Cox 2008).

7.2.2 Pharmacological Assays and Bioassays for Validating Traditional Medicinal Knowledge

Pharmacological assays and bioassays are essential instruments for confirming conventional medical knowledge by supplying data on the bioactivity and effectiveness of conventional treatments (Darwish 2006). Roles of these assays in the validation process include as follows.

7.2.2.1 Pharmacological Assays

Pharmacological assays are laboratory procedures used to gauge how chemicals, such as plant extracts or isolated molecules, affect certain biological targets or functions. Pharmacological assays are used to investigate the interactions of traditional medicines with relevant biological systems in order to evaluate their potential therapeutic capabilities (Darwish 2006). Receptor-binding assays, enzyme inhibition assays, cell-based assays, and organ bath assays are just a few examples of the several pharmacological assays that may be utilized. Receptor-binding tests, for instance, may ascertain if a plant chemical interacts with certain receptors in the body, which is essential for comprehending its pharmacological effects (Darwish 2006).

7.2.2.2 Bioassays

Bioassays are biological tests that evaluate a substance's impact on living things or tissues in order to determine its biological activity or potency. Because they capture the comprehensive and multifaceted nature of conventional treatments, bioassays are particularly pertinent for confirming traditional medical knowledge (Yu and Yang 2017). They take into account how various elements of a cure interact with one another and with biological systems as a whole. Ex vivo testing utilizing isolated tissues, in vivo tests using complete organisms, and in vitro tests using cell cultures are all examples of bioassays. For instance, employing inflammatory markers in cell cultures or testing a plant extract's capacity to reduce inflammation in animal models, bioassays can assess a plant extract's anti-inflammatory properties (Bolton and Gillett 2019; Mukherjee et al. 2016; Hughes et al. 2011). The process of validating traditional medicinal knowledge using pharmacological assays and bioassays typically involves the following steps:

Selection of Bioassay Models The focus of interest for the traditional treatment is replicated by the researchers' choice of suitable biological models or systems.

Preparation of Test Samples Traditional treatments or their constituent parts are produced for testing, frequently in the form of extracts or isolated substances.

Assay Execution In a controlled laboratory setting, the specified bioassays or pharmacological assays are carried out to evaluate the effects of the test substances on the selected biological models.

Data Analysis The bioactivity, potency, and mode of action of the drugs under examination are identified by statistical analysis of the data collected from the tests.

Comparison with Traditional Knowledge To confirm or deny the purported therapeutic capabilities, the assay findings are contrasted with conventional medical knowledge. Positive outcomes demonstrate the success of the conventional treatment.

7.2.3 Statistical Tools for Data Analysis and Interpretation

For data analysis and interpretation in a variety of sectors, such as research, industry, healthcare, and social sciences, statistical techniques are crucial. They assist analysts and researchers in making meaning from data, observing patterns, drawing conclusions, and making appropriate assessments (Taipalus et al. 2022). Commonly used statistical tools for data analysis and interpretation include as follows.

7.2.3.1 Descriptive Statistics

Measures of Central Tendency The average or usual value in a dataset may be described with the use of tools like mean, median, and mode.

Measures of Dispersion Tools that help understand the spread or variability of data include range, variance, standard deviation, and interquartile range.

Frequency Distributions The distribution of data values is summarized using frequency tables, bar charts, and histograms.

7.2.3.2 Inferential Statistics

Hypothesis Testing Analysis of variance (ANOVA) and chi-squared tests, besides other methods, are used to determine if observed differences or connections in data are statistically significant.

Regression Analysis To investigate the correlations between variables and to generate predictions, linear and nonlinear regression models are utilized.

Correlation Analysis The strength and direction of relationships between variables are evaluated using tools like the Pearson correlation coefficient and Spearman rank correlation.

ANOVA (Analysis of Variance) ANOVA analyzes a dataset to see whether there are significant variations between group means.

Nonparametric Tests When data do not conform to the hypotheses of parametric testing, these tests, including the Mann-Whitney U test and the Wilcoxon signed-rank test, are applied.

7.2.3.3 Data Visualization

Box Plots These offer a visual depiction of the data distribution, together with outliers and quartiles.

Scatterplots Scatterplots are used for depicting the relationship between two continuous variables.

Bar Charts Categorical data is displayed using bar charts.

Heatmaps In order to represent values within a matrix, color intensity is used in heatmaps to visualize data.

Time-Series Analysis Time-series data analysis and forecasting techniques include exponential smoothing, autoregressive integrated moving average (ARIMA) models, and moving averages.

Cluster Analysis Methods like k-means clustering and hierarchical clustering assist to find patterns and put related data points together.

Principal Component Analysis (PCA) Complex datasets are easier to understand because PCA decreases the dimensionality of data while preserving crucial information.

Statistical Software Statistical analysis is frequently carried out using software programs like R, SAS, SPSS, Python (with libraries like NumPy, pandas, and SciPy), and Excel.

Confidence Intervals and Margin of Error These instruments aid in estimating the degree of uncertainty surrounding sample statistics and population characteristics.

Statistical Tests for Relationships To evaluate connections between categorical variables, techniques including logistic regression, Fisher's exact test, and chi-squared tests are employed.

Data Mining and Machine Learning Predictive modeling and pattern recognition employ sophisticated statistical methods including decision trees, random forests, and support vector machines. The type of data, the goals of the study, and the questions being asked have an impact on the statistical methods that are employed. To draw reliable findings and make decisions based on data, competent statistical evaluation and interpretation are important (Wu et al. 2021; National Research Council 1988). Data collection and analysis methods used for medicinal plants are depicted in Fig. 7.1.

7.2.4 Comparative Approaches in Ethnopharmacology

The systematic comparison of traditional medical practices, knowledge, and resources among other cultures, regions, or ecosystems is a key component of comparative techniques in ethnopharmacology. These methods are crucial for comprehending the variety, resemblances, and distinctive characteristics of ancient medical systems (Pirintsos et al. 2022; Gonfa et al. 2020). Key aspects and methods of comparative approaches in ethnopharmacology include as follows.

7.2.4.1 Cross-Cultural Comparison

Comparative ethnopharmacology studies how other cultures employ plants, animals, and other natural resources for healthcare purposes. It looks for similarities and contrasts in these uses. Through ethnobotanical surveys, interviews, and field investigations, researchers gather information from various cultures or groups (Saslis-Lagoudakis et al. 2011). They then compare this data to find similar beliefs and methods.

7.2.4.2 Regional Comparison

In order to compare the traditional medical knowledge and practices of nearby or culturally similar cultures, researchers may concentrate on certain geographic locations. Conducting interviews with regional experts and people in the community, recording folk cures, and examining local plant consumption trends are some of the ways used to obtain data (Negi and Abdul Azeez 2022).

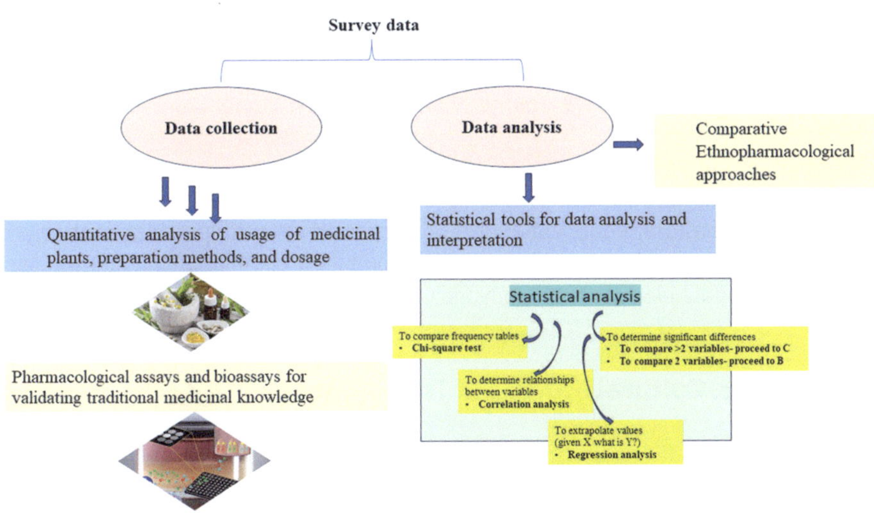

Fig. 7.1 Data collection and analysis methods used for medicinal plants

7.2.4.3 Phytochemical and Pharmacological Comparisons

Comparative phytochemical and pharmacological studies analyze the bioactive elements and pharmacological results of plants utilized in various traditional medical systems. Comparing the chemical profiles, bioactivity, and therapeutic potential of isolated chemicals or plant extracts from multiple sources is a common research task (Xing et al. 2021). Identifying common therapeutic drugs or distinctive chemical components can be made easier by following this approach.

7.2.4.4 Comparative Analysis of Traditional Knowledge Systems

This strategy entails analyzing the composition and arrangement of conventional medical information systems. Researchers can evaluate how various societies categorize diseases, make diagnoses, and recommend treatments. This can shed light on the philosophical foundations and social setting of conventional medicine (Grandgirard et al. 2002).

7.2.4.5 Ecosystem and Biogeographical Comparisons

Comparative research may examine the relationship between traditional medical practices, the local ecosystems and biodiversity. Researchers look at how local environmental elements—such as climate, geology, flora, and fauna diversity—influence the choice and use of medicinal resources (Maroyi and Cheikhyoussef 2015).

7.2.4.6 Comparative Ethnopharmacological Surveys

A thorough understanding of traditional medical practices can be obtained via large-scale studies that span several nations or cultures. Researchers perform studies in various communities, gathering data on medicinal plant species, methods of preparation, dosage, and therapeutic applications. It is possible to analyze this data to find trends and differences (Kadir et al. 2012).

7.2.4.7 Interdisciplinary Collaboration

To contribute a variety of skills to the subject, comparative ethnopharmacology frequently involves collaboration among pharmacologists, ethnobotanists, ecologists, and anthropologists.

Overall, comparative methods in ethnopharmacology assist in improving conventional medical knowledge, providing interesting drug development leads, and promoting intercultural understanding. These methods encourage a comprehensive understanding of traditional medical practices and their ecological and cultural surroundings (Mahdizadeh et al. 2015; Nancarrow et al. 2013).

7.3 Cross-Cultural Studies of Traditional Medicine Practices

The systematic analysis of how various cultures and communities use natural resources, such as plants, animals, minerals, and rituals, for healing, and healthcare is a key component of cross-cultural studies of traditional medical practices (Thatte and Gogtay 2018; Virapongse and Picheansoonthon 2005). These studies seek to comprehend the variety of conventional medical systems, find parallels, and delve into the historical, cultural, and ecological influences on these practices.

7.3.1 Comparative Analysis of Medicinal Plant Properties and Phytochemical Profiles

The systematic investigation and comparison of numerous aspects of plants used in traditional medicine across diverse cultures or areas is a comparative study of medicinal plant qualities and phytochemical profiles (Jagdeep and Sen 2013). This method is useful for appreciating the variety of medicinal plants, finding prospective therapeutic ingredients, and comparing and contrasting how they are used (Agidew 2022; Yu et al. 2021). The key steps and components of a comparative analysis include as follows.

7.3.1.1 Selection of Medicinal Plants

The investigations cover a variety of medicinal plants from various civilizations or geographical areas (Santhan 2020). These plants are frequently selected based on the relevance of their historic therapeutic usage.

7.3.1.2 Documentation of Medicinal Properties

Each chosen plant's common applications, therapeutic indications, and cultural settings are all part of the knowledge that researchers gather on its traditional medicinal characteristics (Hidayat et al., 2023). This might entail fieldwork, speaking with native healers, and reading up on ethnobotanical literature.

7.3.1.3 Phytochemical Analysis

Bioactive chemicals from plant samples are extracted and analyzed as part of phytochemical analysis. These chemicals are identified and measured using a variety of methods including mass spectrometry, spectrophotometry, and chromatography (e.g., HPLC) (Mitra et al. 2021; "Bioact. Phytochem. Drug Discov. to Prod. Dev.," 2020).

Specific phytochemicals such alkaloids, flavonoids, terpenoids, phenolic compounds, and essential oils are targeted for identification and quantification in each plant.

7.3.1.4 Bioactivity Assessment

Plant extracts or isolated chemicals can have their bioactivity assessed using in vitro and in vivo bioassays. In order to ascertain possible therapeutic effects, this involves assessing their impact on certain biological targets, such as enzymes or receptors (Les et al. 2021; Brusotti et al. 2014).

Tests for antibacterial, anti-inflammatory, antioxidant, anticancer, and other pharmacological actions can be included in bioassays.

7.3.1.5 Comparative Analysis

Researchers examine the chosen plants' bioactivity, phytochemical compositions, and traditional therapeutic characteristics across various societies or geographical areas. This investigation identifies similarities and variances in historic applications, common phytochemical components, and probable pharmacological activity (Nwali et al. 2018).

7.3.1.6 Identification of Potential Lead Compound

Researchers may find individual phytochemicals or groupings of compounds through comparative study that are regularly linked to specific pharmacological qualities or therapeutic outcomes. These substances may be used as lead compounds in the creation of pharmaceuticals (Ingle et al. 2017).

7.3.1.7 Cultural and Ecological Context

The research may also take into account the ecological, environmental, and cultural elements that affect the choice and application of medicinal plants in various geographical areas (Hamilton 2004). This clarifies why some plants are favored in particular cultural environments.

7.3.1.8 Data Interpretation

The outcomes of the comparison investigation are interpreted to offer perceptions of the connections between botanical traits, phytochemical profiles, and conventional medical wisdom.

7.3.1.9 Publication and Knowledge Sharing

The results are frequently shared with pertinent stakeholders or published in scholarly journals, which advances knowledge of medicinal plants and their potential uses in contemporary healthcare (Ahmad and Karim 2019).

The therapeutic potential of traditional medicinal plants may be discovered by comparing their qualities and phytochemical profiles. This method is also useful for finding novel drug development leads and fostering the preservation of ecological and cultural information.

7.3.2 Ecological and Geographical Comparisons of Medicinal Plant Use

Understanding how various cultures and ecosystems impact the choice and utilization of medicinal plants requires ecological and geographical comparisons of plant usage (Zhang et al. 2022). Key aspects to consider when making such comparisons include as follows.

7.3.2.1 Biodiversity and Ecosystems

Unique plant species compositions can be found in many habitats. Depending on the local flora and fauna, there may be a wide range of therapeutic plants available

(Beyene 2016). For instance, compared to dry deserts, tropical rainforests are noted for their tremendous plant variety and large number of therapeutic plant species.

7.3.2.2 Cultural Practices

The native flora and its therapeutic benefits are well-known to indigenous and traditional civilizations. These practices frequently have a direct connection to the environment (Patwardhan et al. 2005). Traditional medical systems have created intricate classifications of therapeutic plants based on their ecological and geographic origins, such as Ayurveda in India and traditional Chinese medicine.

7.3.2.3 Climate and Seasonal Variation

The availability of medicinal plants may be impacted by climatic and seasonal fluctuations. Some plants might only grow in certain regions or under certain climatic conditions at various periods of the year.

According to the season, different plants may be used in areas with unique seasons to treat various illnesses.

7.3.2.4 Cultural Exchange and Trade

The dissemination of knowledge about medicinal plants across regions and continents has been facilitated by historical trade routes and cultural interchange.

For instance, the Silk Road made it easier for people in Asia, the Middle East, and Europe to share information and medicinal plants.

7.3.2.5 Economic and Market Factors

The usage of medicinal herbs might also be influenced by economic concerns. Rare or exotic plants may occasionally have a high value and be traded worldwide.

The overharvesting of the desired rare or exotic plants can result to habitat degradation.

7.3.2.6 Local Beliefs and Spiritual Practices

The usage of medicinal herbs is directly related to spiritual and religious beliefs in several cultures. Geographical and ecological factors can drastically alter these practices. While some plants are just utilized for therapeutic purposes, others are revered and employed in ceremonies (Inta et al., 2008).

7.3.2.7 Scientific Research and Conservation

Geographical and ecological research can assist in highlighting regions with a high variety of plant life and the possibility for new medicinal plant discoveries. Understanding the ecological distribution of important medicinal plant species may frequently guide conservation efforts.

7.3.2.8 Regulatory and Legal Considerations

Medicinal plants may have different legal standing in different countries. The trading of some species of medicinal plants is governed by international agreements like the Convention on International Trading in Endangered Species of Wild Fauna and Flora (CITES).

7.3.2.9 Climate Change and Habitat Loss

The availability of medicinal plants can be significantly impacted by climatic changes and habitat degradation. For sustainable usage and conservation, these changes must be closely monitored.

7.3.2.10 Ethnobotanical Studies

Ethnobotanists research human-plant interactions, including the ecological, geographic, and cultural dimensions of plant-based medicine. Their work contributes to the preservation of traditional knowledge and informing conservation initiatives (Gonfa et al. 2020).

7.3.3 Integration of Traditional Knowledge with Modern Medical Practices

The practice of combining ancient healing techniques with modern medical procedures is known as "integrative medicine" or "complementary and alternative medicine (CAM)," and it incorporates the benefits of both traditional medical practices with cutting-edge scientific research (Opoku et al., 2013). This integration can have several benefits.

7.3.3.1 Holistic Approach

Integrative medicine adopts a holistic approach to healthcare, taking into account a patient's mental, emotional, and spiritual health in addition to their physical symptoms. This holistic viewpoint is frequently emphasized in traditional therapeutic approaches.

7.3.3.2 Cultural Sensitivity

Respecting cultural diversity and the information passed down through the generations requires integrating traditional knowledge (Gradellini et al. 2021). It acknowledges that other civilizations have insightful viewpoints on health and healing.

7.3.3.3 Patient-Centered Care

Patient-centered care, in which patients have an active role in making healthcare decisions, is frequently emphasized in integrative medicine (Moeta et al., 2019). Shared decision-making and intimate patient-practitioner connections are frequent features of traditional therapeutic systems.

7.3.3.4 Diverse Treatment Options

A greater variety of therapy alternatives are offered by integrative medicine. When required, patients can receive traditional medical care, but they also have access to alternative medicines whenever that is the best course of action (Tabish 2008).

7.3.3.5 Prevention and Wellness

Ayurveda and traditional Chinese medicine are two examples of traditional medical systems that put a strong emphasis on well-being maintenance and illness prevention. These ideas are incorporated into contemporary healthcare through integrative medicine.

7.3.3.6 Personalized Medicine

Treatments in traditional systems are frequently customized to fit the individual's specific needs and constitution. Integrative medicine aims to personalize care by taking into consideration the unique health and lifestyle characteristics of each patient (Goetz and Schork 2018).

7.3.3.7 Safety and Efficacy

Utilizing empirically supported practices and scientific analysis, integrative medicine aims to assess the efficacy and safety of conventional treatments. By doing this, therapies are made to be both secure and efficient.

7.3.3.8 Combating Chronic Diseases

Integrative medicine offers a wider variety of tools and ways to manage complicated health challenges, making it particularly useful in controlling chronic diseases (Roy 2010).

7.3.3.9 Reduced Side Effects

Compared to some pharmacological interventions, some traditional therapies, including acupuncture and herbal cures, may have less adverse effects. These treatments can be used in accordance with integrative medicine to reduce adverse effects.

7.3.3.10 Cross-Cultural Exchange

Integrative medicine encourages cross-cultural communication and cooperation between Eastern and Western medical specialists. Increased mutual understanding and creativity might arise from this discussion (He et al. 2023). Challenges in integrating traditional knowledge with modern medicine:

Lack of Scientific Validation It might be challenging to evaluate the safety and efficacy of certain traditional medicines because they lack thorough scientific support.

Standardization Regional and professional differences in traditional remedies are common. Controlling standards and quality might prove challenging.

Regulatory Issues Traditional medicine is governed quite differently in every nation, which generates uncertainties regarding the reliability and safety of the products (Yuan et al. 2016; Sankaramourthy et al. 2021).

Integration in Medical Education To guarantee that healthcare professionals are aware about both conventional and contemporary treatments, integrative medicine necessitates reforms in medical education.

Cost Considerations When it comes to cost, integrative medicine can be higher in cost if it's not covered under insurance.

7.4 Utilization of Quantitative Approaches in Ethnopharmacology

Utilizing mathematical and statistical techniques, quantitative insights in ethnopharmacology analyze and comprehend the therapeutic characteristics of plants and other natural items utilized in traditional medicine by many civilizations.

7.4.1 Chemical Profiling

Quantitative examination of the chemical components of medicinal plants, such as the detection and measurement of active substances (such as alkaloids, flavonoids, and terpenoids), can be done by using methods like gas chromatography-mass spectrometry (GC-MS), high-performance liquid chromatography (HPLC), high-performance thin layer chromatography (HPTLC), etc. (Pham et al. 2020).

7.4.2 Bioactivity Testing

Calculation of the IC_{50} values (concentration at which 50% of a biological impact is obtained) in cell culture or in vivo models can be done to quantify the biological activity of plant extracts or compounds (Chik et al. 2015).

7.4.3 Dose-Response Relationships

Dose-response relationships can be assessed by examining the connection between a drug's therapeutic impact and the dose of a medicinal plant or component. This may be used to choose the best dose (Schmeda-Hirschmann and Yesilada 2005).

7.4.4 Pharmacokinetics

Pharmacokinetics of the ethnopharmacological medicinal plants can be studied by examining how phytochemicals are absorbed, distributed, metabolized, and excreted

(ADME) throughout the human body to provide a quantitative knowledge of how they are handled (Sun et al. 2020).

7.4.5 Synergy and Combinatorial Effects

Synergistic effects of medicinal drugs can be determined by mixing several plant extracts or chemicals that might improve medicinal results. Utilizing techniques like the combination index (CI) and fractional inhibitory concentration index (FICI), this may be quantitatively examined (Wei et al. 2021).

7.4.6 Network Pharmacology and Molecular Docking

Studying the interactions between various chemicals in medicinal plants and their targets in the human body using network analysis and systems biology can help to provide a quantitative knowledge of the mechanisms of action. Also, the molecular docking studies are helpful for determination of phytoconstituents' binding with specific genes or proteins (Taha et al. 2020).

Some examples of quantitative approaches in ethnopharmacology are tabulated in Table 7.1.

7.5 Advantages of Quantitative Insights in Ethnopharmacology

7.5.1 Objectivity and Reproducibility

In ethnopharmacological research, quantitative approaches bring accuracy (Leonti 2022; Süntar 2020). Data collection by researchers may be standardized, which increases reproducibility and reduces the chance of bias in the findings.

7.5.2 Statistical Analysis

Robust statistical analysis is possible with quantitative data. This makes it possible for researchers to find patterns, connections, and correlations in the data, giving them a stronger foundation for correlating with the conclusions (Noyes et al. 2019).

7.5.3 Comparative Studies

Comparing the usage of medicinal plants across many cultures, geographical regions, and timeframes is made possible by quantitative insights. This comparison method can highlight patterns and similarities in conventional medical procedures.

7.5.4 Identification of Key Species

Quantitative analysis may be performed to determine which species of medicinal plants are mostly utilized in a certain culture or society (Süntar 2020). Further investigation and conservation efforts can be guided by this knowledge.

7.5.5 Dose-Response Relationships

To establish dose-response relationships and to assist selection of the right dosage of medicinal plants for therapeutic effects while minimizing potential adverse effects, quantitative data can be employed.

7.5.6 Prioritizing Plant Species

Based on quantitative data, researchers can categorize plant species for additional pharmacological and phytochemical investigations (Leonti 2022). This may result in the identification of novel medicinal substances or medications.

7.5.7 Ethnobotanical Conservation

By identifying species that are widely utilized and in danger of being overharvested or having their habitats destroyed, quantitative insights can help with the conservation of medicinal plants (Leonti 2022; Dapar et al. 2020).

7.6 Limitations of Quantitative Insights in Ethnopharmacology

7.6.1 Cultural and Contextual Factors

The cultural and environmental intricacies of conventional therapeutic practices may be overlooked by quantitative methodologies (Leonti 2022; Dapar et al. 2020). Understanding the holistic character of conventional medicine frequently requires consideration of these factors.

7.6.2 Language and Terminology

It might be difficult to translate traditional knowledge into quantitative data since various cultures may have their own terminology and notions for medicinal plants (Panthee 2020).

Table 7.1 Some examples of quantitative approaches in ethnopharmacology

Quantitative approach	Name of plant	Ethnopharmacological relevance	Method used	Compounds assessed	References
Chemical profiling	*Tinospora cordifolia*	Anticancer and immunomodulatory activity	HPTLC	N-formylannonain, 11-hydroxymustakone, and yangambin	Bala et al. (2015)
	Solanum xanthocarpum	Anti-inflammatory and immunomodulatory activity	HPTLC	Solamargine, solasodine, solasonine	Srivastava et al. (2022)
	Synergistic combination (*Withania somnifera* and *Myristica fragrans*)	Neuroprotective effect	TLC-bioautography	Sitoindosides VII, VIII, quercetin, isopelletierine and withanolide S (*Withania somnifera*), Eugenol, methyl eugenol, myristic acid, galbacin, and β-sitosterol (*Myristica fragrans*)	Khan et al. (2021)
	Rubia cordifolia	Anti-tyrosinase effect	TLC-bioautography	Purpurin	Insaf et al. (2023)
	Picrorhiza kurroa	Hepatoprotective effect	UPLC-MS	Picroside I, picroside II, apocyanin	Zahiruddin et al. (2017)
	Andrographis paniculata	Anticancer effect	HPTLC	Andrographolide	Malik et al. (2024)
	Pterocarpus marsupium	Antidiabetic effect	HPTLC-bioautography	Epicatechin, quercetin, naringenin, vanillic acid	Irfan Dar et al. (2022)

(continued)

Table 7.1 (continued)

Bioactivity testing	**Name of plant**	**Ethnopharmacological relevance**	**Type of extract**	**Type of study**	**References**
	Inonotus obliquus	Anti-tumor effect	Aqueous	Antibacterial activity	Glamočlija et al. (2015)
	Oroxylum indicum (L.) Kurz	Effects on glucose and lipid homeostasis	Ethyl acetate	Anti-obesity activity	Mangal et al. (2017)
	Blechnum orientale Linn.	Wound healing	Aqueous	Antibacterial activity	Lai et al. (2016)
	Glycyrrhiza glabra and Honey	Immunomodulatory activity	–	Metabolomics analysis	Kong et al. (2022)
	Pistacia integerrima	Antibacterial activity	Methanolic	Antibacterial activity	Abass et al. (2023)
Pharmacokinetics and metabolomics	**Name of plant**	**Ethnopharmacological relevance**	**Method used**	**Compounds assessed**	**References**
	Picrorhiza kurroa	Hepatoprotective effect	UPLC-MS	Picroside I, picroside II, apocyanin	Zahiruddin et al. (2017)
	Ganoderma lucidum P. Karst	Anticancer effect	UPLC-MS	Ganoderic acid	Ansari et al. (2022)
	–	Anticonvulsant	GC-MS	Piperine	Parveen et al. (2015)

7.6.3 Reductionism

Quantitative analysis has a tendency to oversimplify complicated traditional healing systems, which leaves out their richness and interconnections and reduces them to single data points.

7.6.4 Bias in Data Collection

Countless the efforts to standardize collection of data, biases may still exist, particularly if researchers are unaware of the cultural and social background of the studied location.

7.6.5 Limited Validation

Quantitative findings might not necessarily be consistent with current pharmacological or clinical studies, and some conventional treatments might not have any supporting evidence.

7.6.6 Ethical Considerations

Quantitative data collection from indigenous groups must be done ethically, taking into account concerns about permission, cultural appropriation, and intellectual property rights.

7.6.7 Limited Coverage

Traditional medical practices cannot be measured completely, especially those that incorporate spiritual or metaphysical components (Dapar et al. 2020; Pirintsos et al. 2022).

7.7 Potential Synergies between Quantitative and Comparative Approaches

Socioscientific, ecological, and ethnopharmacological quantitative and comparative methods can work together to produce a deeper and more comprehensive knowledge of complex processes (Borgstede and Scholz 2021; Nwali et al. 2018). Potential synergies between these two approaches include as follows.

7.7.1 Enhanced Data Precision

With the use of measurements, numerical analysis, and statistical tools, quantitative techniques deliver exact data. Comparisons between various groups, areas, or time periods may be made more robust when this quantitative precision is combined with comparative methodologies. Findings are more trustworthy and generalizable as a result of this synergy.

7.7.2 Pattern Recognition

Within databases, quantitative data may highlight patterns and trends. This is expanded upon by comparative analysis, which finds patterns in several datasets. Researchers might get deeper insights by comparing quantitative data from different sources or settings to find similarities and contrasts.

7.7.3 Cross-Cultural Understanding

Comparative studies that use quantitative data can close the gap between various cultural and traditional practices. They make it possible to compare and contrast practices, beliefs, or results in a methodical manner. Comparing quantitative indicators across cultures offers a common language.

7.7.4 Ecosystem Analysis

Species diversity, population dynamics, and ecological processes may all be quantified in ecology. By comparing these quantitative metrics across multiple ecosystems, comparative techniques enable researchers to gain a better understanding of various factors affecting biodiversity and ecosystem health.

7.7.5 Healthcare Research

Quantitative data may be used to evaluate the effectiveness of treatments or interventions in the healthcare industry. Researchers can then analyze treatment outcomes across various patient groups or healthcare systems using comparative analysis, which results in recommendations that are supported by the available data.

7.7.6 Phylogenetics and Evolutionary Biology

Researchers can get a better understanding of the evolution of species features by combining quantitative phylogenetic analysis with comparative methods (Gluckman et al. 2016). Through this synergy, it is possible to examine how features vary between species and how they relate to ecological and environmental variables.

7.7.6.1 Policy and Decision-Making

Quantitative data can help in the development of evidence-based policy. The adoption of best practices can then be facilitated by comparative analysis, which can assist policymakers in understanding how policies and initiatives have functioned in various locations or under various circumstances.

7.7.6.2 Cultural Ecology

Quantitative data can be utilized in anthropology and ethnography to measure cultural practices or traditions. The evaluation of these practices across many cultures and historical periods is therefore made possible by comparative techniques, aiding in our knowledge of how cultures change through time (Nwali et al. 2018).

7.7.6.3 Conservation Biology

The evaluation of species vulnerability and conservation priority across different areas is made possible by the combination of quantitative ecological data and comparative analysis. Effective conservation efforts are guided by this synergy.

7.7.6.4 Social Sciences

Quantitative data may reveal trends in economic activity and human behavior in sociology and economics. Researchers can then examine these patterns across societies or time periods using comparative analysis, which sheds information on the social and economic variables driving behavior.

7.8 Incorporation of Omics Technologies and Computational Methods

Numerous scientific fields, such as ethnopharmacology, ecology, and medicine, have been transformed by the inclusion of omics technologies (genomics, transcriptomics, proteomics, metabolomics, etc.) and computational methodologies (Subramanian et al. 2020; Maria 2011; Pant and Garg 2018). These approaches can be integrated to enhance research in these fields as follows.

7.8.1 Genomic and Metagenomic Analysis

The genetic makeup of medicinal plants may be determined by genomic sequencing, providing information on the production of bioactive substances. The significance of microorganisms in influencing the phytochemical composition of medicinal plants and their ecological interactions can be revealed by metagenomic study of soil and plant microbiomes.

7.8.2 Transcriptomics

Transcriptomic research can shed light on the patterns of gene expression in medicinal plants and assist in identifying important metabolic routes for secondary metabolites (Maria 2011). Comparative transcriptomics can explain variations in medicinal characteristics by illuminating changes in gene expression between plant populations or species utilized in traditional medicine.

7.8.3 Proteomics

Proteins in medicinal plants, such as the enzymes necessary for the manufacture of secondary metabolites, may be identified and quantified using proteomic analysis. Understanding how environmental conditions affect the expression of proteins in plants is an additional advantage of proteomics.

7.8.4 Metabolomics

The thorough characterization of plant metabolites, including bioactive substances, is made possible through metabolomics (Okada et al. 2010). Comparative metabolomics can show differences in metabolite profiles among various plant populations or in various ecological settings.

7.8.5 Phylogenomics

Reconstructing the evolutionary history of medicinal plants using genomes and phylogenetics can help us understand the evolution and diversity of their therapeutic characteristics.

7.8.6 Bioinformatics and Computational Biology

When analyzing data, computational approaches can help identify potential drug targets in plant genomes. The discovery of novel bioactive chemicals and the prediction of their pharmacological characteristics can both be aided by machine learning and data mining.

7.8.7 Network Analysis

Networks of genes, proteins, and metabolites may be built using omics data, showing intricate relationships and processes related to therapeutic properties. Key nodes or biomarkers linked to treatment benefits can be found via network-based research (Navabharath et al. 2022).

7.8.8 Pharmacogenomics

In ethnopharmacology, integrating genomic data with patient data might result in personalized medicine strategies that let patients choose their own therapies based on their unique genetic profiles.

7.8.9 Conservation and Biodiversity Assessment

Assessment of the genetic diversity of medicinal plant species using genomic and metagenomic methods can help with conservation efforts. Computer modeling helps forecast how changes in the environment will affect plant populations.

7.8.10 Drug Discovery

Drug development can be sped up by using computational docking and virtual screening to anticipate the interactions between bioactive plant chemicals and target proteins.

7.8.11 Ethnopharmacological Data Integration

By offering mechanistic insights into the bioactivity of medicinal plants and verifying traditional usage, omics data can supplement ethnopharmacological knowledge (Pant and Garg 2018).

Researchers can better understand the chemical, genetic, and ecological properties of medicinal plants by utilizing omics technologies and computational methods. This will help with drug discovery, conservation efforts, and the evidence-based integration of conventional knowledge into contemporary healthcare.

7.9 Ethical Considerations in Cross-Cultural Research and Knowledge Sharing

To guarantee respect, equality, and sensitivity towards the cultures, communities, and individuals involved, ethical issues in cross-cultural research and information exchange are crucial (Honan et al. 2013). Key ethical principles and considerations include as follows.

7.9.1 Informed Consent

The informed consent of every research subject should be obtained, making sure they are aware of the study's objectives, procedures, and any risks and rewards. Permission is a continuous process and participants are always free to revoke it.

7.9.2 Cultural Sensitivity

Be respectful of the communities' or individuals' cultural customs, traditions, and beliefs. To negotiate cultural sensitivities and customs effectively, seek the advice of local authorities or community leaders.

7.9.3 Community Engagement and Collaboration

During the course of the study process, engaging with the local communities is important along with their inclusion in the planning, gathering, and analysis of the study's data. Working together with community people ensures that the research reflects their goals and needs.

7.9.4 Beneficence and Non-maleficence

It should be made sure the advantages of the research exceed the risks. An effort should be made to minimize any negative effects and maximize community advantages. It should be cautiously taken care of to not abuse or hurt study participants or the communities they live in (Broesch et al. 2020).

7.9.5 Privacy and Confidentiality

The confidentiality and privacy of study participants should be ensured. Confidential information that could hurt people or communities should not be disclosed. Data should be anonymized before being used in publications.

7.9.6 Resource Sharing and Collaboration

The communities and people who are taking part in the study should be informed of the research findings and the generated results. Collaboration with the regional institutions and researchers should be made to advance knowledge transfer and capacity-building.

7.9.7 Intellectual Property and Ownership

Intellectual property and knowledge ownership should be discussed in an ethical and open manner. Traditional knowledge may be possessed by a group as a whole or by particular people or organizations.

7.9.8 Equitable Access to Benefits

It should be made sure that all parties, including the groups and people who provided their expertise, are equally dispersed of the financial advantages or benefits resulting from the research.

7.9.9 Avoiding Harmful Stereotypes and Exploitation

Promoting preconceptions and using cultural customs and traditional knowledge for one's own or other people's benefit should be avoided. Information and research results should be presented in a polite and truthful manner.

7.9.10 Respecting Vulnerable Populations

Close attention should be paid to populations that are at risk, such as marginalized or indigenous peoples. It should be made sure that research does not make existing vulnerabilities worse.

7.9.11 Transparency and Accountability

All aspects of the research process should be kept transparent, including funding sources, conflicts of interest, and any study restrictions. One should be accountable for doing research in an ethical manner and sharing knowledge in a responsible manner.

7.9.12 Institutional Review and Ethical Oversight

The appropriate institutional review boards or ethics committees should be consulted for their ethical approval. Moral norms and principles established by the appropriate organizations should be followed.

7.9.13 Cultural Competence

Researchers' cultural sensitivity and proficiency in cross-cultural research should be enhanced (Broesch et al. 2020). This involves instruction on acceptable behavior as well as training in cultural diversity.

7.9.14 Long-Term Engagement

Interaction with the communities or people being researched should be continued for long-term engagement. Ethics responsibilities go beyond the study project, particularly when it comes to knowledge transfer and long-term gains.

7.9.15 Local Capacity-Building

Local research capacity should be developed by ensuring that communities have the resources and information needed to do research on their own terms (Ekmekci and Arda 2017; Broesch et al. 2020).

7.10 Conclusion

In conclusion, the incorporation of quantitative insights into ethnopharmacology is an important step towards a greater comprehension of conventional medical procedures and their potential to advance contemporary healthcare. We have revealed the rich tapestry of traditional knowledge, showing the many chemical compositions and therapeutic potentials of medicinal plants, via meticulous data collecting, statistical analysis, and the inclusion of cutting-edge technology. In addition to bridging the gap between ancient therapeutic practices and modern science, these quantitative techniques also open the door for evidence-based improvements in drug development, the preservation of medicinal flora, and the promotion of culturally competent healthcare practices. As we continue to investigate the quantitative aspects of ethnopharmacology, we set out on a journey that will hopefully facilitate both our understanding of science and our understanding of the deep wisdom that is ingrained in traditional medical systems all over the world.

References

Abass S, Dar MI, Zahiruddin S, Beg MA, Nagar P, Jan B, Husain SA, Ahmad S (2023) Phytochemical and antibacterial analysis of Pistacia integerrima: an integrated in vitro and in silico approach. Process Biochem 132:236

Adams A, Cox AL (2008) Questionnaires, in-depth interviews and focus groups Open Research Online The Open University' s repository of research publications Questionnaires, in-depth interviews and focus groups Book Chapter. Res. Methods Hum. Comput. Interact. 17–34

Agidew MG (2022) Phytochemical analysis of some selected traditional medicinal plants in Ethiopia. Bull Natl Res Cent 46:87

Ahmad F, Karim M (2019) Impacts of knowledge sharing: a review and directions for future research. J Work Learn 31:207–230

Ansari MHR, Khan W, Parveen R, Saher S, Ahmad S (2022) Pharmacokinetic, metabolomic, and stability assessment of ganoderic acid H based triterpenoid enriched fraction of Ganoderma lucidum P. Karst. Metabolites 12:97

Bala M, Pratap K, Verma PK, Singh B. and Padwad Y (2015) Validation of ethnomedicinal potential of Tinospora cordifolia for anticancer and immunomodulatory activities and quantification of bioactive molecules by HPTLC. Journal of ethnopharmacology, 175, pp.131–137

Beyene B, Beyene B. and Deribe H (2016) Review on application and management of medicinal plants for the livelihood of the local community. Journal of Resources Development and Management, 22(1), pp.33–39

Bolton D. and Gillett G (2019) The biopsychosocial model of health and disease: New philosophical and scientific developments (p. 149). Springer Nature

Borgstede M. and Scholz M (2021) Quantitative and qualitative approaches to generalization and replication–A representationalist view. Frontiers in psychology, 12, p.605191

Broesch T, Crittenden AN, Beheim BA, Blackwell AD, Bunce JA, Colleran H, Hagel K, Kline M, McElreath R, Nelson R.G. and Pisor AC (2020) Navigating cross-cultural research: methodological and ethical considerations. Proceedings of the Royal Society B, 287(1935), p.20201245

Brusotti G, Cesari I, Dentamaro A, Caccialanza G, Massolini G (2014) Isolation and characterization of bioactive compounds from plant resources: the role of analysis in the ethnopharmacological approach. J Pharm Biomed Anal 87:218–228

Chakrabarty SP, Kaur R (2021) A primer to traditional knowledge protection in India: the road ahead. Liverp Law Rev 42:401–427

Chik WI, Zhu L, Fan LL, Yi T, Zhu GY, Gou XJ, Tang YN, Xu J, Yeung WP, Zhao ZZ and Yu ZL (2015) Saussurea involucrata: A review of the botany, phytochemistry and ethnopharmacology of a rare traditional herbal medicine. Journal of ethnopharmacology, 172, pp.44–60

Dapar MLG, Alejandro GJD, Meve U and Liede-Schumann S (2020) Quantitative ethnopharmacological documentation and molecular confirmation of medicinal plants used by the Manobo tribe of Agusan del Sur, Philippines. Journal of ethnobiology and ethnomedicine, 16, pp.1–60

Darwish IA (2006) Immunoassay methods and their applications in pharmaceutical analysis: basic methodology and recent advances. Int J Biomed Sci 2:217–235

Ekmekci PE, Arda B (2017) Interculturalism and informed consent: respecting cultural differences without breaching human rights. Cultura (lasi) 14:159–172

Ghorbani A, Farzaneh Naghibi MM (2005) Ethnobotany, Ethnopharmacology and drug discovery. Iran J Pharm Sci 2:1–23

Glamočlija J, Ćirić A, Nikolić M, Fernandes Â, Barros L, Calhelha RC, Ferreira IC, Soković M and Van Griensven LJ (2015) Chemical characterization and biological activity of Chaga (Inonotus obliquus), a medicinal "mushroom". Journal of ethnopharmacology, 162, pp.323–332

Gluckman P, Beedle A, Buklijas T, Low F, Hanson M (2016) Human evolution and the origins of human diversity. Princ Evol Med:131–158

Goetz LH, Schork NJ (2018) Personalized medicine: motivation, challenges, and progress. Fertil Steril 109:952–963

Gonfa N, Tulu D, Hundera K. and Raga D (2020) Ethnobotanical study of medicinal plants, its utilization, and conservation by indigenous people of Gera district, Ethiopia. Cogent Food & Agriculture, 6(1), p.1852716

Gradellini C, Gómez-Cantarino S, Dominguez-Isabel P, Molina-Gallego B, Mecugni D. and Ugarte-Gurrutxaga MI (2021) Cultural competence and cultural sensitivity education in university nursing courses. A scoping review. Frontiers in psychology, 12, p.682920

Grandgirard J, Poinsot D, Krespi L, Nénon JP, Cortesero AM (2002) Four varieties of comparative analysis. Entomol Exp Appl 103:239–248

Hadi MA, José Closs S (2016) Ensuring rigour and trustworthiness of qualitative research in clinical pharmacy. Int J Clin Pharm 38:641–646
Hamilton AC (2004) Medicinal plants, conservation and livelihoods. Biodiversity & Conservation, 13, pp.1477–1517
Harvey AL (2008) Natural products in drug discovery. Drug Discov Today 13:894–901
He C, Zhao X, Yao R, Xiao P (2023) Food-medicine can promote cross-culture communication between East and West. Chin Herb Med 15:3–5
Heywood VH (2011) Ethnopharmacology, food production, nutrition and biodiversity conservation: towards a sustainable future for indigenous peoples. J Ethnopharmacol 137:1–15
Hidayat S, Subositi D, Batubara I, Munawaroh E, Afandi S. and Cahyaningsih R (2023) Scientific databases for conservation of medicinal plants. In Medicinal Plants: Biodiversity, Biotechnology and Conservation (pp. 265–287). Singapore: Springer Nature Singapore.
Holmes AGD (2023) The design and use of questionnaires in educational research: a new (student) researcher guide. Innovare J Educ:1–5
Honan E, Hamid MO, Alhamdan B, Phommalangsy P, Lingard B (2013) Ethical issues in cross-cultural research. Int J Res Method Educ 36:386–399
Hughes JP, Rees SS, Kalindjian SB, Philpott KL (2011) Principles of early drug discovery. Br J Pharmacol 162:1239–1249
Ingle KP, Deshmukh AG, Padole DA, Dudhare MS, Moharil MP, Khelurkar VC (2017) Phytochemicals: extraction methods, identification and detection of bioactive compounds from plant extracts. J Pharmacogn Phytochem 6:32–36
Insaf A, Parveen R, Srivastava V, Samal M, Kha M. and Ahmad S (2023) TLC–MS-Bioautographic Identification of Antityrosinase Compounds and Preparation of a Topical Gel Formulation from a Bioactive Fraction of an RSM-Optimized Alcoholic Extract of Rubia Cordifolia L. stem. Journal of AOAC International, 106(6), pp.1598–1607
Inta A, Shengji P, Balslev H, Wangpakapattanawong P. and Trisonthi C (2008) A comparative study on medicinal plants used in Akha's traditional medicine in China and Thailand, cultural coherence or ecological divergence?. Journal of ethnopharmacology, 116(3), pp.508–517
Irfan Dar M, Qureshi MI, Zahiruddin S, Abass S, Jan B, Sultan A, Ahmad S (2022) In silico analysis of PTP1B inhibitors and TLC-MS bioautography-based identification of free radical scavenging and α-amylase inhibitory compounds from heartwood extract of Pterocarpus marsupium. ACS Omega
Jagdeep S, Sen TS (2013) Comparative analysis of medicinal plants for their antimicrobial potential and phytoconstituents screening. Int J Drug Dev Res 5:447–453
Kadir MF, Bin Sayeed MS, Mia MMK (2012) Ethnopharmacological survey of medicinal plants used by indigenous and tribal people in Rangamati, Bangladesh. J Ethnopharmacol 144:627–637
Khan MA, Srivastava V, Kabir M, Samal M, Insaf A, Ibrahim M, Zahiruddin S. and Ahmad S (2021) Development of synergy-based combination for learning and memory using in vitro, in vivo and TLC-MS-bioautographic studies. Frontiers in Pharmacology, 12, p.678611
Kong S, Li P, Verpoorte R, Wang J, Zhu C, Dai Y and Chen S (2022) Synergistic mechanism for the bioactivity fortification of licorice by honey. Journal of Ethnopharmacology, 289, p.115048
Kruse W, Eggert-Kruse W, Rampmaier J, Runnebaum B, Weber E (1991) Dosage frequency and drug-compliance behaviour—a comparative study on compliance with a medication to be taken twice or four times daily. Eur J Clin Pharmacol 41:589–592
Kumar D, Kumari S, Seth T, Prajapat K, Reddy R (2019) Indigenous traditional knowledge on folk medicinal plants. J Pharmacogn Phytochem 8:2409–2412
Lai JCY, Lai HY, Rao NK and Ng SF (2016) Treatment for diabetic ulcer wounds using a fern tannin optimized hydrogel formulation with antibacterial and antioxidative properties. Journal of ethnopharmacology, 189, pp.277–289
Leonti M (2022) The relevance of quantitative ethnobotanical indices for ethnopharmacology and ethnobotany. J Ethnopharmacol 288:115008
Leonti M, Weckerle CS (2015) Quantitative and comparative methods in ethnopharmacology. Ethnopharmacol:29–40

Les F, Cásedas G, López V (2021) Bioactivity of medicinal plants and extracts. Biology (Basel) 10:2021–2023
Mahdizadeh M, Heydari A, Karimi Moonaghi H (2015) Clinical interdisciplinary collaboration models and frameworks from similarities to differences: a systematic review. Glob J Health Sci 7:170–180
Malik Z, Parveen R, Zahiruddin S, Gautam G, Husain SA and Ahmad S (2024) HPTLC Stability Indicating Analytical Method of Andrographolide and 5-fluorouracil with Network Pharmacology Analysis against Cancer. Combinatorial Chemistry & High Throughput Screening, 27(6), pp.894–909
Mangal P, Khare P, Jagtap S, Bishnoi M, Kondepudi KK and Bhutani KK (2017). Screening of six Ayurvedic medicinal plants for anti-obesity potential: An investigation on bioactive constituents from Oroxylum indicum (L.) Kurz bark. Journal of Ethnopharmacology, 197, pp.138–146
Maria (2011) Omics technologies, data and bioinformatics principles. Methods Mol Biol 719:61779
Maroyi A, Cheikhyoussef A (2015) A comparative study of medicinal plants used in rural areas of Namibia and Zimbabwe. Indian J Tradit Knowl 14:401–406
Mitra S, Naskar N and Chaudhuri P (2021) A review on potential bioactive phytochemicals for novel therapeutic applications with special emphasis on mangrove species. Phytomedicine plus, 1(4), p.100107
Moeta M, Mogale RS, Mgolozeli SE, Moagi MM and Pema-Bhana V (2019) Integrating African traditional health knowledge and practices into health sciences curricula in higher education: An Imbizo Approach. International Journal of African Renaissance Studies-Multi-, Inter-and Transdisciplinarity, 14(1), pp.67–82
Mukherjee P, Harwansh R, Bahadur S, Banerjee S, Kar A (2016) Evidence based validation of Indian traditional medicine—way forward. World J Tradit Chin Med 2:48–61
Nancarrow SA, Booth A, Ariss S, Smith T, Enderby P and Roots A, (2013) Ten principles of good interdisciplinary team work. Human resources for Health, 11, pp.1–11
National Research Council (1988) Methods of data collection, representation, and analysis. Behav Soc Sci:167–181
Navabharath M, Srivastava V, Gupta S, Singh SV and Ahmad S (2022) Ursolic Acid and Solasodine as Potent Anti-Mycobacterial Agents for Combating Paratuberculosis: An Anti-Inflammatory and In Silico Analysis. Molecules, 28(1), p.274
Negi DP, Abdul Azeez EP (2022) Diminishing traditional methods and inaccessible modern healthcare: the dilemma of tribal health in India. J Health Res 36:867–877
Noyes J, Booth A, Moore G, Flemming K, Tunçalp Ö. and Shakibazadeh E, (2019) Synthesising quantitative and qualitative evidence to inform guidelines on complex interventions: clarifying the purposes, designs and outlining some methods. BMJ global health, 4(Suppl 1), p.e000893
Nwali ON, Idoko A, Okolie JE, Ezeh E, Ugwudike PO, Rita ON, Ezenwali MO, Odo IA, Ani PN and Okolie SO (2018) Comparative analysis of the phytochemical compositions of leaf, stem-bark and root of Azadirachta Indica (neem). Universal Journal of Pharmaceutical Research
Okada T, Mochamad Afendi F, Altaf-Ul-Amin M, Takahashi H, Nakamura K, Kanaya S (2010) Metabolomics of medicinal plants: the importance of multivariate analysis of analytical chemistry data. Curr Comput Aided Drug Des 6:179–196
Opoku JK, Addai-Mensah P. and Wiafe F (2013) Traditional and Modern Medicine: A survey of views on its integration in Ghana. Medicine, 4(2), pp.256–257
Pant M, Garg S (2018) Applications of computational methods in manufacturing processes. Adv Appl Manuf Eng:191–229
Panthee D (2020) Uses and limitations of quantitative research in English language education Dinesh. Malaysian Palm Oil Council 21, 1–9
Parveen B, Pillai KK, Tamboli ET and Ahmad S (2015) Effect of piperine on pharmacokinetics of sodium valproate in plasma samples of rats using gas chromatography-mass spectrometry method. Journal of Pharmacy and Bioallied Sciences, 7(4), pp.317–320
Patwardhan B, Warude D, Pushpangadan P, Bhatt N (2005) Ayurveda and traditional Chinese medicine: a comparative overview. Evidence-Based Complementary and Alternative Medicine, 2(4), pp. 465–473

Periwal V, Bassler S, Andrejev S, Gabrielli N, Typas A, Patil KR (2020) Bioactivity assessment of natural compounds using machine learning models based on drug target similarity. bioRxiv. 2020.11.06.371112
Pham GN, Nguyen TTT. and Nguyen-Ngoc H (2020) Ethnopharmacology, phytochemistry, and pharmacology of Syzygium nervosum. Evidence-Based Complementary and Alternative Medicine, 2020(1), p.8263670
Pirintsos S, Panagiotopoulos A, Bariotakis M, Daskalakis V, Lionis C, Sourvinos G, Karakasiliotis I, Kampa M and Castanas E (2022) From traditional ethnopharmacology to modern natural drug discovery: A methodology discussion and specific examples. Molecules, 27(13), p.4060
Rodriguez S, Hug C, Todorov P, Moret N, Boswell SA, Evans K, Zhou G, Johnson NT, Hyman BT, Sorger PK and Albers MW (2021) Machine learning identifies candidates for drug repurposing in Alzheimer's disease. Nature communications, 12(1), p.1033
Roy R (2010) Integrative medicine to tackle the problem of chronic diseases. J Ayurveda Integr Med 1:18–21
Sailaja V, Madhu M, Neeraja V (2016) Quantitative phytochemical analysis of some medicinal plant seed by using various organic solvent. J Pharmacogn Phytochem 5:30–34
Sankaramourthy D, Subramanian K, Sadras SR (2021) Safety and regulatory issues on traditional medicine entrusted drug discovery. Evid Based Valid Tradit Med A Compr Approach:589–603
Santhan P (2020) A field study on Indian medicinal plants. J Med Plants Stud 8:198–205
Saslis-Lagoudakis CH, Williamson EM, Savolainen V, Hawkins JA (2011) Cross-cultural comparison of three medicinal floras and implications for bioprospecting strategies. J Ethnopharmacol 135:476–487
Saxena KG, Rao KS (2023) Traditional knowledge systems and sustainable development. Tradit Ecol Knowl Resour Manag Asia:15–26
Schmeda-Hirschmann G. and Yesilada E (2005) Traditional medicine and gastroprotective crude drugs. Journal of ethnopharmacology, 100(1-2), pp.61–66
Srivastava V, Navabharath M, Gupta S, Singh SV and Ahmad S (2022) Exploration of Solanum xanthocarpum Schrad. & Wendl. against mycobacterium avium subspecies paratuberculosis and assessment of its immunomodulatory and anti-inflammatory potential. Pharmaceuticals, 15(11), p.1367
Subramanian SV, Smith GD, Subramanyam M (2006) Indigenous health and socioeconomic status in India. PLoS Med 3:1794–1804
Subramanian I, Verma S, Kumar S, Jere A and Anamika K (2020) Multi-omics data integration, interpretation, and its application. Bioinformatics and biology insights, 14, p.1177932219899051
Sun Y, Yang AWH and Lenon GB (2020) Phytochemistry, ethnopharmacology, pharmacokinetics and toxicology of Cnidium monnieri (L.) Cusson. International journal of molecular sciences, 21(3), p.1006
Süntar I (2020) Importance of ethnopharmacological studies in drug discovery: role of medicinal plants. Phytochem Rev 19:1199–1209
Tabish SA (2008) Complementary and alternative healthcare: is it evidence-based? Int J Health Sci (Qassim) 2:V–IX
Taha KF, Khalil M, Abubakr MS and Shawky E (2020) Identifying cancer-related molecular targets of Nandina domestica Thunb. by network pharmacology-based analysis in combination with chemical profiling and molecular docking studies. Journal of ethnopharmacology, 249, p.112413
Taipalus T, Isomöttönen V, Erkkilä H. and Äyrämö S (2022) Data analytics in healthcare: A tertiary study. SN Computer Science, 4(1), p.87
Thatte UM, Gogtay NJ (2018) Traditional medicine: a review of work in India (2012-2017). Proc Indian Natl Sci Acad 84:267–279
Tugume P, Kakudidi EK, Buyinza M, Namaalwa J, Kamatenesi M, Mucunguzi P and Kalema J (2016) Ethnobotanical survey of medicinal plant species used by communities around Mabira Central Forest Reserve, Uganda. Journal of ethnobiology and ethnomedicine, 12, pp.1–28
Virapongse A, Picheansoonthon C (2005) Researching traditional medicine: a review and evaluation of objectives and methodologies. J R Inst Thail 30:17–19

Wei MM, Zhao SJ, Dong XM, Wang YJ, Fang C, Wu P, Song GQ, Gao JN, Huang ZH, Xie T and Zhou JL, (2021). A combination index and glycoproteomics-based approach revealed synergistic anticancer effects of curcuminoids of turmeric against prostate cancer PC3 cells. Journal of Ethnopharmacology, 267, p.113467

Wu WT, Li YJ, Feng AZ, Li L, Huang T, Xu AD. and Lyu J, (2021). Data mining in clinical big data: the frequently used databases, steps, and methodological models. Military Medical Research, 8, pp.1–12

Xing B, Li S, Yang J, Lin D, Feng Y, Lu J, Shao Q (2021) Phytochemistry, pharmacology, and potential clinical applications of saffron: a review. J Ethnopharmacol 281:114555

Yu B, Yang H (2017) Evaluation of different estimation methods for accuracy and precision in biological assay validation. PDA J Pharm Sci Technol 71:297–305

Yu M, Gouvinhas I, Rocha J and Barros AI (2021) Phytochemical and antioxidant analysis of medicinal and food plants towards bioactive food and pharmaceutical resources. Scientific reports, 11(1), p.10041

Yuan H, Ma Q, Ye L and Piao G (2016) The traditional medicine and modern medicine from natural products. Molecules, 21(5), p.559

Zahiruddin S, Khan W, Nehra R, Alam MJ, Mallick MN, Parveen R and Ahmad S (2017) Pharmacokinetics and comparative metabolic profiling of iridoid enriched fraction of Picrorhiza kurroa–An Ayurvedic Herb. Journal of ethnopharmacology, 197, pp.157–164

Zhang MX, Chen Y, Guo JX, Zhang R, Bi YQ, Wei XX, Niu H, Zhang CH. and Li MH (2022) Complex ecological and socioeconomic impacts on medicinal plant diversity. Frontiers in Pharmacology, 13, p.979890

Unveiling Plant-Based Healing Wisdom Through Ethnobotany and Medicinal Ethnopharmacology

8

Niharika, Mukta Satsangi, Shahid Umar, Akbar Ali, Bushra Parveen, and Sayeed Ahmad

Niharika and Mukta Satsangi contributed equally with all other contributors.

Niharika
Department of Botany, School of Chemical and Life Sciences, Jamia Hamdard, New Delhi, India

Department of Pharmacognosy & Phytochemistry, School of Pharmaceutical Education & Research, Centre of Excellence in Unani Medicine (Pharmacognosy & Pharmacology), Bioactive Natural Product Laboratory (BNPL), Jamia Hamdard, New Delhi, India

M. Satsangi
Department of Botany, Dayalbagh Educational Institute, Dayalbagh, Agra, Uttar Pradesh, India

Department of Pharmacognosy & Phytochemistry, School of Pharmaceutical Education & Research, Centre of Excellence in Unani Medicine (Pharmacognosy & Pharmacology), Bioactive Natural Product Laboratory (BNPL), Jamia Hamdard, New Delhi, India

S. Umar
Department of Botany, School of Chemical and Life Sciences, Jamia Hamdard, New Delhi, India

A. Ali
Department of Botany, Dayalbagh Educational Institute, Dayalbagh, Agra, Uttar Pradesh, India

B. Parveen
Department of Pharmacology, School of Pharmaceutical Education & Research, Jamia Hamdard, New Delhi, India

S. Ahmad (✉)
Department of Pharmacognosy & Phytochemistry, School of Pharmaceutical Education & Research, Centre of Excellence in Unani Medicine (Pharmacognosy & Pharmacology), Bioactive Natural Product Laboratory (BNPL), Jamia Hamdard, New Delhi, India

© The Author(s), under exclusive license to Springer Nature Singapore Pte Ltd. 2024
M. Nandave et al. (eds.), *Ethnopharmacology and OMICS Advances in Medicinal Plants Volume 1*, https://doi.org/10.1007/978-981-97-2367-6_8

Abstract

Ethnobotany and medical ethnopharmacology, the interdisciplinary fields, explore the intricate relationship between ancient wisdom and contemporary knowledge. The close relationship between humans and plants has been a cornerstone of cultural, medicinal, and ecological aspects throughout history. This book chapter highlights the rich tapestry of theoretical foundations of ethnobotany and ethnopharmacology woven across diverse cultures. We discuss the wealth of ethnobotanical and ethnopharmacological knowledge to explore how different communities have cultivated, and harnessed local medicinal plants for their ritualistic, dietary, medicinal, and cultural needs. Moreover, it illuminates ethnopharmacological knowledge and its relevance in modern science for developing herbal-based drugs and products. It also addresses the challenges and ethical considerations of researching and compiling conventional plant knowledge. In conclusion, it advocates the invaluable insights derived from the convergence fusion of traditional knowledge and modern pharmacological exploration, paving a path towards holistic healthcare approaches, and safeguarding our rich botanical heritage.

Keywords

Medicinal plants · Ethnobotany · Ethnopharmacology · Phytoconstituents Pharmacological activities

8.1 Introduction

For millennia, medicinal plants have played an integral role in human life. They are known for their essential medicinal and therapeutic potential in curing human ailments around the world since ancient times, thus promoting healing and well-being (Lawal et al. 2022). Traditional systems of medicine such as Ayurveda, Unani, Siddha, and traditional Chinese medicine (TCM) view health as a holistic concept, where the balance and harmony of the body, mind, and spirit are crucial (Ahmad et al. 2022; Patwardhan et al. 2005). The use of medicinal plants plays a significant role in restoring and maintaining this delicate equilibrium.

The fields of ethnobotany and medical ethnopharmacology are the intersection of nature's healing powers and modern science in traditional medicine and contemporary healthcare (Matejić et al. 2020). These fields delve into the deep connection between humans and plants, uncovering the invaluable knowledge that indigenous communities have held for centuries (Martinez et al. 2023). Studying the diverse cultures and the use of medicinal plants can provide a deeper appreciation for the intricate relationship between humans and nature. It can reveal the vast array of botanical treasures that have been used for centuries to alleviate ailments, boost immunity, and improve overall health. Furthermore, exploring traditional practices alongside modern scientific research can lead to innovative approaches to drug discovery and the development of natural remedies. This integration of knowledge and techniques can provide a comprehensive understanding of the therapeutic potential of medicinal plants, paving the way for the development of natural remedies grounded in centuries of empirical knowledge and collaboration between different cultures and scientific communities.

8.1.1 Ethnobotany

The term ethnobotany was first used by an American botanist Johan W. Harshberger in 1896 "to the study of plants used by primitive and aboriginal people" (Ahmad et al. 2019). It focuses on the multifaceted and close relationships between the locals and plants, as well as the ecosystems they inhabit (McClatchey et al. 2009). Ethnobotanical knowledge was introduced by Dr. Janaki Ammal (Indian botanist) in the official program of the Economic-Botany Botanical Survey of India (Kolkata) in 1954. Dr. S. K. Jain (an Indian ethnobotanist) started a rigorous field survey among the tribal areas in middle India in 1960 (Sharma and Jain 2021). From prehistoric times, people belonging to different cultures have used and perceived local plants for various purposes such as medicine, food, cosmetics, fuel, agriculture, rituals, and ceremonies (Kumar Srivastava 2018).

Ethnobotany is the scientific study or field of research of indigenous knowledge and cultural value of plants prevailing in many different cultures. Thus, it encompasses the cultural significance of plants, highlighting their roles in daily life. Indigenous societies have a deep understanding of the local flora and recognize the unique qualities and potential uses of each plant species (Ndhlovu et al. 2023). They have mastered the art of using plants for their therapeutic properties and other uses. Ethnobotany is not limited to the medical realm; it also focuses on plant collection, documentation, and analysis. This relationship promotes the preservation of traditional knowledge and helps to conserve plant diversity (Obakiro et al. 2020).

8.1.2 Medical Ethnopharmacology

Medical ethnopharmacology, on the other hand, is a subfield of ethnopharmacology that focuses on the therapeutic potential of plants (Das et al. 2022). It explores the traditional and indigenous knowledge of the use of medicinal plants and the medical purposes of active constituents isolated from them which are used in folk medicine (Bouafia et al. 2021). Medical ethnopharmacologists strive to bridge the gap between traditional medicine and modern science by assessing the safety and efficacy of traditional medicines and identifying prospective drug development.

8.1.3 Dynamic Interactions Between Ethnobotany and Medical Ethnopharmacology

Ethnobotany and medical ethnopharmacology interact dynamically in a way like a two-way bridge (Pardo-De-Santayana et al. 2015). One side of the coin (ethnobotany) plays an important role in cultivation practices, conservation, and sustainable use of medicinal plants for their continued availability. Moreover, such practices help in safeguarding indigenous cultures and their unique insights into plant-based medicine. The other side of the coin is the wealth of information about medicinal plants that have been conserved by traditional healers and indigenous people. This historical knowledge serves as the basis for ethnobotanical study and is a useful tool for identifying possible therapeutic targets and directing contemporary pharmacological studies (Leonti et al. 2020). In contrast, modern science supports ethnobotanical practices with rigor, validation, and modern analytical methods (Saini et al. 2023). It extracts valuable pharmacological ingredients from medicinal plants and aids in confirming the safety and effectiveness of conventional treatments (Malik et al. 2023). By working together, traditional knowledge is preserved and incorporated into modern treatment (Lone et al. 2015; Shah and Bhat 2019). Few examples are summarized in Table 8.1.

Table 8.1 List of plants with ethnobotanical and ethnopharmacological uses

Botanical name	Common name	Ethnobotanical use	Ethnopharmacological use	References
Allium sativum	Garlic	Treatment of infections, high blood pressure, and high cholesterol	Antibacterial, antiviral, antihyperlipidemic activity	Ansary et al. (2020), Bayan et al. (2014)
Aloe barbadensis	Aloe vera	Treatment of burns, wounds, and skin conditions	Anti-inflammatory, antimicrobial, wound healing activity	Nalimu et al. (2021)
Boswellia serrata	Indian frankincense	Treatment of arthritis and inflammation	Anti-inflammatory, analgesic activity	Siddiqui (2011), Chaudhary and Kumari (2021)
Echinacea purpurea	Purple coneflower	Treatment of the common cold and flu	Immunomodulatory, antiviral activity	Burlou-Nagy et al. (2022), de Oliveira et al. (2022)
Ginkgo biloba	Ginkgo	Treatment of Alzheimer's disease and memory loss	Neuroprotective, antioxidant activity	Singh et al. (2019)
Panax ginseng	Ginseng	Treatment of fatigue, stress, and erectile dysfunction	Adaptogenic, immunomodulatory, aphrodisiac activity	Jin et al. (2019), Sengupta and Dutta (2022)
Silybum marianum	Milk thistle	Treatment of liver disease	Antioxidant, hepatoprotective activity	Jiang et al. (2022)
Asparagus racemosus	Satawar	Weakness, lactation in women, spermatorrhea, pain	Immunomodulator, galactagogue effect, aphrodisiac activity, antidiarrheal effect	Kaushik et al. (2023), Shaha and Bellankimath (2017)

Table 8.1 (continued)

Botanical name	Common name	Ethnobotanical use	Ethnopharmacological use	References
Ficus benghalensis L.	Banyan	Asthma, diabetes, ulcer protection, leprosy treatment	Antidiabetic, anti-inflammatory, wound healing, antidepressant effects, antioxidant	Logesh et al. (2023), Murugesu et al. (2021)
Piper longum L.	Laung	Cough, asthma, fever, indigestion, cold, toothache, headache	Stomachic, anti-asthmatic, antibronchitis, aphrodisiac	Gani et al. (2019)
Rauvolfia serpentina L.	Indian snakeroot	Headache, fever, malaria, snakebite, joint pain, mental disease	Antipsychotic, antihypertensive	Kumar et al. (2022), Abnabi (2019), Mukherjee et al. (2019)
Solanum nigrum L.	Black nightshade	Skin diseases, laxative, ulcers, asthma, sore throat, toothache, dermatitis	Acute nephritis, cancers, urethritis, leukorrhea	Keerthana et al. (2022), Eskandari et al. (2019)
Camellia sinensis	Green tea	Refreshing and stimulating effect	Antioxidant	Chao et al. (2022)
Curcuma longa	Turmeric	Used as spice, inflammation, joint pain	Anti-inflammatory, antioxidant	Srivastava et al. (2022), Fuloria et al. (2022)

8.1.4 Value of Traditional Knowledge

Herbal therapy has a deep cultural and historical association with many cultures. Many indigenous groups have kept their age-old medicinal procedures alive by passing down knowledge from generation to generation (Kachmar et al. 2021). Herbal therapy is becoming more popular as a method of treatment as people want to embrace holistic wellness and regain their ancestral roots. Furthermore, recent scientific findings have shed light on the pharmacological properties of a variety of therapeutic plants (Aye et al. 2019). The traditional medicinal plant-based knowledge from different communities or tribals forms the foundation stone for medical ethnopharmacological research. The bioactive compounds isolated from medicinal plants are evaluated for their safety and efficacy (Kebede et al. 2021). It incorporates the accumulated knowledge of many centuries and offers a holistic method of healing that considers not only one's physical health but also the cultural and spiritual facets of well-being (Courric et al. 2023). Finding novel possibilities for phytopharmaceutical research, traditional knowledge acts as an inspiration source for drug discovery (El-Amir et al. 2022).

The purpose of this chapter is to emphasize the importance of traditional knowledge for modern science. It emphasizes the requirement to respect and recognize indigenous populations' contributions to herbal medicine. By establishing a mutually beneficial interaction between medical ethnopharmacology and ethnobotany, doors are open to the possibility of finding new treatments, strengthening healthcare, and promoting cultural understanding. Additionally, this chapter will include specific examples of medicinal plants, their traditional uses, as well as their pharmacological characteristics, and their prospective current medical applications. Thus, medical ethnopharmacology and ethnobotany are two disciplines that focus on the investigation of conventional medical practices and the application of plant-based treatments for an array of ailments.

8.2 Roots of Traditional Medicine

The plant-based healing practices are the oldest form of medicine and an important part of the healthcare system for treating various human ailments and best due to fewer side effects (Gaonkar and Hullatti 2020). Herbal medicines have been used for thousands of years in cultures all over the world, including traditional Chinese medicine (TCM), Ayurveda in India, and Native American healing techniques in the modern period (Dong 2013; Yuan et al. 2016). Prehistoric periods, like Rigveda (4000–1500 BC), Atharvaveda (1500 BC), Upanishads (1000–600 BC), and Mahabharata and Puranas (700–400 BC), have claimed the use of plant species for medicinal purposes (Sharma and Jain 2021). In India, Ayurveda has been practiced for more than 5000 years. *Charaka Samhita* and the *Sushruta Samhita*, the popular Ayurvedic literature have explained how herbal medicines treat a variety of health problems, enhance well-being, and restore harmony and balance (Yao et al. 2023). Enormous indigenous knowledge of medicinal plants exists in different cultures. Different cultures all over the world have explored and used the therapeutic benefits of medicinal plants. Different parts of medicinal plants have been extensively utilized as herbal remedies since ancient times by tribals or inhabitants residing in different parts of the world.

8.3 Integrating Ethnobotany and Medical Ethnopharmacology across Different Cultures

Various regions around the globe are endowed with rich sources of medicinal plant biodiversity having a diverse range of cultivated and wild plant species.

8.3.1 Indigenous Communities: Keepers of their Belief Systems

Based on Jose R. Martinez Cobo's study Indigenous communities, and peoples, are those that perceive themselves as distinctive from other groups of the civilizations that currently rule those territories or parts of them because of their heritage with pre-invasion and pre-colonial civilization that originated on their territory. Following their cultural patterns, social institutions, and legal system, they are currently non-dominant sectors of society that are utterly convinced about conserving, developing, and passing on to subsequent generations their ancestral lands and cultural beliefs as the basis of their continued existence as people (Mohd Salim et al. 2023). Indigenous people are groups belonging to different cultures. They are inextricably related to their own beliefs and upholding social characteristics owned to their understanding and cultural beliefs and experiences related to traditional ecological knowledge inherited from their natural surroundings (Mohamed Shaffril et al. 2020). The rituals, practices, and beliefs are important to their way of life, and thus, such groups are considered as the keeper of their belief systems.

Different tribal groups or indigenous communities usually reside in various remote areas of the world and depend solely on natural resources (wild plants) primarily for their livelihoods and survival, particularly for food and medicine. Such areas are rich in cultural heritage. With time, they develop a special bond with their surrounding environment. As such groups have been living in their area for a long time; they exhibit a deep understanding and knowledge of their surrounding native flora. Moreover, they possess invaluable knowledge regarding sustainable practices for the management of natural resources. Currently, the estimated total population of indigenous people worldwide is 476 million. They have shown different ethnobotanical uses of plants in their natural surroundings. Most of the traditional healers practice in rural and forest areas. Furthermore, tribals utilize wild medicinal plants for the treatment of various ailments or make their use in ethnomedicinal preparations (Sen et al. 2011).

8.3.2 Unveiling Ancient Wisdom Via Ethnobotanical and Ethnopharmacological Studies

Ethnobotanists work closely with indigenous and local communities. Exploring the relationship between people and plants through ethnobotany allows us to appreciate the deep-rooted connections that have shaped societies throughout history (Gahamanyi et al. 2021; Leonti 2022). They are involved in learning directly from their cultural traditional knowledge and practices about plants, including their uses, cultivation, conservation, and cultural importance (Gonfa et al. 2020). They explore how different civilizations and communities connect with plants for spiritual, cultural, and medical reasons and have relied on nature for sustenance and healing over the centuries (Pathak 2020). This field of study, which merges elements of anthropology and botany, investigates the deep-rooted link between indigenous cultures and the plant kingdom. Ethnobotany, dating back thousands of years, has its origins in ancient civilizations that recognized the medicinal and cultural value of plants (Kalle et al. 2022). From the ancient Egyptians who considered plants as sacred entities to the indigenous tribes of the Amazon rainforest who possess a profound knowledge of the healing properties of their surroundings, the history of ethnobotany is as varied as the cultures that practiced it (Obakiro et al. 2020). They observed the behavior of animals and learned from their interactions with various plants, gradually accumulating knowledge about their properties and potential uses (Katara et al. 2023). Ethnobotanical and ethnopharmacological data are collected by interviews and discussions with tribal healers. In earlier times, the practitioners and healers passed their knowledge verbally based on trial-and-error methods and empirical observations. The only literature available is in the form of vernacular trivial ethnographic publications (Petkeviciute et al. 2010). Ethnopharmacological studies shed light on the efficacy and safety of these traditional remedies, offering a connection between ancient wisdom and modern medicine (Pullaiah et al. 2016).

The knowledge ingrained in different ethnic groups and cultures and intricately linked to the cultural identities of ethnic groups is passed down through generations connecting indigenous relationships with heritage, beliefs, and livelihood and forming the basis of traditional medicine systems that still exist today (Jütte et al. 2017). Table 8.1 shows ethnobotanical and ethnopharmacological uses of some of the common medicinal plants.

8.3.3 Importance of Ethnobotanical and Medical Ethnopharmacological Studies

This kind of research is noteworthy, and the data collected by ethnobotanical and ethnopharmacological studies is of paramount importance as it helps in safeguarding the traditional knowledge possessed by indigenous groups. Through such studies, the history of herbal medicine can be traced back. The preservation of this knowledge guarantees that these communities can preserve their cultural history, which is essential for their self-esteem and togetherness.

Over millennia, indigenous cultures have established sustainable practices that allow them coexist harmoniously with natural surroundings (Shaheen et al. 2023). This includes understanding which plants can be harvested when they can be harvested, and in what quantities without causing harm to the environment. It is critical to document these sustainable practices through ethnobotanical research to maintain ecosystems and biodiversity. Medical ethnopharmacology and ethnobotany can aid in the development of more environment-friendly ways to utilize plant resources (Tresca et al. 2020).

Documenting traditional knowledge through ethnobotanical research can also help protect the intellectual property rights of indigenous communities (Goel and Srikanth 2023). This is particularly important when their knowledge is at risk of being exploited or patented without their consent. The traditional system of medicines uses medicinal plant species belonging to different taxonomic groups, where whole plants, plant parts, or extracts are used. Living in surroundings rich in biodiversity and utilizing plants for various human ailments, indigenous tribes developed extensive knowledge of the medicinal characteristics of plants. This understanding has the potential for the development of new therapeutic drugs (Porras et al. 2021). Traditional plant-based therapies have given rise to many pharmaceutical drugs.

Ethnobotanical study leads to the discovery of new medicinal compounds having potent therapeutic potential for further use in modern medicine (Lautié et al. 2020). This is called ethnopharmacology, the field which highlights the importance of traditional knowledge in modern medicine. Ethnobotanical research provides a link between traditional practices and modern science. It enables researchers to scientifically validate the traditional knowledge and claims exhibited by indigenous groups regarding the efficacy of medicinal plants (Packer et al. 2019). The significance of indigenous knowledge in medical ethnopharmacology lies in its potential to enhance modern medicine. Scientific research often validates the effectiveness of traditional remedies discovered by indigenous communities, leading to the development of new drugs and therapies (Pirintsos et al. 2022). Traditional healers consider not only the symptoms but also the underlying causes of diseases, addressing imbalances within the body and the surrounding environment. This comprehensive approach aligns with the growing demand for personalized and integrative medicine in today's world.

8.4 Integration of Traditional Knowledge with Modern Science for Holistic Healthcare

Medicinal plants have been working as a panacea since ancient times till date. According to the World Health Organization (WHO), traditional herbal medicine serves more than 80% of the world's population's primary healthcare needs directly or indirectly (Shaheen et al. 2023). Plant-based medicines have rationalized the use of therapeutic agents to alleviate human agony (Kumar et al. 2018). Herbal remedies, an essential part of alternative and complementary medicine, are known to be the safest approach due to fewer side effects on human health. There has been a considerable increase in the demand for and consumption of herbal drugs and products in recent years, resulting in the expansion of the natural product industry on a global scale (Bommakanti et al. 2023).

With time, the traditional plant-based remedies followed by different indigenous groups and cultures across the world have played a vital role and foundation in the modern healthcare system through drug development and discovery. Indigenous people follow the practice of using plant-based remedies without knowing any scientific reason (Saha et al. 2022).

Ethnobotanical and medical ethnopharmacological research studies are the important key players in confluence of the traditional knowledge/wisdom with modern medicine (contemporary science). Ethnopharmacologists collect and document data from indigenous groups and try to find active phytoconstituents posing therapeutic benefits. It helps us to better understand how indigenous people associated with plants, and with their tremendous traditional knowledge, novel medications, and treatments, can be developed using this information derived from centuries of experience (Suárez 2019; Tresca et al. 2020). Only a small portion of the world's estimated 300,000–500,000 plant species have had their therapeutic potential investigated.

The goal of ethnopharmacological research is to determine the specific phytoconstituents showing the medicinal effect by isolating and characterizing them using sophisticated analytical techniques. In some cases, active phytoconstituents are isolated from medicinal plants and synthesized in a laboratory to form standardized doses. Further, the isolated constituents are screened for their biological efficacy through in vitro and in vivo experiments. During clinical trials, they need to follow strict ethical guidelines and provide compelling proof of their efficacy and safety. Moreover, modification and optimization of phytochemicals from plants can lead to the development of more effective and safer drugs. Also, traditional plant-based medicines can provide therapies for treating diseases for which modern remedies are limited or nonexistent.

With the growing interest in herbal medicines, pharmacological studies are conducted and focused by researchers and health professionals to scientifically validate, uncover, and comprehend underlying mechanisms of action and provide evidence for the safety and efficacy of traditional remedies used by ethnic groups. Modern scientific methods study the pharmacological properties of active constituents and their mechanism of action against diseases and disorders. Compounds with intriguing bioactivity are developed into potential medicines leading to the emergence of pristine pharmaceuticals. Computational drug development has been increasingly important in recent decades, owing to lower risks, time, cost-effectiveness, and resources when compared to traditional experimental procedures (Pirintsos et al. 2022). The artificial intelligence and availability of collected data resulted in computational drug discovery with special influence on natural products and novel drug compound identification.

It is essential to have collaboration among various traditional healers and practitioners, healthcare professionals, and researchers to incorporate herbal medicine into modern healthcare. By collaborating with indigenous healers and incorporating their knowledge, researchers can tap into a vast resource of untapped potential (Atanasov et al. 2021). Moreover, indigenous knowledge promotes a holistic approach to healthcare that recognizes the interconnectedness of the physical, mental, and spiritual aspects of healing. Here comes the major role of ethnobotanist's research to preserve and record the usage of therapeutic plants in indigenous cultures. Researchers from the scientific field and ethnobotanists can collaborate to bridge the knowledge gap between the two. We embrace both the practical knowledge of conventional healing systems and the rigors of contemporary scientific study.

8.4.1 Success Stories in Utilizing Ethnobotanical Knowledge for Modern Medicine

There are numerous case studies of ethnobotanical and medical ethnopharmacological research leading to the creation of new medications and treatments.

8.4.1.1 Case Study I

The Amazon rainforest's native *Cinchona officinalis* L. served as the initial source of the antimalarial medication *quinine*. The traditional knowledge of indigenous peoples who used the bark of the cinchona tree to treat malaria served as the foundation for the discovery of quinine named by *Pierre Joseph Pelletier* and *Joseph Caventou* (Gachelin et al. 2017). Spanish conquistadors introduced quinine to Europe in the sixteenth century. *Quinine* swiftly replaced other malaria treatments as the go-to option and helped countless people. One of the most significant antimalarial medications accessible today is quinine which is still used to treat malaria.

8.4.1.2 Case Study II

The cancer medication *paclitaxel*, which comes from the Pacific yew tree or *Taxus brevifolia*, is a further example. As part of a US government program to test plants for potential anticancer action, *paclitaxel* was found in the 1960s. The American Indians who employed the *T. brevifolia* tree's bark to treat a range of illnesses, including cancer, served as the basis for the discovery of *paclitaxel* (Wang and Du 2018). It is used to treat a few cancers, such as lung, ovarian, and breast cancer. *Paclitaxel* has saved the lives of countless people, and it is a testament to the power of ethnobotanical and medical ethnopharmacological research.

8.4.1.3 Case Study III

A Chinese native plant called sweet wormwood or *Artemisia annua* is the source of the antimalarial drug artemisinin. Chinese scientist *Tu Youyou*, who was researching traditional Chinese medicine, made the discovery of *artemisinin* in the 1970s. *Tu Youyou* discovered from ancient Chinese medical writings that *Artemisia annua* had been used for generations to treat malaria. She started to extract substances from the *A. annua* plant and test them in the lab for anti-plasmodial action. She ultimately determined that the most active substance was *artemisinin* (Septembre-Malaterre et al. 2020). The most significant antimalarial medication in the world now is *artemisinin*. Millions of lives have been spared, thanks to its usage in treating malaria in more than 80 countries. The most efficient form of treatment for malaria is thought to be artemisinin-based combination treatments (ACTs). She also received the *Nobel Prize* in Physiology/Medicine for her discovery (Ekiert et al. 2021; Media 2015).

8.4.1.4 Case Study IV

Cancer medications vincristine and vinblastine are made from the *Catharanthus roseus* (periwinkle), which is indigenous to Madagascar. A group of researchers from the National Cancer Institute (NCI) made the discoveries of vincristine and vinblastine in the 1960s (Škubník et al. 2021). The researchers were looking for potential anticancer properties in plants from all around the world. They discovered that the *C. roseus* plant had been utilized in Madagascar by traditional healers to cure various illnesses, including cancer. The *C. roseus* plant's extracts were examined in the lab for anticancer properties by the researchers. They ultimately determined that vincristine and vinblastine were the most effective substances. Today, diseases, including pediatric leukemia, lymphoma, and testicular cancer, are treated with vincristine and vinblastine. Many kids and adults have had their lives saved by them (Kumar 2016; Moha 2017).

8.4.1.5 Case Study V

One of the most used painkillers in the world is aspirin. In the nineteenth century, it was initially isolated from the willow tree or *Salix alba*. But before it was isolated from the willow tree, *aspirin* had been used for generations to alleviate pain and inflammation. Willow bark was utilized by the ancient Egyptians to cure inflammation and pain. *S. alba* was also utilized as a treatment for several illnesses by the Greeks and Romans. A treatise on the use of willow bark to alleviate pain and fever was written in the eighteenth century by British physician Edward Stone. The use of *S. alba* bark for therapeutic purposes became more widespread in Stone's paper (Abdelaziz et al. 2019; Desborough and Keeling 2017).

German chemist Felix Hoffmann extracted salicylic acid from willow bark in the nineteenth century. *Aspirin's* active component is salicylic acid. Hoffmann also created a method for producing salicylic acid, which allowed for the mass production of *aspirin*. It is used to treat several ailments, such as fever, inflammation, and discomfort. Additionally, *aspirin* has been demonstrated to lower the risk of heart attack and stroke (Moon 2022).

8.4.1.6 Case Study VI

The *Digitalis purpurea*, a native of Europe, is the source of the cardiac medicine *digitalis*. In the eighteenth century, English physician *William Withering* discovered digitalis. Traditional healers in England informed Withering that heart illness had long been treated using the foxglove plant (Bessen 1986). *D. purpurea* plant constituents were isolated by withering and then examined in a lab to see how they affected the heart. He ultimately determined that the most active substance was digitalis. Congestive heart failure and atrial fibrillation are two diseases that are now treated with digitalis. It is a cardiac drug that has been around for a long time and is very effective (Kreis 2017; Maes et al. 2022).

8.4.1.7 Case Study VII

The *Rauwolfia serpentina* plant, which is indigenous to India, is the source of the hypertension medication *reserpine* (Jahan et al. 2022). A group of NCI scientists made the discovery of reserpine in the 1950s. The researchers were looking for potential antihypertensive properties in plants from all around the world. They discovered that the *R. serpentina* plant had been utilized for millennia to treat high blood pressure by local healers in India. The chemicals were taken out of the *Rauwolfia serpentina* and put to the test in a lab for their ability to lower blood pressure. The *Rauwolfia serpentina* dried root, which had been used for generations in India to cure fever, mania, and snakebites, was where it was originally isolated in 1952 (Khan et al. 2018).

8.4.1.8 Case Study VIII

The European native *Galanthus nivalis* plant is the source of the medication galantamine, which is used to treat Alzheimer's disease (Naguib et al. 2020). A group of researchers from the Bulgarian Academy of Sciences made the discovery of galantamine in the 1950s. The researchers were looking for possible anticholinesterase action in plants from all around the world. Acetylcholine, a neurotransmitter that plays a role in memory and learning, is broken down by the enzyme acetylcholinesterase (Metz and Pavlov 2021). The Bulgarian traditional healers told the scientists that memory loss had long been treated with the *Galanthus nivalis*. The chemicals were taken out of the *G. nivalis* plant and evaluated in the lab for anticholinesterase action. *Galantamine* was ultimately found to be the most active substance (Fadhlullah et al. 2021). Today, galantamine is used to treat Alzheimer's disease in mild to moderate cases.

8.4.1.9 Case Study IX

Plants, including *Berberis aquifolium* and *Berberis vulgaris*, contain the alkaloid *berberine* (Samadi et al. 2020). Traditional Chinese medicine has long utilized berberine to treat a range of illnesses, including infections, diarrhea, and dysentery. The therapeutic potential of berberine has recently attracted the attention of researchers. There are several possible health advantages of berberine, including anti-inflammatory properties, anti-microbiological action, antidiabetic action, and anti-cancer properties. Currently, type 2 diabetes, Alzheimer's disease, and cancer are among the conditions for which berberine is being researched as a potential treatment (Imenshahidi and Hosseinzadeh 2016; Nuralın and Gürü 2022).

8.4.1.10 Case Study X

Turmeric (*Curcuma longa* L.), a spice that is frequently used in Indian cooking, has a substance called *curcumin*. Since ancient times, traditional Indian medicine has utilized *C. longa* to treat a wide range of conditions, including infections, inflammation, and discomfort. The therapeutic potential of curcumin has recently attracted the attention of researchers (Ayati et al. 2019; Singh et al. 2020b). Numerous possible health advantages of *curcumin* have been demonstrated, including inflammatory-reduction capacity, antioxidant function, anticancer properties, and a neuroprotective effect. Disorders, such as cancer, Alzheimer's disease, and arthritis, are currently being studied concerning curcumin as a potential therapeutic (Hewlings and Kalman 2017).

The case studies presented here emphasize the importance of collaboration among scientists, researchers, and indigenous communities to uncover the healing properties hidden within nature. By acknowledging and respecting the traditional knowledge passed down through generations, we can access a vast reservoir of potential medicinal resources. As our knowledge of ethnobotany and medical ethnopharmacology grows, it is crucial to recognize the significance of preserving traditional plant knowledge and the ecosystems that support these invaluable resources. The success stories shared here serve as an inspiration and remind us of the rich heritage we have yet to fully explore in our ongoing pursuit of better healthcare solutions. Thus, with the help of ethnopharmacological research, traditional remedies are integrated with modern healthcare, thus representing a significant interplay between cultural knowledge and scientific progress. This approach broadens the spectrum of therapeutic possibilities. Moreover, it also respects indigenous culture's heritage and wisdom for the development of evidence-based medicine.

8.5 Uncovering Ecological Wisdom and Cultural Preservation through Ethnobotany

The widespread utilization and importance of some of the commercially valuable medicinal plant species have led to their ruthless collection (unsustainable collection practices) and overexploitation by plant gatherers and traders, consequently leading to loss of biodiversity and a decline in their population to fulfill the increase in herbal drug demand (Targe et al. 2022). Market demand-driven harvesting occurs without any concern about the conservation of natural flora. This problem mainly arises because of a lack of knowledge about forest management. Due to exponential growth in human population and urbanization, a rapid loss in natural habitats occurs leading to loss of diversity and risk of local medicinal plant extinctions, making use of medicinal plants in an unsustainable way. Even the altered patterns of climate parameters can disrupt medicinal plants (Rana et al. 2020). Thus, with suitable conservation methods, the preservation of natural habitats is important.

Indigenous groups are in close association with their surrounding environment. Ethnobotanical studies document detailed knowledge of local ecosystems and the interactions between plant species as people residing in these areas have a thorough awareness of plant ecology, including species interdependence and their responsibilities in preserving ecological equilibrium. Indigenous groups are aware of sustainable harvesting practices to ensure the continuous availability of valuable plants (Dasgupta et al. 2023; Singh et al. 2020a; Priyadarshini and Abhilash 2019). Such practices in today's time will help in conservation and sustainable resource management by providing valuable insights into sustainable farming practices. With time they have developed sustainable agriculture methods to preserve their surroundings for the future.

The World Bank has initiated many programs for indigenous groups to increase their participation awareness. With government agencies, they have conducted projects to help ethnic groups. Implementation of sustainable harvesting methods by promoting cultivation and certification programs will help alleviate the burden on the wild plant population (Liu et al. 2021). Various in situ (natural reserves, nurseries) and ex situ (seed banks, botanical gardens) conservation strategies have also been developed for the conservation of plant diversity.

Overexploitation and habitat loss result in loss of many plant species. With time, even the indigenous knowledge about plants is disappearing. To conserve biodiversity and for sustainable and equitable use of plants, quantitative ethnobotany comes into role. The importance of ethnobotanical studies is enhanced by incorporating quantitative methods for the collection, processing, and interpretation of ethnobotanical data. In 1987, the concept of quantitative ethnobotany was coined by Prance and co-workers. It is defined as "the application of quantitative techniques to the direct analysis of contemporary plant use data" (Hussain et al. 2018).

8.6 Integration and Bridging the Gap between Ethnobotany and Ethnopharmacology

By combining indigenous knowledge with modern scientific methods, innovative and sustainable treatments can be developed. Bioinformatics plays a crucial role in advancing natural product research, drug discovery, and ethnobotany by integrating omics technologies with traditional knowledge. The synergy between omics technologies, such as genomics, proteomics, transcriptomics, and metabolomics, enhances the understanding of plant-derived natural products for drug development, thereby contributing to modern drug discovery (Pirintsos et al. 2022). The omics technology has significantly advanced ethnobotany and medical ethnopharmacology by identifying bioactive compounds from plants. It uses pharmacophylogenomics to identify the genes responsible for producing bioactive compounds, ethno-directed screening to screen plant extracts based on their traditional use in medicine, and computational chemistry to predict the biological activity of compounds (Fig. 8.1).

Additionally, the combination of modern technologies with traditional ethnobotanical knowledge can help elucidate the phylogenetic relationships between plant species, facilitating the identification of potential bioactive compounds (Gras et al. 2021). These bioinformatics-assisted omics studies provide a comprehensive approach to drug discovery and bioprospecting by developing new drugs from natural sources (Leonti 2022). The application of bioinformatics in the context of ethnobotany and natural product research has significant reciprocal benefits by bridging the gap between traditional and modern science (Sharma and Sarkar 2013).

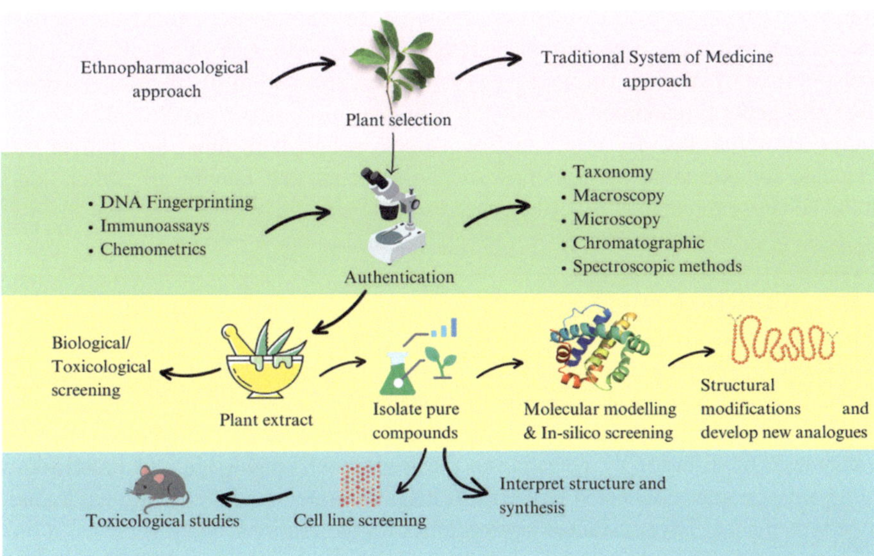

Fig. 8.1 Approaches from botanical sources to modern drug discovery and development process

8.7 Challenges and Ethical Considerations in Ethnobotanical Research

Ethnobotanical research is an intriguing field that explores the relationship between plants and cultures. However, like any scientific pursuit, it has its challenges and ethical considerations. Understanding these challenges is critical to conducting responsible and meaningful research in ethnobotany and medical ethnopharmacology.

One of the primary challenges in ethnobotanical research is preserving indigenous knowledge and cultural heritage. Many traditional healing practices and plant uses have been passed down through generations within specific communities (Casas et al. 2023). When conducting research, it is essential to approach communities with respect, ensuring their participation and consent. Researchers must establish trusting relationships and acknowledge the intellectual property rights of indigenous peoples.

Another significant challenge is ensuring the sustainable use of plant resources. Ethnobotanical research often highlights the medicinal properties of various plants, leading to increased demand. However, overexploitation can endanger plant species and disrupt delicate ecosystems. Researchers must strive for sustainable practices, promoting conservation and collaborating with local communities to ensure responsible harvesting and cultivation techniques (Eldeen et al. 2016).

When it comes to sharing research findings, ethical considerations are of utmost importance. The open-access movement has made it easier to spread knowledge, but it should be done responsibly. Researchers must maintain a balance between sharing valuable information and protecting the intellectual property rights of indigenous communities (Dapar and Alejandro 2020). Prior informed consent, proper attribution, and collaborative agreements are essential for maintaining ethical standards in the field. In addition, ethnobotanical research carries the risk of biopiracy, which poses ethical dilemmas. Biopiracy refers to the exploitation of traditional knowledge and genetic resources without fair compensation or benefit-sharing with the source communities. To avoid such practices, researchers must be vigilant and advocate for equitable partnerships and agreements that benefit all stakeholders involved (Puspangadan et al. 2016).

8.8 Conclusion

Medical ethnopharmacology and ethnobotany are topics that are expanding quickly and have the potential to significantly improve human health and well-being. We can find new sources of medicinal plants and create novel medications and therapies for a wide range of diseases by researching the traditional knowledge and practices of indigenous peoples connected to plants. It helps us understand the traditional knowledge systems that have sustained communities for generations, and it serves as a link between traditional practices and modern medicine. By exploring the healing properties of plants, we can gain a greater appreciation for our shared human heritage and the importance of preserving and respecting the natural world. The research focuses on developing new drugs and therapies using medicinal plants and further enhancing our knowledge of the molecular mechanisms of action of plant-based medicines. It also studies the interactions between plant-based medicines and the human body to identify any potential side effects and drug interactions. However, navigating the inherent challenges and ethical considerations in this field is essential. By conducting research responsibly, respecting indigenous knowledge, promoting sustainability, and ensuring fair benefit-sharing, we can unlock the potential of ethnobotany and medical ethnopharmacology while preserving cultural heritage and biodiversity. This can help develop personalized treatment plans for patients, ensuring that the medication is tailored to their individual needs. Ethnobotany and medical ethnopharmacology have the potential to make significant contributions to human health and well-being in the coming years. Through enhanced collaboration between researchers and indigenous communities, advancements in plant discovery and drug development technologies, and a greater appreciation for traditional knowledge, we can anticipate the emergence of novel drugs and therapies in this field.

References

Abdelaziz HK, Saad M, Pothineni NVK, Megaly M, Potluri R, Saleh M, Kon DLC, Roberts DH, Bhatt DL, Aronow HD, Abbott JD, Mehta JL (2019) Aspirin for primary prevention of cardiovascular events. J Am Coll Cardiol. https://doi.org/10.1016/j.jacc.2019.03.501

Abnabi MN (2019) Therapeutics, phytochemistry and pharmacology of Rauvolfia Serpentina (Asrol): an important Unani drug. Int J Sci Res, ISSN: 2319–7064

Ahmad I, Ahmad MSA, Hussain M, Hameed M (2019) Ethnobotanical aspects of some traditional medicinal plants:1–19. https://doi.org/10.1007/978-981-13-9566-6_1

Ahmad S, Katiyar CK, Ulrich-Merzenich GS, Mukherjee PK (2022) Editorial: metabolomics and Ethnopharmacology in the development of herbal and traditional medicine. Front Pharmacol 13:851023. https://doi.org/10.3389/FPHAR.2022.851023/BIBTEX

Ansary J, Forbes-Hernández TY, Gil E, Cianciosi D, Zhang J, Elexpuru-Zabaleta M, Simal-Gandara J, Giampieri F, Battino M (2020) Potential health benefit of garlic based on human intervention studies: a brief overview. Antioxidants. https://doi.org/10.3390/antiox9070619

Atanasov AG, Zotchev SB, Dirsch VM, Orhan IE, Banach M, Rollinger JM, Barreca D, Weckwerth W, Bauer R, Bayer EA, Majeed M, Bishayee A, Bochkov V, Bonn GK, Braidy N, Bucar F, Cifuentes A, D'Onofrio G, Bodkin M et al (2021) Natural products in drug discovery: advances and opportunities. Nat Rev Drug Discov 20(3):200–216. https://doi.org/10.1038/s41573-020-00114-z

Ayati Z, Ramezani M, Amiri MS, Moghadam AT, Rahimi H, Abdollahzade A, Sahebkar A, Emami SA (2019) Ethnobotany, phytochemistry and traditional uses of curcuma spp. and pharmacological profile of two important species (C. longa and C. zedoaria): a review. Curr Pharm Des. https://doi.org/10.2174/1381612825666190402163940

Aye MM, Aung HT, Sein MM, Armijos C (2019) A review on the phytochemistry, medicinal properties and pharmacological activities of 15 selected Myanmar medicinal plants. Molecules 24:2. https://doi.org/10.3390/MOLECULES24020293

Bayan L, Koulivand PH, Gorji A (2014) Garlic: a review of potential therapeutic effects. Avicenna J Phytomed 4:1

Bessen HA (1986) Therapeutic and toxic effects of digitalis: William Withering, 1785. J Emerg Med. https://doi.org/10.1016/0736-4679(86)90048-X

Bommakanti V, Puthenparambil Ajikumar A, Sivi CM, Prakash G, Mundanat AS, Ahmad F, Haque S, Prieto MA, Rana SS (2023) An overview of herbal nutraceuticals, their extraction, formulation, therapeutic effects and potential toxicity. Separations 10(3):177. https://doi.org/10.3390/SEPARATIONS10030177

Bouafia M, Amamou F, Gherib M, Benaissa M, Azzi R, Nemmiche S (2021) Ethnobotanical and ethnomedicinal analysis of wild medicinal plants traditionally used in Naâma, Southwest Algeria. Vegetos 34(3):654–662. https://doi.org/10.1007/S42535-021-00229-7/TABLES/2

Burlou-Nagy C, Bănică F, Jurca T, Vicaş LG, Marian E, Muresan ME, Bácskay I, Kiss R, Fehér P, Pallag A (2022) Echinacea purpurea (L.) Moench: biological and pharmacological properties—a review. Plan Theory. https://doi.org/10.3390/plants11091244

Casas A, Blancas J, Vibrans H (2023) Perspectives of the ethnobotanical research in Mexico. https://doi.org/10.1007/978-3-319-77089-5_57-1

Chao J, Chen TY, Pao LH, Deng JS, Cheng YC, Su SY, Huang SS (2022) Ethnobotanical survey on bitter tea in Taiwan. Front Pharmacol. https://doi.org/10.3389/fphar.2022.816029

Chaudhary G, Kumari I (2021) Boswellia serrata ROXB. EX COLEBR. (Salai): an Ayurvedic herb with anti-inflammatory potential. Int J Pharm Sci Rev Res. https://doi.org/10.47583/ijpsrr.2021.v69i01.024

Courric E, Brinvilier D, Couderc P, Ponce-Mora A, Méril-Mamert V, Sylvestre M, Pelage JH, Vaillant J, Rousteau A, Bejarano E, Cebrian-Torrejon G (2023) Medicinal plants and plant-based remedies in Grande-Terre: an ethnopharmacological approach. Plan Theory 12(3):654. https://doi.org/10.3390/PLANTS12030654

Dapar M, Alejandro G (2020) Ethnobotanical studies on indigenous communities in The Philippines: current status, challenges, recommendations and future perspectives. J Complement Med Res. https://doi.org/10.5455/jcmr.2020.11.01.51

Das SK, Das B, Jena AB, Pradhan C, Sahoo G, Dandapat J (2022) Therapeutic potential and ethnopharmacology of dominant mangroves of Bhitarkanika National Park, Odisha, India. Chem Biodivers 19(3):e202100857. https://doi.org/10.1002/CBDV.202100857

Dasgupta R, Dhyani S, Basu M, Kadaverugu R, Hashimoto S, Kumar P, Johnson BA, Takahashi Y, Mitra BK, Avtar R, Mitra P (2023) Exploring indigenous and local knowledge and practices (ILKPs) in traditional Jhum cultivation for localizing sustainable development goals (SDGs): a case study from Zunheboto District of Nagaland, India. Environ Manag 72(1):147–159. https://doi.org/10.1007/S00267-021-01514-6/METRICS

de Oliveira BG, Santos LFF, da Costa MC, Bastos RW, Do Carmo PHF, de Assis Santos D, Pianetti GA, César IC (2022) Antimicrobial and immunomodulatory activities of dried extracts of Echinacea Purpurea. Braz J Pharm Sci. https://doi.org/10.1590/s2175-97902022e21026

Desborough MJR, Keeling DM (2017) The aspirin story—from willow to wonder drug. Br J Haematol. https://doi.org/10.1111/bjh.14520

Dong J (2013) The relationship between traditional Chinese medicine and modern medicine. Evid Based Complement Alternat Med 2013. https://doi.org/10.1155/2013/153148

Ekiert H, Świątkowska J, Klin P, Rzepiela A, Szopa A (2021) Artemisia annua-importance in traditional medicine and current state of knowledge on the chemistry, biological activity and possible applications. Planta Med. https://doi.org/10.1055/a-1345-9528

El-Amir M, Hegazy F, Pare PW, Ibrahim MAA, Najmi A, Javed SA, Al Bratty M, Alhazmi HA (2022) Modern approaches in the discovery and development of plant-based natural products and their analogues as potential therapeutic agents. Molecules 27(2):349. https://doi.org/10.3390/MOLECULES27020349

Eldeen IMS, Effendy MAW, Tengku-Muhammad TS (2016) Ethnobotany: challenges and future perspectives. Res J Med Plant. https://doi.org/10.3923/rjmp.2016.382.387

Eskandari M, Assadi M, Shirzadian S, Mehregan I (2019) Ethnobotanical study and distribution of the solanum section solanum species (Solanaceae) in Iran. J Med Plants

Fadhlullah M, Prasetyati SB, Marleni NR, Mahmud (2021) The application of different heat processing technique on eel (Monopterus albus) galantine. IOP Conf Ser Earth Environ Sci. https://doi.org/10.1088/1755-1315/674/1/012003

Fuloria S, Mehta J, Chandel A, Sekar M, Rani NNIM, Begum MY, Subramaniyan V, Chidambaram K, Thangavelu L, Nordin R, Wu YS, Sathasivam KV, Lum PT, Meenakshi DU, Kumarasamy V, Azad AK, Fuloria NK (2022) A comprehensive review on the therapeutic potential of Curcuma longa Linn. in relation to its major active constituent curcumin. Front Pharmacol. https://doi.org/10.3389/fphar.2022.820806

Gachelin G, Garner P, Ferroni E, Tröhler U, Chalmers I (2017) Evaluating Cinchona bark and quinine for treating and preventing malaria. J R Soc Med. https://doi.org/10.1177/0141076816681421

Gahamanyi N, Munyaneza E, Dukuzimana E, Tuyiringire N, Pan CH, Komba EVG (2021) Ethnobotany, ethnopharmacology, and phytochemistry of medicinal plants used for treating human diarrheal cases in Rwanda: a review. Antibiotics. https://doi.org/10.3390/antibiotics10101231

Gani HMO, Hoq MO, Tamanna T (2019) Ethnomedicinal, phytochemical and pharmacological properties of Piper longum (Linn). Asian J Med Biol Res. https://doi.org/10.3329/ajmbr.v5i1.41038

Gaonkar VP, Hullatti K (2020) Indian traditional medicinal plants as a source of potent antidiabetic agents: a review. J Diabetes Metab Disord 19(2):1895–1908. https://doi.org/10.1007/S40200-020-00628-8/METRICS

Goel S, Srikanth N (2023) Local health traditions and preservation of national heritage: emphasis on intellectual property rights protection, benefit sharing and mainstreaming. 307–323. https://doi.org/10.1007/978-981-19-4286-0_20

Gonfa N, Tulu D, Hundera K, Raga D (2020) Ethnobotanical study of medicinal plants, its utilization, and conservation by indigenous people of Gera district, Ethiopia. Cogent Food Agric 6:1. https://doi.org/10.1080/23311932.2020.1852716

Gras A, Hidalgo O, D'ambrosio U, Parada M, Garnatje T, Vallès J (2021) The role of botanical families in medicinal ethnobotany: a phylogenetic perspective. Plan Theory. https://doi.org/10.3390/plants10010163

Hewlings SJ, Kalman DS (2017) Curcumin: a review of its effects on human health. Food Secur. https://doi.org/10.3390/foods6100092

Hussain W, Ullah M, Dastagir G, Badshah L (2018) Quantitative ethnobotanical appraisal of medicinal plants used by inhabitants of lower Kurram, Kurram agency, Pakistan. Avicenna J Phytomed 8(4):313. /pmc/articles/PMC6204146/

Imenshahidi M, Hosseinzadeh H (2016) Berberis vulgaris and berberine: an update review. Phytother Res. https://doi.org/10.1002/ptr.5693

Jahan N, Hussian F, Ayub AR, Ilyas M, Khan MA, Manzoor R, Khan WA (2022) Isolation and characterization of flavonoids from roots of Rauvolfia serpentina and evaluation of their hypotensive potential through angiotensin-converting enzyme (ACE) inhibition mode of action. Chem Pap. https://doi.org/10.1007/s11696-022-02238-5

Jiang G, Sun C, Wang X, Mei J, Li C, Zhan H, Liao Y, Zhu Y, Mao J (2022) Hepatoprotective mechanism of Silybum marianum on nonalcoholic fatty liver disease based on network pharmacology and experimental verification. Bioengineered. https://doi.org/10.1080/21655979.2022.2037374

Jin Y, Cui R, Zhao L, Fan J, Li B (2019) Mechanisms of Panax ginseng action as an antidepressant. Cell Prolif. https://doi.org/10.1111/cpr.12696

Jütte R, Heinrich M, Helmstädter A, Langhorst J, Meng G, Niebling W, Pommerening T, Trampisch HJ (2017) Herbal medicinal products—evidence and tradition from a historical perspective. J Ethnopharmacol. https://doi.org/10.1016/j.jep.2017.06.047

Kachmar MR, Naceiri Mrabti H, Bellahmar M, Ouahbi A, Haloui Z, El Badaoui K, Bouyahya A, Chakir S (2021) Traditional knowledge of medicinal plants used in the northeastern part of Morocco. Evid Based Complement Alternat Med 2021. https://doi.org/10.1155/2021/6002949

Kalle R, Pieroni A, Svanberg I, Sõukand R (2022) Early citizen science action in ethnobotany: the case of the folk medicine collection of Dr. Mihkel Ostrov in the territory of present-day Estonia, 1891–1893. Plan Theory. https://doi.org/10.3390/plants11030274

Katara K, Chaudhari K, Pandya H, Maitreya B (2023) The review on ethno-botanical concept, history and traditional knowledge of indigenous communities of India and its threats. Int Assoc Biol Comput Dig. https://doi.org/10.56588/iabcd.v2i1.120

Kaushik P, Ahlawat P, Singh K, Singh R (2023) Chemical constituents, pharmacological activities, and uses of common ayurvedic medicinal plants: a future source of new drugs. Adv Tradit Med. https://doi.org/10.1007/s13596-021-00621-3

Kebede T, Gadisa E, Tufa A (2021) Antimicrobial activities evaluation and phytochemical screening of some selected medicinal plants: a possible alternative in the treatment of multidrug-resistant microbes. PLoS One 16(3):e0249253. https://doi.org/10.1371/JOURNAL.PONE.0249253

Keerthana SD, Philip A, Abdul R, Rupesh Kumar M, Syed SA, Syed A, Bharathi DR (2022) Ethnopharmacology of solanum nigrum: a review. World J Curr Med Pharm Res. https://doi.org/10.37022/wjcmpr.v4i4.223

Khan ZA, Shahzad SA, Anjum A, Bale AT, Naqvi SAR (2018) Synthetic approaches toward the reserpine. Synth Commun. https://doi.org/10.1080/00397911.2018.1434546

Kreis W (2017) The foxgloves (digitalis) revisited. Planta Med. https://doi.org/10.1055/s-0043-111240

Kumar A (2016) Vincristine and vinblastine: a review. Int J Med Pharm Sci

Kumar Srivastava A (2018) Significance of medicinal plants in human life. Synthesis of Medicinal Agents from Plants, 1–24. https://doi.org/10.1016/B978-0-08-102071-5.00001-5

Kumar S, Singh A, Singh B, Maurya R, Kumar B (2018) Structural characterization and quantitative determination of bioactive compounds in ethanolic extracts of Boerhaavia diffusa L. by liquid chromatography with tandem mass spectrometry. Sep Sci Plus 1(9):588–596. https://doi.org/10.1002/SSCP.201800056

Kumar S, Kumari D, Singh B (2022) Genus Rauvolfia: a review of its ethnopharmacology, phytochemistry, quality control/quality assurance, pharmacological activities and clinical evidence. J Ethnopharmacol. https://doi.org/10.1016/j.jep.2022.115327

Lautié E, Russo O, Ducrot P, Boutin JA (2020) Unraveling plant natural chemical diversity for drug discovery purposes. Front Pharmacol 11:510516. https://doi.org/10.3389/FPHAR.2020.00397/BIBTEX

Lawal IO, Rafiu BO, Ale JE, Majebi OE, Aremu AO (2022) Ethnobotanical survey of local Flora used for medicinal purposes among indigenous people in five areas in Lagos State, Nigeria. Plants 11(5):633. https://doi.org/10.3390/PLANTS11050633/S1

Leonti M (2022) The relevance of quantitative ethnobotanical indices for ethnopharmacology and ethnobotany. J Ethnopharmacol. https://doi.org/10.1016/j.jep.2022.115008

Leonti M, Casu L, de Oliveira Martins DT, Rodrigues E, Benítez G (2020) Ecological theories and major hypotheses in ethnobotany: their relevance for ethnopharmacology and pharmacognosy in the context of historical data. Rev Bras 30(4):451–466. https://doi.org/10.1007/S43450-020-00074-W

Liu H, Burkhart EP, Chen VYJ, Wei X (2021) Promotion of in situ Forest farmed American ginseng (Panax quinquefolius L.) as a sustainable use strategy: opportunities and challenges. Front Ecol Evol 9:652103. https://doi.org/10.3389/FEVO.2021.652103/BIBTEX

Logesh R, Vivekanandarajah Sathasivampillai S, Varatharasan S, Rajan S, Das N, Pandey J, Prasad Devkota H (2023) Ficus benghalensis L. (Moraceae): a review on ethnomedicinal uses, phytochemistry and pharmacological activities. Curr Res Biotechnol. https://doi.org/10.1016/j.crbiot.2023.100134

Lone PA, Bhardwaj AK, Bahar FA (2015) Study of indigenous/traditional medicinal plant knowledge-an Endeavour towards new drug discovery. Afr J Tradit Complement Altern Med 12(2):73–95. https://doi.org/10.4314/AJTCAM.V12I2.14

Maes KR, Depuydt P, Vermassen J, De Paepe P, Buylaert W, Lyphout C (2022) Foxglove poisoning: diagnostic and therapeutic differences with medicinal digitalis glycosides overdose. Acta Clin Belg. https://doi.org/10.1080/17843286.2020.1773652

Malik Z, Parveen R, Zahiruddin S, Gaurav, Husain SA, Ahmad S (2023) HPTLC stability indicating analytical method of Andrographolide and 5-fluorouracil with network pharmacology analysis against cancer. Comb Chem High Throughput Screen 26. https://doi.org/10.2174/1386207326666230609104038

Martinez JL, Maroyi A, Wagner ML (2023) Ethnobotany: ethnopharmacology to bioactive compounds. Ethnobotany: Ethnopharmacology to Bioactive Compounds, 1–240. https://doi.org/10.1201/9781003323969

Matejić JS, Stefanović N, Ivković M, Živanović N, Marin PD, Džamić AM (2020) Traditional uses of autochthonous medicinal and ritual plants and other remedies for health in eastern and southeastern Serbia. J Ethnopharmacol 261:113186. https://doi.org/10.1016/J.JEP.2020.113186

McClatchey WC, Mahady GB, Bennett BC, Shiels L, Savo V (2009) Ethnobotany as a pharmacological research tool and recent developments in CNS-active natural products from ethnobotanical sources. Pharmacol Ther 123(2):239–254. https://doi.org/10.1016/J.PHARMTHERA.2009.04.002

Media N (2015) The 2015 Nobel prize in physiology or medicine—press release. Nobelprize.Org

Metz CN, Pavlov VA (2021) Treating disorders across the lifespan by modulating cholinergic signaling with galantamine. J Neurochem. https://doi.org/10.1111/jnc.15243

Moha R (2017) Chemotherapy medication of vincristine and vinblastine. Cancer Res Cell Ther. https://doi.org/10.31579/2640-1053/007

Mohamed Shaffril HA, Ahmad N, Samsuddin SF, Samah AA, Hamdan ME (2020) Systematic literature review on adaptation towards climate change impacts among indigenous people in the Asia Pacific regions. J Clean Prod 258. https://doi.org/10.1016/J.JCLEPRO.2020.120595

Mohd Salim J, Anuar SN, Omar K, Tengku Mohamad TR, Sanusi NA (2023) The impacts of traditional ecological knowledge towards indigenous peoples: a systematic literature review. Sustain For 15(1):824. https://doi.org/10.3390/SU15010824

Moon JS (2022) The tale of aspirin: a weeping willow to soothe your sorrows. Anesthesiology. https://doi.org/10.1097/ALN.0000000000004384

Mukherjee E, Gantait S, Kundu S, Sarkar S, Bhattacharyya S (2019) Biotechnological interventions on the genus Rauvolfia: recent trends and imminent prospects. Appl Microbiol Biotechnol. https://doi.org/10.1007/s00253-019-10035-6

Murugesu S, Selamat J, Perumal V (2021) Phytochemistry, pharmacological properties, and recent applications of ficus benghalensis and ficus religiosa. Plan Theory. https://doi.org/10.3390/plants10122749

Naguib S, Bernardo-Colón A, Cencer C, Gandra N, Rex TS (2020) Galantamine protects against synaptic, axonal, and vision deficits in experimental neurotrauma. Neurobiol Dis. https://doi.org/10.1016/j.nbd.2019.104695

Nalimu F, Oloro J, Kahwa I, Ogwang PE (2021) Review on the phytochemistry and toxicological profiles of Aloe vera and Aloe ferox. Futur J Pharm Sci. https://doi.org/10.1186/s43094-021-00296-2

Ndhlovu PT, Asong JA, Omotayo AO, Otang-Mbeng W, Aremu AO (2023) Ethnobotanical survey of medicinal plants used by indigenous knowledge holders to manage healthcare needs in children. PLoS One 18:3. https://doi.org/10.1371/JOURNAL.PONE.0282113

Nuralın L, Gürü M (2022) Berberis Vulgaris fruit: determination of phenolic compounds in extracts obtained by supercritical CO2 and Soxhlet methods using HPLC. Food Anal Methods. https://doi.org/10.1007/s12161-021-02136-8

Obakiro SB, Kiprop A, Kowino I, Kigondu E, Odero MP, Omara T, Bunalema L (2020) Ethnobotany, ethnopharmacology, and phytochemistry of traditional medicinal plants used in the management of symptoms of tuberculosis in East Africa: a systematic review. Trop Med Health 48:1. https://doi.org/10.1186/S41182-020-00256-1

Packer J, Turpin G, Ens E, Venkataya B, Hunter J (2019) Building partnerships for linking biomedical science with traditional knowledge of customary medicines: a case study with two Australian indigenous communities. J Ethnobiol Ethnomed 15(1):1–11. https://doi.org/10.1186/S13002-019-0348-6/TABLES/1

Pardo-De-Santayana M, Quave CL, Sõukand R, Pieroni A (2015) Medical ethnobotany and ethnopharmacology of Europe. Ethnopharmacol:343–355. https://doi.org/10.1002/9781118930717.CH29

Pathak MM (2020) Mapping ethnobotany research in India reviews and mini reviews. Ethnobot Res Appl. https://www.academia.edu/44772684/Mapping_ethnobotany_research_in_India_Reviews_and_Mini_reviews

Patwardhan B, Warude D, Pushpangadan P, Bhatt N (2005) Ayurveda and traditional Chinese medicine: a comparative overview. Evid Based Complement Alternat Med 2(4):465. https://doi.org/10.1093/ECAM/NEH140

Petkeviciute Z, Savickiene N, Savickas A, Bernatoniene J (2010) (PDF) Urban ethnobotany study in Samogitia region, Lithuania. J Med Plants Res 4:1. https://www.researchgate.net/publication/285032172_Urban_ethnobotany_study_in_Samogitia_region_Lithuania

Pirintsos S, Panagiotopoulos A, Bariotakis M, Daskalakis V, Lionis C, Sourvinos G, Karakasiliotis I, Kampa M, Castanas E (2022) From traditional ethnopharmacology to modern natural drug discovery: a methodology discussion and specific examples. Molecules 27(13):4060. https://doi.org/10.3390/MOLECULES27134060

Porras G, Chassagne F, Lyles JT, Marquez L, Dettweiler M, Salam AM, Samarakoon T, Shabih S, Farrokhi DR, Quave CL (2021) Ethnobotany and the role of plant natural products in antibiotic drug discovery. Chem Rev 121(6):3495–3560. https://doi.org/10.1021/ACS.CHEMREV.0C00922/SUPPL_FILE/CR0C00922_SI_001.XLSX

Priyadarshini P, Abhilash PC (2019) Promoting tribal communities and indigenous knowledge as potential solutions for the sustainable development of India. Environ Dev 32:100459. https://doi.org/10.1016/J.ENVDEV.2019.100459

Pullaiah T, Krishnamurthy KV, Bahadur B (2016) Ethnobotany of India. Ethnobotany of India. https://doi.org/10.1201/9781315366371

Puspangadan P, George V, Ijinu TP (2016) Ethnobotany, ethnopharmacology, bioprospecting, and patenting. Ethnobotany of India https://doi.org/10.1201/9781315366371

Rana SK, Rana HK, Ranjitkar S, Ghimire SK, Gurmachhan CM, O'Neill AR, Sun H (2020) Climate-change threats to distribution, habitats, sustainability and conservation of highly traded medicinal and aromatic plants in Nepal. Ecol Indic 115:106435. https://doi.org/10.1016/J.ECOLIND.2020.106435

Saha A, Marma KKS, Rashid A, Tarannum N, Das S, Chowdhury T, Afrin N, Chakraborty P, Emran M, Hamidullah Mehedi HM, Hussain MI, Barua A, Mistry SK (2022) Risk factors associated with self-medication among the indigenous communities of Chittagong Hill tracts, Bangladesh. PLoS One 17(6):e0269622. https://doi.org/10.1371/JOURNAL.PONE.0269622

Saini AK, Zahiruddin S, Newton KG, Sawant L, Mitra R, Kumar Rai R, Ahmad S (2023) Validation and standardization of gallic acid and ellagic acid in Quercus Infectoria, Terminalia Chebula, and Pistacia Integerrima. J AOAC Int 106(5):1305–1312. https://doi.org/10.1093/JAOACINT/QSAD070

Samadi P, Sarvarian P, Gholipour E, Asenjan KS, Aghebati-Maleki L, Motavalli R, Hojjat-Farsangi M, Yousefi M (2020) Berberine: a novel therapeutic strategy for cancer. IUBMB Life. https://doi.org/10.1002/iub.2350

Sen S, Chakraborty R, De B, Devanna N (2011) An ethnobotanical survey of medicinal plants used by ethnic people in West and South district of Tripura, India. J For Res 22(3):417–426. https://doi.org/10.1007/S11676-011-0184-6/METRICS

Sengupta P, Dutta S (2022) Panax ginseng as reproductive medicine in male infertility: with a brief focus on "herb-drug" interaction. Chem Biol Lett 2022, 9(1), 279

Septembre-Malaterre A, Rakoto ML, Marodon C, Bedoui Y, Nakab J, Simon E, Hoarau L, Savriama S, Strasberg D, Guiraud P, Selambarom J, Gasque P (2020) Artemisia annua, a traditional plant brought to light. Int J Mol Sci. https://doi.org/10.3390/ijms21144986

Shah S, Bhat JA (2019) Ethnomedicinal knowledge of indigenous communities and pharmaceutical potential of rainforest ecosystems in Fiji Islands. J Integr Med 17(4):244–249. https://doi.org/10.1016/J.JOIM.2019.04.006

Shaha P, Bellankimath A (2017) Pharmacological profile of Asparagus racemosus: a review. Int J Curr Microbiol App Sci. https://doi.org/10.20546/ijcmas.2017.611.144

Shaheen S, Harun N, Ijaz R, Mukhtar N, Ashfaq M, Bibi F, Ali M, Abbas Z, Khalid Z (2023) Sustainability issues in conservation of traditional medicinal herbs and their associated knowledge: a case study of district Lahore, Punjab, Pakistan. Sustain For 15(9):7343. https://doi.org/10.3390/SU15097343

Sharma BK, Jain AK (2021) Ethnobotanical studies of Sikkim: a review. Int Res J Biol Sci 10(4):34–44. www.isca.me

Sharma V, Sarkar IN (2013) Bioinformatics opportunities for identification and study of medicinal plants. Brief Bioinform. https://doi.org/10.1093/bib/bbs021

Siddiqui MZ (2011) Boswellia serrata, a potential antiinflammatory agent: an overview. Indian J Pharm Sci. https://doi.org/10.4103/0250-474X.93507

Singh SK, Srivastav S, Castellani RJ, Plascencia-Villa G, Perry G (2019) Neuroprotective and antioxidant effect of Ginkgo biloba extract against AD and other neurological disorders. Neurotherapeutics. https://doi.org/10.1007/s13311-019-00767-8

Singh RK, Kumar A, Singh A, Singhal P (2020a) Evidence that cultural food practices of Adi women in Arunachal Pradesh, India, improve social-ecological resilience: insights for sustainable development goals. Ecol Process 9(1):1–19. https://doi.org/10.1186/S13717-020-00232-X/TABLES/6

Singh R, Upadhyay SK, Tuli HS, Singh M, Kumar V, Yadav M, Aggarwal D, Kumar S (2020b) Ethnobotany and herbal medicine: some local plants with anticancer activity. Bull Pure Appl Sci Botany. https://doi.org/10.5958/2320-3196.2020.00009.9

Škubník J, Pavlíčková VS, Ruml T, Rimpelová S (2021) Vincristine in combination therapy of cancer: emerging trends in clinics. Biology. https://doi.org/10.3390/biology10090849

Srivastava BBL, Ripanda AS, Mwanga HM (2022) Ethnomedicinal, Phytochemistry and antiviral potential of turmeric (Curcuma longa). Compounds. https://doi.org/10.3390/compounds2030017

Suárez ME (2019) Medicines in the forest: ethnobotany of wild medicinal plants in the pharmacopeia of the Wichí people of Salta province (Argentina). J Ethnopharmacol. https://doi.org/10.1016/j.jep.2018.10.026

Targe K, Lyngdoh S, Bussmann RW, Singh B, Correspondence A, Singh Adhikari B (2022) Conservation linkages of rare and endangered medicinal plants used in the traditional health care system in Pin Valley National Park, Himachal Pradesh. Ethnobot Res Appl 24:1–22. https://doi.org/10.32859/era.24.36.1-22

Tresca G, Marcus O, Politi M (2020) Evaluating herbal medicine preparation from a traditional perspective: insights from an ethnopharmaceutical survey in the Peruvian Amazon. Anthropol Med. https://doi.org/10.1080/13648470.2019.1669939

Wang L, Du GH (2018) Paclitaxel. Natural small molecule drugs from plants. https://doi.org/10.1007/978-981-10-8022-7_89

Yao R, Heinrich M, Zhang B, Wei X, Qi Y, Gao W (2023) Single botanical drugs in the Ayurvedic pharmacopoeia of India—a quantitative ethnobotanical analysis. Front Pharmacol. https://doi.org/10.3389/fphar.2023.1136446

Yuan H, Ma Q, Ye L, Piao G (2016) The traditional medicine and modern medicine from natural products. Molecules 21:5. https://doi.org/10.3390/MOLECULES21050559

A Current Perspective of Ethnopharmacology in India

9

Amrita Mishra and Arun K. Mishra

Abstract

Ethnopharmacology is the scientific study of traditional and folk medicines which includes social and cultural context, analysis of constituents, investigation of pharmacological activities, and evaluation of physiological and clinical impact of medicinal substance. The field of ethnopharmacology has played a crucial role in establishing the therapeutic potential of natural chemicals by documenting the advantageous effects of many plants. These substances have been extensively studied and recognized as highly significant resources for the development of pharmaceutical drugs. The current chapter gives glimpse of traditional systems of medicines practiced in India, their preclinical and clinical research, pre- and post-COVID-19 scenario and challenges.

Keywords

Traditional system of medicine · Indian Systems of Medicine · CoVID-19 · Market analysis

9.1 Introduction

Natural products refer to substances that are derived from natural sources, such as plants, animals, or minerals. These goods are not synthesized or chemically modified in a laboratory setting. The field of research traces its roots back to the practice

A. Mishra (✉)
Department of B.Pharm (Ayu), Delhi Pharmaceutical Sciences and Research University, New Delhi, India

A. K. Mishra
SOS School of Pharmacy, IFTM University, Moradabad, India

© The Author(s), under exclusive license to Springer Nature Singapore Pte Ltd. 2024

M. Nandave et al. (eds.), *Ethnopharmacology and OMICS Advances in Medicinal Plants Volume 1*, https://doi.org/10.1007/978-981-97-2367-6_9

of ethnopharmacology. Throughout the course of history, plants possessing medicinal attributes have been utilized, and contemporary scientific advancements have, in several instances, successfully isolated, identified, and comprehensively elucidated the pharmacological mode of action (MOA) of these active substances. The field of research concerned with the scientific examination of traditional medicines is commonly referred to as "ethnopharmacology." In every human community, a traditional medicine system is present, wherein the utilization of flowering plants plays a prominent role in the procurement of resources for medicinal purposes. The field of ethnopharmacology has played a crucial role in establishing the therapeutic potential of natural chemicals by documenting the advantageous effects of many plants. Throughout history, various people have utilized natural goods in their original state or following the crude extraction of their active components. These substances have been extensively studied and recognized as highly significant resources for the development of pharmaceutical drugs. India possesses a diverse and extensive legacy of ethnomedicine systems that have played a significant role in promoting the well-being of a substantial portion of its population. This chapter looks at various traditional systems of medicine that are practiced in India, focusing on their present research status, market demand, and the limitations encountered in this area of study.

9.2 Traditional Systems of Medicines in India

It is widely acknowledged that traditional medical systems have historically played a crucial role in meeting the demands of the global healthcare system. Currently, they are engaged in this activity and are expected to maintain their substantial influence in the times to come. Indian Systems of Medicine refer to medical practices that are thought to have originated in India or have been assimilated into Indian culture from other countries (Prasad 2002). India is renowned for being home to six recognized medical systems. The aforementioned practices encompass Ayurveda, Siddha, Unani, Yoga, Naturopathy, and Homoeopathy.

9.2.1 Ayurveda

The term "Ayurveda" derives from the Sanskrit language and translates to "the Science of Life." The establishment and enunciation of the foundational and practical principles of Ayurveda are considered to have occurred around 1500 BC. The Atharvaveda has a collection of 114 hymns that pertain to the development of remedies for diverse ailments. The fourth component of the four significant collections of knowledge, collectively referred to as the Vedas, serves as the fundamental basis of Indian civilization. Over the course of history, a total of eight distinct specialities and two major schools have emerged as a result of the accumulation and development of knowledge. There were two distinct schools of education in ancient times. The first was a medical school referred to as "DhanvantriSampradaya," with

"Sampradaya" denoting a tradition. The second was a surgical school known as "AtreyaSampradaya" in literary works (Ramachandra Rao 1987).

The Charaka Samhita and the Sushruta Samhita, belonging to the domains of medicine and surgery, respectively, both possess distinct compilations that represent their respective schools. The former comprises several chapters that encompass various aspects of medicine and related subjects. This book provides a comprehensive analysis of about 600 drugs derived from plant, animal, and mineral sources. The primary focus of the Sushruta Samhita revolves around several aspects of fundamental surgical theory and principles. This text provides a comprehensive discussion on the applications of over 100 distinct surgical instruments, including but not limited to scalpels, scissors, forceps, and specula. The utilization of vegetables and deceased animals serves as a means to demonstrate dissection and surgical methodologies. The study delves into other facets of other surgical-related domains, such as anatomy, embryology, toxicology, and therapeutics, encompassing detailed accounts of about 650 medicines (Mondal 2018).

Ayurveda was the main healthcare system in India for centuries, but after the repeated attacks and Mughal invasions, it started fading. In British India also, it was considered as nonscientific knowledge. Ayurveda regained popularity in the early twentieth century as a result of the national freedom movement. Even in British India, representatives of the people began requesting appropriate actions to expand Ayurveda along scientific lines. The thought for the resuscitation of Traditional Systems of Medicine gathered force after India attained independence from British rule in 1947. For the purpose of providing health care to the nation's citizens, the systems received official recognition and joined the national healthcare network (Sharma 2011).

Ayurveda is one of the main healthcare systems that is essential for meeting the primary healthcare needs of the nation, and the Indian government has taken a number of steps to strengthen its standing in that regard. Ayurvedic hospitals and colleges have sprung up all around. The Central Institute of Research in Indigenous System of Medicine (CIRISM) was established in 1955 to address R&D needs. In 1956, a Post Graduate Training Center of Ayurveda was established to provide postgraduate education. In 1967, Gujarat Ayurved University was established at Jamnagar in the Gujarat State. In 1972, the Central Council of Indian Medicine (CCIM) was set up to regulate education and regulation.

The Ministry of Health and Family Welfare originally established the Department of Indian Systems of Medicine and Homeopathy (ISM&H) in 1995. The Department of AYUSH replaced ISM&H as its official title. The Modi administration transformed the division into a formal ministry in 2014. The development of education, research, and dissemination of traditional medical systems in India is the responsibility of the Ministry of AYUSH, a ministry of the Government of India. The term AYUSH was created using the names of the alternative medical practices that the ministry covers: Ayurveda, Yoga and Naturopathy, Unani, Siddha, and Homeopathy (Mukherjee et al. 2007).

In Ayurveda, drug therapy has advanced significantly. Up to 8000 recipes for creating pharmacological combination can be found in at least 70 books. The medicines that are used come from a variety of plants, animals, and minerals. It is recommended to use a variety of plant preparations to increase the host's overall resistance. These medications, also known as Rasayana, Jeevanya, and Balya medicines, all improve tissue resilience to illness. Ayurveda emphasizes that one must follow a regulated diet, sleep, and sex in order to promote and maintain good health and prevent disease. It also prescribes the observation of certain principles, including daily, nightly, seasonal, and ethical routines. Ayurveda is so more than just a medical science; it is also a way of life (Che et al. 2017).

9.2.2 Siddha

In various regions of South India, particularly in the state of Tamil Nadu, the Siddha system of medicine is practiced. Although it shares many similarities with Ayurveda, it still possesses its own unique identity. The Tamil civilization has grown to be intimately associated with this system. The word "Siddha" derives from the word "Siddhi," which implies accomplishment. In the fields of medicine, yoga, and tapa (meditation), Siddhars were the ones who attained supreme knowledge. It is a well-known fact that South India, particularly the banks of the rivers Cauvery, Vaigai, Tamiraparani, etc., was home to a highly advanced civilization before the arrival of the Aryans. The medical approach that is popular in this society appears to be the forerunner of the Siddha system of medicine that is used today. As time went on, it interacted with other medical streams, completing and enhancing them while also enriching itself. In contrast to ancient Ayurveda, which was primarily based on remedies of vegetable origin, the materia medica of the Siddha system of medicine heavily relies on drugs of metal and mineral origin. The development of Siddha medicine, yoga, and philosophy was credited to 18 Siddhars, according to tradition. However, the material they produced isn't entirely accessible. According to the well-known humility of the ancient Indian Acharyas (preceptors), it is unknown who wrote numerous literary works of outstanding worth. Another custom was to credit one's instructor, sponsor, or even a famous scholar of the day as the author of one's work. This has made it incredibly challenging to determine who the true author of many classics is (Narayanaswamy 1975).

9.2.3 Unani

Unani medicine traces its origins back to ancient Greece. The establishment of it (460–377 BC) is attributed to Hippocrates, a renowned physician and philosopher. Galen, a prominent figure in the field, made significant contributions to the continued advancement of the subject. Aristotle (384–322 BC) laid the groundwork for the field of anatomy and physiology. Dioscorides, a prominent physician during the first century AD, exerted a substantial impact on the development of pharmacology, specifically in relation to the utilization of plant-based remedies. The second stage of growth was witnessed by Egypt and Persia (modern-day Iran). As per the information provided on the official website of the Department of Indian Systems of Medicine and Homoeopathy (http://www.indianmedicine.nac.in), it is said that the ancient Egyptians possessed a sophisticated pharmaceutical system and showed proficiency in formulating several dosage forms such as oils, powders, ointments, and alcohol (Ishtiyaq 1983).

Around 1350 AD, the Arabs played a key role in the introduction of Unani medicine to India. Zia Mohd Masood Rasheed Zangi was the first recognized Hakim (Physician). Hakim M. Shareef Khan (1725–1807), a distinguished physician best known for his book *IlajulAmraz*, and Akbar Mohd Akbar Arzani (about 1721 AD), authors of the books *QarabadinQadri* and *Tibbe Akbar*, are two of the well-known physicians who played a key role in the development of the system. A well-known figure among Indian unani doctors of the twentieth century was Hakim Ajmal Khan (1864–1927). He had a complex personality; in addition to being a doctor, he was also a scientist, a politician, and a freedom warrior. He played a key role in the foundation of the Unani and Ayurvedic College in Delhi's Karol Bagh. As a dedicated researcher, he oversaw numerous investigations on Rauwolfiaserpentina, the plant that produces many well-known alkaloids like reserpine and Ajamaloon. Hakim Kabeeruddin (1894–1976), who translated 88 Unani texts from Arabic and Persian into Urdu, is another outstanding contributor. The first school of unani medicine was founded in the unified India in 1872 as Oriental College in Lahore. The establishment of numerous institutions followed. Following independence, numerous government organizations interested in the growth of ISM provided Unani with a boost. There are currently around 30 universities that provide degree programs in unani medicine, and about 20,000 doctors graduate each year on average. Around 177 hospitals are present. In 1983, in partnership with the Karnataka government, a National Institute of Unani Medicine was founded in Bangalore, Karnataka state, to meet both academic and R & D needs. The leading organization engaged in R & D activities is the Central Council for Research in Unani Medicine (CCRUM).

9.2.4 Homoeopathy

The discovery of homoeopathy can be attributed to Dr. Christian Friedrich Samuel Hahnemann (1755–1843), a German physician, during the latter part of the eighteenth century. The therapeutic system of medicine is based on the notion of "SimiliaSimilibusCurentur," which translates to "let likes be treated by likes." Homeopathy is a therapeutic approach that involves the administration of medicinal substances capable of inducing symptoms comparable to those experienced by a healthy individual, so mimicking the natural ailment that it aims to treat in the afflicted patient. The treatment strategy not only encompasses a comprehensive perspective but also takes into account the unique individual qualities of each patient. The principle of the "law of similars" was previously articulated by Hippocrates and Paracelsus. However, Dr. Hahnemann, although living in an era with little knowledge of current laboratory techniques, was able to establish this concept on a scientific basis.

The introduction of homeopathy in India occurred when German missionaries and physicians began disseminating homeopathic medications among the indigenous population. Homoeopathy was introduced to India in 1839, when Dr. John Martin Honigberger achieved an appropriate treatment outcome for Maharaja Ranjit Singh's vocal cord paralysis. After relocating to Kolkata, formerly known as Calcutta, Dr. Honigberger established himself as a renowned medical practitioner specializing in the treatment of cholera. Subsequently, Dr. M.L. Sirkar, a highly distinguished physician of his era, commenced his professional engagement with the field of homeopathy. In the year 1868, he assumed the role of editor for the inaugural edition of the "Calcutta Journal of Medicine," which was the first Homoeopathic Journal. The establishment of the "Calcutta Homoeopathic Medical College" in 1881 was a significant milestone in the field of medicine, with the participation of eminent physicians such as Dr. P.C. Mujumdar and Dr. D.N. Roy. Dr. Lahiri, along with Dr. B. K. Sarkar and several other individuals, played crucial roles in their personal endeavors to establish homoeopathy as a recognized and respected profession. They have gained significant recognition for their substantial contributions to the advancement of Homoeopathy, not only within the region of West Bengal but also on a national scale.

The Indian government has consistently exerted efforts toward the advancement and progress of Homoeopathy and other traditional systems of medicine, namely, Ayurveda, Yoga, Naturopathy, Unani, Siddha, and Sowa Rigpa. These systems are together recognized as AYUSH. The establishment of an institutional framework for Homoeopathy at both the national and state levels has been achieved as a result of the persistent endeavors of the government. The current state of infrastructure in the field of homeopathic medicine is notable, comprising of 195 undergraduate and 43 postgraduate colleges dedicated to this discipline. These institutions are responsible for providing quality university education in homeopathy. Additionally, there is a regulatory mechanism in place to ensure the standards of these educational institutions. Furthermore, there exists an autonomous research council consisting of 22 institutes and units, which contribute to the advancement of knowledge in homeopathy. The number of registered homeopathic practitioners is significant, with a total of 283,840 individuals practicing in this field. Moreover, there are strict regulations in place to ensure the safety of homeopathic drugs, with 403 drug manufacturing units adhering to these regulations (Ghosh 2010).

9.2.5 Yoga and Naturopathy

Yoga has evolved over time to become both an art and a science that encompasses various aspects of life. Maharishi Patanjali is widely recognized as the progenitor of the ancient discipline of Yoga. The practice of Yoga fosters a state of harmony with the natural world, thereby contributing to the prevention of diseases, the promotion of health, and the management of many lifestyle-related problems, particularly those of a psychosomatic and stress-related nature. The term "Yoga" has a literal interpretation as "Yuj," which signifies the act of uniting. This phenomenon facilitates the integration of individual awareness with the broader framework of universal consciousness. The practice of yoga encompasses a series of eight steps known as "Ashtanga Yoga." The sequence of steps in the practice of yoga includes Yama, which refers to constraints; Niyama, which pertains to observances; Asana, which involves physical postures; Pranayama, which focuses on controlling breathing; Pratyahara, which entails withdrawing from worldly objects; Dharana, which involves concentration; Dhyana, which refers to meditation; and Samadhi, which signifies the unification of mind and body. Yoga is commonly associated with a holistic approach to maintaining a healthy lifestyle, achieved via the practice of Yogic Asanas, Pranayama, and other techniques. These practices are believed to promote overall well-being and offer long-lasting solutions for a variety of maladies. Furthermore, yoga is often integrated with other therapeutic systems, creating a comprehensive approach to wellness. The foundation of yoga philosophy is rooted in the Panchakosha doctrine, which encompasses the five layers of existence known as annamaya kosha, pranamaya kosha, manomaya kosha, vijnanamaya kosha, and anandamayakosha. This statement underscores the significance of every aspect of our being.

Naturopathy has been strongly championed by Mahatma Gandhi, widely recognized as the Father of the Nation. He advocated and propagated Nature Cure as the optimal solution for the betterment of India, considering its geographical, cultural, social, and economic circumstances. Nature Cure is a holistic approach to human well-being that emphasizes the alignment of individuals with the constructive principles of nature across the physical, mental, and moral dimensions. According to the principles of naturopathy, the human body is considered to be composed of the five fundamental elements known as Panchamahabootas, which include akash (space), vayu (air), prithvi (earth), aap (water), and teja (sun—heat and light). The equilibrium of these constituents is associated with well-being, while their disequilibrium is linked to poor health. Naturopathy aims to incorporate a comprehensive range of evidence-based drugless therapies, offering individuals a holistic approach to health care that excludes the use of chemical or herbal drugs. This is particularly pertinent in contemporary times, as lifestyle diseases have reached nearly epidemic levels, despite the improvements in modern science and specialized health care.

9.3 Research

The inclusion of scientific evidence is crucial in substantiating the justification behind the utilization of traditional systems formulations in health care, as it serves to both improve and protect the cultural legacy. The Indian subcontinent's diverse geographical landscapes and biodiversities have played a significant role in the utilization of various plant species and natural resources for healthcare purposes. This has contributed to the development of the Materia Medica in traditional systems of medicine. The investigation of plant species utilized in Ayurveda and other Indian Systems of Medicine (ISM) holds potential for the discovery of novel therapeutic approaches in the treatment of various ailments (Choudhary 2002). Methods such as high-throughput screening, phytochemical profiling, quality controls, standardization of raw materials and finished products, clinical trials, herbal therapeutics, pharmacokinetics, and herbal pharmacovigilance can contribute to substantiating the rationale for employing these systems and optimizing the potential advantages of natural resources. The Government of India, specifically the Ministry of Health and Family Welfare, has implemented various initiatives aimed at the advancement and safeguarding of these cultural heritages. The Department of AYUSH, which encompasses the fields of Ayurveda, Siddha, Unani, and Homeopathy, is responsible for overseeing the governance of education and research within these systems. The Central councils established by AYUSH Ministry continuously evolved in research and promotion of these systems of medicine. The annual report of year 2022–2023 shared by the Ministry shows the contribution in the field of medicinal plant research, drug standardization research and pharmaceutical research, preclinical studies, and clinical research encompassing wide range of topics (Table 9.1). Government is providing various kinds of funding support to the researchers to promote these systems.

Other than research councils, a large number of people from scientific community are involved in research in the field of ethanopharmacology. The area of interest varies from literature survey for ethnomedicines to pharmacological evaluation, analytical profiling to latest metabolomics, and genomics studies. Table 9.2 contains some details about the latest publications done by Indian scientific community in the field of ethnopharmacology in 2023. Here also we can see the wide variety of topics from formulations pharmacology, toxicity, gene expression, clinical studies, etc.

The research in the field of ethnopharmacology has been an area of thrust from the last several years. From year 2011 to 2020, India has published 7159 publications in the field of ethnopharmacology research. This data is huge, and if we go in details, we can see that the numbers of publication has improved from 751 publications from 2011 to 982 in 2020. These publications include preclinical as well as clinical research data from India. These data indicates the increasing interest of researchers in ethnopharmacology in India (Chaman et al. 2022).

Table 9.1 AYUSH annual report 2022–23

	Medicinal plant research	Drug standardization research and pharmaceutical research	Preclinical studies	Clinical research
Central Council for Research in Ayurvedic Sciences	• Completed Projects-04 • In Progress-43 • NMPB Project in Progress-03 • Medico-EthnoBotanical Survey tours-36 (for drug collection). • Herbarium sheets-815. • Raw drug collected 2315 kg (approx.)	• Completed projects—13 • In progress—28 • Collaborative research studies in progress for Pharmaceutical Research—02	• Completed projects—07 • In progress—33 • Collaborative research studies in progress—9	• Clinical Research Projects completed 13 (iron deficiency anemia, abnormal uterine bleeding, urolithiasis, dandruff, leukorrhea, psoriasis, Amajirna, eczema, Grahani, and external wounds) • In progress-17 (urolithiasis, cognitive deficit, dry age-related macular degeneration syndrome, occupational stress among IT professionals, iron deficiency anemia, occupational stress among nurses, prehypertension, fistula in Ano, alcohol dependence, gout, obesity, czema, leukorrhea, nonalcoholic fatty liver disease (NAFLD) and subclinical hypothyroidism) • Collaborative Clinical Research: Completed—7 (Ayush D for prediabetic subjects; Ayush D for type II diabetes mellitus and AYUSH-SL for patients of chronic lymphedema, validation and reliability testing of Ayurveda Diagnostic Methods, Reproductive and Child Health project for AnteNatal Care, Ayush CCT and Raj Yoga Meditation on clinical recovery and postoperative cardiothoracic surgeries and Ayush-SS Granules in insufficient lactation) • In Progress—12 (Carctrol S for Cancer Patients, Ayush A for Bronchial Asthma, Ayush-LND for abnormal uterine bleeding, PTK for hepato-protective as add-on therapy in the patients of tuberculosis on ATT, ayurvedic interventions in ADHD, SarpangandhaMishran for hypertension, gut bacterial modulation induced by Virechana in the healthy adults, Marma therapy in lumbar disc herniation with radiculopathy, Ayush M-3 in the management of migraine, Ayush-GMH for nonalcoholic fatty liver disease, primary knee osteoarthritis and effect of intranasal oil instillation (PratimarshaNasya) on nasal barrier function among healthy individuals).

(continued)

Table 9.1 (continued)

	Medicinal plant research	Drug standardization research and pharmaceutical research	Preclinical studies	Clinical research
Central Council for Research in Unani Medicine	Completed projects—05 In progress –01 Surveys for "Collection of medicinal plants"—07 ("Documentation and Digitization of Herbarium specimens") are going on Projects on "Cultivation of medicinal plants"—04 "Maintenance of drug samples"—05	Completed projects—20 (Drug Standardization) In Progress –15, Monographs revised—124	Completed-Acute and repeated dose (28-day) oral toxicity studies on six Unani formulations (MajoonPiyaz, TriyaqeAfayee, MajoonUshba, KhamireGaozabaanSaada, QurseMullayin and Cap. Mubarek) In Progress-Acute, and sub-acute (28-days repeated dose) toxicity on SharbatEjaz, SharbatUnnab, Sharbat toot Siyah, ItrifalShahatra, SufoofDamaHaldiWala, and QursAsfar	Validation studies on safety and efficacy in progress—50 Completed-Jawarish-e- Bisbasa in Saman-e-Mufrit (central obesity), KhamiraBanafsha in Sual–e-Yabis (dry cough) and Majoone-Piyaz in Surat-e-Inzaal (premature ejaculation)
Central Council for Research in Siddha	Completed projects—04	Completed projects—07 (Drug Standardization Research)	Completed projects—14 projects related (preclinical studies)	Completed projects —14 projects related (clinical studies)

Central Council for Research in Homoeopathy	• Pharmacopoeia activities—21. • Collection of medicinal plants—42. • Cultivation of medicinal plants—104	• Drug monographs for HPI completed—51, revision of drug monographs—21 drugs • Chemo profiling studies: 08. • Safety evaluation of homoeopathic medicines: arsenic album, ferrum phos and phosphorus (acute oral and subacute oral toxicity studies the acute and subacute toxicity study) • Completed evaluation of immunomodulatory effect of homoeopathic medicines: arsenic album, ferrum phos and phosphorus • Completed evaluation of antinociceptive and anti-inflammatory effect of homoeopathic medicines: *Calendula officinalis* • Completed evaluation of safety of homeopathic mother tinctures and their potencies on zebrafish embryonic development: Hyocyamus nigre, Janosiaasoka, Sarsaparilla • Completed evaluation of pharmacological effect of homoeopathic drugs in zebrafish models: *Artemisia vulgaris*, *Cicuta virosa*, *Cuprum metallicum*, Belladonna	22 preclinical studies were undertaken during the reported period, out of which 13 studies were in collaboration with other institutes and 09 were in-house studies	During the reported period, ten clinical research studies were continued from previous years, out of which four studies were concluded. Four new research studies were initiated. The screening/enrolment/follow-up of the ongoing studies is continued during the reporting year

Table 9.2 Latest trends in research (2023)

	Traditional system	Formulation/drug	Activity/clinical studies	Reference
1.	Ayurvedic formulations	*Dhanwantaram Kashaya*, *Saraswatarishta*	*Drosophila* models of Huntington's and Alzheimer's disease	Sharma et al. (2023)
2.	Ayurvedic formulations	*Ashwagandha* (*Withania somnifera* L. Dunal)	Ethnological validation of ghrita as *Vajikarana Rasayana:* In silico, in vitro, *and* in vivo *approach*	Gurav et al. (2023)
3.	Ayurvedic drug	*Moringa oleifera* Lam. leaves	Constipation of rats induced by low-fiber-diet: A proteomics study	Li et al. (2024)
4.	Ayurvedic drug	*Clerodendrum viscosum* leaves	Lead-induced neurotoxicity through upregulation of BDNF-Akt-Nrf2 pathway in mice	
5.	Ayurvedic drug	*Fritillaria cirrhosa* D. Don (Syn: *Fritillaria roylei* Hook. bulbs	Antiplasmodial activity of the bulbs and UPLC-IM-Q-TOF-MS/MS-based biochemometric approach for the identification of marker compounds	Bora et al. (2023)
6.	Ayurvedic formulation	Febrojith: polyherbal formulation	Ameliorating effect of on acute and chronic inflammatory model via inhibiting inflammation and oxidative stress	Ratheesh et al. (2023)
7.	Ayurvedic herbo-metallic formulation	*Nityanand Rasa*	Subchronic oral toxicity	Kshirsagar et al. (2023)
8.	Unani formulation	Sharbat-e-bazoori Motadil	Ameliorative effect of on TNF-α, IL-1β, and caspase-3 expression in kidneys of wistar rats against sodium fluoride induced oxidative stress	Khan et al. (2024)
9.	Unani formulation	2-phenyl ethyl alcohol isolated from the spent floral distillate of damask rose (*Rosa damascena* Mill.)	Vasorelaxant property of and its possible mechanism	Singh et al. (2023)
10.	Siddha formulation	Kabasura Kudineer	Immunomodulation(miRNA-15/IL-10Rα) by suppressing oxidative stress	Behera et al. (2023)
11.	Siddha formulation	Abha Guggulu	Anti-arthritic activity in osteoarthritis patients	Hedaoo et al. (2023)
12.	Yoga	Integrated yoga regimen	Prefrontal oxygenation and working memory in patients with type 2 diabetes mellitus	Kaligal et al. (2023)

(continued)

Table 9.2 (continued)

	Traditional system	Formulation/drug	Activity/clinical studies	Reference
13.	Ayurvedic formulation	AYUSH 64	Coadministration of as an adjunct to standard of care in mild and moderate COVID-19	Chopra et al. (2023)
14.	Unani system	Hot fomentation (Takmīd-e-haar)	Pain alleviation in nonspecific low back pain	Khan et al. (2023)
15.	Ayurveda	Sacred fig (Ficusreligiosa)	Genomic and transcriptomic analysis	Ashalatha et al. (2023)
16.			Genome-wide DNA methylation profiling after Ayurveda intervention to bronchial asthmatics identifies differential methylation in several transcription factors with immune process related function	Bhat et al. (2023)
17.	Ayurveda	Literature synthesis	Ayurgenomics in Antiviral therapy	Quazi and Pathak (2023)

9.4 Post-COVID-19 Scenario

The global impact of the COVID-19 pandemic has been observed over the entirety of the human population. The initial instance of COVID-19 in India can be attributed to an imported case from Wuhan, China, on January 30, 2020, which was subsequently identified in Kerala (Sahasranaman and Kumar 2020). As of December 12, 2020, the mortality rate of COVID-19 in India is at 1.45%. During the pandemic, the severe acute respiratory syndrome-related coronavirus (SARS-CoV-2) posed the greatest level of risk. The scientific community as a whole has redirected its research efforts toward the study of COVID-19. Various strategies were implemented to mitigate the spread of infection, such as isolation measures, among others. The management strategy for COVID-19 involves providing supportive care, which is complemented by the administration of a mix of broad-spectrum antibiotics, antiviral medications, corticosteroids, and convalescent plasma (Yang et al. 2020). Researchers are diligently engaged in the development of efficacious medicines.

As of October 18, 2020, a significant number of clinical trials, over 3611, have been conducted or are now in progress, focusing on COVID-19. Among these trials, over 100 specifically investigate the efficacy of complementary medicines. Furthermore, new trials continue to emerge regularly, reflecting the escalating global incidence of COVID-19 cases. The medications under investigation encompass a variety of pharmaceuticals, including repurposed flu therapies, unsuccessful ebola drugs, and malaria treatments that were initially formulated some decades ago (Lythgoe and Middleton 2020). In such circumstances, the general populace exhibited a tendency to turn to conventional medical systems as a means of preventive health care. The Ministry of AYUSH has also released recommendations pertaining to the utilization of traditional medicines for the purpose of preventative care. Numerous established traditional AYUSH formulations, renowned for their immune-modulating properties, have been utilized for an extensive period in the treatment of respiratory illnesses and allergy problems. The Ministry of AYUSH, a governmental body in India, has compiled a list of formulations and endorsed their utilization as a prophylactic measure in areas designated as red zones or containment zones, as well as for those involved in combating the COVID-19 pandemic. A considerable number of these treatments are currently undergoing clinical trials in people diagnosed with COVID-19. Various formulations, including AYUSH Kwath, SamshamaniVati, AYUSH-64, AgasthayaHareetaki, Anuthaila, Triyaq-e-Araba, Roghan-e-Baboona, Arq-e-Ajeeb, Khamira-e-Banafsha, as well as individual herbs such as *Allium sativum* L. (Garlic), *Cinnamomum verum* J. Presl. (Cinnamon), *Curcuma longa* L. (turmeric), *Phyllanthus emblica* L. (Amla), *Piper nigrum* L. (black pepper), were employed for the management of mild to moderate symptoms and self-care purposes. The growing preference of individuals for ethnomedicines has stimulated study in the subject of ethnopharmacology, a trend that persists to the present day (Ahmad et al. 2021).

9.5 Market Share and Demand

The market size of the Ayush sector has experienced significant expansion following the move of the Department of AYUSH to the Ministry of AYUSH in 2014. The Ayush manufacturing sector witnessed a significant growth over a span of 7 years, as indicated by the latest study conducted by the Research and Information System for Developing Countries (RIS) in 2020. In 2014–2015, the industry's value stood at Rs. 21,697 crores (USD 2.85 Bn), whereas the study estimates the industry's size to have reached Rs. 1,37,800 crores (USD 18.1 Bn) in 2020. This represents a six-fold increase in value during the aforementioned 7-year period. In a similar vein, an initial examination of the Revenue Information System (RIS) reveals a total revenue of Rs 1,66,797 crore in the Ayush Service sector (Annual Report, Ministry of AYUSH, 2022–23).

Prior to the COVID-19 pandemic, the ayurveda market experienced an average yearly growth rate ranging from 15% to 20%. During the previous fiscal quarter, numerous enterprises, both large and small, have experienced growth ranging from 50% to 90%. The incorporation of Ayurveda as a comprehensive and organic healthcare system is anticipated to yield favorable outcomes within the market. In the period spanning from March of the previous year to the present, there has been a notable increase in the consumer demand for honey, Chyawanprash, and turmeric inside Ayurvedic retail establishments. Specifically, honey has experienced a surge of 45%, Chyawanprash has witnessed a substantial rise of 85%, and turmeric has observed a growth of 40%. The surge in demand for these items has mostly been driven by the endorsements from the Ministry of AYUSH as a means to combat the coronavirus.

According to the Union Ministry of Ayush, there has been a 2.67% increase in the value of exports of Ayush and herbal products in the fiscal year 2022–2023. In contrast, imports have experienced a growth of 8.6% when compared to the previous financial year.

The financial year 2022–2023 had a total export value of $628.5 million for Ayush and herbal items from India, representing a little increase from the previous fiscal year's similar 12-month period, which recorded exports valued at $612.1 million. According to statistics provided by the India Exim Bank Research and the Directorate General of Commercial Intelligence and Statistics (DGCI&S), there was an increase of 8.6 percent in imports, rising from $111.1 million in the fiscal year 2021–2022 to $120.7 million in the fiscal year 2022–2023, as reported by the Ministry.

9.6 Challenges

- From the perspective of ethnopharmacology investigators and regulatory agencies, the evaluation of the safety and effectiveness of complex mixes, some of which contain minerals/metallic components, should be approached differently compared to traditional biomedical medications, primarily because of their historical use in humans. The ethical standards of 2000, established by the Indian Council of Medical Research (ICMR), classified traditional formulations into three distinct groups to provide regulatory advice. The World Health Organization (WHO) has acknowledged the principle of excluding phase I trials for formulations that provide strong evidence of their previous utilization in humans. The aforementioned information was explicitly articulated in the guidance document of 2005. The Indian Council of Medical Research (ICMR) included certain pertinent elements from the aforementioned article into the revised edition of its ethical principles from 2006. These elements were specifically included in the part pertaining to the clinical assessment of traditional Ayurveda, Siddha, Unani (ASU) treatments, and medicinal plants. By considering the contents of both publications, the subsequent sections present viable foundational perspectives for a standardized methodology in conducting research on traditional medicine/herbal or herbo-mineral formulations.
- The guidelines or regulations pertaining to chemical medications and traditional formulations cannot be assumed to be uniform. It is important to acknowledge that a plant material or product may have chemicals that have not been fully defined, and these constituents may potentially offer therapeutic benefits by working along with the recognized active constituent. One notable illustration of this phenomenon is the Indian identification of reserpine from Sarpagandha (Rauwolfia serpentina), a botanical species commonly known as Indian snakeroot. This discovery proved to be a potent antihypertensive agent; yet, its widespread usage was hindered by the significant adverse effect of inducing depression. Nevertheless, when the extract containing additional elements is utilized, this particular negative effect is not observed. It continues to be employed

in Ayurveda to this day. The plant has multiple active principles that possess an intrinsic system for maintaining balance. However, this equilibrium is disrupted when a single active ingredient is extracted and purified to create a novel pharmaceutical medicine.
- There is lack of authentic source of crude drugs for the preparation of traditional formulations to be used in clinical trials. The quality of the crude material plays important role in the pharmacological properties of the formulation, and it is a continuous challenge in this field of research.
- According to the guidelines provided by the Indian Council of Medical Research (ICMR), it is recommended to do restricted toxicity studies in two animal species if ASU formulations are intended for administration for a duration exceeding 3 months, if toxicity has been described in the literature, or if a large multicentric phase III trial is planned. However, researchers tend to adhere to the rules established by the Organization for Economic Co-operation and Development (OECD) for conducting toxicity studies on animals in order to ensure safety and enhance global recognition.
- The paramount concern in health care revolves around the assurance of the safety of any substances used by individuals for medical reasons. The efficiency of the material that is consumed is considered a minor factor. Numerous proponents of traditional medicines have the belief that their consumption is generally harmless; however, certain conjectures are derived from substantial practical knowledge acquired in the field. Medicinal substances are frequently ingested through both nonprescription and prescription medications in various regions across the globe. The Indian traditional systems of medicine must prioritize quality in order to align with international standards. This phenomenon is expected to result in widespread international recognition and foster additional advancements in this particular domain. There are additional issues, as previously mentioned, pertaining to uniformity and conduction of both clinical and nonclinical trials for traditional medicines. However, these challenges can be effectively addressed through the formulation and implementation of appropriate policies.

9.7 Summary and Conclusion

Indian traditional systems of medicines have been there from thousands of years and are still the first choice of major population for their primary healthcare needs. Ethnopharmacology has played an important role in scientifically establishing the therapeutic efficacy of these drugs. In India, these systems are believed to be safe and effective. There is notable growth in the field of research, marked demand and awareness. Government is also committed to preserve and expand this knowledge treasure. The global acceptance of these systems is also increasing after COVID-19 pandemic. The WHO Global Centre for Traditional Medicine is being established in Jamnagar under the Ministry of AYUSH. This would be the first and only global outposted center (office) for traditional medicine across the globe. This indicates the increasing reputation of India in the field of traditional systems of medicines across the globe. Now the emphasis should be given to the quality, standardization, systematic ethnopharmacological evaluations to meet the global standards if we wish to continue this growth.

References

Ahmad S, Zahiruddin S, Parveen B, Basist P, Parveen A, Gaurav PR, Ahmad M (2021) Indian medicinal plants and formulations and their potential against COVID-19-preclinical and clinical research. Front Pharmacol 11:578970. https://doi.org/10.3389/fphar.2020.578970

Ashalatha KL, Arunkumar KP, Gowda M (2023) Genomic and transcriptomic analysis of sacred fig (Ficusreligiosa). BMC Genomics 24(1):197. https://doi.org/10.1186/s12864-023-09270-z

Behera J, Kumar MP, Femela AI, Senguttuvan G, Ramasamy MS (2023) miRNA-15/IL-10Rα axis promotes KabasuraKudineer (Indian traditional Siddha formulation) induced immunomodulation by suppressing oxidative stress. J Ethnopharmacol 305:116032. https://doi.org/10.1016/j.jep.2022.116032

Bhat S, Rotti H, Prasad K, Kabekkodu SP, Saadi AV, Shenoy SP, Joshi KS, Nesari TM, Shengule SA, Dedge AP, Gadgil MS, Dhumal VR, Salvi S, Satyamoorthy K (2023) Genome-wide DNA methylation profiling after Ayurveda intervention to bronchial asthmatics identifies differential methylation in several transcription factors with immune process related function. J Ayurveda Integr Med 14(2):100692. https://doi.org/10.1016/j.jaim.2023.100692

Bora PS, Agrawal P, Kaushik NK, Puri S, Sahal D, Sharma U (2023) Antiplasmodial activity of the bulbs of Fritillariacirrhosa D.Don (Syn: Fritillariaroylei Hook.): UPLC-IM-Q-TOF-MS/MS-based biochemometric approach for the identification of marker compounds. J Ethnopharmacol 310:116389. https://doi.org/10.1016/j.jep.2023.116389

Chaman SM, Kappi M, Ahmed KKM (2022) Ethnopharmacology research: a Scientometric assessment of Indian publications during 2011 to 2020. J Pharmacol Pharmacother 13(1):48–58. https://doi.org/10.1177/0976500X221082839

Che C-T, George V, Ijinu TP, Pushpangadan P, Andrae-Marobela K (2017) Chapter 2—traditional medicine. In: Badal S, Delgoda R (eds) Pharmacognosy. Academic Press, pp 15–30., ISBN 9780128021040. https://doi.org/10.1016/B978-0-12-802104-0.00002-0

Chopra A, Tillu G, Chuadhary K, Reddy G, Srivastava A, Lakdawala M, Gode D, Reddy H, Tamboli S, Saluja M, Sarmukaddam S, Gundeti M, Raut AK, Rao BCS, Yadav B, Srikanth N, Patwardhan B (2023) Co-administration of AYUSH 64 as an adjunct to standard of care in mild and moderate COVID-19: a randomized, controlled, multicentric clinical trial. PLoS One 18(3):e0282688. https://doi.org/10.1371/journal.pone.0282688

Choudhary B (2002) The new international seed treaty: promises and prospects for food security. Curr Sci 83:366–369

Ghosh AK (2010) A short history of the development of homeopathy in India. Homeopathy 99(2):130–136. https://doi.org/10.1016/j.homp.2009.10.001

Gurav S, Wanjari M, Bhole R, Raut N, Prasad S, Saoji S, Chikhale R, Khanal P, Pant A, Ayyanar M, Gurav N (2023) Ethnological validation of Ashwagandha (Withaniasomnifera L. Dunal) ghrita as 'VajikaranaRasayana': in-silico, in-vitro and in-vivo approach. J Ethnopharmacol 304:116064. https://doi.org/10.1016/j.jep.2022.116064

Hedaoo M, Patil-Bhole T, Sharma R, Mahajan M (2023) Exploratory quasi-experimental study of anti-arthritic activity of Ayurvedicpolyherbal formulation, AbhaGuggulu in osteoarthritis patients. Drug Metab Pers Ther 38(3):281–288. https://doi.org/10.1515/dmpt-2022-0187

Ishtiyaq A (1983) Kulliyat-e-Asri, New Public Press Delhi, 34–83

Kaligal C, Kanthi A, Vidyashree M, Krishna D, Raghuram N, HongasandraRamarao N, Deepeshwar S (2023) Prefrontal oxygenation and working memory in patients with type 2 diabetes mellitus following integrated yoga: a randomized controlled trial. Acta Diabetol 60(7):951–961. https://doi.org/10.1007/s00592-023-02085-0. Epub 2023 Apr 4

Khan F, Ansari AN, Nayab M (2023) Application of hot medicated fomentation for pain alleviation in non-specific low back pain—a randomized controlled clinical trial. J Bodyw Mov Ther 35:196–201. https://doi.org/10.1016/j.jbmt.2023.04.075

Khan MU, Basist P, Gaurav, Zahiruddin S, Reddy N, Penumallu AS (2024) Ameliorative effect of traditional polyherbal formulation on TNF-α, IL-1β and Caspase-3 expression in kidneys of wistar rats against sodium fluoride induced oxidative stress. J Ethnopharmacol 318(Part A):116900. https://doi.org/10.1016/j.jep.2023.116900

Kshirsagar SR, Kumari M, Bajad SM, Kumar MJM, Saxena SN, Kumari I (2023) Assessment of sub-chronic oral toxicity of Nityanand rasa: an ayurvedicherbo-metallic formulation. J Ethnopharmacol 312:116494. https://doi.org/10.1016/j.jep.2023.116494

Li Z, Li X, Tang S, Gao Q, Li C, Chen P, Yue X, Fu R, Huang X, Zhang Y, Yang H, Yang B (2024) Moringa oleifera Lam. Leaf improves constipation of rats induced by low-fiber-diet: a proteomics study. J Ethnopharmacol 318(Part A):116936. https://doi.org/10.1016/j.jep.2023.116936

Lythgoe MP, Middleton P (2020) Ongoing clinical trials for the management of the COVID-19 pandemic. Trends Pharmacol Sci 41(6):363–382. https://doi.org/10.1016/j.tips.2020.03.006

Mondal S (2018) UNIT-II Basic principles of treatment in Ayurvedic systems of medicine. DOI: 10.13140/RG.2.2.32019.58404

Mukherjee PK, Venkatesh M, Kumar V (2007) An overview on the development in regulation and control of medicinal and aromatic plants in the Indian system of medicine. Bol Latinoam Caribe Plant Med Aromat 6(4):129–136

Narayanaswamy V (1975) Introduction to the siddha system of medicine. Research Institute of Siddha Medicine, T. Nagar, Madras (Chennai)

Prasad LV (2002) In: Roy CR, Muchatar RU (eds) Indian system of medicine and homoeopathy traditional medicine in Asia. WHO- Regional Office for South East Asia, New Delhi, pp 283–286

Quazi SH, Pathak SS (2023) Ayurgenomics in anti-viral therapy: a literature synthesis. Cureus 15(2):e34649. https://doi.org/10.7759/cureus.34649

Ramachandra Rao SK (1987) Encyclopedia of Indian medicine, vol 2. Dr. P.V. Parameshvara Charitable trust, Bangalore

Ratheesh M, Jose SP, Sheethal S, Sindhu A, Sandya S, Asish A (2023) Ameliorating effect of polyherbal formulation (Febrojith) on acute and chronic inflammatory model via inhibiting inflammation and oxidative stress. J Ethnopharmacol 317:116824. https://doi.org/10.1016/j.jep.2023.116824

Sahasranaman A, Kumar N (2020) Network structure of COVID-19 spread and the lacuna in India's testing strategy. SSRN Electron J. https://doi.org/10.2139/ssrn.3558548

Sharma PV (2011) Ayurveda kaVaijnanika Itihasa. Chaukhambha Orientalia, Varanasi, p 606

Sharma S, Singh G, Patwardhan K (2023) Effects of two Ayurvedic formulations, Dhanwantaram Kashaya and Saraswatarishta on life history parameters and toxic aggregates in Drosophila models of Huntington's and Alzheimer's disease. J Ethnopharmacol 305:116080. https://doi.org/10.1016/j.jep.2022.116080

Singh MK, Savita K, Singh S, Mishra D, Rani P, Chanda D, Verma RS (2023) Vasorelaxant property of 2-phenyl ethyl alcohol isolated from the spent floral distillate of damask rose (Rosa damascena mill.) and its possible mechanism. J Ethnopharmacol 313:116603. https://doi.org/10.1016/j.jep.2023.116603

Yang Y, Islam MS, Wang J, Li Y, Chen X (2020) Traditional Chinese medicine in the treatment of patients infected with 2019-new coronavirus (SARS-CoV-2): a review and perspective. Int J Biol Sci 16(10):1708–1717. https://doi.org/10.7150/ijbs.45538

Databases of Medicinal Plants: An Update

10

Gandhar Pradhan and Yogesh A. Kulkarni

Abstract

There are various traditional and herbal medicines available as a solution for the treatment of various disorders such as cancer, diabetes, Alzheimer's disease, rheumatoid arthritis, Parkinson's disease, and their associated comorbidities. These medicines are composed of various forms such as powder, extract of medicinal plant/s. Authentic information regarding the properties and geographical location of the medicinal plants in a particular region plays important role in medicinal plant research. Thus, many countries are focusing on the compilation and curation of online databases and compendia to serve as a platform for researchers to acquire adequate knowledge with respect to medicinal plants in a specific zone. The present chapter aims to recognize the details of the work carried out by various countries by providing systematic information of the traditional knowledge databases of medicinal plants developed by researchers from different parts of the world.

Keywords

Medicinal plants · Databases · Traditional knowledge · TKDL · BSI · OSADHI

10.1 Introduction

Traditional systems of medicine have long been used for treatment of various ailments in many countries such as India, China, Arabia, etc. (Ikram et al. 2015; Gureje et al. 2015). The prevalent systems in India include the Ayurveda (Mukherjee et al.

G. Pradhan · Y. A. Kulkarni (✉)
Shobhaben Pratapbhai Patel School of Pharmacy & Technology Management, SVKM's Narsee Monjee Institute of Management Studies (NMIMS) Deemed to be University, Mumbai, India

© The Author(s), under exclusive license to Springer Nature Singapore Pte Ltd. 2024
M. Nandave et al. (eds.), *Ethnopharmacology and OMICS Advances in Medicinal Plants Volume 1*, https://doi.org/10.1007/978-981-97-2367-6_10

2017), the Siddha system (Thas 2008), the Unani system (Itrat 2020), and the Homeopathic System (Jonas 2003). These systems primarily rely upon the use of medicinal plants as a cure for various diseases and ailments right from the prehistoric era. The Chinese system of medicine also involves the use of medicinal plants growing in the local region as a cure for all various diseases (Wang et al. 2019). China is working on developing an integration of its traditional system of medicine with the Western system of medicine in order to cope with the emerging era of precision medicine (Wang and Zhang 2017). Thus, the importance of traditional systems of medicine knows no bounds even after several years of its existence.

Medicinal plants have been reported to be the sources of various primary and secondary metabolites with an array of pharmacological activities such as antihypertensive (Sultana and Asif 2017), anticancer (Alami Merrouni and Elachouri 2021), analgesic (Lee et al. 2021), antimicrobial (Cowan 1999), antiviral (Ali et al. 2021), antioxidant (Zafar et al. 2023), etc. Research projects have been undertaken with respect to the identification of sources of potential drug like candidates against specific diseases from medicinal plants (Guzman and Molina 2018).

Thus, there is a need of developing a system for understanding of the properties and potential of various medicinal plants. A single database or multiple databases for each country focusing on the medicinal plants should be made available in that respective region.

10.2 Databases Used in India

10.2.1 Traditional Knowledge Digital Library (TKDL)

The Traditional Knowledge Digital Library is one of the most widely utilized databases of medicinal plants in India which was launched to satisfy the needs of 70% of the population of the country dependent on traditional medicine. TKDL has been able to successfully address the barriers of language and format by giving a scientific structure to the traditional texts of Ayurveda, Siddha, Unani, Sowa Rigpa, and Yoga. All these texts have been translated into various international languages such as English, Japanese, French, German, and Spanish with the help of information technology tools such as Traditional Knowledge Resource Classification. This unique system of classification allows the embodiment of 200 subgroups under A61K 36/00 in the International Patent Classification which is instrumental in modifying the efficiency of the search of patent applications related with the knowledge of the traditional systems of medicine. The basis of this database lies in the books of the Indian system of medicine that are within the open domain and can be accessed by any individual or organization. Thus, this database can bridge the gap between the books and the International Patent Examiners with over 3.6 lakh transcribed formulations/practices (Kalbande and Suradkar 2021).

10.2.2 The Medicinal Plant Database of the Botanical Survey of India

This database has a variety of information with respect to the scientific name, family, vernacular name, medicinal uses, location of species, and images of herbarium specimen of all the medicinal flora and fauna prevalent in India. It is extremely dynamic with weekly updates and 1915 species along with an additional upcoming 1000 species. Other digital resources such as virtual herbarium, Journal Nelumbo, etc. are provided by the Botanical Survey of India to extend the knowledge of the viewer with respect to the appropriate data regarding any medicinal plant species of interest. The Indian Virtual Herbarium contains a set of information of the family, genus, species, author citation, subspecies, variety (if any), collector, collection number, collection date, herbarium region, localities, plant description, habitat, and comments (if applicable) of a species along with the JPEG image. The virtual herbarium can be accessed at https://bsi.gov.in/page/en/virtual-herbarium. (Accessed 22 Oct 23) Journal Nelumbo is specialized service provided by the Botanical Survey of India with the aim of facilitating botanical research in the country. It provides a center of attraction for researchers from all over the globe working in the areas of plant taxonomy, biosystematics, ethno- and economic botany, phytogeography, endemism, etc. The journal is available at https://nelumbo-bsi.org/index.php/nlmbo/aj (Accessed 22 Oct 2023).

10.2.3 Indian Medicinal Plants, Phytochemistry, and Therapeutics Database

It is developed by the Institute of Mathematical Sciences, Chennai, hosting a data set of around 1742 Indian medicinal plants, 9596 phytochemicals, 1124 therapeutic uses constituting 27,074 plant-phytochemical associations, and 11,514 plant-therapeutic associations (Mohanraj et al. 2018). Indian medicinal plants have been linked with 974 traditional Indian medicinal formulations. An in silico library of 9596 phytochemicals with two-dimensional (2D) and three-dimensional (3D) chemical structures along with their predicted absorption, distribution, metabolism, excretion (ADME), and toxicity has been curated in the database (Vivek-Ananth et al. 2023). The drug likeness of all the compounds has been predicted with the help of Lipinski's rule of five (Lipinski et al. 2001), Oral PhysChem Score (Traffic Lights) (Lobell et al. 2006), GlaxoSmithKline's 4/400 (Gleeson 2008), Pfizer's 3/75 (Hughes et al. 2008), Veber rule (Veber et al. 2002), Egan rule (Egan et al. 2000), etc. Thus, the database has successfully enabled the application of cheminformatics in order to streamline the process of natural product-based drug discovery.

10.2.4 Online Structural and Analytics-Based Database for Herbs of India (OSADHI)

This database is classified as one of the largest manually curated databases based on documentation, digitization, deposition, and data science by the Council of Scientific Education and Research, Lucknow. The current version has anticipated the data related with 21,238 medicinal plants of India comprising of 6959 unique plant species. The geographical locations of 27,440 phytochemicals along with 2477 therapeutic uses of these unique plants have been articulated. The vernacular names of 343 plant families have been mentioned in this database. The two-dimensional chemical structures of 22,314 phytochemicals along with their physicochemical properties, drug-likeness properties, and predicted ADMET properties have been provided in the database. The major features of this database have been summarized in Table 10.1.

10.2.5 Consumer Unity and Trust Society (CUTS) Database on Medicinal Plants

This database, developed by the CUTS Centre for International Trade, Economics, and Environment is well equipped with the data of medicinal plants from the Himalayan region, Arunachal Pradesh, Meghalaya, and Uttaranchal. The features of the plant species described include common name, family, plant parts used, medicinal properties/uses, type of species, estimated domestic demand, market price, extent of availability, the presence or absence in the Red Data Book, and the distribution of a particular species in the various states. It contains the data of 35 most important medicinal plant species in the states based on environmental concerns and the potential to generate income through cultivation.

10.2.6 Compendium of Antiviral Medicinal Plants of Northeast India

Developed by the Institute of Bioresources and Sustainable Development (IBSD), Imphal, Manipur and Botanical Survey of India, Kolkata, this compendium focuses

Table 10.1 Features of OSADHI database

Feature	Components
Traditional knowledge	Vernacular names, plant parts used, therapeutic use, and taxonomy
Geographical classification	Distribution of medicinal plants throughout the states of India and Union Territories
Phytochemicals	SMILES, InChIKey, IUPAC, 2D and 3D structures of phytochemicals
Chemoinformatic analysis	Physicochemical properties, ADMET, classification, etc.

only on the medicinal plants showing antiviral activity predominantly found in the Northeast region of India. It provides a detailed account of all the 70 plants in terms of scientific classification, vernacular names, synonyms, botanical description, geographical distribution, location, plant material of interest, medicinal/therapeutic uses, phytochemical profile, pharmacological activities, toxicity and safety, clinical studies, contraindications, precautions, adverse reactions, and marketed formulations.

10.2.7 eFloraofIndia

This is one of the largest noncommercial databases covering all the plants of the Indian Subcontinent, maintained by the members of the efloraofIndia Google group. It has a data of around 13,000 species with over 3,00,000 pictorial representations of each, out of which, 2,50,000 have been portrayed alongside the genera and family pages to enhance identification.

10.2.8 Indian Ocean Rim Association (IORA) Medicinal Plants Database

It is maintained by the Ministry of External Affairs, Government of India, and contains an array of knowledge with regard to the botanical information, geographical distribution and regional names, traditional uses in local communities, pharmacological properties, herbal formulations, status in the global markets, patents, and cultivation of the medicinal plants from the IORA recognized countries only.

A summary of the different databases prevalent in India is presented in Table 10.2.

10.3 Databases Used in the USA

10.3.1 Dr. Duke's Phytochemical and Ethnobotanical Databases

Maintained by the US Department of Agriculture, it provides various types of reports based on different features of the phytochemical in question. Plant reports document the taxonomy information, details of the chemicals, and activities of the medicinal plants. The chemical reports document an array of plants and their activities having an interaction with a particular chemical. The biological activity reports document the information about the chemicals and plants linked with an activity. The ethnobotany plant reports document a list of uses, geographical range, and references for a plant in the ethnobotanical database. The ethnobotany use report gives a list of plant linked with a specific use in the ethnobotanical database. The database is available at: https://phytochem.nal.usda.gov/. Accessed 22 Oct 2023.

Table 10.2 Summary of medicinal plant databases of India

Name of the database	Maintained by	Link
Traditional Knowledge Digital Library	Council of Scientific Education and Research (CSIR)	https://www.tkdl.res.in/tkdl/langdefault/common/Home.asp?GL=Eng. Accessed 22 Oct 2023
The Medicinal Plant Database of the Botanical Survey of India	Botanical Survey of India, Kolkata	https://bsi.gov.in/page/en/medicinal-plant-database. Accessed 22 Oct 2023
Indian Medicinal Plants, Phytochemistry and Therapeutics Database	The Institute of Mathematical Sciences, Chennai	https://doi.org/10.1021/acsomega.3c00156. Accessed 22 Oct 2023
Online Structural and Analytics Based Database for Herbs of India	Council of Scientific Education and Research (CSIR)	https://www.neist.res.in/osadhi/fsearch.php. Accessed 22 Oct 2023
Consumer Unity and Trust Society (CUTS) Database on Medicinal Plants	CUTS Centre for International Trade, Economics and Environment	https://www.cuts-international.org/pdf/Database-fullreport.pdf. Accessed 22 Oct 2023
Compendium of Antiviral Medicinal Plants of Northeast India	• Institute of Bioresources and Sustainable Development (IBSD), Imphal, Manipur • Botanical Survey of India, Kolkata	https://ibsd.gov.in/ibsd/home/index.php/research/view?id=Compendium. Accessed 22 Oct 2023
eFloraofIndia	eFloraofIndia Google Group	https://efloraofindia.com/home/. Accessed 22 Oct 2023
IORA Medicinal Plants Database	Ministry of External Affairs, Government of India	https://iccmp.in/about.php. Accessed 22 Oct 2023

10.4 Databases Used in Chile

10.4.1 Rizoma

This database is maintained by the Centre for Agroecological and Botanical Studies, Valparaíso, and contains 1380 use records of 736 native Chilean vascular plant species constituting 399 genera and 128 families (Cordero et al. 2022). It provides a detailed account of the geographic distribution, phytogeographic origin, life form, life span, and use category of the medicinal plants. The online version provides information on the mode of use, common names, and photographs of each species. Thus, this database has emerged as a depot of traditional knowledge by conserving the local, biological, and cultural diversity of the Chilean flora and fauna. It is available at: https://ceab-rizoma.com/database/. Accessed 22 Oct 2023.

10.5 Databases Used in the UK

10.5.1 Garden of Medicinal Plants

It is maintained by the Royal College of Physicians, UK, and houses over 1000 medicinal plants currently or previously cultured in the garden along with the basic information regarding their name, geographical distribution, and reason for growing and their medicinal uses. It can be accessed at: https://garden.rcplondon.ac.uk/. Accessed 22 Oct 2023.

10.6 Databases Used in Japan

10.6.1 NIBIOHN Comprehensive Medicinal Plant Database

Developed by the Research Centre for Medicinal Plant Resources, NIBIOHN, Japan, this database acts as a storehouse for most of the medicinal flora and fauna found in the region of Japan. The information regarding any plant can be accessed through a search involving the common name of the plant or scientific name of the plant or name of the crude drug or compound or sample which would be at the interest of the researchers. It is available at: http://mpdb.nibiohn.go.jp/mpdb-bin/top.cgi?lang=en. Accessed 22 Oct 2023.

10.6.2 WAKANYAKU Database

It is maintained by the Institute of Natural Medicine, Japan, and provides information related to the reviews on natural medicines, natural compounds isolated and identified at the institute along with their LC-MS profiles. It also publishes the research involving the use of any of the crude drugs mentioned in the *Honzotsukan*, a book written in the Edo period. It can be accessed at: https://www.inm.u-toyama.ac.jp/en/database/. Accessed 22 Oct 2023.

10.6.3 Traditional Medical and Pharmaceutical Database

It is developed by the Institute of Natural Medicine and stores all sorts of scientific data on crude drugs, kampo formulas, and natural compounds. The published results of genetic, chemical, and biological analysis tests carried out on the abovementioned categories are well articulated in this database. Thus, the standardization of natural products is facilitated thereby aiding the process of new drug discovery. It is available at: https://dentomed.toyama-wakan.net/index_en. Accessed 22 Oct 2023.

10.6.4 The Database of Ethnomedicines in the World

Maintained by the Museum of Materia Medica, this database has all the scientific information with regard to the samples of the crude drugs stored in the museum. It includes the drug from different traditional systems of medicine such as the Traditional Japanese and Chinese System, Unani System, Thai System, Ayurveda, and Indonesian system, respectively. It can be accessed through: https://ethmed.toyama-wakan.net/SearchEn/. Accessed 22 Oct 2023.

10.6.5 Kampo DB

Developed by the Institute of Natural Medicine, this database provides information related to the interactions between natural compounds and their target proteins predicted with the help of molecular docking and machine learning techniques. It also gives a detailed account about the pathways and biological processes which are most likely to be altered by the kampo formulas and crude drugs. It is available at: https://wakanmoview.inm.u-toyama.ac.jp/kampo/. Accessed 23 Oct 2023.

10.6.6 Plant Genome Database Japan

It is maintained by the National Bioscience Database Center of the Japan Science and Technology Agency and is composed of three sections, namely, the plant resource, the ortholog database and marker, QTL, and others. The plant resource section deals with the search of a medicinal plant from the citrus origin and consists of the information regarding the name of the plant, variety, classification, and the English Name, respectively. It also provides information related to the full-length cDNA clones, seeds, and cultured cell lines of arabidopsis, physcomitrella, poplar, and cassava along with the cereals, forage crops, and legumes. The ortholog database gives a detailed account of the genetic information of the plant species which can be accessed through an amino acid or keyword information search. The marker and QTL section includes an array of information related to the sequence-tagged DNA markers thereby enhancing a more accurate and precise genome comparisons between different species. It can be accessed at: http://pgdbj.jp/index.html?ln=en. Accessed 23 Oct 2023.

10.6.7 Chinese Medicinal Herbal Classics "Authentic Herbs" Database

It is maintained by the Museum of Materia Medica and provides the Japanese translation of the original text of "Shennong Bencaojing" which is a drug book containing 1700 items including medicinal plants, animals, minerals, edible fruits, rice grains, vegetables, and poultry along with their clinical efficacy. Thus, this database has given birth to a rich and precise source of information regarding pharmaceuticals and botany by focusing on both the medicinal and food effects of the items listed. It is available at: https://ethmed.toyama-wakan.net/Honzou/. Accessed 23 Oct 2023.

10.7 Databases Used in China

10.7.1 Chinese Herbal Medicine Database

Developed by the Hong Kong Polytechnic University, this database acts as a storehouse of authentic information about nearly 400 herbal medicines and crude drugs used in China. It gives a summary of source, basic nature, action, usage, and safety of the Chinese herbs along with their images (decoction pieces or Chinese materia medica). Decoction pieces refer to the Chinese herbs which have been cut into pieces, while materia medica refers to the dried herbs which have not been cut into specific shapes. This database acts as a one-stop solution for researchers across the globe to access information regarding Chinese Herbal Medicines thereby facilitating the learning of Chinese Traditional Medicine. It is available at: https://herbaltcm.sn.polyu.edu.hk/. Accessed 23 Oct 2023.

10.8 Databases Used in Singapore

10.8.1 Singapore Herbarium Online

It is developed by the Singapore Botanic Gardens and allows the researchers to access authenticated information related to the specimen data and images of the species stored in the Singapore Herbarium. This herbarium stores around 750,000 specimens from different regions such as Malesian region (Peninsular Thailand, Malaysia, Singapore, Brunei, Indonesia, the Philippines, and New Guinea) and nearby areas (East Asia, mainland SE Asia, the Southwest Pacific), with the most elaborate collections from Singapore and Peninsular Malaysia. The search query can be processed with the help of the name of the family, genus, or species, respectively. It can be accessed at: https://herbaria.plants.ox.ac.uk/bol/sing. Accessed 23 Oct 2023.

10.9 Miscellaneous Databases

10.9.1 WHO Monographs of Selected Medicinal Plants

It is maintained by the World Health Organization, Geneva, and provides a detailed account of the plants which are scientifically well explored in all the WHO regions to demonstrate the safety and efficacy of the phytochemicals present in these plants

(World Health Organization et al. 2006). The purpose of designing the monographs was to establish scientific evidence for the safety, efficacy, and quality control/quality assurance of the medicinal plants, to provide a model to assist the Member States to develop their own monographs or formularies, and to enable exchange of information amongst the Member States. It is available at: https://apps.who.int/iris/handle/10665/42052. Accessed 30 July 2023.

10.10 Conclusion

The importance of having a common integrative platform to supply information regarding most of the medicinal plants in a particular region knows no bounds. India is the country with a great "herbal heritage" due to which many Indian states have initiated the process of development of databases related to medicinal plants in terms of their phytochemicals, taxonomy, phytochemicals, and therapeutic uses.

The world is realizing the value of traditional and herbal medicines due to the side effects of various synthetic drugs which can be easily overcome with the aid of natural products. Many countries are aiming to be "bio-rich" which in simple terms means having an abundance of local medicinal plants sufficing the need for an effective therapeutic approach toward the management of various existing and upcoming diseases.

Thus, to facilitate the conduction of further research around natural products, traditional and herbal medicines, there is a need to undertake many projects focused on the development of a common traditional knowledge database of medicinal plants which would enlighten researchers with the detailed knowledge of each and every medicinal plant growing all over the world.

References

Alami Merrouni I, Elachouri M (2021) Anticancer medicinal plants used by Moroccan people: ethnobotanical, preclinical, phytochemical and clinical evidence. J Ethnopharmacol 266:113435. https://doi.org/10.1016/j.jep.2020.113435

Ali SI, Sheikh WM, Rather MA, Venkatesalu V, Muzamil Bashir S, Nabi SU (2021) Medicinal plants: treasure for antiviral drug discovery. Phytother Res 35:3447–3483. https://doi.org/10.1002/ptr.7039

Cordero S, Abello L, Gálvez F (2022) Rizoma: a new comprehensive database on traditional uses of Chilean native plants. Biodivers Data J 10:e80002. https://doi.org/10.3897/BDJ.10.e80002

Cowan MM (1999) Plant products as antimicrobial agents. Clin Microbiol Rev 12:564–582. https://doi.org/10.1128/CMR.12.4.564

Egan WJ, Merz KM, Baldwin JJ (2000) Prediction of drug absorption using multivariate statistics. J Med Chem 43:3867–3877. https://doi.org/10.1021/jm000292e

Gleeson MP (2008) Generation of a set of simple, interpretable ADMET rules of thumb. J Med Chem 51:817–834. https://doi.org/10.1021/jm701122q

Gureje O, Nortje G, Makanjuola V, Oladeji BD, Seedat S, Jenkins R (2015) The role of global traditional and complementary systems of medicine in the treatment of mental health disorders. Lancet Psychiatry 2(2):168–177. https://doi.org/10.1016/S2215-0366(15)00013-9

Guzman E, Molina J (2018) The predictive utility of the plant phylogeny in identifying sources of cardiovascular drugs. Pharm Biol 56:154–164. https://doi.org/10.1080/13880209.2018.1444642

Hughes JD, Blagg J, Price DA, Bailey S, Decrescenzo GA, Devraj RV, Ellsworth E, Fobian YM, Gibbs ME, Gilles RW, Greene N, Huang E, Krieger-Burke T, Loesel J, Wager T, Whiteley L, Zhang Y (2008) Physiochemical drug properties associated with in vivo toxicological outcomes. Bioorg Med Chem Lett 18:4872–4875. https://doi.org/10.1016/j.bmcl.2008.07.071

Ikram RRR, Abd Ghani MK, Abdullah N (2015) An analysis of application of health informatics in traditional medicine: a review of four traditional medicine systems. Int J Med Inform 84(11):988–996. https://doi.org/10.1016/j.ijmedinf.2015.05.007

Itrat M (2020) Methods of health promotion and disease prevention in Unani medicine. J Educ Health Promot 9. https://doi.org/10.4103/jehp.jehp_618_19

Jonas W (2003) A critical overview of homeopathy. Ann Intern Med 139(8):W-76. https://doi.org/10.7326/0003-4819-138-5-200303040-00009

Kalbande D, Suradkar P (2021) Traditional knowledge digital library: a Magic bullet in the war against biopiracy. Library Philosophy and Practice (e-journal)

Lee JH, Kim N, Park S, Kim SK (2021) Analgesic effects of medicinal plants and phytochemicals on chemotherapy-induced neuropathic pain through glial modulation. Pharmacol Res Perspect 9:e00819. https://doi.org/10.1002/prp2.819

Lipinski CA, Lombardo F, Dominy BW, Feeney PJ (2001) Experimental and computational approaches to estimate solubility and permeability in drug discovery and development settings. Adv Drug Deliv Rev 46:3–26. https://doi.org/10.1016/s0169-409x(00)00129-0

Lobell M, Hendrix M, Hinzen B, Keldenich J, Meier H, Schmeck C, Schohe-Loop R, Wunberg T, Hillisch A (2006) In silico ADMET traffic lights as a tool for the prioritization of HTS hits. ChemMedChem 1:1229–1236. https://doi.org/10.1002/cmdc.200600168

Mohanraj K, Karthikeyan BS, Vivek-Ananth RP, Chand RPB, Aparna SR, Mangalapandi P, Samal A (2018) IMPPAT: a curated database of Indian medicinal plants, phytochemistry and therapeutics. Sci Rep 8:4329. https://doi.org/10.1038/s41598-018-22631-z

Mukherjee PK, Harwansh RK, Bahadur S, Banerjee S, Kar A, Chanda J, Biswas S, Ahmmed SM, Katiyar CK (2017) Development of Ayurveda–tradition to trend. J Ethnopharmacol 197:10–24. https://doi.org/10.1016/j.jep.2016.09.024

Sultana S, Asif HM (2017) Review: medicinal plants combating against hypertension: a green antihypertensive approach. Pak J Pharm Sci 30:2311–2319

Thas JJ (2008) Siddha medicine—background and principles and the application for skin diseases. Clin Dermatol 26(1):62–78. https://doi.org/10.1016/j.clindermatol.2007.11.010

Veber DF, Johnson SR, Cheng H-Y, Smith BR, Ward KW, Kopple KD (2002) Molecular properties that influence the oral bioavailability of drug candidates. J Med Chem 45:2615–2623. https://doi.org/10.1021/jm020017n

Vivek-Ananth RP, Mohanraj K, Sahoo AK, Samal A (2023) IMPPAT 2.0: an enhanced and expanded phytochemical atlas of Indian medicinal plants. ACS Omega 8:8827–8845. https://doi.org/10.1021/acsomega.3c00156

Wang WJ, Zhang T (2017) Integration of traditional Chinese medicine and Western medicine in the era of precision medicine. J Integr Med 15(1):1–7. https://doi.org/10.1016/S2095-4964(17)60314-5

Wang L, Wang Z, Ma Q, Fang G, Yang J (2019) The development and reform of public health in China from 1949 to 2019. Glob Health 15(1):1–21. https://doi.org/10.1186/s12992-019-0486-6

World Health Organization, Plants, W.C. on S.M., WHO Consultation on Selected Medicinal Plants (2006) (2nd : 1999 : Ravello-Salerno, I., WHO Consultation on Selected Medicinal Plants (3rd : 2001 : Ottawa, O.), WHO Consultation on Selected Medicinal Plants (4th : 2005 : Salerno-Paestum, I.. WHO monographs on selected medicinal plants. World Health Organization. https://iris.who.int/handle/10665/42052

Zafar F, Asif HM, Shaheen G, Ghauri AO, Rajpoot SR, Tasleem MW, Shamim T, Hadi F, Noor R, Ali T, Gulzar MN, Nazar H (2023) A comprehensive review on medicinal plants possessing antioxidant potential. Clin Exp Pharmacol Physiol 50:205–217. https://doi.org/10.1111/1440-1681.13743

Role of Genome-Based Barcoding in Quality Control and Standardization of Medicinal Plants

11

S. Shreedevasena and Krishnan P. Abhijith

Abstract

DNA barcoding has emerged as a potential tool in medicinal plant research, providing a robust and accurate method for species identification. Its application in the authentication and classification of medicinal plants addresses challenges associated with traditional methods like morphological identification. Utilizing specific DNA regions, such as barcode sequences from the chloroplast genome (e.g., *rbcL*, *matK*, and *trnH-psbA*) and nuclear ribosomal DNA (ITS), facilitates discrimination among closely related species, enhancing the reliability of plant identification. The significance of DNA barcoding lies in its ability to overcome limitations associated with morphological variations, seasonal changes, and the presence of processed plant materials in medicinal products. DNA barcoding plays a crucial role in quality control within the herbal medicine industry, ensuring the safety and efficacy of medicinal plant products. This technique aids in detecting adulteration, contamination, or substitution, thereby safeguarding consumers and promoting regulatory compliance. Moreover, DNA barcoding contributes to biodiversity conservation by facilitating the identification of endangered or threatened medicinal plant species. As the global demand for medicinal plants continues to rise, the integration of DNA barcoding in both research and industry holds promise for advancing authenticity, traceability, and sustainability in the medicinal plant sector. This chapter underscores the transformative impact of DNA barcoding on medicinal plant science, providing a

S. Shreedevasena (✉)
ICAR- Directorate of Medicinal and Aromatic Plant Research, Anand, Gujarat, India
e-mail: Shreedevasena.s@icar.gov.in

K. P. Abhijith
School of Crop Improvement, ICAR- Indian Agricultural Research Institute Assam, Dhemaji, Assam, India

© The Author(s), under exclusive license to Springer Nature Singapore Pte Ltd. 2024
M. Nandave et al. (eds.), *Ethnopharmacology and OMICS Advances in Medicinal Plants Volume 1*, https://doi.org/10.1007/978-981-97-2367-6_11

molecular foundation for precision in identification, quality assurance, and conservation efforts.

Keywords
DNA barcodes · Medicinal plants · Biodiversity conservation · Adulteration · Global trade

11.1 Introduction

India boasts a rich heritage of Ayurvedic medicinal plants that enjoy significant international market demand. The expansion of the global pharmaceutical industry has heightened the demand for these plants in Indian markets, with India contributing 7–8% to the world biodiversity of medicinal plants. International trade and exchange of medicinal plants in India amount to a substantial turnover of $1.6–$1.8 billion (Mishra et al. 2016). Concerns regarding the adulteration of herbal plants in the global market have prompted organizations such as the American Spice Trading Association (ASTA) and the European Spice Association (ESA) to establish minimal quality requirements for phytochemical specifications, authentication, and standardization of medicinal and aromatic plants for worldwide food markets to mitigate unfair trading practices (Kala 2005). According to the World Health Organization (WHO), the total international market value of the pharmaceutical industry is currently US$62 billion, expected to grow to US$5 trillion by 2050 (Yadav and Misra 2012). In recent years, DNA barcoding has emerged as a valuable molecular technique to complement traditional taxonomic expertise for rapid species identification and biodiversity inventories (Laiou et al. 2013). China is taking the lead in DNA barcoding research of medicinal plants, establishing a DNA barcode database (Lone et al. 2019). Reports on DNA barcoding of Indian medicinal plants are also available to enhance quality standardization in the global herbal market (Techen et al. 2014).

DNA barcoding represents a burgeoning development in medicinal plants, primarily employed for quality control and standardization in the pharmaceutical industry to curb the substitution and adulteration of medicines (Kress and Erickson 2012). Given the precarious status of many medicinal plants nearing extinction, DNA barcodes serve as crucial tags for detection and identification purposes. The escalating demand has prompted numerous small farmers, medical businesses, and entrepreneurs to shift their focus to medicinal crop farming, resulting in challenges related to adulteration when resources are limited (Chen et al. 2010). Despite this, numerous wild medicinal and aromatic plants (MAPs) with potential therapeutic significance remain unidentified.

DNA barcoding, a standard molecular universal identification technique, enables the characterization of various species and the detection of adulteration in plant genotypes (Dhanya and Sasikumar 2010). DNA barcoding involves using small DNA fragments for taxonomy identification. The universal standard region employed for identification is termed the DNA barcode. Barcoding finds extensive application in studying phylogeny, analyzing inter- and intra-diversity, and aiding in the classification, identification, and distinction of wild, cultivar, and adulterant plants. This article underscores the crucial role of DNA barcoding in identifying and characterizing morphologically indistinguishable MAPs, facilitating the analysis of differences between closely related taxa and species.

11.2 Plant Authentication Approaches

The organoleptic methods (identification by the senses: taste, sight, smell, touch), macroscopic and microscopic methods (Identification by form, colour, texture), and chemical profiling (e.g., TLC, HPLC-UV, HPLC-MS) are among the conventional methods of medicinal plant identification (Barbuto et al. 2010). The related species cannot be easily identified in processed products using either method, either, as the former requires specialized personnel for macroscopic and microscopic tests. Chemical profiling or markers-based identification is the recently used authentication in medicinal plants which is highly influenced by the plant physiology, storage conditions, and other external factors (Hollingsworth et al. 2009). Studies on the systematics and phylogeny of aromatic plants have typically been based on physical traits (Singh 2019). However, during the past few years, steady advancements in molecular biology provided a number of novel methods that can be used to characterize the unusual chemical makeup of related cultivars and study the relationships between these species (Marieschi et al. 2009). The DNA barcoding conditions, namely, a minimum sequence length of 500 bp and more than three organisms per single species, have been proven by 13,761 species (Joshi et al. 2004). The plant-based markers like plastidial (*rpoB*, *rpoC1*, *rbcL*, and *matK*) genes; intergenic plastidial spacers (*trnH-psbA*, *atpF-atpH*, and *psbK-psbI*) and the nuclear internal transcribed spacers have are the efficient DNA barcoding regions (Chase et al. 2007; Fazekas et al. 2008). The major markers used for the plant identification are *matK*, *rbcL*, *trnH-psbA*, *ITS*, *trnL-F*, 5S-rRNA, and 18S-rRNA. Presently, the characterization of critical and complex plant groups is identified with *matK* and *rbcL* combinations to increase their efficiency in identification (Hebert et al. 2003).

The DNA-based approaches provide high reliability and are not affected by external factors; therefore, developing DNA-based markers for medicinal plants is very important for proper authentications. These molecular markers are effective; however despite this, they are not always able to distinguish between distinct species or cultivars, and they are frequently not universally useable on a wide panel of taxa, because they were created on specific genera or species (Yang et al. 2012). DNA barcoding, which analyzes a short, universal DNA region, has recently been presented as a universal tool for species identification. The main premise of the discrimination method is that species can be successfully distinguished from one another if the sequence variation of the DNA barcode across species is higher than within species. In practice, a DNA sequence from such a standardized DNA region can be created from a small tissue sample and compared to a library of reference sequences belonging to documented species, allowing for rapid and reliable taxonomic recognition. DNA barcoding opens up new avenues for medicinal plant identification, which is valuable not only for clarifying taxonomic issues within the family but also for investigating commercial elements of spice traceability from field to market (Newmaster et al. 2013).

11.3 DNA Barcoding of Medicinal Plants

Hebert et al. proposed the DNA barcoding approach for species identification in 2003. The Barcode loci, or DNA sections used for identification, can distinguish between closely related species and identify new cryptic species (Hollingsworth et al. 2011). Different barcode loci are utilized for the purpose depending on the taxon and complexity of the species (Fig. 11.1). To find more effective barcodes in plant species, researchers have looked at chloroplast and nuclear genomes (Joshi et al. 2004). Major standard single-locus candidate barcodes have been proposed in the recent two decades: *ITS*, *ITS2*, *matK*, *rbcL*, *psbA-trnH*, and *trnL-trnF*, which discriminate plant species with high accuracy. However, it was discovered that a single barcode was insufficient for identifying all plants, necessitating the usage of multi-locus DNA barcodes (Kress and Erickson 2012). The Plant Working Group of the Consortium for the Barcoding of Life (CBOL) recommended a combination of *matK* and *rbcL* loci to improve species discrimination accuracy (Liang and Hilu 1996).

The advancement of sequencing technology permitted the advancement of DNA barcoding from identifying a single herb in medicinal plants to concurrently detecting numerous herbs in an herbal medicine cocktail, hence encouraging the selection and use of DNA barcoding (Marichamy et al. 2014). The development of

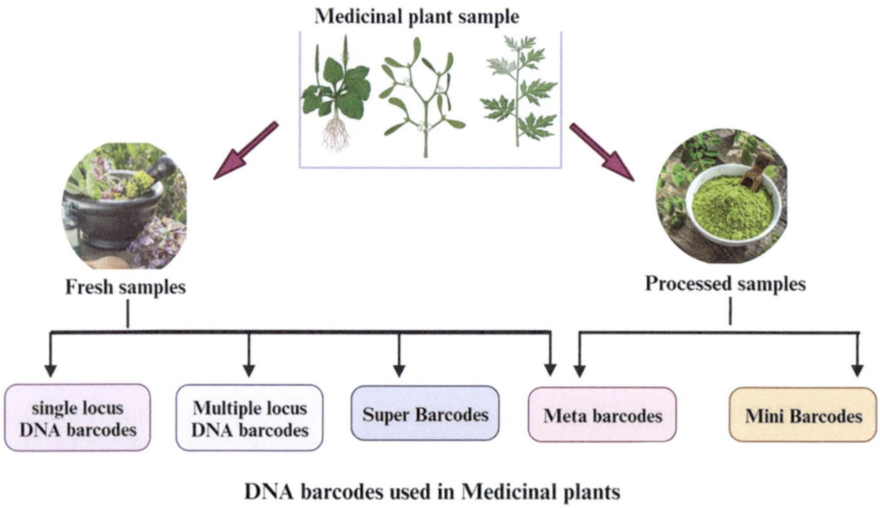

Fig. 11.1 Schematic representation of different barcoding used in medicinal plants

next-generation sequencing (NGS) and third-generation sequencing technologies has accelerated the development of DNA barcode-derived technologies for identifying herbal medicine species. Super-barcoding, meta-barcoding, and mini-barcoding are examples of modern DNA barcode derivative techniques (Fig. 11.1) (Chen et al. 2023). The mini-barcoding can identify species from highly degraded DNA, while meta-barcoding can be used to analyze species richness in a sample including a variety of species (li et al. 2015). The super-barcoding based on the plant chloroplast genome can be used to determine species-relatedness (Park et al. 2020). These three DNA barcoding-based technologies have expanded the applications and made DNA barcoding more realistic. Data mining and analysis tools have increased the use of DNA barcoding-based technologies that may effectively identify biological systems in herbal medicines (Anabat et al. 2020).

The seven key plant barcodes that have received the greatest attention in standard single locus DNA barcoding are ITS and ITS2, *rbcL*, *matK*, *trnL-trnF*, *psbA-trnH*, *ycf1*, and *rpoC* (Ganie et al. 2019). Among the barcodes, ITS and ITS2 are predominantly booming and more utilized barcodes. ITS2 region can differentiate not only plant taxa from distinct plant families but also closely related taxa at the genus and species levels. When a full-length ITS is not available, ITS2 could be utilized as an alternative mini-barcoding method (Sheidai et al. 2019). The second most common barcodes are *rbcL*, *matK*, and *trnL-trnF*, probably because they can be utilized as multi-locus DNA barcodes. Several genes, including *atpF-atpH*, *ndhF-rpl32*, and *psbK-psbI*, are candidates for barcoding. When employing various barcodes for related species, conflicting findings have been discovered; however, a single locus barcode does not provide the evolutionary distinctions required to distinguish related species (Tripathi et al. 2013). Multi-locus DNA barcoding is preferable when accurate discrimination and sufficient genetic information are required (Liu et al. 2011). According to Kress and Erickson (2007), ITS + *psbA-trnH* has the ability to discriminate against a wide range of plant species. Chen et al. (2010) first proposed the ITS2 sequence as a universal barcode for identifying medicinal plants and the ITS2 + *psbA-trnH* combination as a DNA barcoding system for identifying botanical medicinal herbs. Super-barcoding is a technique that uses entire plastid genome sequencing as a barcode to accurately identify closely related species (Wu et al. 2021). In order to identify closely related groups, super-barcoding uses high-throughput next-generation sequencing to produce a high copy of a reliable sequence (Kane and Cronk 2008). Super-barcoding improves the density and phylogenetic coverage of the entire plastid genome sequence and is predicted to detect traditional herbal medications accurately (Krawczyk et al. 2018).

The rise of meta-barcoding, a new DNA barcoding-based method for rapidly and concurrently identifying several species in a single environmental sample, has emerged with the advent of next-generation sequencing technologies. Meta-barcoding is used to identify plant species using entire but degraded DNA extracted from an environmental sample. It also has the potential to identify plant species variation within processed commodities, which is crucial for confirming the validity of therapeutic plant products. Due to the typical DNA degradation in processed samples, obtaining full-length sequencing data with regular standard barcodes is difficult. This constraint can be overcome with mini-DNA barcoding technology (Meusnier et al. 2008). Mini-barcoding can identify various species by utilizing incomplete, relatively short sequences from normal DNA barcodes, which is important for damaged DNA preservation. It increases the accuracy of species identification overall (Little 2014).

The CBOL Plant Working Group evaluated seven chloroplast genomic regions and proposed the 2-locus *matK* + *rbcL* plant barcode in an international conference since matK provides high resolution but less universality, whereas rbcL provides high universality but less species resolution (Tali et al. 2019). To achieve higher discrimination in closely related species, CBOL Group proposed to add the nuclear ITS to the *matK* + *rbcL* combination. The ITS2 sequence has been identified as a universal barcode for medicinal plant identification, and the ITS2 + *psbA-trnH* combination is recommended as a DNA barcoding system for identifying botanical medicinal herbs (Taberlet et al. 2012). The strengths and limitations of the barcodes and their derivative barcoding technology, including single-locus barcoding, multi-locus barcoding, super-barcoding, meta-barcoding, and mini-barcoding, were discussed (Table 11.1).

11.4 Application of DNA Barcodes in Medicinal Plants

DNA barcoding has revolutionized the field of medicinal plant research by providing a reliable, standardized, and efficient method for identifying species. Its multifaceted applications have significant implications for the pharmaceutical industry, traditional medicine, biodiversity conservation, and global trade regulations (Fig. 11.2). It involves analyzing specific DNA regions to accurately identify and authenticate medicinal plant species. They also prevent misidentification and substitution, ensuring the efficacy and safety of herbal medicines. DNA barcodes aid in distinguishing closely related species that might have similar morphological features enable accurate taxonomic classification, crucial for identifying specific plant traits and chemical constituents. DNA barcoding can confirm the botanical origins of plants used in traditional medicine, bridging indigenous knowledge with modern science. DNA barcode helps in identifying threatened or endangered medicinal plant species, aiding conservation efforts. Contributes to understanding the genetic diversity and distribution of medicinal plants. It also assists in legal cases involving illegal trade, ensuring the authenticity of traded plant materials and provides evidence for enforcing regulations on the sustainable use of medicinal plants.

The efficiency of a DNA barcoding approach to commercial spice recognition is influenced by the presence of taxonomic challenges as well as evidence of hybridization events within specific families. DNA barcoding is a good endeavor to confirm HMP authenticity, which may assist reputable brands in creating a reputation and ensuring consumer safety. The DNA barcoding methods and databases help to detect the misidentification; however, this technique varies with the tissue type and marker locus. DNA barcodes like plastid markers and nuclear markers like rbcL, matK, and ITS2 were used in the investigation of herbal medicines sold in markets.

Table 11.1 Strengths and limitations of barcode loci

Barcode loci	Strength	Limitation	Taxonomic units	References
ITS1, ITS2	Generality, simplification in usage provides high copy number, shows interspecific variability and intraspecific uniformity	Provide incomplete concerted evolution as well as difficulties of amplification and sequencing Heterogeneity is an issue for ITS2 due to concerted evolution, which may lead to inaccurate or misleading results	Family, genus, species, subspecies	Chen et al. (2010), Yao et al. (2010)
matK	High sequence variation with efficiency rates, PCR amplification, with suitable sequence length, accurate discrimination between plants Intra and inter-specific divergence distinction in the barcoding gap.	While matK generally shows higher variability than other markers like rbcL, there are cases of low intraspecific variation, especially in certain plant groups. This can limit the ability to discriminate between closely related species	Family, genus, species	Wu et al. (2021)
rbcL	High primer versatility, easy amplification and alignment, and high discrimination power	Relatively low interspecific identification power Poor discrimination of closely related species	Family, genus	Ming et al. (2017)
psbA-trnH	Relatively high amplification of good length with high success rates	Manual correction is required for sequence analysis which difficult to compare different genera and species	Family, genus, species, subspecies	Pang et al. (2012), Liu et al. (2011)
trnL–trnF	High nucleotide conversion rate, give high genetic variation with more systematic taxonomic information	Mononucleotide repeats can impact sequencing reads in some taxa	Genus, species, subspecies	Tripathi et al. (2013)
Super-barcoding	Higher resolution and better versatility Accurate discrimination of subspecies	High cost and the requirement for high quality and quantity of DNA Large next-generation sequence data generated which requires skillful technicians	Genus, species, subspecies	Liu et al. (2014)

(continued)

Table 11.1 (continued)

Barcode loci	Strength	Limitation	Taxonomic units	References
Meta-barcoding	Ability to identify every species in a complex sample or processed mixtures simultaneously Provides high coverage sequence data Ability to reveal plant species diversity within processed products	Limited by the PCR success rate Difficult in creating taxonomic reference libraries Sequencing errors in high-throughput sequencing	Species, subspecies	Govender et al. (2022)
Mini-barcoding	Due to the shorter molecular markers of mini-barcode, different physicochemical technologies can be combined to identify herbal samples rapidly	Only few nucleotides provide limited taxonomic discrimination Limited resolution Sufficient reference sequence data is needed for accurate species identification which mainly depends on the acceptable resolution	Family, genus, species, subspecies	Wang et al. (2022)

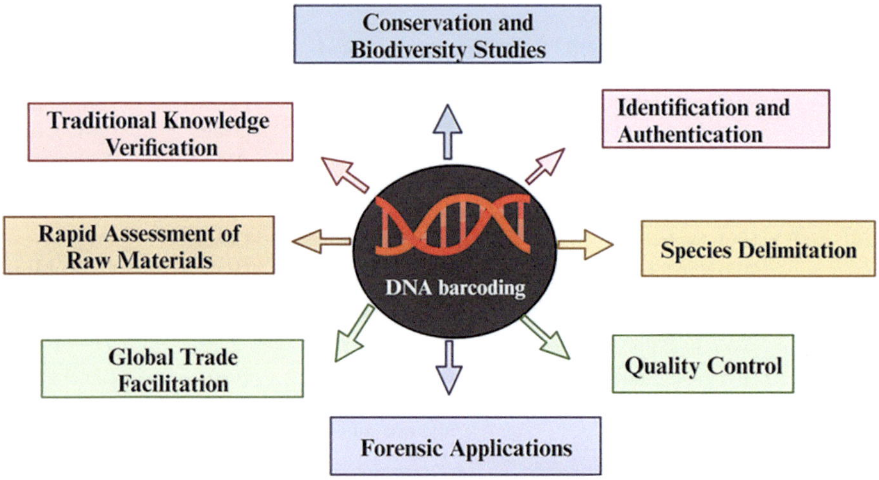

Fig. 11.2 Application of DNA barcoding in medicinal plants

ITS2 alone was accurate enough to match 88% of ITS2 sequences to the predicted species, whereas only 59% of sequenced rbcL and 45% of matK could match. To assure quality, regulatory agencies may suggest ITS2 as a viable DNA barcode to herbal medicine makers and merchants (Michel et al. 2016).

The samples from six different genera (*Mentha*, *Ocimum*, *Origanum*, *Salvia*, *Thymus*, and *Rosmarinus*) were amplified and sequenced with universal primers to identify four barcodes (*rpoB*, *rbcL*, *matK*, and *trn-psbA*). The non-coding plastid gene *trnHpsbA* intergenic region appears to be the best marker for molecular species identification, followed by *matK*, with interspecific genetic distance from 0% to 7% and 0% to 5%, respectively (De Mattia et al. 2011). Punarnava (*Boerhavia diffusa*) is India's traditional indigenous medicinal herb. The DNA barcoding of *Boerhavia diffusa* aids in the reliable identification of closely related species, hence improving medicine quality and biosafety. For the phylogenetic study of four *Boerhavia* species, bar code candidates containing nuclear ribosomal DNA regions ITS, ITS1, ITS2, and the chloroplast plastid gene *psbA-trnH* were used. Among the bar codes, the phylogenetic tree built with the ITS sequence separated one species from the others. As a result, ITS and ITS1 could be possible candidate locations for recognizing and authenticating herbal medicinal plants (Selvaraj et al. 2012). Similarly, *Epimedium elatum*, a rare medicinal plant noted for its high concentration of Epimedin B, C, and icariin, is confined to the gloomy woodland area of India's northwestern Himalayas. Three plastid genes (*matK*, rbcL, and *trnH-psbA*) and one nuclear region gene (ITS) were partially sequenced to create a unique barcode as well as examine nucleotide diversity and evolutionary relationships among five geographically distinct *E. elatum* accessions (Lone et al. 2019).

MatK and ITS sequencing revealed a diverse haplotype and single nucleotide polymorphism profile. Both primers created a dependable and effective method of detecting nucleotide sequence diversity. ITS and *matK* satisfied the Consortium of Barcode of Life's ideal DNA barcode qualities and can thus be recommended as excellent barcode markers for the identification of medicinal plants. The combination of morphological observation with DNA barcoding ensures the correct identification of wild plants of medicinal value and also focuses on public safety (Chen et al. 2023).

11.5 Challenges of Barcoding

11.5.1 Levels of Genomic Variation Can Play Tricks with Barcoding!

The barcoding relies on the genomic variation available between the species. A DNA sequence that is variable enough to distinguish various species yet flanked by a highly conservative genomic region defines an idea barcode. Finding such regions is not so easy, especially in the case of plants (Rubinoff et al. 2006). There are two challenges associated with this aspect. First designing a universal primer which can amplify the region of interest in high resolution across the species is a complex task (Singh et al. 2012), and secondly, such a single locus barcode often fails to harbor sufficient genetic variation to distinguish across all the species, especially species which belong to closely associated taxonomic groups (Ficetola et al. 2010; Singh et al. 2012; Li et al. 2015). Therefore, it's always advisable to go with multiple markers which target the various genomic regions to capture accurate genetic diversity rather than solely relying on a single barcode (Nazar et al. 2022).

11.5.2 The Quality of DNA Matters!

Extraction of pure and high molecular weight DNA is crucial in the successful implementation of various genomic studies, and the same requirement applies to DNA barcoding also. However medicinal plants often contain significant amounts of secondary metabolites like polysaccharides, polyphenols, alkaloids, and flavonoids, which can hamper in isolation of high-quality DNA. This case is much more critical in the case of samples of processed medicinal plants where the DNA is often degraded and its extraction is quite challenging. However, to tackle, this various DNA extraction methods for plants with high levels of secondary metabolites and polysaccharides have developed (Sahu et al. 2012). Kazi et al. (2013) have demonstrated a PCR-based identification technique to identify the presence of St. John's Wort (*Hypericum perforatum* L.) from various herbal products including capsules, tablets, and tinctures. However, Särkinen et al. (2012) reported that the amplification of DNA fragments from degraded DNA templates was not uniform, and amplification of shorter fragments has more success rate than that of longer fragments.

11.5.3 Erroneous Identification in the Cases of Hybridization, Introgression, and Cryptic Species

In cases of hybridization between closely related species or introgression (gene flow) between species, the barcode sequences of the hybrid or introgressed individuals may not match the expected barcode of either parent species. This can lead to misidentification if the barcode is used as the sole identifier. Since hybrids contain genetic material from two or more parent species, leading to mixed DNA sequences that may not match any reference barcode in the database, this can cause also ambiguities in the results. Another issue frequently arising in this regard is hybrid DNA which may contain variations in primer binding sites or target regions, making PCR amplification less efficient or leading to a failed amplification.

When the barcoding is based on the plastid genome, hybridization can create much more complications. For instance, when two species get hybridized, the nuclear genome of both species will be passed on to the progeny, whereas the cytoplasm of only one species gets passed on, and eventually after several generations of backcrosses, the progenies will retain the nuclear genome of one species and plastid genome of another species. In this situation, barcoding will lead to an erroneous conclusion (Cowan and Fay 2012).

Cryptic species are species that are morphologically similar but genetically distinct. These species may share similar or identical barcode sequences, leading to misidentification or failure to distinguish between them (Nazar et al. 2022). One such example is the case of mint (*Mentha spicata*) where several cryptic species are identified through a detailed phylogenetic analysis (Heylen et al. 2021).

11.5.4 Polyploidy

Polyploid plants have multiple copies of genes, including barcode regions. This can lead to complications in amplification and sequencing, as different gene copies might have variations that affect the accuracy of barcoding (Barcaccia et al. 2015). Another issue that might arise in the case of polyploid is each gene may have different alleles due to mutations, leading to sequence variation within the same species. This can make it challenging to define a single representative sequence for barcoding. In case of allopolyploids (where the genomes are homoeologous), sequences are similar but are distinct copies of genes from different ancestral genomes. These sequences can complicate the PCR reaction if the barcode is derived from such genomic regions.

In the case of barcoding based on the plastid genome, further complications occur. In most cases, recently formed autopolyploids will have the same plastid genome as the parental diploid (Cowan and Fay 2012). Plastid genomes are typically maternally inherited in plants, which mean that the plastid genome of an autopolyploid will be inherited from the maternal parent. Since autopolyploids arise from the same species, the maternal parent's plastid genome remains unchanged (Crosby and Smith 2012). This will make it difficult to distinguish between the diploid progenitor and the autopolyploid with a barcode alone. Similar constraints will be faced in the case of an allopolyploid where its plastid genome will be similar to only one of the progenitors.

11.6 Conclusion

The Indian subcontinent is blessed with its abundant heritage of diverse ayurvedic medicinal plants that has caught worldwide attention which generates a large sum of revenue. However, the absence of robust quality control measures can easily lead to adulteration and substitution which will ultimately pull down the integrity of native medicinal products in the market. Genome-based barcoding addresses this major challenge in the herbal medicine industry to aid in the identification and authentication of medicinal plants. This approach benefits both the indigenous herbal medicinal industry and the global pharmaceutical industry in maintaining the integrity and quality of medicinal plants and their products. Attempts toward developing DNA-based markers, like *matK*, *rbcL*, *ITS*, etc., have very well succeeded and proved their potential in identifying and characterizing medicinal plant species. The advent of next-generation sequencing and novel genomic tools and techniques such as super-barcoding and meta-barcoding has expanded the applications of DNA barcoding, enabling the simultaneous detection of multiple plant components in complex mixtures. Even though DNA barcoding is a promising technology, it still faces some lacunas. Issues such as extraction of high-quality DNA from processed plant samples, issues posed by genetic variation, erroneous identification in the cases of hybridization and polyploidy, introgression, and cryptic species need to be addressed, and appropriate solutions and technologies can revamp this technology to be foolproof. Therefore a thorough refinement in the technology as well as exponential growth of comprehensive reference databases of barcoding will bring more utility of DNA barcoding toward identifying, verifying, and upholding the quality of medicinal plants. However, implementing DNA barcoding for the reliable identification of medicinal plants demands more than technology. A collaborative effort with the participation of encompassing researchers, regulators, and the herbal medicine industry can play a pivotal role in endorsing this technology widely thereby ensuring the safety of consumers. This will also provide an opportunity for brands to establish their reputations in the field.

References

Anabat MM, Riahi H, Sheidai M, Koohdar F (2020) Population genetic study and barcoding in Iran saffron (Crocus sativus L.). Ind Crop Prod 143:111915

Barbuto M, Galimberti A, Ferri E, Labra M, Malandra R, Galli P et al (2010) DNA barcoding reveals fraudulent substitutions in shark seafood products: the Italian case of "palombo"(Mustelus spp.). Food Res Int 43:376–381

Barcaccia G, Lucchin M, Cassandro M (2015) DNA barcoding as a molecular tool to track down mislabeling and food piracy. Diversity 8:2

Chase MW, Cowan RS, Hollingsworth PM, Van Den Berg C, Madriñán S, Petersen G et al (2007) A proposal for a standardised protocol to barcode all land plants. Taxon 56:295–299

Chen S, Yao H, Han J, Liu C, Song J, Shi L et al (2010) Validation of the ITS2 region as a novel DNA barcode for identifying medicinal plant species. PLoS One 5:e8613

Chen S, Yin X, Han J, Sun W, Yao H, Song J et al (2023) DNA barcoding in herbal medicine: retrospective and prospective. J Pharm Anal 13:431

Cowan RS, Fay MF (2012) Challenges in the DNA barcoding of plant material. In: Sucher N, Hennell J, Carles M (eds) Plant DNA fingerprinting and barcoding. Methods in molecular biology, vol 862. Humana Press. https://doi.org/10.1007/978-1-61779-609-8_3

Crosby K, Smith DR (2012) Does the mode of plastid inheritance influence plastid genome architecture? PLoS One 7. https://doi.org/10.1371/journal.pone.0046260

De Mattia F, Bruni I, Galimberti A, Cattaneo F, Casiraghi M, Labra M (2011) A comparative study of different DNA barcoding markers for the identification of some members of Lamiacaea. Food Res Int 44:693–702. https://doi.org/10.1016/j.foodres.2010.12.032

Dhanya K, Sasikumar B (2010) Molecular marker based adulteration detection in traded food and agricultural commodities of plant origin with special reference to spices. Curr Trends Biotechnol Pharm 4:454–489

Fazekas AJ, Burgess KS, Kesanakurti PR, Graham SW, Newmaster SG, Husband BC et al (2008) Multiple multilocus DNA barcodes from the plastid genome discriminate plant species equally well. PLoS One 3:e2802. https://doi.org/10.1371/journal.pone.0002802

Ficetola GF, Coissac E, Zundel S, Riaz T, Shehzad W, Bessière J et al (2010) An in silico approach for the evaluation of DNA barcodes. BMC Genomics 11:1–10. https://doi.org/10.1186/1471-2164-11-434

Ganie AH, Tali BA, Khuroo AA, Reshi ZA, Nawchoo IA (2019) Impact assessment of anthropogenic threats to high-valued medicinal plants of Kashmir Himalaya, India. J Nat Conserv 50:125715. https://doi.org/10.1016/j.jnc.2019.125715

Govender A, Singh S, Groeneveld J, Pillay S, Willows-Munro S (2022) Experimental validation of taxon-specific mini-barcode primers for metabarcoding of zooplankton. Ecol Appl 32:e02469. https://doi.org/10.1002/eap.2469

Hebert PDN, Cywinska A, Ball SL, DeWaard JR (2003) Biological identifications through DNA barcodes. Proc R Soc B Biol Sci 270:313–321. https://doi.org/10.1098/rspb.2002.2218

Heylen OCG, Debortoli N, Marescaux J, Olofsson JK (2021) A revised phylogeny of the mentha spicata clade reveals cryptic species. Plan Theory 10:819. https://doi.org/10.3390/plants10040819

Hollingsworth ML, Andra Clark A, Forrest LL, Richardson J, Pennington RT, Long DG et al (2009) Selecting barcoding loci for plants: evaluation of seven candidate loci with species-level sampling in three divergent groups of land plants. Mol Ecol Resour 9:439–457. https://doi.org/10.1111/j.1755-0998.2008.02439.x

Hollingsworth PM, Graham SW, Little DP (2011) Choosing and using a plant DNA barcode. PLoS One 6:e19254. https://doi.org/10.1371/journal.pone.0019254

Joshi K, Chavan P, Warude D, Patwardhan B (2004) Molecular markers in herbal drug technology. Curr Sci 87:159–165

Kala CP (2005) Current status of medicinal plants used by traditional Vaidyas in Uttaranchal state of India. Ethnobot Res Appl 3:267. https://doi.org/10.17348/era.3.0.267-278

Kane NC, Cronk Q (2008) Botany without borders: barcoding in focus. Mol Ecol 17:5175–5176. https://doi.org/10.1111/j.1365-294X.2008.03972.x

Kazi T, Hussain N, Bremner P, Slater A, Howard C (2013) The application of a DNA-based identification technique to over-the-counter herbal medicines. Fitoterapia 87:27–30. https://doi.org/10.1016/j.fitote.2013.03.001

Krawczyk K, Nobis M, Myszczyński K, Klichowska E, Sawicki J (2018) Plastid super-barcodes as a tool for species discrimination in feather grasses (Poaceae: Stipa). Sci Rep 8:1924. https://doi.org/10.1038/s41598-018-20399-w

Kress WJ, Erickson DL (2007) A two-locus global DNA barcode for land plants: the coding rbcL gene complements the non-coding trnH-psbA spacer region. PLoS One 2:e508. https://doi.org/10.1371/journal.pone.0000508

Kress WJ, Erickson DL (2012) DNA barcodes: methods and protocols (pp. 3-8). Humana Press

Laiou A, Mandolini LA, Piredda R, Bellarosa R, Simeone MC (2013) DNA barcoding as a complementary tool for conservation and valorisation of forest resources. ZooKeys 365:197–213. https://doi.org/10.3897/zookeys.365.5670

Li X, Yang Y, Henry RJ, Rossetto M, Wang Y, Chen S (2015) Plant DNA barcoding: from gene to genome. Biol Rev Camb Philos Soc 90:157–166. https://doi.org/10.1111/brv.12104

Liang H, Hilu KW (1996) Application of the *matK* gene sequences to grass systematics. Can J Bot 74:125–134. https://doi.org/10.1139/b96-017

Little DP (2014) A DNA mini-barcode for land plants. Mol Ecol Resour 14:437–446. https://doi.org/10.1111/1755-0998.12194

Liu C, Liang D, Gao T, Pang X, Song J, Yao H et al (2011) PTIGS-IdIt, a system for species identification by DNA sequences of the *psbA-trnH* intergenic spacer region. In: BMC bioinformatics. Springer, pp 1–7. https://doi.org/10.1186/1471-2105-12-S13-S4

Liu J, Shi L, Han J, Li G, Lu H, Hou J et al (2014) Identification of species in the angiosperm family Apiaceae using DNA barcodes. Mol Ecol Resour 14:1231–1238. https://doi.org/10.1111/1755-0998.12262

Lone SA, Hassan QP, Gupta S (2019) Development of DNA barcode for rapid identification of Epimedium elatum (Morren & Decne) from northwestern Himalayas in India. J Appl Res Med Aromat Plants 13:100205. https://doi.org/10.1016/j.jarmap.2019.100205

Marichamy K, Kumar NY, Ganesan A (2014) Sustainable development in exports of herbals and ayurveda, siddha, unnani, homeopathy (Ayush) in India. Sci Park 1:1–6

Marieschi M, Torelli A, Poli F, Sacchetti G, Bruni R (2009) RAPD-based method for the quality control of Mediterranean oregano and its contribution to pharmacognostic techniques. J Agric Food Chem 57:1835–1840. https://doi.org/10.1021/jf8032649

Meusnier I, Singer GAC, Landry JF, Hickey DA, Hebert PDN, Hajibabaei M (2008) A universal DNA mini-barcode for biodiversity analysis. BMC Genomics 9:1–4. https://doi.org/10.1186/1471-2164-9-214

Michel CI, Meyer RS, Taveras Y, Molina J (2016) The nuclear internal transcribed spacer (ITS2) as a practical plant DNA barcode for herbal medicines. J Appl Res Med Aromat Plants 3:94–100. https://doi.org/10.1016/j.jarmap.2016.02.002

Ming S, Gang-Qiang DONG, Zhang YQ, Xia LIU, Wei SUN (2017) Identification of processed Chinese medicinal materials using DNA mini-barcoding. Chin J Nat Med 15(7):481–486

Mishra P, Kumar A, Nagireddy A, Mani DN, Shukla AK, Tiwari R et al (2016) DNA barcoding: an efficient tool to overcome authentication challenges in the herbal market. Plant Biotechnol J 14:8–21. https://doi.org/10.1111/pbi.12419

Nazar N, Howard C, Slater A, Sgamma T (2022) Challenges in medicinal and aromatic plants DNA barcoding—lessons from the Lamiaceae. Plan Theory 11:137. https://doi.org/10.3390/plants11010137

Newmaster SG, Grguric M, Shanmughanandhan D, Ramalingam S, Ragupathy S (2013) DNA barcoding detects contamination and substitution in North American herbal products. BMC Med 11:1–13. https://doi.org/10.1186/1741-7015-11-222

Pang X, Liu C, Shi L, Liu R, Liang D, Li H et al (2012) Utility of the trnH-psbA intergenic spacer region and its combinations as plant DNA barcodes: a meta-analysis. PLoS One 7:e48833. https://doi.org/10.1371/journal.pone.0048833

Park HS, Jayakodi M, Lee SH, Jeon JH, Lee HO, Park JY et al (2020) Mitochondrial plastid DNA can cause DNA barcoding paradox in plants. Sci Rep 10:6112. https://doi.org/10.1038/s41598-020-63233-y

Rubinoff D, Cameron S, Will K (2006) Are plant DNA barcodes a search for the Holy Grail? Trends Ecol Evol 21:1–2. https://doi.org/10.1016/j.tree.2005.10.019

Sahu SK, Thangaraj M, Kathiresan K (2012) DNA extraction protocol for plants with high levels of secondary metabolites and polysaccharides without using liquid nitrogen and phenol. ISRN Mol Biol 2012:1–6. https://doi.org/10.5402/2012/205049

Särkinen T, Staats M, Richardson JE, Cowan RS, Bakker FT (2012) How to open the treasure chest? Optimising DNA extraction from herbarium specimens. PLoS One 7. https://doi.org/10.1371/journal.pone.0043808

Selvaraj D, Shanmughanandhan D, Sarma RK, Joseph JC, Srinivasan RV, Ramalingam S (2012) DNA barcode ITS effectively distinguishes the medicinal plant Boerhavia diffusa from ITS adulterants. Genom Proteom Bioinform 10:364–367. https://doi.org/10.1016/j.gpb.2012.03.002

Sheidai M, Tabaripour R, Talebi SM, Noormohammadi Z, Koohdar F (2019) Adulteration in medicinally important plant species of Ziziphora in Iran market: DNA barcoding approach. Ind Crop Prod 130:627–633. https://doi.org/10.1016/j.indcrop.2019.01.025

Singh G (2019) Plant systematics: an integrated approach (4th edn). CRC Press. https://doi.org/10.1201/9780429289521

Singh H, Parveen I, Raghuvanshi S, Babbar SB (2012) The loci recommended as universal barcodes for plants on the basis of floristic studies may not work with congeneric species as exemplified by DNA barcoding of Dendrobium species. BMC Res Notes 5:1–11. https://doi.org/10.1186/1756-0500-5-42

Taberlet P, Coissac E, Pompanon F, Brochmann C, Willerslev E (2012) Towards next-generation biodiversity assessment using DNA metabarcoding. Mol Ecol 21:2045–2050. https://doi.org/10.1111/j.1365-294X.2012.05470.x

Tali BA, Khuroo AA, Nawchoo IA, Ganie AH (2019) Prioritizing conservation of medicinal Flora in the Himalayan biodiversity hotspot: an integrated ecological and socioeconomic approach. Environ Conserv 46:147–154. https://doi.org/10.1017/S0376892918000425

Techen N, Parveen I, Pan Z, Khan IA (2014) DNA barcoding of medicinal plant material for identification. Curr Opin Biotechnol 25:103–110. https://doi.org/10.1016/j.copbio.2013.09.010

Tripathi AM, Tyagi A, Kumar A, Singh A, Singh S, Chaudhary LB et al (2013) The internal transcribed spacer (ITS) region and trnhH-psbA are suitable candidate loci for DNA barcod-

ing of tropical tree species of India. PLoS One 8:e57934. https://doi.org/10.1371/journal.pone.0057934
Wang G, Liu Y, Bai X, Cao P, Pang X, Han J (2022) Identification and poisoning diagnosis of Aconitum materials using a genus-specific nucleotide signature. Ecotoxicol Environ Saf 237:113539. https://doi.org/10.1016/j.ecoenv.2022.113539
Wu L, Wu M, Cui N, Xiang L, Li Y, Li X et al (2021) Plant super-barcode: a case study on genome-based identification for closely related species of Fritillaria. Chin Med (United Kingdom) 16:1–11. https://doi.org/10.1186/s13020-021-00460-z
Yadav M, Misra S (2012) Sustainable development: a role for market information systems for non-timber forest products. Sustain Dev 20:128–140. https://doi.org/10.1002/sd.470
Yang JY, Jang SY, Kim H-K, Park SJ (2012) Development of a molecular marker to discriminate Korean Rubus species medicinal plants based on the nuclear ribosomal DNA internal transcribed spacer and chloroplast trnL-F intergenic region sequences. J Korean Soc Appl Biol Chem 55:281–289. https://doi.org/10.1007/s13765-012-1044-6
Yao H, Song J, Liu C, Luo K, Han J, Li Y et al (2010) Use of ITS2 region as the universal DNA barcode for plants and animals. PLoS One 5:e13102. https://doi.org/10.1371/journal.pone.0013102

Future Directions and Emerging Trends in Medicinal Plant DNA Barcoding

12

Aftab Alam, Mohammad Usman, Bushra Parveen, and Sayeed Ahmad

Abstract

The analysis of genome sequences in various plant species has uncovered genes linked to different traits and their inheritance patterns. This exploration has facilitated the mapping of complex traits associated with diverse phenotypes. Recent discoveries demonstrate the utility of DNA barcoding in categorizing unknown specimens, authenticating phytomedicines, and conserving plant diversity. However, no individual common DNA barcode exists for entire plant sets, necessitating a relative examination of plant barcode loci to select the most appropriate candidates for authenticating specific medicinal plant families. Both chloroplast and nuclear regions are utilized as universal barcodes for verifying phytomedicines. Advancements in genomics, particularly next-generation sequencing, have enabled the complete sequencing of plastomes, termed super-barcodes, significantly enhancing traditional ethno-

A. Alam · B. Parveen (✉)
Centre of Excellence in Unani Medicine (Pharmacognosy and Pharmacology) and Bioactive Natural Product Laboratory, SPER, Jamia Hamdard, New Delhi, India

Department of Pharmacology, SPER, Jamia Hamdard, New Delhi, India

M. Usman
Department of Pharmacognosy and Phytochemistry, SPER, Jamia Hamdard, New Delhi, India

Universal Biotech, Delhi, India

S. Ahmad (✉)
Centre of Excellence in Unani Medicine (Pharmacognosy and Pharmacology) and Bioactive Natural Product Laboratory, SPER, Jamia Hamdard, New Delhi, India

Department of Pharmacognosy and Phytochemistry, SPER, Jamia Hamdard, New Delhi, India

© The Author(s), under exclusive license to Springer Nature Singapore Pte Ltd. 2024
M. Nandave et al. (eds.), *Ethnopharmacology and OMICS Advances in Medicinal Plants Volume 1*, https://doi.org/10.1007/978-981-97-2367-6_12

botanical knowledge and promoting safer usage of phytomedicines. The current focus is on using DNA barcoding to ensure the integrity and authenticity of herbal products, defending consumers from potential health hazards due to product replacement or adulteration. Developing a DNA database via next-generation sequencing, and prioritizing conserved regions, holds significant promise for precise identification of medicinal plants.

Keywords

DNA barcoding · Bar-HRM technique · DNA mini-barcoding · Meta-barcoding · DNA Barcoding techniques

Abbreviations

BINs	Barcode index numbers
BOLD	Barcode of life data system
bp	Base pairs
CBG	Centre for Biodiversity Genomics
CBOL	Consortium for the Barcode of Life
COI	Cytochrome c oxidase subunit I
DBW	DNA Barcoding Workshop
DDBJ	DNA Data Bank of Japan
dsDNA	Double-stranded DNA
ENA	European Nucleotide Archive
GBIF	Global Biodiversity Information Facility
HP-TLC	High-performance thin-layer chromatography
HRM	High-resolution melting technology
iBOL	International Barcode of Life Project
LAMP	Loop-mediated isothermal amplification
LC-MS	Liquid chromatography-mass spectrometry
MMDBD	Medicinal Materials DNA Barcode Database
NCBI	National Center for Biotechnology Information
NGS	Next-generation sequencing
OTU	Operational taxonomic units (s)
PCR	Polymerase chain reaction
RPA	Recombinase polymerase amplification
TLC	Thin-layer chromatography
WGS	Whole-genome shotgun

12.1 Introduction

For centuries, medicinal plants have served a vital function in traditional medicine, offering a rich source of natural remedies to address a wide range of health issues. These botanical resources have served as the basis for developing numerous therapeutic compounds used globally. However, due to the increasing demand for herbal remedies, dietary supplements, and natural healthcare products, it has become essential to establish precise methods for identifying, quality control, and tracking medicinal plant materials, which are crucial to confirm the safety and effectiveness of these materials within traditional (herbal) medicine systems (Yuan et al. 2016).

The traditional practice of medicinal plants is deeply rooted in culture and history. Yet, the modernization of health care and the globalization of herbal product markets have presented a series of challenges, primarily concerning the quality and safety of medicinal plants. Adulteration, contamination, and mislabeling are among the issues that jeopardize the integrity of traditional medicine practices and the well-being of consumers (Jităreanu et al. 2022).

In the modern age of genomics and advanced molecular techniques, the integration of traditional wisdom with cutting-edge science presents a promising solution. Genome-based barcoding, originally designed for identifying species in ecological and biodiversity studies, is now emerging as a revolutionary tool in medicinal plant research. Scientists can use this approach to target specific regions that are extremely preserved of the plant genome, allowing for the foundation of distinct genomic signatures for each species which, in turn, enables accurate and dependable species identification, even when the botanical material undergoes processing or mixed with other botanical ingredients (Mosa et al. 2019).

The use of genome-based barcoding has the potential to redefine our approach to authenticating and controlling the quality of medicinal plants. It offers a robust and precise method for confirming the botanical identity of plant materials, thus mitigating issues related to substitution and contamination. Additionally, it plays a pivotal role in preserving endangered plant species, which ensure their sustainable utilization in traditional medicine (Malik et al. 2019).

While DNA sequencing techniques have been around since the mid-twentieth century, they were initially constrained by their labor-intensive, time-consuming, and expensive nature. These limitations rendered DNA sequencing impractical for species identification (Heather and Chain 2016).

The history of DNA barcoding is a captivating journey that has significantly impacted the field of biological identification. DNA barcoding, utilizing short and standardized DNA sequences as unique species identifiers, has undergone continuous evolution (Gostel and Kress 2022a). It was initially introduced by Canadian scientist Paul Hebert in the early 2000s, proposing a method for distinguishing between species based on specific DNA regions. The cytochrome c oxidase subunit 1 (COI) gene was chosen as the common animal barcode due to its combination of conserved areas and interspecies diversity. These qualities made it suitable for sequencing and amplification. Over time, DNA barcoding has expanded to encompass plants and fungi, becoming an invaluable tool in biodiversity research. It aids in identifying new species, differentiating cryptic species, and assessing ecosystem diversity. Challenges and debates emerged, especially regarding the selection of barcode genes, but the technique has continued to evolve alongside advances in sequencing technology and bioinformatics. Today, DNA barcoding finds applications in various scientific fields, contributing to the understanding and conservation of biodiversity (Rodrigues et al. 2017).

This chapter aims to explore the rapidly evolving landscape of genome-based barcoding in the context of medicinal plants. It will delve into the methodologies, applications, and advantages of this innovative approach, shedding light on its potential to revolutionize traditional medicine, enhance consumer safety, and promote the sustainable use of botanical resources. Furthermore, the chapter will discuss the current challenges and future opportunities, including the need for comprehensive DNA barcode databases, global standardization, and regulatory integration.

In an increasingly interconnected and health-conscious world, genome-based barcoding holds the promise of preserving the ancient wisdom of traditional medicine while aligning it with the precision and accountability of modern science. This chapter endeavors to provide a comprehensive and insightful overview of this transformative approach, guiding researchers, practitioners, and policy-makers toward a more secure and scientifically informed future for medicinal plants in traditional medicine.

12.2 Techniques of DNA Barcoding: A Genome-Based Tool

Various DNA sequencing techniques are crucial in biological and ecological research. Sanger sequencing is precise and used for species identification. Next-generation sequencing (NGS) simultaneously analyzes multiple genes, which is essential in metagenomics. Mini-barcodes offer rapid, cost-effective species identification. High-resolution melting (HRM) analysis swiftly detects DNA sequence variations, which is useful in food authentication. DNA microarrays enable high-throughput screening for biodiversity and disease detection. Meta-barcoding is valuable for broad-scale species identification in ecosystems, aiding environmental monitoring. Details are given in Table 12.1.

12.2.1 The DNA Barcoding Technique Integrated with Other Technologies

Several analytical methods, including chromatography techniques, have been effectively combined with barcode markers. These methods include TLC, HPTLC, HPLC, and LC-MS. These approaches have proven to be valuable in the identification of bioactive constituents within medicinal plants and herbal products (Mahima et al. 2022a; Palhares et al. 2015).

Table 12.1 DNA barcoding techniques with descriptions, targets, advantages, and applications

Technique	Method description	Targeted genomic region	Advantages	Common applications	Source reference
Sanger sequencing	Conventional DNA sequencing method	Cytochrome c oxidase subunit 1 (COI) gene	Well-established and accurate	Species identification, biodiversity research	Deagle et al. (2014)
NGS	High-throughput sequencing technologies	Various genes (e.g., COI, rbcL, 16S rRNA)	Simultaneous analysis of multiple genes	Metagenomics, phylogenetics	Sanschagrin and Yergeau (2014), Goodwin et al. (2016)
Mini-barcodes	Short DNA sequences for rapid analysis	Short segments of COI	Rapid identification, cost-effective	Field-based species identification	Elbrecht and Leese (2017)
High-resolution melting (HRM) analysis	Detection of DNA sequence variations	COI, rbcL, and other genes	Swift and cost-effective species identification	Food authentication, disease diagnosis	Chatzoglou et al. (2023)
DNA microarrays	Detection of species-specific sequences	Multiple target genes	High-throughput screening, multiplexing	Biodiversity assessment, disease detection	Sánchez-Pla (2014)
Meta-barcoding	Sequencing environmental DNA for species ID	Multiple marker genes	Broad-scale species identification in ecosystems	Environmental monitoring, microbiome analysis	Topstad et al. (2021)

12.2.2 DNA Mini-Barcoding

"DNA mini-barcoding" is a variation of the traditional DNA barcoding method that employs shorter and more specific genetic regions utilized for species identification based on DNA. Unlike standard DNA barcoding, which typically depends on relatively long DNA sequences, such as the mitochondrial COI gene, to differentiate species, DNA mini-barcoding uses shorter DNA sequences, often only a fraction of the standard barcode region's length (Bingpeng et al. 2018).

Mini-barcodes have been developed to address issues related to sample degradation, particularly in cases of poor DNA preservation. These mini-barcodes, usually around 200–300 base pairs in length, can be increased further through amplification efficiently than full-length barcodes due to their reduced size. However, mini-barcodes lack a standardized reference database and may not provide the same.

Level of taxonomic discrimination as full-length barcodes because of their limited length (Shokralla et al. 2015).

Consequently, mini-barcodes may be more suitable for genus or family-level identification rather than universal species identification (Gao et al. 2019a; Guo et al. 2022).

12.2.3 Bar-HRM Technique

The utilization of high-resolution melting (HRM) technology in genotyping medicinal plants marks a recent and promising technological advancement. This approach examines various factors associated with DNA base sequences, such as length, G.C. content, and variations in base complementarity, to analyze the melting curve of amplified genetic fragments. HRM, similar to other fluorescent PCR techniques, closely monitors nucleic acid melting curves using particularly saturated stains like LC Green and Eva Green. This comprehensive analysis facilitates the categorization or genotyping of the test populations considering distinct variations in melting curve shapes. The method offers numerous advantages, including high throughput, sensitivity, specificity, repeatability, ease of use, and cost-effectiveness.

Additionally, its enclosed tube operation significantly decreases the hazard of contamination. Lately, in the past few years, the HRM technique has gained traction in genotyping medicinal plants, especially when combined with conventional DNA barcode markers, known as Bar-HRM technology. The implementation of the Bar-HRM technique, along with its high-temperature uniformity and resolution, enhances the precision in distinguishing single base differences, thereby refining the specification of medicinal plants (Yu et al. 2021). A compilation of 24 papers published between 2011 and 2019 integrated Bar-HRM technology, involving 2 review articles (Sun et al. 2016). Bar-HRM technology streamlines the identification process by eliminating the need for sequence-specific probes and sequencing. Various barcode sequences such as matK, trnH- ITS, rbcL, rpoC, psbA, and trnL, among others, have been united with HRM technology to effectively differentiate and identify medicinal plants. This strategy has demonstrated accuracy, reliability, and speed in authenticating market samples of raw materials, highlighting its suitability for high-throughput assays (Buddhachat et al. 2015). Species authentication using Bar-HRM holds the potential to foster consumer faith and enhance the quality of herbal products (Osathanunkul et al. 2015; Yu et al. 2021).

12.2.4 Meta-Barcoding

Meta-barcoding, employing universal PCR primers, enables the simultaneous amplification of multiple DNA barcodes, fostering the identification of a diverse array of species within a single environmental sample. This approach significantly enhances the efficiency of species identification, providing a comprehensive assessment of biodiversity from a solitary environmental specimen. Through the concurrent amplification of various DNA barcodes, meta-barcoding offers a detailed insight into the multifaceted species composition within a given environmental context. In studies involving traditional Chinese medicinal plants (CMPs), ITS2 or other plant barcodes are often preferred due to their high identification efficiency across various species. However, meta-barcoding research generates a substantial amount of sequence data, making the traditional Sanger sequencing technique impractical for this purpose (Gao et al. 2019b).

12.2.5 Loop-Mediated Isothermal Amplification (LAMP)

The LAMP technology has found extensive application in detecting numerous pathogenic organisms related to human diseases, confirming meat authenticity, and identifying genetically engineered organisms from its inception. This innovative method has been widely employed across various domains due to its versatility and effectiveness in detecting diverse pathogens associated with human illnesses, verifying the legitimacy of meat products, and discerning genetically modified organisms, marking a significant milestone in its inception. This advanced molecular biology method offers rapid and precise testing, especially in settings lacking the infrastructure for PCR facilities (Gao et al. 2019a). LAMP utilizes a set of four to six primers, each binding to various segments of the specific DNA target sequence, assisted by the DNA polymerase derived from *Bacillus stearothermophilus*. Positive PCR results are detected through observable changes, such as a color shift from translucent to lively green or increased turbidness due to magnesium pyrophosphate accumulation. The LAMP technique doesn't necessitate specialized equipment or molecular methods.

LAMP has found utility in species identification within herbal medicine. A method derived from LAMP, pioneered by Sasaki and Nagumo, concentrated on the sequence of the trnK gene to differentiate between *C. aromatic* and *C. longa*. The improved ability to distinguish between different species using DNA barcodes, specifically the ITS2 sequence, makes LAMP a swift method for verifying the authenticity of plants. By contrasting authentic and adulterated ITS2 sequences, specific LAMP primers can be crafted for the precise identification of *Crocus sativus*, *Hedyotis diffusa*, *Dendrobium officinale*, and *Taraxacum formosanum*. Furthermore, the amalgamation of LAMP with barcoding analysis offers the capability to identify the toxic additive *Aristolochia fangchi* (Chen et al. 2023).

12.2.6 Recombinase Polymerase Amplification (RPA)

RPA is an isothermal amplification process renowned for its exceptional sensitivity and selectivity. Operating within a temperature range from 37 to 42 °C, RPA demands nominal sample preparation and is capable of amplification of DNA target copies as minimal as 1–10 within less than 20 min (Wu et al. 2019). RPA is versatile, amplifying various targets, together with RNA, dsDNA, ssDNA, and miRNA, derived from various organisms and samples. It can be performed in solution phase, solid phase, and bridge amplification formats. Various ways to detect, like using endpoint lateral flow strips and real-time fluorescence detection, have been effectively paired with RPA, making the process more human-friendly and accessible (Lobato and O'Sullivan 2018).

RPA is recognized for its remarkable attributes, including its straightforwardness, heightened acuteness, selectivity, suitability with multiplexing, quick amplification, and capacity to operate at a sustained low temperature without necessitating an early denaturation step or numerous primers. Its overall versatility positions RPA ideally for extensive utilization in kits and assays, especially for point-of-need applications, particularly in resource-limited settings. This technology offers a combination of affordability, acuteness, specificity, swiftness, durability, and equipment-free operation, aligning closely with the ASSURED (Affordable, Sensitive, Specific, User-friendly, Rapid, Equipment-free, and Delivered) criteria (Daher et al. 2016).

12.3 Methodology

DNA barcoding is a potent technique for identifying plant species and their extracts, inclusive of phytopharmaceuticals (Thakur et al. 2021). This method effectively employs concise genetic markers to swiftly and precisely identify organisms, ensuring a rapid and accurate identification process (Hays and Fagan 2016). The DNA barcoding procedure includes extracting high-quality genomic DNA, conducting PCR amplification, sequencing the resulting amplicons, and then performing subsequent bioinformatic analysis (see Fig. 12.1). Compared to traditional methods, DNA barcoding is widely recognized as a fast, precise, and cost-effective approach that commonly provides species-level resolution, making it valuable for authenticating medicinal herbs (Hays and Fagan 2016).

DNA barcodes are stored in the sequence library of the Barcode of Life Database (BOLD, www.barcodeoflife.org), managed by the Consortium for the Barcode of Life (CBOL). This database is particularly useful for DNA analysis in processed samples and products (Geary and Bubela 2019).

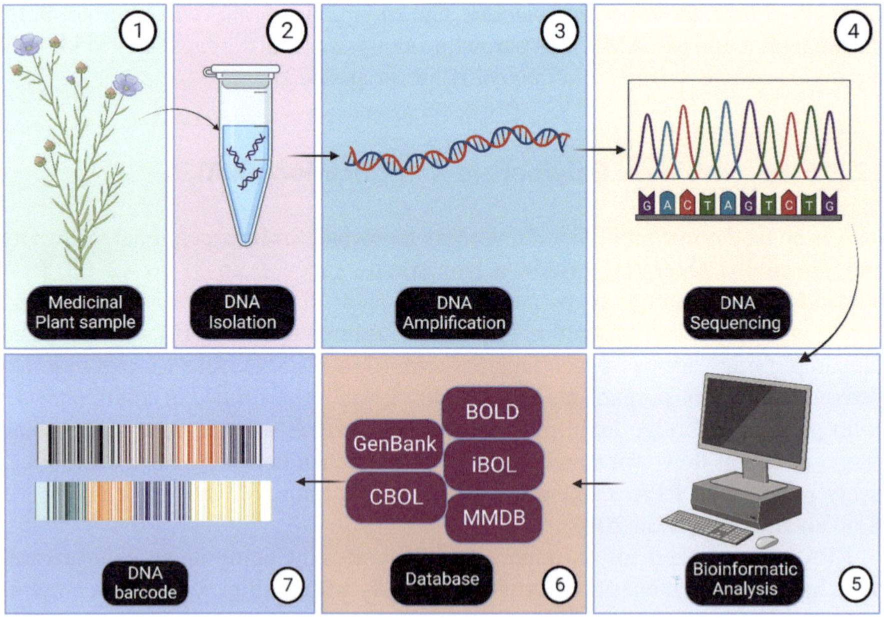

Fig. 12.1 Illustrating the fundamental process of DNA barcoding for medicinal plants

12.3.1 Database

Creating DNA barcode databases dedicated to authenticating medicinal plants is crucial for maintaining updated, well-organized, and easily accessible information for the global scientific community. Several primary databases focused on collecting, managing, and examining DNA barcodes across various organisms are pivotal in this endeavor. These databases include (1) the BOLD, (2) CBOL, (3) GenBank®, and (4) the International Barcode of Life Project (iBOL). They actively promote the utilization of DNA barcoding to identify and classify Earth's biodiversity. Currently, the Medicinal Materials DNA Barcode Database (MMDBD) stands as a unique repository storing DNA barcodes of medicinal plants documented in the Chinese Pharmacopoeia (Wong et al. 2018).

12.3.1.1 Consortium for the Barcode of Life (CBOL)

The CBOL was introduced in May 2004 and has since expanded to include over 120 organizations from 45 countries. CBOL's primary mission is to facilitate the establishment of international research collaborations necessary to create a wide-ranging DNA barcode library for all eukaryotic life over the next two decades. It has already launched the first global initiatives aimed at providing barcode coverage for all bird and fish species by 2012 (Marshall 2005). While these two projects are expected to produce approximately 0.5 million records, a complete barcode library for the animal kingdom is projected to consist of around 100 million records, nearly double the size of GenBank at the time (52 million sequence records as of March 7, 2006).

The extensive amount of data involved emphasizes the requirement for enterprise-level software to aid the distinctive elements of DNA barcoding, which led to the creation of the BOLD system. This system includes crucial features like maintaining a continuous connection between a barcode sequence and its originating specimen, offering a secure platform for storing, organizing, and searching records, and guaranteeing accessibility to the entire biodiversity community. Establishing and enforcing data standards is another critical requirement. To address these challenges, CBOL introduced discussions with significant genomics repositories such as the NCBI, biodiversity organizations like the GBIF, leading barcoding centers, and diverse taxonomic communities. These collaborative efforts resulted in the formulation of formal guidelines that records must adhere to receive a barcode designation. These guidelines outline specific criteria stating that gene sequences must originate from a selected gene region, ensuring they meet stringent quality standards. Additionally, the sequences should be identifiable to a specific specimen held in a major collection, facilitating taxonomic verification and enhancing the authenticity of the collection (Miller 2008; Ratnasingham and Hebert 2007).

12.3.1.2 Barcode of Life Data Systems (BOLD)

The BOLD system, developed by Ratnasingham and Hebert in 2013, is a cloud-based platform designed for storing and analyzing data (Ratnasingham and Hebert 2013). It is managed by the CBG in Canada and involves four main components: (1) data portal, (2) educational portal, (3) registry of BINs, and (4) comprehensive platform for gathering and analyzing data. Researchers are encouraged to use BOLD for efficient management of their DNA barcoding projects.

The BOLD Handbook, accessible on the BOLD website underneath the "Resources" section, and guidance provided on the "DNA Barcoding Workshop (DBW) website" offer detailed instructions on tasks like setting up a BOLD account, organizing specimen data, starting projects, and incorporating records and images into the platform. Following these steps is crucial for successfully uploading DNA barcode sequences onto BOLD and utilizing the BINs platform effectively. It's important to note that BOLD maintains the confidentiality of the user's work, including records of specimens and sequences, accessible only to the user and their collaborators unless the user chooses to share it publicly (Wilson et al. 2019).

After the upload process, DNA barcodes that meet specific length requirements of "200 bp" and strict quality standards (minimal proportion of Ns) are spontaneously grouped into operational taxonomic units (OTUs) known as BINs. These groupings are established using advanced single linkage analysis through the extensive BOLD library and have shown significant correlation with species identified through traditional taxonomic methods like morphology. As a result, these alphanumeric designations can serve as alternative taxonomic labels when complete Linnaean assignments are unavailable (Wilson et al. 2019).

12.3.1.3 International Barcode of Life Project (iBOL)

The sixth iBOL Conference took place in Guelph, Canada, from August 18 to 21, 2015, under the theme "Barcodes to Biomes." This conference served as a pivotal forum for unveiling the modern advancements in DNA barcoding study and its widespread applications (Adamowicz 2015). It functioned as a central hub that united researchers worldwide, facilitating the exchange of groundbreaking ideas and fostering collaborative initiatives. Notably, all abstracts, whether plenary or contributed, presented during the conference were published in the journal *Genome*.

In a comparative analysis between the sixth conference and the third held in Mexico City in 2009, Arzanlou highlighted ten notable trends evident in the contributed abstracts. A prominent trend observed was the escalating focus on diverse socioeconomic applications stemming from DNA barcoding. These applications encompassed a wide range, including studies related to farming pests, invasive and isolated species, carriers of diseases, monitoring ecosystem health, and ensuring the legitimacy of seafood items and herbal remedies in various markets. Furthermore, there was a discernible surge in interest in utilizing barcoding and meta-barcoding methods for conducting dietary assessments and exploring intricate food networks that spanned multiple trophic levels.

This increased emphasis regarding the socioeconomic uses of DNA barcoding represents a significant shift in research focus, indicating its growing practical relevance across various sectors and industries. This expansion underscores the adaptability of DNA barcoding, demonstrating its potential to address real-world challenges and make substantive contributions to diverse fields beyond the traditional realms of scientific inquiry (Arzanlou 2013).

12.3.1.4 GenBank

GenBank®, found at http://www.ncbi.nlm.nih.gov, is a vast repository containing openly accessible nucleotide sequences linked to approximately 260,000 officially recognized species. These sequences are mainly acquired by submissions made by individual laboratories and large-scale sequencing initiatives, which include whole-genome shotgun (WGS) projects and efforts in sampling of environment-related materials. Submission of data is typically accomplished using web-based tools like BankIt or stand-alone programs such as Sequin. The GenBank team allocates accession numbers after receiving data. The database upholds an ongoing data substitution with the ENA and the DDBJ, ensuring comprehensive global coverage.

GenBank can be easily accessed via the NCBI Entrez retrieval system, consolidating data from significant DNA and protein sequence databases. It includes supplementary information concerning taxonomy, genomes, mapping, protein structures, domains, and biomedical journal literature available through PubMed. Users have the capability to perform sequence similarity searches utilizing BLAST, which spans across GenBank and various other sequence databases. The GenBank database is updated with every other month releases and updates provided on a daily basis, accessible via FTP. For access to GenBank and its associated retrieval and analysis facilities, users can start at the NCBI home page available on www.ncbi.nlm.nih.gov (Benson et al. 2012).

12.3.1.5 MMDBD

The Medicinal Plant DNA Barcoding Database (MMDBD) is an online repository established in 2010, accessible at https://rdccm.cuhk.edu.hk/mherbsdb/. This repository serves as a comprehensive archive for DNA sequences acquired through DNA barcoding technology, specifically focusing on sequences associated with medicinal plants recorded in the Chinese Pharmacopoeia, Herbal Medicine Compendium of the US Pharmacopoeia, and Dietary Supplements Compendium. Additionally, the MMDBD includes data on commonly found adulterants related to these medicinal plant species. Through its provision of DNA sequences and associated information, the database aids researchers and practitioners in verifying and identifying medicinal plants, ensuring the authenticity and standard of herbal products and supplements. Its concentration on prominent pharmacopeial plant species signifies its relevance in supporting traditional medicine practices, while its emphasis on DNA barcoding contributes significantly to scientific research, quality control, and safety assessment within the realm of herbal-based products and remedies (Gesto-Borroto et al. 2021). The sequence files are regularly updated, and as of December 16, 2020, MMDBD houses 65,520 DNA sequences representing 2225 species.

This platform offers extensive details about medicinal materials, encompassing their names, medical descriptions, information from the pharmacopoeia, biological classifications, and their status in compliance with the Convention on International Trade in Endangered Species of Wild Fauna and Flora. It also incorporates data sourced from the "International Union for Conservation of Nature's Red List of Threatened Species," where applicable. Notably, this database allows users to search for DNA barcodes in herbal medicine using scientific names in Latin or Chinese names (Wong et al. 2018).

12.4 Application

12.4.1 Protection of Endangered Species

The ongoing decline in biodiversity on a global scale is a concerning trend, resulting in the endangered status of numerous species. This reduction stems largely from the unlawful utilization of living organisms, particularly targeting unusual and valuable animals and plants. This illicit activity poses a severe threat to the survival of various species, exacerbating the already critical situation faced by many ecosystems worldwide. This issue is of global concern and is particularly intense in certain regions, notably in Africa, where it has led to the endangerment of many species, significantly impacting ecological balance and the environment. DNA barcoding has appeared as a vital tool in the conservation of endangered species, enabling rapid and precise identification of species, which is essential for the management and protection of these species (Thinh et al. 2021).

DNA barcoding serves as a crucial tool in identifying and combating the unlawful trade of threatened species (Din Dipita et al. 2022). In response to these challenges, a recent initiative has been launched, aiming to establish a comprehensive global strategic DNA barcoding project. The primary aim of this project is to establish an extensive barcode library specifically focusing on threatened species. This ambitious endeavor seeks to strengthen efforts in protecting endangered organisms by enhancing identification methods and facilitating the enforcement of regulations aimed at conserving biodiversity (Gostel and Kress 2022b).

12.4.2 Evaluations of Regional Biodiversity

DNA barcoding investigations play a dual role in supporting crucial local and regional biodiversity assessments, particularly in floristics. In areas with limited knowledge of species diversity, collecting and studying specimens are vital for both identifying species and conserving them. Contemporary field expeditions employ various methods for species collection and documentation, combining traditional natural history specimens, photographs, ecological notes, and newer genetic vouchers intended for genetic and genomics research (Funk et al. 2017). Integrating genetic vouchers and DNA barcode sequencing into standard species inventories not only aids in building the global barcode reference database but also often reveals previously unknown cryptic diversity (Gostel and Kress 2022b).

DNA barcode studies play a significant role in supporting local and regional biodiversity assessments, including floristics. Especially in areas characterized by rich biodiversity and inadequate understanding of species diversity, explorations reliant on collections and inventory studies play a pivotal role in both α taxonomy and conservation efforts. Contemporary field expeditions use various methods for species collection: conventional specimens from natural history, photos, ecological observations, and new genetic vouchers. These genetic vouchers aid genetic and genomic research, enriching biodiversity documentation (Funk et al. 2017).

Additionally, they often lead to the revelation of previously unrecognized cryptic diversity (Gostel and Kress 2022b), which underscores the pivotal role that DNA barcoding plays in advancing our understanding of biodiversity and conservation efforts.

12.4.3 Preservation of Natural Resources

DNA barcoding has multifaceted applications, extending beyond disease prevention and the control of illegal hunting of endangered species. It also finds relevance in the realms of agriculture and forestry. Many countries that heavily depend on natural resources are facing challenges related to the excessive exploitation of agricultural, marine, and forestry resources. This exploitation has resulted in the decline and potential extinction of multiple species, ultimately causing ecological and biological differences. DNA barcoding is instrumental in the conservation of assets from nature, offering several vital applications. It enables the identification of endangered species, thereby contributing to conservation initiatives. Through the identification of diverse species, researchers can enhance their understanding of biodiversity and the distinct roles each species fulfills within ecosystems. Additionally, DNA barcoding aids in the identification of invasive species and the monitoring of their propagation. Swift identification allows for proactive measures to thwart these species from disrupting the delicate equilibrium of ecosystems (Safhi et al. 2022).

12.4.4 DNA Barcoding in Pharmacognosy

Understanding the composition of raw materials obtained from animals and plants is crucial, particularly concerning the early detection and prevention of counterfeit drugs. DNA barcoding has appeared as an efficacious method for swiftly and accurately identifying components within medicinal raw materials. Its notable applications in this domain are evident in extensive literature discussing the incorporation of DNA barcoding protocols into national pharmacopeias. This integration ensures reliable and efficient screening of plant raw materials (Chen et al. 2014), and even in industrial-scale applications (Sgamma et al. 2017).

A compelling case study highlights the practical application of DNA barcoding in detecting counterfeit medicinal plants prevalent in the Iranian marketplace. Notably, plants like *Ziziphora clinopodioides* Lam. and *Z. tenuior* L., bearing close morphological resemblances and similar odors to species such as "*Thymus kotschyanus* Boiss. & Hohen., *Th. vulgaris* L., *Thymus. daenensis* Celak., and *Th. trautvetteri* Klokov & Des.-Shost," were subject to DNA barcoding analysis. The findings revealed that all examined samples contained impurities originating from other plant species (Sheidai et al. 2019).

Furthermore, another illustrative example of DNA barcoding's application in herbalism involves the identification of potentially harmful plants that might be present in medicinal raw materials. As an example, an extensive assessment covering 106 medicinal plants (different species) from the Chinese pharmacopeia, along with associated toxic plants, demonstrated the effectiveness of ITS2 sequences in detecting adequate polymorphism. This polymorphism aided in distinguishing toxic plants and other impurities among the examined medicinal plant species, showcasing the utility of DNA barcoding in ensuring the safety and authenticity of medicinal materials (Liu et al. 2019).

Furthermore, various studies have showcased the effectiveness of ITS2 sequences in distinguishing closely related species of medicinal plants that share morphological similarities but differ in chemical composition. In a particular study, the ITS2 barcode successfully distinguished between the rhizomes of *Panax vietnamensis* var. fuscidicus K. Komatsu, S. Zhu & S.Q. Cai, and *Panax notoginseng* F.H. Chen, one of which is employed in pharmaceutical drug preparation (Yang et al. 2018). A similar approach aided in distinguishing phenotypically similar species within the genera Uncaria and Mucuna and in identifying roots and rhizomes of *Rheum tanguticum Rheum officinale* Baill., *Rh. palmatum* L., and Maxim. from the Polygonaceae family (Shadrin 2021).

These instances underscore the critical role of DNA barcoding in ensuring the authenticity and safety of medicinal raw materials (Dawan and Ahn 2022).

12.5 Challenges

In recent times, the use of plant-based traditional medicines has been on the rise, with an anticipated 80% of the global population turning to herbal products for their well-being and healthcare needs. To ensure the safety of consumers and effectively identify medicinal plants and herbal products, the DNA barcoding technique has proven to be a valuable tool. However, a comprehensive analysis of existing literature reveals certain limitations associated with this technique, and it's important to note that there isn't a single efficient barcode suitable for all plant groups (Welz et al. 2018).

A significant drawback of DNA barcoding is its vulnerability to fluctuations in the quality of template DNA, primer selection, the impact of PCR procedures on herbal products, and the likelihood of DNA sample contamination due to additives. Additionally, DNA availability can be compromised or degraded during manufacturing processes, which may involve extensive heat treatment, irradiation, exposure to ultraviolet light, or extractive distillation. In cases where DNA is absent from processed products, DNA barcoding is not a viable option (Fazekas et al. 2009). Furthermore, DNA barcoding can be challenged by the presence of multiple species within herbal products, which can introduce bias in PCR outcomes. To mitigate this bias, approaches such as conducting parallel PCR or cloning specific PCR products into a vector to target particular species have been employed (Newmaster et al. 2013).

Another obstacle in the domain of DNA barcoding stems from the disruption caused by secondary plant metabolites, which encompass polysaccharides, tannins, alkaloids, and polyphenols. These metabolites may compromise the quality of DNA, and sequencing success can be reduced. Under such circumstances, the use of meta-barcoding and high-throughput sequencing techniques can help address these issues. In the future, enhancing DNA barcoding perspectives involves the incorporation of biological reference material and anatomical studies in conjunction with sequence data. This combined approach aims to provide users with a higher level of authenticity when it comes to the identification of medicinal plants and herbal products. Future directions in DNA barcoding include the development of a novel mini-barcode sequence library, the establishment of a BRM library, and a focus on anatomical studies to bolster herbal drug authentication (Mahima et al. 2022b).

12.6 Conclusion and Future Outlook

This book chapter offers a comprehensive overview of the evolution of DNA barcoding concerning medicinal plants. It encompasses diverse facets of DNA barcoding, discussing established methodologies as well as cutting-edge and emerging technologies within this field. DNA barcoding is widely acknowledged for its contribution to classifying medicinal plants, detecting substitutes and adulterants, and overseeing the pharmaceutical market. Established methods have demonstrated effectiveness in identifying and utilizing the majority of plant raw materials and minimally processed products within the medicinal materials market. However, challenges persist, particularly in the identification of finely processed products like tablets, oral liquids, pills, and injections, where rapid and standardized identification methods are still lacking, which is particularly evident in the complex botanical compositions of Indian patent medicines.

The establishment of distinct DNA barcodes for medicinal plants is intricately linked to advancements in genome sequencing. Despite significant strides in next-generation and third-generation sequencing technologies, the exploration of chloroplast (or plastid) genomics in medicinal plants remains incomplete on a comprehensive scale. There is an ongoing need for extensive sequencing of chloroplast genomes in various medicinal plants to facilitate the development of new, precise DNA barcodes for enhanced identification and classification purposes. Historically, DNA barcoding technology has focused on marker selection and sequence database construction. While universal plant DNA barcodes remain a subject of debate, the standard markers of rbcL, matK, trnH-psbA, and ITS are widely employed in most applications. International DNA barcode databases are sustained by organizations like BOLD in Canada and China's independent DNA barcode database. The continual progress and extensive utilization of technologies such as super-barcoding and meta-barcoding have broadened the spectrum of the DNA barcode toolkit.

In the realm of identification within traditional Chinese medicine, the utilization of plant DNA barcoding technology is anticipated to expand significantly, showing potential for more extensive and profound applications in the future. In summary, DNA barcoding holds significant promise for the field of medicinal plants, offering valuable contributions to the modernization of traditional medicines worldwide.

References

Adamowicz SJ (2015) International barcode of life: evolution of a global research community. Genome 58(5):151–162

Arzanlou M (2013) DNA barcoding: a new tool with wide array of applications. Res Mol Med 1(2):1–2. https://doi.org/10.18869/ACADPUB.RMM.1.2.1

Benson DA, Cavanaugh M, Clark K, Karsch-Mizrachi I, Lipman DJ, Ostell J, Sayers EW (2012) GenBank. Nucleic Acids Res 41(D1):D36–D42

Bingpeng X, Heshan L, Zhilan Z, Chunguang W, Yanguo W, Jianjun W (2018) DNA barcoding for identification of fish species in the Taiwan Strait. PLoS One 13(6):e0198109. https://doi.org/10.1371/JOURNAL.PONE.0198109

Buddhachat K, Osathanunkul M, Madesis P, Chomdej S, Ongchai S (2015) Authenticity analyses of Phyllanthus amarus using barcoding coupled with HRM analysis to control its quality for medicinal plant product. Gene 573(1):84–90. https://doi.org/10.1016/J.GENE.2015.07.046

Chatzoglou E, Tsaousi N, Apostolidis AP, Exadactylos A, Sandaltzopoulos R, Giantsis IA, Gkafas GA, Malandrakis EE, Sarantopoulou J, Tokamani M, Triantaphyllidis G, Miliou H (2023) High-resolution melting (HRM) analysis for rapid molecular identification of Sparidae species in the Greek fish market. Genes 14(6):1255. https://doi.org/10.3390/GENES14061255/S1

Chen S, Pang X, Song J, Shi L, Yao H, Han J, Leon C (2014) A renaissance in herbal medicine identification: from morphology to DNA. Biotechnol Adv 32(7):1237–1244. https://doi.org/10.1016/J.BIOTECHADV.2014.07.004

Chen S, Yin X, Han J, Sun W, Yao H, Song J, Li X (2023) DNA barcoding in herbal medicine: retrospective and prospective. J Pharm Anal 13(5):431–441. https://doi.org/10.1016/j.jpha.2023.03.008

Daher RK, Stewart G, Boissinot M, Bergeron MG (2016) Recombinase polymerase amplification for diagnostic applications. Clin Chem 62(7):947–958. https://doi.org/10.1373/CLINCHEM.2015.245829

Dawan J, Ahn J (2022) Application of DNA barcoding for ensuring food safety and quality. Food Sci Biotechnol 31(11):1355–1364

Deagle BE, Jarman SN, Coissac E, Pompanon F, Taberlet P (2014) DNA metabarcoding and the cytochrome c oxidase subunit I marker: not a perfect match. Biol Lett 10(9):20140562. https://doi.org/10.1098/RSBL.2014.0562

Din Dipita A, Missoup AD, Tindo M, Gaubert P (2022) DNA-typing improves illegal wildlife trade surveys: tracing the Cameroonian bushmeat trade. Biol Conserv 269:109552. https://doi.org/10.1016/J.BIOCON.2022.109552

Elbrecht V, Leese F (2017) Validation and development of COI metabarcoding primers for freshwater macroinvertebrate bioassessment. Front Environ Science 5(APR):237020. https://doi.org/10.3389/FENVS.2017.00011/BIBTEX

Fazekas AJ, Kesanakurti PR, Burgess KS, Percy DM, Graham SW, Barrett SCH, Newmaster SG, Hajibabaei M, Husband BC (2009) Are plant species inherently harder to discriminate than animal species using DNA barcoding markers? Mol Ecol Resour 9:130–139

Funk VA, Gostel M, Devine A, Kelloff CL, Wurdack K, Tuccinardi C, Radosavljevic A, Peters M, Coddington J (2017) Guidelines for collecting vouchers and tissues intended for genomic work (Smithsonian Institution): botany best practices. Biodivers Data J 5(5):1. https://doi.org/10.3897/BDJ.5.E11625

Gao Z, Liu Y, Wang X, Wei X, Han J (2019a) DNA mini-barcoding: a derived barcoding method for herbal molecular identification. Front Plant Sci 10:987
Gao Z, Liu Y, Wang X, Wei X, Han J (2019b) DNA mini-barcoding: a derived barcoding method for herbal molecular identification. Front Plant Sci 10:987. https://doi.org/10.3389/FPLS.2019.00987
Geary J, Bubela T (2019) Governance of a global genetic resource commons for non-commercial research: a case-study of the DNA barcode commons. Int J Commons 13(1):1
Gesto-Borroto R, Medina-Jimenez K, Lorence A, Villarreal ML (2021) Application of DNA barcoding for quality control of herbal drugs and their phytopharmaceuticals. Rev Bras 31:127–141
Goodwin S, McPherson JD, McCombie WR (2016) Coming of age: ten years of next-generation sequencing technologies. Nat Rev Genetics 17(6):333–351. https://doi.org/10.1038/nrg.2016.49
Gostel MR, Kress WJ (2022a) The expanding role of DNA barcodes: indispensable tools for ecology, evolution, and conservation. Diversity 14:213. https://doi.org/10.3390/D14030213
Gostel MR, Kress WJ (2022b) The expanding role of DNA barcodes: indispensable tools for ecology, evolution, and conservation. Diversity 14(3):213. https://doi.org/10.3390/D14030213
Guo M, Yuan C, Tao L, Cai Y, Zhang W (2022) Life barcoded by DNA barcodes. Conserv Genet Resour 14(4):351–365
Hays CG, Fagan CM (2016) Conservation biology, evolution and encyclopedia of Evolution Biology. Elsevier, Amsterdam, pp 347–353. https://doi.org/10.1016/B978-0-12-800049-6.00301-2
Heather JM, Chain B (2016) The sequence of sequencers: the history of sequencing DNA. Genomics 107(1):1. https://doi.org/10.1016/J.YGENO.2015.11.003
Jităreanu A, Trifan A, Vieriu M, Caba IC, Mârțu I, Agoroaei L (2022) Current trends in toxicity assessment of herbal medicines: a narrative review. Processes 11(1):83. https://doi.org/10.3390/PR11010083
Liu M, Li XW, Liao BS, Luo L, Ren YY (2019) Species identification of poisonous medicinal plant using DNA barcoding. Chin J Nat Med 17(8):585–590. https://doi.org/10.1016/S1875-5364(19)30060-3
Lobato IM, O'Sullivan CK (2018) Recombinase polymerase amplification: basics, applications and recent advances. TrAC Trends Anal Chem 98:19–35. https://doi.org/10.1016/J.TRAC.2017.10.015
Mahima K, Sunil Kumar KN, Rakhesh KV, Rajeswaran PS, Sharma A, Sathishkumar R (2022a) Advancements and future prospective of DNA barcodes in the herbal drug industry. Front Pharmacol 13:947512. https://doi.org/10.3389/FPHAR.2022.947512
Mahima K, Sunil Kumar KN, Rakhesh KV, Rajeswaran PS, Sharma A, Sathishkumar R (2022b) Advancements and future prospective of DNA barcodes in the herbal drug industry. Front Pharmacol 13:947512. https://doi.org/10.3389/FPHAR.2022.947512/BIBTEX
Malik S, Priya A, Babbar SB (2019) Employing barcoding markers to authenticate selected endangered medicinal plants traded in Indian markets. Physiol Mol Biol Plants 25(2):327. https://doi.org/10.1007/S12298-018-0610-8
Marshall E (2005) Will DNA bar codes breathe life into classification? American Association for the Advancement of Science, Washington, DC
Miller SE (2008) Consortium for the barcode of life: support for consortium activities and secretariat office
Mosa KA, Gairola S, Jamdade R, El-Keblawy A, Al Shaer KI, Al Harthi EK, Shabana HA, Mahmoud T (2019) The promise of molecular and genomic techniques for biodiversity research and DNA barcoding of the Arabian peninsula Flora. Front Plant Sci 9:372140. https://doi.org/10.3389/FPLS.2018.01929/BIBTEX
Newmaster SG, Grguric M, Shanmughanandhan D, Ramalingam S, Ragupathy S (2013) DNA barcoding detects contamination and substitution in north American herbal products. BMC Med 11:1–13
Osathanunkul M, Madesis P, De Boer H (2015) Bar-HRM for authentication of plant-based medicines: evaluation of three medicinal products derived from Acanthaceae species. PLoS One 10(5):e0128476. https://doi.org/10.1371/JOURNAL.PONE.0128476

Palhares RM, Drummond MG, Brasil DSAFB, Cosenza GP, Brandão MDGL, Oliveira G (2015) Medicinal plants recommended by the world health organization: DNA barcode identification associated with chemical analyses guarantees their quality. PLoS One 10(5):e0127866. https://doi.org/10.1371/JOURNAL.PONE.0127866

Ratnasingham S, Hebert PDN (2007) Bold: the barcode of life data system. Mol Ecol Notes 7(3):355–364. https://doi.org/10.1111/J.1471-8286.2007.01678.X. http://www.barcodinglife.org

Ratnasingham S, Hebert PDN (2013) A DNA-based registry for all animal species: the barcode index number (BIN) system. PLoS One 8(7):e66213. https://doi.org/10.1371/JOURNAL.PONE.0066213

Rodrigues MS, Morelli KA, Jansen AM (2017) Cytochrome c oxidase subunit 1 gene as a DNA barcode for discriminating Trypanosoma cruzi DTUs and closely related species. Parasit Vectors 10(1):1–18. https://doi.org/10.1186/S13071-017-2457-1/TABLES/3

Safhi FA, Alshamrani SM, Fiteha YG, Abd El-Moneim D (2022) DNA barcoding of endangered and rarely occurring plants in Faifa Mountains (Jazan, Saudi Arabia). Agriculture 12(11):1931

Sánchez-Pla A (2014) DNA microarrays technology: overview and current status. Compr Anal Chem 63:1–23. https://doi.org/10.1016/B978-0-444-62651-6.00001-5

Sanschagrin S, Yergeau E (2014) Next-generation sequencing of 16S ribosomal RNA gene amplicons. JoVE 90:e51709

Sgamma T, Lockie-Williams C, Kreuzer M, Williams S, Scheyhing U, Koch E, Slater A, Howard C (2017) DNA barcoding for industrial quality assurance. Planta Med 83(14/15):1117–1129

Shadrin DM (2021) DNA barcoding: applications. Russ J Genet 57:489–497

Sheidai M, Tabaripour R, Talebi SM, Noormohammadi Z, Koohdar F (2019) Adulteration in medicinally important plant species of Ziziphora in Iran market: DNA barcoding approach. Ind Crop Prod 130:627–633. https://doi.org/10.1016/J.INDCROP.2019.01.025

Shokralla S, Hellberg RS, Handy SM, King I, Hajibabaei M (2015) A DNA mini-barcoding system for authentication of processed fish products. Sci Rep 5:15894. https://doi.org/10.1038/SREP15894

Sun W, Jian LJ, Xiong C, Zhao B, Chen SL (2016) The potential power of bar-HRM technology in herbal medicine identification. Front Plant Sci 7(MAR2016):185807. https://doi.org/10.3389/FPLS.2016.00367/BIBTEX

Thakur VV, Tripathi N, Tiwari S (2021) DNA barcoding of some medicinally important plant species of Lamiaceae family in India. Mol Biol Rep 48(4):3097–3106. https://doi.org/10.1007/S11033-021-06356-3

Thinh BB, Doudkin RV, Chac LD, Chinh HV, Hoi QV, Lien NT (2021) Identification of Tinomiscium petiolare from Vietnam using the DNA barcode. Proc Appl Bot Genet Breed 182(2):114–122. https://doi.org/10.30901/2227-8834-2021-2-114-122

Topstad L, Guidetti R, Majaneva M, Ekrem T (2021) Multi-marker DNA metabarcoding reflects tardigrade diversity in different habitats. Genome 64(3):217–231

Welz AN, Emberger-Klein A, Menrad K (2018) Why people use herbal medicine: insights from a focus-group study in Germany. BMC Complement Altern Med 18:1–9

Wilson JJ, Sing KW, Jaturas N (2019) DNA barcoding: bioinformatics workflows for beginners. In: Encyclopedia of bioinformatics and computational biology, vol 1–3. Elsevier, Amsterdam, pp 985–995. https://doi.org/10.1016/B978-0-12-809633-8.20468-8

Wong TH, But GWC, Wu HY, Tsang SSK, Lau DTW, Shaw PC (2018) Medicinal materials DNA barcode database (MMDBD) version 1.5-one-stop solution for storage, BLAST, alignment and primer design. Database (Oxford) 2018(2018):bay112. https://doi.org/10.1093/DATABASE/BAY112

Wu L, Ye L, Wang Z, Cui Y, Wang J (2019) Utilization of recombinase polymerase amplification combined with a lateral flow strip for detection of Perkinsus beihaiensis in the oyster Crassostrea hongkongensis. Parasit Vectors 12(1):360. https://doi.org/10.1186/S13071-019-3624-3

Yang J, Dong L, Wei G, Hu H, Zhu G, Zhang J, Chen S (2018) Identification and quality analysis of Panax notoginseng and Panax vietnamensis var. fuscidicus through integrated DNA barcoding and HPLC. Chin Herb Med 10(2):177–183. https://doi.org/10.1016/j.chmed.2018.03.008

Yu J, Wu X, Liu C, Newmaster S, Ragupathy S, Kress WJ (2021) Progress in the use of DNA barcodes in the identification and classification of medicinal plants. Ecotoxicol Environ Saf 208:111691. https://doi.org/10.1016/j.ecoenv.2020.111691
Yuan H, Ma Q, Ye L, Piao G (2016) The traditional medicine and modern medicine from natural products. Molecules 21(5):559. https://doi.org/10.3390/MOLECULES21050559

Conservation of High-Value Medicinal Plants Through Agronomical Approaches

13

Pooja Oli, Kuldeep Joshi, Shailaja Punetha, Manish Sharma, Vineet Kaswan, and Mithilesh Singh

Abstract

Medicinal plants have played an essential role in the formulation of modern pharmaceutical and nutraceutical products. With the increasing global population, the demand for healthcare services, including pharmaceutical products derived from medicinal plants, also increases. The increased demand for medicinal plant raw materials has led to their overexploitation from their natural habitat and placed many of them under the threat category. Therefore, the development of conservation strategies is crucial to maintaining the genetic diversity, sustainable harvesting, and preservation of traditional practices associated with medicinal plant species. Agronomic practices have emerged as a potential tool for promoting the sustainable management and conservation of therapeutic plant resources. This chapter provides an overview of the importance of medicinal plants and various agronomic practices that can be employed to conserve medicinal plants, propagation and cultivation, sustainable harvesting, and natural habitat management. By adopting agronomic practices, long-term viability and

P. Oli · S. Punetha (✉) · M. Singh
G.B. Pant National Institute of Himalayan Environment, Kosi-Katarmal, Almora, Uttarakhand, India

K. Joshi
Centre for GMP Extraction Facility, National Institute of Pharmaceutical Education and Research, Guwahati, Assam, India

M. Sharma
Pulses Research Station, Sardarkrushinagar Dantiwada Agricultural University, Banaskantha, Gujarat, India

V. Kaswan
College of Basic Science and Humanities, Sardarkrushinagar Dantiwada Agricultural University, Banaskantha, Gujarat, India

© The Author(s), under exclusive license to Springer Nature Singapore Pte Ltd. 2024

M. Nandave et al. (eds.), *Ethnopharmacology and OMICS Advances in Medicinal Plants Volume 1*, https://doi.org/10.1007/978-981-97-2367-6_13

availability of medicinal plants can be ensured while preserving their germplasm and ecological integrity.

Keywords

Pharmaceutical · Cultivation · Genetic diversity · Sustainable · Overharvesting Conservation · Propagation · Natural habitat · Germplasm

Abbreviations

FYM	Farmyard manure
GAP	Good agricultural practices
MPs	Medicinal plants
NMPB	National Medicinal Plant Board
NPK	Nitrogen phosphorus potassium

13.1 Introduction

The Indian Himalayan Region is well-known for its rich plant diversity, boasting about 1748 plant species of known medicinal value, and around 60–80% of the world's population depends on these plants for their primary health care (Samant et al. 1998). These medicinal plant species (MPs) are used in traditional medicine systems like Ayurveda, Unani, and Siddha, and various pharmacological investigations have confirmed the presence of active compounds in these plants that have the potential to play an active role in modern drug development (Dhar et al. 2002). These medicinally active compounds, such as alkaloids, polyphenols, tannins, glycosides, flavonoids, and terpenoids, have diverse biological activities that can be harnessed in the treatment of various diseases such as cancer, kidney disorders, asthma, piles, skin diseases, etc. (Akinyemi et al. 2018; Samant et al. 2011). Approximately 25% of medicines were formulated using plant-derived drugs that are essential components of modern pharmacopeia (Phondani et al. 2016; Pathak and Das 2013). These MPs are extensively used by the local communities for primary health care, thus preserving the traditional knowledge as well as sustaining their livelihood through the cultivation and trade of these plants (Bhatt et al. 2021). The high demand for MPs raw material in the pharmaceutical and nutraceutical industries has led to the overharvesting of these plants from the wild and pushed many of them to the verge of extinction (Phondani et al. 2010).

Moreover, the collection of these MPs from wild sources also poses challenges like adulteration that can affect the quality of the drug developed (Amujoyegbe et al. 2012). According to reports, the existing loss of plant species is between 100 and 1000 times higher than the expected natural extinction rates, which highlights the impact of plant extinctions on future drug development (Pimm et al. 1995). Based on the available data from the International Union for Conservation of Nature (IUCN) and the World Wide Fund (WWF), there are an estimated 50,000–80,000 flowering plants used for medicinal purposes, and about 15,000 are facing the risk of extinction due to overharvesting in countries like India, China, Kenya, Uganda, Nepal, and Tanzania (Ross 2005; Heywood and Iriondo 2003; Hamilton 2008; Zerabruk and Yirga 2012). Good agricultural practices (GAP) should be developed in accordance with agronomic techniques to provide a comprehensive framework from production to postharvesting in order to facilitate the sustainable use of MPs. Agronomic approaches refer to the implementation of scientific principles in agriculture, i.e., collection of suitable crop varieties, soil management, nutrient requirement, pest and weed control, crop monitoring, precision agriculture, best harvesting time, etc. (Fig. 13.1). The National Medicinal Plant Board (NMPB) of India has developed best cultivation practices for a few MPs, for example, *Aconitum balfourii*, *Hedychium spicatum*, *Nardostachys grandiflora*, *Swertia chirata*, *Valeriana jatamansi*, and *Zanthoxylum armatum* (NMPB 2008, 2014). Therefore, it is essential to develop promising agronomic approaches

Fig. 13.1 Good agricultural cultivation practices (GACPs) for medicinal plants

for high-value medicinal plants that might offer a path toward sustainable resource management, biodiversity preservation, economic opportunities, as well as the promotion of traditional knowledge.

13.2 Agronomic Approaches for Medicinal Plant Cultivation and Conservation

Agronomic approaches are used in the agriculture sector to increase the productivity of MPs in different manners, such as maintaining the wild genetic diversity, homogenous material, high yields, and fulfilling the rising demand for raw materials. These agronomic methods also allow better species identification, quality assurance, genetic advancement, land development, crop management, and the transfer of technology to rural people (Madukwe et al. 2002; Kala et al. 2006). In the Himalayan region, MPs are cultivated in significantly fewer numbers due to the unavailability of plant material, lack of knowledge about good agriculture practices, and market availability (Purohit 1997; Nautiyal et al. 1997). Agronomic approaches are used in Northern India to cultivate almost 20 medicinal plant species, for example, *Withania somnifera*, *Hedychium spicatum*, *Nardostachys grandiflora*, *Tinospora cordifolia*, *Zanthoxylum armatum*, *Curcuma caesia*, *Rheum emodi*, *Swertia chirata*, and *Valeriana jatamansi* (NMPB 2008). Some MPs have a variety of medicinal benefits, such as *Withania somnifera*, which has been shown to have antitumor, antibacterial, immunomodulatory, anticonvulsant, antipyretic, and antihelminthic qualities. Similar to this, *Hedychium spicatum* rhizomes are used to treat inflammations, bronchitis, vomiting, and dyspepsia. The agro-techniques employed in MPs having diverse medicinal properties are discussed in Table 13.1 (NMPB 2008, 2014). However, many MPs are still neglected in terms of agro-techniques and systematic cultivation practices, and this might be due to the unavailability of plant material for propagation (Rao et al. 2004). It is crucial to take action to slow the rate of

Table 13.1 Scientific cultivation practices of selected medicinal plants for large-scale production

S. No.	Medicinal plants	Part used	Therapeutic uses	Propagation material	Propagule rate per hectare	Sowing time	Optimum spacing	Fertilizer concentration per hectare		Yield per hectare	Diseases and pest
								FYM	NPK ratio (kg)		
1.	*Andrographis paniculata*	Dried leaves and tender shoots	Malaria, fever, cough, inflammation, dysentery, dyspepsia, hemorrhoids, and diarrhea	Seeds	650–750 gm seeds	September	15 × 15 cm	20 t	75:75:50	2.5 tons (t) (whole plant)	Not reported
2.	*Angelica glauca*	Roots/rhizome	Anorexia, flatulent colic, bronchitis, and spasms	Seeds and rootstock splits	6.2 kg seeds	April and may	45 × 45 cm	15–20 t	–	593–600 kg (whole plant)	Not reported
3.	*Aquilaria malaccensis*	Wood and oil	Diarrhea, indigestion, wounds, injuries, pain, etc.	Seeds	4500 plants	April–June	3 × 3 m	20 t	60:60:40	–	*Heortia vitessoides*
4.	*Aristolochia indica*	Leaves and roots	Anti-inflammatory, antiperiodic, diuretic, antiarthritic, antibilious	Seeds	30,000 seedlings	August–September	60 × 60 cm	10 t	25:60:100	640 kg (whole plant)	Leaf blight, *Pachliopta aristolochiae*
5.	*Bergenia ciliata*	Rhizome	Antidiuretic, astringent, tonic, and anti-inflammatory	Seeds and rhizome	18–20 quintals rhizome	July	10 × 10 cm	35 t	–	7.0–7.2 t (rhizomes)	Not reported
6.	*Catharanthus roseus*	Roots and leaves	Blood dysentery, scabies, hypotensive, antiviral, and sedative	Seeds	2.5 kg seeds		45 × 30 cm	10 t	–	1.5 t (leaves), 0.5 t (roots)	Little leaf and die-back disease
7.	*Centella Asiatica*	Leaves and whole plant	Nervine tonic, anti-leprosy, anti-leprosy, diuretic, asthma, fever, insomnia, and in mental disorders	Seeds and suckers	300 kg rooted suckers	February–march	45 × 45 cm	20 t	100:50:50	10–12 t (whole plant)	Not reported
8.	*Chlorophytum arundinaceum*	Roots	Renal calculus, diabetes, and leukorrhea	Seeds and roots tubers	45,000–50,000 seedlings	March	60 × 30 cm	20 t	90:30:30	500–800 kg (roots)	Not reported`
9.	*Cineraria maritima*	Leaves	Eye infections, blood congestion, conjunctivitis, and cataract removal	Seeds and stem cuttings	–	December	30 × 30 cm	10 t	–	1.5–2.0 t (leaves)	Not reported
10.	*Clerodendrum serratum*	Roots and leaves	Antitoxic, styptic, antiseptic, and antiseptic	Stem cuttings	–	May	2.0 × 2.0 m	–	75:75:75 (g)	1.5–2.0 t (whole plant)	Sooty larval and mealy bugs
11.	*Cryptolepis buchananii*	Roots and mature stem	In fever, skin diseases and leprosy	Seeds	10 kg	April	60 × 60 cm	10 t	130:100:60 (kg)	17.74 t (herbage)	Blue beetle attack on leaves

(continued)

Table 13.1 (continued)

S. No.	Medicinal plants	Part used	Therapeutic uses	Propagation material	Propagule rate per hectare	Sowing time	Optimum spacing	Fertilizer concentration per hectare		Yield per hectare	Diseases and pest
								FYM	NPK ratio (kg)		
12.	*Eclipta alba*	Whole plant	Edema, liver disorders, rheumatic joint pains, fever, stimulate appetite and improve digestion	Seeds and stem cuttings	450–500 gm seeds	February–march	20 × 20 cm	15 t	30:40:20 (kg)	8 t (whole plant)	Insect attack
13.	*Embelia ribes*	Berries and roots	Headache, headache, insomnia, epilepsy, and in rhinitis	Seeds	–	June–July	1.0 × 1.0 m	5–10 t	–	190–200 kg (seeds)	Not reported
14.	*Evolvulus alsinoides*	Whole plant	Blood purifier and cures bleeding piles	Seeds	190 gm seeds	June–July	25 × 25 cm	10 t	–	18.6 quintal (herbage)	Not reported
15.	*Fumaria parviflora*	Whole plant	Cures indigestion, fever, urinary discharges, vomiting, leprosy, fatigue, biliousness, and thirst	Seeds	4–5 kg seeds	November	30 × 15 cm	12–15 t	40:50:20 (kg)	2.5–3.0 t (whole plant)	Not reported
16.	*Glycyrrhiza glabra*	Roots and stolon	Antiviral, anti-inflammatory, mild expectorant, demulcent, and in treating peptic ulcers	Stolon cuttings	–	July	60 × 45 cm	10 t	–	7 t (roots)	Panash mosaic
17.	*Habenaria intermedia*	Tuber	Age sustaining	Tubers	250,000 tubers	May	20 × 20 cm	15 t	–	550–600 kg (tubers)	White grub attack in tubers
18.	*Leonotis nepetifolia*	Whole plant	In burns, ringworm, rheumatism, rheumatism, antidermatophytic, anticancerous, and antihelminthic	Seeds	20 kg	June	30 × 30 cm	5 t	100: 80: 60 (kg)	27 t (whole plant)	Leaf rust
19.	*Malaxis music era*	Bulb	In fever, bleeding, dysentery, seminal weakness, emaciation, and in conditions of pitta and Vata	Bulbs	1,11,100 bulbs	October–November	30 × 30 cm	25 t	–	60 kg (bulbs)	Not reported
20.	*Mucuna pruriens*	Seeds	In Parkinson's disease and rheumatic ailments	Seeds	–	June	1 m × 0.75 m	10–20 t	75:50:50 (kg)	2.5 to 3.0 t (seeds)	Collar rot and leaf-eating hairy caterpillar
21.	*Paederia scandens*	Whole plant	In paraplegia, sciatica, and rheumatism	Vegetative cutting	22,400 cuttings	August and September	60 × 60 cm	15–20 t	100:50:50 (kg)	1.2 t	Not reported

22.	*Phyllanthus amarus*	Whole plant	Treating jaundice and hepatitis-B	Seeds	4 kg seed	June	25 × 25 cm	20 t	–	2–3 t (herbs)	Powdery mildew disease
23.	*Piper longum*	Dried spikes and roots	Used in dyspepsia, diuretic plexopathy, asthma, fever, joint pain, snake bite, night blindness, and liver inflammation	Stem/vine cutting	–	March and April	60 × 60 cm	20 t	–	100–150 kg (fruits)	Phytophthora leaf, stem rot, and anthracnose
24.	*Pluchea lanceolata*	Leaves	In edema, muscular pain, muscular pains, analgesic, antipyretic, and in rheumatism	Seeds	5000 plants	October–march	90 × 120 cm	–	–	40 t (herbs)	Not reported
25.	*Premna mucronata*	Roots and its bark	To cure fever, neurologic problems, and stomachache. They also have anti-inflammatory properties	Seeds	100 gm seed	June	4.5 × 4.5 m	5–10 kg per pit	90:100:80 (g)	6–8 kg (roots)	Termites attack, aphids, and thrips
26.	*Psoralea corylifolia*	Seeds	Treating leprosy, psoriasis, leukoderma, and other skin diseases	Seeds	–	March–April	60 × 30 cm	10 t	60: 60: 30 (kg)	1.0–1.2 t (seeds)	Powdery mildew
27.	*Roscoea purpurea*	Tuberous roots	Treatment of leukorrhea and impotence	Tubers	–	April	15 × 10 cm	25–30 t	–	1.2–1.4 t (tubers)	Not reported
28.	*Salvadora persica*	Roots	Used in dental caries, skin infection, conjunctiva, stimulant, gastropathy, and gingivostomatitis	Seeds	15 gm seeds	June	5 × 5 m	–	30:20:15 (kg)	200 kg (roots)	Not reported
29.	*Solanum surattense*	Whole plant	Helpful in bronchitis, thirst, urinary concretions, fever, and asthma	Seeds	1.25 to 2.5 kg seeds	June	60 × 45 cm	1 kg	30:40:20 (kg)	16–20 t (whole plant)	Not reported
30.	*Tecomella undulata*	Bark	Anti-leucorrhoea, anti-syphilitic, used in diabetes and liver diseases	Seeds	–	July–august	3 × 4 m	–	750: 600:300 (g)	–	Not reported
31.	*Tribulus terrestris* Linn.	Whole plant	Diuretic, emollient, anodyne, styptic, astringent, depurative, anthelmintic, verminosis, etc.	Seeds	1–2.0 kg seeds	February–march	90 × 120 cm	10 t	60:40:40 (kg)	2.88 t (whole plant)	Not reported
32.	*Withania somnifera*	Roots, leaves and seeds	Antitumor, antibacterial, antibacterial, immunomodulatory, anticonvulsant, antipyretic, and anthelmintic	Seeds	–	July–august	60 × 60 cm	–	25:25:25	0.5–0.7 t (roots), 30–40 kg (seeds)	Seedling blight, seed rotting, die-back, leaf curl tobacco, and crutches broom

(continued)

Table 13.1 (continued)

S. No.	Medicinal plants	Part used	Therapeutic uses	Propagation material	Propagule rate per hectare	Sowing time	Optimum spacing	Fertilizer concentration per hectare		Yield per hectare	Diseases and pest
								FYM	NPK ratio (kg)		
33.	*Abroma augustum*	Roots, roots bark, stem, and leaves	Used in intrauterine and menstrual disorders	Seeds	150–200 g seeds	March and April	1 × 1 m	10 t	30:40:20	1.95 quintals (roots bark)	Damping off
34.	*Aconitum balfourii*	Tuberous roots	Used as diaphoretic, diuretic, analgesic, febrifuge, anti-inflammatory, anti-rheumatic, antipyretic, and vermifuge	Seeds and tuber	2 kg seeds	October–November or march–April	45 × 45 cm	25–30 quintals	–	450 kg (roots)	Not reported
35.	*Aconitum heterophyllum*	Tuberous roots	Used as an expectorant, febrifuge, anthelmintic, antidiarrheal, antiemetic, and anti-inflammatory	Seeds, tubers, and leaves	1.5 kg seeds	February–March (lower altitudes) May–June (alpine sites)	30 × 30 cm	150 quintals	–	518 kg (whole plant)	Not reported
36.	*Alpinia galanga*	Rhizome and seeds	Used in sore throat and respiratory congestion	Rhizome	5.5 t rhizome	February–April	30 × 30 cm	5 t	–	23.93 t (rhizome)	*Colletotrichum* spots
37.	*Alstonia scholaris*	Stem bark, leaves, latex, and flowers	Treatment of fever, dyspepsia, skin diseases, liver complaints, chronic diarrhea, and dysentery	Seeds	250 g seeds	March–April	5 × 5 m	–	–	2.5 kg (bark)	*Colletotrichum goleosporioides*, *Sordaria humana*
38.	*Asparagus racemosus*	Tubers and leaves	Cure diseases of blood, kidney, liver, scalding urine, rheumatism, and gonorrhea	Seeds and rhizomes	7 kg seeds	June	15 × 15 cm	10 t	60:40:40	30–35 kg (seeds)	Not reported
39.	*Bacopa monnieri*	Whole plant	Improve mental alertness and treatment of insomnia	Shoot cuttings	40,000 propagules	June–august	20 × 20 cm	10 t	30:50:40	22.5 t (whole plant)	Not reported
40.	*Baliospermum montanum*	Roots, leaves, and seeds	Treatment of dropsy, jaundice, jaundice, cathartic, and in rheumatism	Seeds and terminal cutting	7–8 kg seeds	June	1.5 × 1.2 m	10 t	75:50:25	2 t (roots)	Not reported
41.	*Caesalpinia sappan*	Wood, leaves, pods, and seeds	Treating contusion and thrombosis	Seeds	600–800 g seeds	April–may	4 × 4 m	6–8 kg	100:50:75	200–250 kg (seeds)	Pod borer and termites attack
42.	*Clerodendrum indicum*	Roots and leaves	Asthma, cough, fever, tumors, and skin diseases	Stem and root cutting	–	March–April	45 × 30 cm	100 quintals	40 N:75P	10–12 quintals (roots)	Termites attack

43.	*Colchicum luteum*	Corms and seed	Treatment of gout, rheumatism, and liver and spleen diseases	Corms	0.5 million corms	October	10 × 20 cm	6 t	–	500–750 kg (corm)	Not reported
44.	*Coleus barbatus*	Roots	Used in epilepsy	Stem cutting	28,000 rooted cuttings	June–July	60 × 60 cm	15 t	30:30:30	250 kg (roots)	Not reported
45.	*Commiphora wightii*	Oleo-gum resin	Treating arthritis, rheumatism, hyperlipidemia, thrombosis, and hypercholesterolemia	Seeds	100 g seeds	July–august	2 × 2 m	1 t	75:130:30 (g)	120–130 kg (oleo-gum resin)	Collar rot disease and termite attack
46.	*Curculigo orchioides*	Tuberous roots and rhizomes	Treating cough, jaundice, asthma, and piles	Tuber	600–750 kg of root segments	May–June	1.5 × 2 cm	20 t	25:15:10	1000–1700 kg (tubers)	Seedling rot and black rot disease
47.	*Curcuma caesia*	Rhizome	To cure jaundice, cough, hiccups, and respiratory disorders	Rhizome	2.2 tons rhizomes	April	30 × 30 cm	5 t	100:80:60	48 t (rhizomes)	Leaf spot and leaf blotch disease
48.	*Curcuma zedoaria*	Rhizomes and leaves	In the cure, jaundice, cough, hiccups, and respiratory disorders	Rhizome	10–12 quintals rhizome	April–June	40 × 20 cm	150 quintals	100:80:60	285 quintals (rhizomes)	Not reported
49.	*Desmodium gangeticum*	Roots and whole plant	Used in postnatal complaints, diarrhea, chronic fever, biliousness, cough, vomiting, and asthma	Seeds	3–4.5 kg seeds	March–April	45 × 45 cm	10 t	20 N:40P	50–55 quintals (herbs)	Not reported
50.	*Dioscorea bulbifera*	Corms and tubers	Treating sore throat boils and swellings, dysentery, piles, and syphilis and acting as an anticancer	Corms and bulbils	80–100 kg corms/ bulbils	April–may	50 × 50 cm	15–20 t	–	12 t (corms)	Not reported
51.	*Gymnema sylvestre*	Leaf	Treatment of urinary disorders and neutralization of excess sugar in the body	Cutting	6700 rooted cuttings	February–march	1 × 1.5 m	10 t	250 kg NPK	1250 kg (leaves)	Aphid
52.	*Hedychium spicatum*	Rhizomes and its essential oil	Treatment of asthma, bronchitis, vomiting, dyspepsia, and inflammations	Seeds and rhizomes	25 quintals rhizomes	April	45 × 30 cm	20 t	100:120:60	115–120 quintals (rhizomes)	Rhizome rot and leaf spot disease
53.	*Hemidesmus indicus*	Roots	Treatment of leprosy, skin diseases, fever, asthma, bronchitis, syphilis, urinary diseases, chronic rheumatism, and leukorrhea	Stem and root cutting	28,000 rooted cuttings	July–September	60 × 60 cm	1–2 kg	–	1.2 t (roots)	Not reported

(continued)

Table 13.1 (continued)

S. No.	Medicinal plants	Part used	Therapeutic uses	Propagation material	Propagule rate per hectare	Sowing time	Optimum spacing	Fertilizer concentration per hectare		Yield per hectare	Diseases and pest
								FYM	NPK ratio (kg)		
54.	Holostemma ada-kodien Schult.	Roots	Treatment of cough, burning sensation, stomach pain, constipation, fever, and tridoshas	Seeds	1.5 kg seeds	May–June	60 × 60 cm	30 t	–	10–15 t (roots)	Aphid attack
55.	*Inula racemosa*	Roots and rhizome	In the cure of cure loss of appetite and stomach problems	Seeds and roots	1 kg seeds	November or early march	50 × 50 cm	15 t	100:100:50	80 quintals (roots)	Not reported
56.	*Leptadenia reticulata*	Leaves, stem, and roots	Treating skin infections, wounds, cardiac disease, bleeding, diuretics, fever, and galactagogue	Stem cutting	5000 plantlets	February–march	2 × 1 m	20 t	80:160:160 (g)	6–7 t (roots)	Powdery mildew disease, aphid, and mite infestation
57.	*Marsdenia tenacissima*	Roots	Treatment of colic and gonorrhea	Seeds and vegetative cuttings	100 seeds	May–June	60 × 60 cm	5 t	–	528 kg (roots)	Not reported
58.	*Microstylis wallichii*	Pseudobulbs and tubers	Cures bleeding diathesis, fever, and burning sensation	Vegetative parts	125,000 bulbs	November or December	20 × 20 cm	40–50 t	–	600–750 kg (tubers)	White grub attack
59.	*Nardostachys grandiflora*	Rhizomes and roots	It treats insomnia, hysteria, convulsions, and vertigo and is also used as a cardiac tonic	Seeds	600 g seeds	May	1 × 1 m	40 quintals	–	835 kg (roots)	Not reported
60.	*Operculina turpethum*	Roots bark, roots, and stem	Astringent, antidiarrheal, diuretic, analgesic, and is used to cure dropsy stomachache	Seeds and vegetative cutting	2 kg seeds	March–June	30 × 30 cm	2 t	–	1500 kg (roots)	Not reported
61.	*Oroxylum indicum*	Roots bark, stem bark, fruits, seeds, and leaves	Treatment of asthma, cold, cough, liver complaints, anemia, and jaundice	Seeds	250 g seeds	March	60 × 60 cm	10 kg	150 N:250P (g)	1 kg (bark)	Not reported
62.	*Picrorhiza kurroa*	Roots and rhizome	Rheumatism, joint pain and paralysis	Seeds and rhizomes	250 g seeds	October–November	20 × 30 cm	10 kg	–	11 quintals (roots)	Powdery mildew disease and aphid attack

63.	*Plumbago zeylanica*	Roots and milky juice	Used in rheumatism, joint pain, and paralysis	Stem cutting and seeds	1–1.5 kg seeds	March–April	50 × 25 cm	6 t	30:40:30	12–18 quintals (roots)	Semilooper larvae and Bihar hairy caterpillar
64.	*Premna integrifolia*	Root, stem bark, and leaves	In the cure of liver complaints, cold, obstinate fevers, flatulence, rheumatism, and neuralgia	Stem cutting	80,000 rooted stem cuttings	April– June	2 × 2 m	10 t	50:50:50	1250 kg (roots)	Not reported
65.	*Pterocarpus marsupium*	Heartwood, bark, leaves, and gum	Treating boils, sores, and other skin diseases	Seeds	2500 rooted cuttings	April	8 × 8 m	–	200 N:150P (g)	750–800 quintals (heartwood)	Seed borer
66.	*Pterocarpus santalinus*	Heartwood and bark	Treating vomiting, skin diseases, leprosy, ulcers, and hemorrhages	Seeds	100 g seeds	March, April, or may	4 × 4 m	25 kg	150:100:100 (g)	15,000 t (heartwood)	Leaf-eating caterpillars
67.	*Rheum emodi*	Roots and rhizomes	Anti-inflammatory drug used in treating ulcers	Seeds and rhizome	1 kg seeds	February	50 × 50 cm	10–15 kg	–	37–66 quintals	*Fusarium* sp. infection
68.	*Rubia cordifolia*	Roots and whole plant	Used in treating blood dysentery, inflammations, and urinogenital disorders	Seeds and roots cutting	600 g seeds	January	60 × 75 cm	100 quintals and 150–200 quintals (low altitude)	30:40:20	3 t (roots)	Not reported
69.	*Saraca asoca*	Stem bark, flowers, seeds	Astringent, uterine sedative, and a uterine tonic	Seeds	2 kg seeds	March	3 × 3 m	10 kg/tree/year applied twice	–	1 t (bark)	Not reported
70.	*Solanum indicum*	Whole plants, roots, and fruits	` treating asthma, chest pain, chronic fever, spasmodic cough, scorpion stings, difficult urination, and worm infestation	Seeds	4 kg seeds	May– June	30 × 30 cm	5 t	–	300 kg (seeds), 800 kg (fruits)	Not reported
71.	*Strychnos nux-vomica*	Dried ripe seeds and stem bark	Used in treating chronic dysentery, paralytic and neuralgic disorders, epilepsy, rheumatic arthritis, and hydrophobia	Seeds	1 kg seeds	December or January	5 × 5 m	10 kg for each pit	–	50–75 kg (seeds)	Not reported

(continued)

Table 13.1 (continued)

S. No.	Medicinal plants	Part used	Therapeutic uses	Propagation material	Propagule rate per hectare	Sowing time	Optimum spacing	Fertilizer concentration per hectare		Yield per hectare	Diseases and pest
								FYM	NPK ratio (kg)		
72.	*Strychnos potatorum*	Seeds, fruits, leaves, bark, and roots	Treating eye problems, leucoderma, kidney complaints, poisoning, etc.	Seeds	200 g seeds	February–march	5 × 5 m	1.5–2.0 kg per plant	–	6750–9000 kg (seeds)	Seed damage due to fungal attack, i.e., *Alternaria*, *fusarium*, *aspergillus*, and *Rhizopus*
73.	*Swertia chirata*	Whole plant	Used in piles treatment, skin diseases, ulcers, and diabetes	Seeds	200 g seeds	October–November	45 × 45 cm	15 t	–	3.75 t (herbage)	Not reported
74.	*Terminalia arjuna*	Stem-bark	Used as an astringent, aphrodisiac, urinary astringent, and expectorant	Seeds	4–10 kg seeds	March	6 × 6 m	10 kg	75:50:30 (g)	500 kg (bark)	Aphid attack
75.	*Terminalia chebula*	Dried immature fruits	Antiseptic, astringent, diuretic, and cardio tonic	Seeds	5 kg seeds	July	6 × 6 m	15 kg	75:30:30 (g)	40–50 kg (fruits)	Termite attack
76.	*Tinospora cordifolia*	Stem, roots, and whole plant	Diuretic, aphrodisiac and used in malaria, fever, urinary disorders, diabetes, rheumatism and dyspepsia	Stem cuttings	2500 cuttings	June–July	3 × 3 m	10 t	75 kg N	1500 kg (stem)	Not reported
77.	*Tylophora indica*	Aerial parts and roots	Treating bronchial asthma and 25–3.75 t	Seeds	250 g of seeds	–	75 × 75 cm	20 t	45:50:40	25–3.75 t (aerial parts)	Insects of *Coleoptera* order affect the crop
78.	*Uraria picta*	Roots and whole plant	Anti-inflammatory, expectorant, and diuretic	Seeds	4–5 kg seeds	April–June	30 × 30 cm	10 t	–	3–4 quintals (roots)	Not reported
79.	*Valeriana jatamansi*	Dried roots and rhizome	Antipyretic, diuretic, hypotensive, nervous unrest, epilepsy, hysteria, and also used in hepatic and nervine tonic	Seeds and rootstock	2.5–3 kg seeds	June	40 × 50 cm	35–40 t	–	35–40 quintals (roots)	Rhizome rot disease
80.	*Vigna trilobata*	Whole plant	Anti-edema, anti-inflammatory, soothing, and also used in urogenital disorders	Seeds	15–20 kg seeds	June	20 × 15 cm	10 t	30:60:40	2.5–5 quintals (seed)	Red hairy caterpillars, galerucid beetles, and aphids damage this crop disease

81.	*Viola serpens*	Whole plant	Antipyretic, diuretic, aperients, diaphoretic, and also used in the cure of bleeding piles, constipation, fever, headache, and skin diseases	Runners	110,000 plantlets/1–1.5 kg seeds	March–April	30 × 30 cm	12 t	–	–	Browning and blotching of the leaves
82.	*Zanthoxylum armatum*	Stem bark, fruits, and seeds	Used as an aromatic tonic in fever and dyspepsia, deodorants, and dental caries	Seeds	2–3 kg seeds	August–September	50 × 50 cm	10–12 t	–	20 quintals (fruits)	Not reported

species extinction, and MPs can help by ensuring that resources are available for the foreseeable future (Negi et al. 2010; Volenzo and Odiyo 2020). The large-scale cultivation of MPs is a significant step toward conservation and fulfilling the demand for herbs for farmer's livelihood in hilly areas (Brodt et al. 2011; Lubbe and Verpoorte 2011).

13.3 Agronomic Practices for Medicinal Plant Cultivation

The agronomical characteristics of MPs focused on propagation materials, planting procedures, land preparation, fertilizer application, optimal spacing, intercropping, irrigation strategies, disease control, and postharvest management. Agronomic approaches might be used to raise the safety and quality of medicinal plant raw resources and final goods. Additionally, the use of such practices could support their conservation and cultivation (NMPB 2008).

13.3.1 Optimal Soil and Climatic Conditions

The soil and climatic condition significantly influences the growth and yield. Fertilizers like farmyard manure (FYM), vermicompost, and chemical fertilizers with the right amounts of nitrogen, phosphorus, and potassium are also used to increase the yield of MP, even though soil already has all the necessary nutrients, organic matter, drainage, moisture level, and pH for healthy MP growth (WHO 2003). Different species prefer different kinds of soil and climatic conditions in order to grow and thrive. For instance, *Costus speciosus* grows well in alluvial soil having subtropical climates (Pandey et al. 2011). Similarly, several medicinal plants, such as *Cineraria maritime*, increase in the tropical and tropical region in medium sandy soil; *Nepeta cataria* is well suited to sandy loam soil; and *Cnicus benedictus* thrives in low and subtropical zones and grows in sandy soil (Rashmi 2002).

13.3.2 Propagation Methods

Conventional and nonconventional methods can propagate plants. Traditional methods include propagation via seeds, rhizomes, vegetative cuttings, layering, grafting, and budding; the nonconventional method involves tissue culture techniques such as nodal culture, embryo culture, haploid culture, meristem culture, callus culture, somatic embryogenesis, etc. together with the appropriate species-specific planting practices for the successful generation of new plant. However, various works have been done by NMPB, India, in the selection of the proper propagule as well as the rate of propagule per hectare for achieving the maximum yield. For example, the bark of *Terminalia arjuna* is widely used for its therapeutic properties, and only 4–10 kg seeds per hectare are required for the production of approximately 500 kg of bark. Similarly, the roots of *Valeriana jatamansi* are helpful in curing many ailments, and 2.5–3 kg of seeds per hectare produced 35–40 quintal roots. Many MPs are propagated by stem cuttings like *Tinospora cordifolia*, *Premna integrifolia*, *Plumbago zeylanica*, *Leptadenia reticulata*, *Coleus barbatus*, *Piper longum*, etc. Likewise, *Bergenia ciliate*, *Alpinia galangal*, *Asparagus racemosus*, *Curcuma caesia*, *Curcuma zedoaria*, etc. are propagated by rhizomes (NMPB 2008). The seeds bear the majority of the MPs, but appropriate methods of seed collection, storage, and sowing are essential for successful germination. Low germination under ideal circumstances is caused by a number of variables, including seed dormancy, seed type, and temperature. However, growth stimulants like gibberellins (GA_3) are the most effective among them for speeding up emergence and germination. Some species like *Mentha piperita*, *Coriandrum sativum*, and *Ocimum basilicum* are important medicinal plants used in various formulations and disease treatments having poor seed germination. Therefore, seeds were treated with growth regulators for fast germination and mass-scale propagation (Elhindi et al. 2016). Similarly, the effect of pre-sowing treatments along with various combinations of plant growth hormones and other chemicals for breaking seed dormancy in *Dioscorea deltoidea* were studied and reported that prechilling treatment at 3–4 °C for 30 days along with gibberellic acid was effective in enhancing seed germination (Bano et al. 2021).

13.3.3 Fertilizers and Irrigation Practices

In order to increase the yield of cultivated plant species, it is crucial to regulate nutrients and water levels in the soil. Fertilizers are the substances that are used to provide essential nutrient to the plants, promote their growth, and ultimately enhance the yield. Many chemical and organic fertilizers are used in the cultivation of MPs, but the overuse of chemical fertilizers imbalances the soil nutrients and reduces soil fertility (Tilman et al. 2002; Aktar et al. 2009). Thus, organic fertilizers such as farm yard manure (FYM) and vermicompost are used in MP cultivation, such as in *Artemisia pallens* that successfully increased its biomass, essential oil, and nutrient uptake (Rao et al. 1998). Similarly, different concentrations of FYM with nitrogen, potassium, and phosphorus are used in the selected MPs by NMPB, which subsequently increased their yield as 20 t of FYM was used in *Andrographis paniculata*, 15–20 t in *Angelica glauca*, 35 t in *Bergenia ciliata*, 40 quintals in *Nardostachys grandiflora*, etc. (NMPB 2008).

Moreover, proper management of plant spacing, as well as balanced fertilizer treatment, can significantly influence growth parameters like plant height, number of branches per plant, leaf area, yield per plant, as well as quality parameters. Gunda et al. (2022) investigated the influence of various combinations of plant spacing options and NPK treatment on *Ocimum basilicum*. They recommended 40 cm × 20 cm and 60 cm × 40 cm plant spacing options along with 150:75:75 and 150:75:75 NPK kg per hectare as the best treatment for achieving the best growth and quality parameters. Similar work was done by Joshi et al. (2023) for the development of agro-techniques for *Paeonia emodi* Royle, in which he investigated the best cultivation practices for *P. emodi* by application of FYM and vermicompost and reported that the pretreatment of a field with vermicompost has significantly increased the biomass and plant secondary metabolites. In contrast, the postharvest agronomic characteristics of *P. emodi* were reported highest from the treated field. Treatment of cow manure with urea increased the yield of *Crocus sativus* by improving the soil parameters such as soil organic matter, soil pH, and soil electrical conductivity (Mohammad et al. 2012).

13.3.4 Weed Management

Weeds are undesired plants that compete with the crop of interest for essential nutrients, water, and sunlight and ultimately decrease the yield. To reduce the adverse effects of weeds, a combination of integrated weed management techniques is employed, including physical, cultural, and chemical methods. In the physical technique, weeding, hoeing, and mulching are done by hand; culturally, clean seeds are planted; and in the chemical method, pendimethalin is used at a rate of 1.5 kg per hectare for *Asparagus racemosus* and 1.5 kg per hectare for *Andrographis paniculata* (Upadhyay et al. 2012). Likewise, *Withania somnifera* is one of the most valuable MP cultivated in India that is adversely affected by weeds. Various chemical and plant-based herbicides were applied by Mohare et al. (2023) to investigate their role in eradicating weeds without affecting the plant, and they reported that a one-time spray of Quizalofop-ethyl at the rate of 50 g per hectare after planting was effective in controlling weeds without affecting the plant.

13.3.5 Sustainable Harvesting Practices

Sustainable harvesting practices ensure the long-term survival of the species, MPs sustainability, and improve the ecological balance. Overharvesting and illegal collection of MPs might result in population decline, loss of genetic diversity, and species extinction. *Persea macrantha* is a tree species that has many medicinal properties and is very well known for its bark, as bark is used for the preparations of incense sticks in India. Still, their population is declining at a fast pace due to its unscientific and overharvesting. In order to address this issue, specific agro-techniques were devised. For instance, the application of auxins, namely, indole-3-butyric acid and 1-naphthaleneacetic acid at a concentration of 1000 ppm, led to an 80% success rate in stem rooting.

Additionally, the use of GA3 resulted in a boost in seed germination, achieving an 80% germination rate, as reported by Praveena et al. 2023. The plant growth stages are crucial to ensure the best quality and potency of the plant materials and the final products derived from them. MPs should be harvested after reaching maturity using NMPB procedures, and the material should be cleaned, dried, and kept in sanitary baskets, dry sacks, trailers, hoppers, or other well-aerated containers (Pandey and Savita 2017).

13.4 Conclusion

Agronomical approaches hold massive potential in safeguarding the diversity and abundance of MPs while addressing the needs of traditional medicine and modern health care. By implementing agronomic practices in an integrated and considerate manner, we can ensure that future generations will continue to benefit from the invaluable gifts that medicinal plants offer to humanity. These agronomic approaches have been developed for a few selected MPs based on their demand and availability to promote their cultivation on a broad scale to meet the pharmaceutical demands for clean and standardized chemical ingredients. However, significant efforts are also required to develop these agronomical approaches for other critical MPs, and the cultivators should follow the species-specific approaches.

Acknowledgments The authors are thankful to the director, GBPNIHE, for their continuous support and motivation.

References

Akinyemi O, Oyewole SO, Jimoh KA (2018) Medicinal plants and sustainable human health: a review. Hortic Int J 2(4):194–195

Aktar MW, Sengupta D, Chowdhury A (2009) Impact of pesticides use in agriculture: their benefits and hazards. Interdiscip Toxicol 2(1):1

Amujoyegbe BJ, Agbedahunsi JM, Amujoyegbe OO (2012) Cultivation of medicinal plants in developing nations: means of conservation and poverty alleviation. Int J Med Aromat Plants 2(2):345–353

Bano H, Rather RA, Bhat JI, Bhat TT, Azad H, Bhat SA, Hamid F, Bhat MA (2021) Effect of pre-sowing treatments using phytohormones and other dormancy breaking chemicals on seed germination of Dioscorea deltoidea wall. Ex Griseb: an endangered medicinal plant species of North Western Himalaya. Eco Env Cons 27:253–260

Bhatt ID, Joshi K, Negi VS, Sekar KC, Rawal RS (2021) Threatened and high value medicinal plants of Kailash sacred landscape-Indian part. G.B. Pant National Institute of Himalayan environment (GBP-NIHE), Kosi-Katarmal, Almora

Brodt S, Six J, Feenstra G, Ingels C, Campbell D (2011) Sustainable agriculture. Nat Educ Knowl 3(1):112–121

Dhar U, Manjkhola S, Joshi M, Bhatt A, Bisht AK, Joshi M (2002) Current status and future strategy for development of medicinal plants sector in Uttaranchal, India. Curr Sci 83(8):956–964

Elhindi KM, Dewir YH, Asrar AW, Abdel-Salam E, El-Din AS, Ali M (2016) Improvement of seed germination in three medicinal plant species by plant growth regulators. HortScience 51(7):887–891

Gunda V, Padma M, Rajkumar M, Cheena J, Vijaya D, Chary DS (2022) Influence of NPK and plant spacing on growth, yield and quality of sweet basil (Ocimum basilicum L.) under Telangana conditions

Hamilton AC (2008) Medicinal plants in conservation and development: case studies and lessons learned. In: Kala CP (ed) Medicinal plants in conservation and development. Plantlife International Publisher, Salisbury, pp 1–43

Heywood VH, Iriondo JM (2003) Plant conservation: old problems, new perspectives. Biol Conserv 113(3):321–335

Joshi K, Jugran AK, Bhatt ID (2023) Agrotechniques development for Paeonia emodi Royle: evaluation of soil composition, biomass, and secondary metabolites. J Soil Sci Plant Nutr 18:1–2

Kala CP, Dhyani PP, Sajwan BS (2006) Developing the medicinal plants sector in northern India: challenges and opportunities. J Ethnobiol Ethnomed 2:1–5

Lubbe A, Verpoorte R (2011) Cultivation of medicinal and aromatic plants for specialty industrial materials. Ind Crop Prod 34(1):785–801

Madukwe MC, Okoli EC, Eze SO. 2002 Analysis and comparison of the agricultural development programme and University agricultural technology transfer systems in Nigeria

Mohammad M, Amiri ME, Sharghi Y (2012) Respond of saffron (Crocus sativus L.) to animal manure application. J Med Plant Res 6(7):1323–1326

Mohare S, Tiwari G, Tripathi N, Rawat A, Jha A, Ramakrishnan RS, Samaiya RK (2023) Comparative efficacy of chemical and plant based herbicides on weed Management in winter Sown Ashwagandha Crop [Withania somnifera (L.) Dunal]. Biol Forum Int J 15(2):444–449

National Medicinal Plants Board (2008) Agro-techniques of selected medicinal plants, vol 1. Department of AYUSH, Ministry of Health and Family Welfare, Government of India, New Delhi. ISBN 978-81-7993-154-7

National Medicinal Plants Board (2014) Agro-techniques of selected medicinal plants, vol 2. Department of AYUSH, Ministry of Health and Family Welfare, Government of India, New Delhi. ISBN 978-81-909121-3-6

Nautiyal BP, Pandey N, Bhatt AB (1997) Analysis of vegetation pattern in an alpine zone in north-west Himalaya: a case study of Garhwal Himalaya with reference to diversity and distribution patterns. Int J Ecol Environ Sci 23(1):49–65

Negi VS, Maikhuri RK, Phondani PC, Rawat LS (2010) An inventory of indigenous knowledge and cultivation practices of medicinal plants in Govind Pashu Vihar wildlife sanctuary, central Himalaya, India. Int J Biodiversity Sci Ecosyst Service Manag 6(3–4):96–105

Pandey AK, Savita R (2017) Harvesting and post-harvest processing of medicinal plants: problems and prospects. Pharm Innov J 6(12):229–235

Pandey A, Gupta S, Yadav KR (2011) AgroTechniques of Costus speciosus: an important endangered medicinal plant. In: National conference on forest biodiversity: earth's living treasure. Uttar Pradesh State Biodiversity Board, Lucknow

Pathak K, Das RJ (2013) Herbal medicine-a rational approach in health care system. Int J Herb Med 1(3):86–89

Phondani PC, Maikhuri RK, Kala CP (2010) Ethnoveterinary uses of medicinal plants among traditional herbal healers in Alaknanda catchment of Uttarakhand, India. Afr J Tradit Complement Altern Med 7(3):195

Phondani PC, Bhatt ID, Negi VS, Kothyari BP, Bhatt A, Maikhuri RK (2016) Promoting medicinal plants cultivation as a tool for biodiversity conservation and livelihood enhancement in Indian Himalaya. J Asia Pac Biodivers 9(1):39–46

Pimm SL, Russell GJ, Gittleman JL, Brooks TM (1995) The future of biodiversity. Science 269(5222):347–350

Praveena S, Jithin J, Jose PA (2023) Development of Agrotechniques and restocking of Persea macrantha—a JIGAT yielding plant for sustainable utilisation. J Trop For Sci 35(3):331–338

Purohit AN (1997) Medicinal plants-need for upgrading technology for trading the traditions. Harvesting Herbs–2000 1997:46–76

Rao EP, Narayana MR, Rao BR (1998) The effect of nitrogen and farm yard manure on yield and nutrient uptake in Davana (Artemisia pollens wall, ex DC). J Herbs Spices Med Plants 5(2):39–48

Rao MR, Palada MC, Becker BN (2004) Medicinal and aromatic plants in agroforestry systems. In: New vistas in agroforestry: a compendium for 1st World Congress of Agroforestry. Springer, Cham, pp 107–122

Rashmi R (2002) Agrotechniques for cultivation of some homoeopathic medicinal plants

Ross IA (2005) Medicinal plants of the world, volume 3: chemical constituents, traditional and modern medicinal uses. Humana Press Incorporated, Totowa

Samant SS, Dhar U, Palni LM (1998) Medicinal plants of Indian Himalaya. Gyanodaya Prakashan, Delhi

Samant SS, Vidyarthi S, Pant S, Sharma P, Marpa S, Sharma P (2011) Diversity, distribution, indigenous uses and conservation of the medicinal plants of Indian Himalayan region used in cancer. J Biodiversity 2(2):117–125
Tilman D, Cassman KG, Matson PA, Naylor R, Polasky S (2002) Agricultural sustainability and intensive production practices. Nature 418(6898):671–677
Upadhyay RK, Baksh H, Patra DD, Tewari SK, Sharma SK, Katiyar RS (2012) Integrated weed management of medicinal plants in India. Int J Med Aromat Plant 1(2):51–56
Volenzo T, Odiyo J (2020) Integrating endemic medicinal plants into the global value chains: the ecological degradation challenges and opportunities. Heliyon 6(9):e04970
World Health Organization (2003) WHO guidelines on good agricultural and collection practices [GACP] for medicinal plants. World Health Organization, Geneva
Zerabruk S, Yirga G (2012) Traditional knowledge of medicinal plants in Gindeberet district, Western Ethiopia. S Afr J Bot 78:165–169

Understanding the Interplay: Medicinal Plants and Biotic Stress

14

S. Shreedevasena, N. Kavya, Namburi Karunakar Reddy, and Pooja S. Patel

Abstract

Medicinal plants, vital sources of therapeutic compounds, are subjected to various environmental stresses, including biotic factors that significantly influence their growth and bioactive compound production. Biotic stress, induced by pathogens such as bacteria, fungi, viruses, and herbivores, poses a substantial threat to the sustainable cultivation of medicinal plants. The interaction between medicinal plants and pathogens triggers a cascade of molecular responses, including the activation of defense-related genes and synthesis of secondary metabolites. Pathogen-induced stress alters the biosynthesis pathways of bioactive compounds, impacting the quality and quantity of medicinal properties in these plants. Understanding the molecular basis of plant-pathogen interactions is crucial for developing effective mitigation strategies. Integrated pest management, genetic engineering for disease resistance, and the use of beneficial microorganisms are among the promising approaches to alleviate biotic stress on medicinal plants. Additionally, promoting agroecological practices and enhancing plant resilience through proper cultural practices can contribute to sustainable cultivation and better yield. This chapter provides a comprehensive overview of the challenges posed by biotic stress on medicinal plants, emphasizing the need for interdisciplinary research to safeguard the global supply of essential therapeutic compounds derived from these valuable botanical resources. Implementing sustainable cultivation practices and innovative mitigation strate-

S. Shreedevasena (✉)
ICAR-Directorate of Medicinal and Aromatic Plant Research, Anand, Gujarat, India
e-mail: Shreedevasena.s@icar.gov.in

N. Kavya · N. K. Reddy · P. S. Patel
University of Agricultural Sciences, GKVK, Bangalore, India

© The Author(s), under exclusive license to Springer Nature Singapore Pte Ltd. 2024

M. Nandave et al. (eds.), *Ethnopharmacology and OMICS Advances in Medicinal Plants Volume 1*, https://doi.org/10.1007/978-981-97-2367-6_14

gies will ensure the continued availability of high-quality medicinal plants for pharmaceutical and healthcare applications.

Keywords

Medicinal plants · Biotic stress · Secondary metabolites · Defense mechanisms · Bioactive compounds

14.1 Introduction

India is known for their most extensive array of well-known medicinal plants, and their age-old traditional knowledge of the medicinal plants gained world market attention after the emergence of COVID-19. The people's faith in herbal medicine creates a greater demand for plant-based drugs for curing disease without side effects, attracting people toward Ayurveda. Most of the medicinal plants are widely distributed in the forest, where indigenous tribes know their traditional medicinal knowledge for treating diseases and also its alternate income-generating source. The cosmetic industry is in the growing sector, which highly depends on the natural ingredients extracted from medicinal plants. The important unexplored medicinal plants still exist, and few plants have low population densities in the wild due to harvesting, and few are in an extinct stage, which needs conservation. Medicinal and aromatic plants (MAPs) play a pivotal role in providing valuable compounds that serve as the foundation for numerous pharmaceuticals, cosmetics, and flavoring agents. However, the cultivation of these plants is constantly threatened by various environmental stresses, among which biotic stress stands out as a significant factor. Biotic stress in medicinal and aromatic plants is primarily instigated by living organisms, including pathogens such as bacteria, fungi, viruses, and herbivores. Understanding the dynamics of these interactions is crucial for ensuring sustainable production and maintaining the quality of bioactive compounds essential for human well-being. The impact of biotic stress on MAPs is multifaceted. Pathogens can compromise plant health by triggering diseases that affect growth, yield, and overall vitality. Furthermore, herbivores, ranging from insects to larger animals, pose direct threats by feeding on these plants, leading to physical damage and altered biochemical profiles. These interactions result in a complex web of responses at the molecular, physiological, and ecological levels within the plants. The study of biotic stress in MAPs encompasses a wide range of research areas, including plant-pathogen interactions, host defense mechanisms, and the synthesis of secondary metabolites. Unraveling the intricacies of these processes is essential not only for understanding the fundamental biology of these plants but also for developing effective strategies to mitigate the impact of biotic stress. Moreover, the diminishing awareness and understanding of the traditional use of lesser-known medicinal plants among younger generations contribute to a rapid decline in indigenous knowledge. Recognizing the ongoing loss of traditional wisdom regarding numerous valuable medicinal plants in the past and the current resurgence of interest, there arises a necessity to reassess practical knowledge. This reassessment is undertaken with the anticipation of fostering the development of the medicinal plants sector.

14.2 Influence of Biotic Stress Signals on Secondary Metabolites in Medicinal Plants

Regular observations and monitoring of the pest population dynamics and disease epidemics in agroecosystems can reveal the economic losses and importance of a particular pest or disease against which plant products can be tried (Mitra and Biswas 2002; Blum et al. 2010; Garg 2011; Tara et al. 2011). Herbivory, mechanic stimulation, and pathogen attack trauma-induced stimuli like rain, hail, wind, sand, and herbivory forces plants to respond. Such mechanical phenomena and interactions with microorganisms can have a direct impact on secondary metabolism because they trigger biochemical responses (Vazquez-Flota et al. 2004; Cipollini 1997; Wurst et al. 2008). The induction amount varies depending on the causative agent and other elements, such as insect saliva containing stimulants (Zheng and Wu 2004; Reymond et al. 2000), which may affect the damaged vegetable organ, or it may cause a broader, more widespread reaction (Korth and Dixon 1997; Neuvonen et al. 1987; Waterman and Mole 1989). In reaction to insect ingestion, which leads to the generation of peptides functioning as protease inhibitors, substances can be synthesized. For instance, these substances may impede the palatability of the plant, acting as a deterrent to herbivores, as observed in the cases of tomatoes and potatoes. Alternatively, these substances might stimulate an increase in the production or accumulation of secondary metabolites, causing animals to avoid consumption (Waterman and Mole 1989; Harborne 1989). Regarding diseases brought on by fungi, bacteria, and viruses, a signalization cascade is produced, and it can result in long-term resistance, with changes in the cellular wall's makeup, the synthesis of phytoalexins, and protein structures like chitin and glucanase all being linked to this resistance. Therefore, when a second stimulus is given, the response will happen quicker than the initial infection, allowing for a more effective response to the new infection and decreasing the organ's or organism's tolerance to further attacks (Waterman and Mole 1989; Graham et al. 1986). The phytoalexins, which are typically derived from phenylpropanoids, terpenoids, and polyethylene, are produced in response to pathogen attacks, as seen on cotton leaves (*Gossypium hirsutum*) (Waterman and Mole 1989; Salisbury and Ross 1991; Hilder et al. 1987; Harborne 1989; Opitz et al. 2008). When subjected to jasmonic acid, a plant hormone recognized for its role in mediating defense reactions against both biotic and abiotic stresses, plants exhibit a notable augmentation in the synthesis of terpenoids and the corresponding storage glands. The increased terpenoid levels observed on cotton leaves in response to both real and simulated herbivory have been established as a universal response guided by jasmonic acid. This suggests that these compounds can collaboratively contribute to the plant's defense mechanisms against herbivores or diseases (Heil and Bostock 2002). It is well known that jasmonic acid regulates biosynthetic pathways and contributes to the signaling mechanism that causes

Nicotiana sylvestris roots to produce more nicotine, which is then distributed to the entire plant; a positive correlation is found between the amount of trauma on the leaves, endogenous jasmonic acid, and nicotine accumulation in the whole plant (Heil and Bostock 2002). Regarding the leaves of the plant *Eriodictyon californicum*, a study on artificial trauma found that the amount of phenolic resins decreased by up to 63% (Opitz et al. 2008). After *Asclepias syriaca* leaves are injured, poisonous cardenolides accumulate quickly, reaching a peak 24 h later. After 148 h, the cardenolides start to decline practically in elemental levels (Malcolm and Zalucki 1996). It was discovered that in pumpkin circuits, such compounds are induced an hour after the trauma and return to the elemental levels in less than 10 h (Mccloud et al. 1995). There are also reports of an increase in the content of tannins (Schultz and Baldwin 1982) and iridoids (Bowers and Stamp 1993) in reaction to the trauma caused by caterpillars besides the accumulation of condensed tannin in quaking aspen (*Populus tremuloides*) after trauma caused by herbivores (Osier and Lindroth 2001). Ajmalicine and vindoline levels in *Catharanthus roseus* plantules increased by over 100% after 48 h of an intentionally induced trauma, although catharanthine content did not alter significantly (Vazquez-Flota et al. 2004). Due to their toxicity, volatile chemicals in many plants can serve as a direct defense or an indirect defense by luring natural enemies of the insects that attack them (Yuan et al. 2008). Thus, biotic stress can also lead to the emission of volatile compounds. For instance, herbivory damage can increase the likelihood that these compounds will be released, which may result from a rupture of previously existing structures where the volatile compounds are stored, such as the trichomes (Holopainen and Gershenzon 2010).

14.3 Biotic Stress

Medicinal plants, nature's pharmacy, offer a treasure trove of potent therapeutic compounds that aid in healing various ailments and promoting the overall well-being of humanity. They are the source of elite raw materials that cater to a wide range of industries, including pharmaceutical, nutraceutical, perfumery, food, and cosmetics. Extracts from these plants are essential in combating pests and diseases in crops (Sofowora et al. 2013; Vaou et al. 2021). The vast global demand for raw materials has led to a surge in the cultivation of medicinal plants in recent years (Zahra et al. 2020). However, the production is hindered by various abiotic and biotic factors, among which the latter is a significant constraint. The biotic factors are animate entities that include fungi, bacteria, viruses, viroid, M.L.O.s, nematodes, insects, weeds, and phanerogamic plant parasites. The adverse effects of these organisms on plants are referred to as biotic stress. Biotic stress profoundly affects medicinal plants, causing reduced yield, compromised quality of economically vital plant parts, and decreased production of essential therapeutic secondary metabolites. Therefore, a comprehensive understanding of significant biotic stresses posing a threat to medicinal crops is critical to developing effective management strategies.

14.4 Disease: An Invisible Threat

The disease is a condition where an organism shows deviation from its normal physiological state, often associated with signs and symptoms. Infectious plant diseases are caused by fungi, bacteria, and viruses, among which fungi are very destructive. Fungi are widespread in the ecosystem, serving as both mutualistic partners and hostile adversaries. Notably, approximately 80% of plant species partake in symbiotic relationships with fungi, thereby influencing nutrient exchange and overall plant health. Nonetheless, this delicate equilibrium occasionally falters, leading to the transformation of these symbiotic allies into potent plant pathogens, inducing a diverse array of distressing symptoms (Zeilinger et al. 2016). Different fungi can attack other plants and plant organs, causing a range of symptoms, such as mildews, leaf spot, blight, anthracnose, rust, damping off, rotting, and wilt, resulting in extensive quantitative and qualitative damage to the plants (Table 14.1). Leaf spot diseases are more widespread in medicinal crops under cultivation, viz., *Chlorophytum*, *Cassia angustifolia*, *Rauwolfia serpentina*, *Aloe barbadense*, *Adhatoda vasica*, *Centella asiatica*, *Commiphora wightii*, *Andrographis paniculata*, *Datura*, *Asparagus*, *Tinospora cordifolia*, *Mentha arvensis*, and *Withania somnifera* (Kalra et al. 2005; Sinha and Chattopadhyay 2011; Rai and Tetrawal 2010; Ramappa Parashurama and Shivanna 2013; Thakur and Harsh 2014). Leaf spots appear as localized lesions consisting of dead and collapsed tissues. Mildews primarily appear on the leaf, stem, and reproductive structures, affecting photosynthesis; severe infection may result in sterility and production of inferior seeds. The consequences are striking, with documented reductions of up to 45% and 20%, respectively, in seed yield and husk yield of isabgol infected by the downy mildew fungus *Peronospora plantaginis* (Gohel and Prajapati 2022). Several other diseases, such as damping off, root rots, and wilts, cause a collapse of the plants, thus severely affecting the crop stand (Pandey 2017). Bacterial diseases are mainly identified by decaying tissues with a water-soaked, spongy, or slimy appearance (Singh et al. 2016).

A large number of medicinal and aromatic plants, including *Mentha* spp., *Andrographis paniculata*, *Hyoscyamus muticus*, *Catharanthus roseus*, *Rauwolfia serpentina*, *Hyoscyamus albus*, *Asparagus*, *Datura*, and *Withania somnifera*, are reported to be infected by viruses (Singh et al. 2016). The infection is manifested as local lesions or spread systemically, producing diverse symptoms, including chlorosis, mosaic patterns, vein clearing, leaf distortion, and necrosis. The severity of symptoms can vary depending on the viral strain, the host plant's susceptibility, and prevailing environmental conditions. In addition to the reduced vigor and survival of plants, viruses also make the host vulnerable to invasion by secondary pathogens (Beute 1970). For example, coinfection of young *Asparagus* plants with *Asparagus I and II virus* increases susceptibility to *Fusarium* wilting. DNA viruses, especially from the *Begomovirus* genera, primarily impact medicinal plant biomass and yield. The rapid evolution of these DNA viruses, coupled with their insect vector (whiteflies) exhibiting polyphagous feeding behavior, has escalated viral disease incidence worldwide (Saeed and Samad 2017). A drastic reduction in the yield of the medicinal herb Kalmegh, showing typical symptoms, yellow vein, upward leaf curling,

Table 14.1 List of diseases infecting medicinal plants

Diseases	Host	Causal organism	References
Downey mildew	*Plantago ovata*	*Peronospora plantaginis*	Gohel and Prajapati (2022)
	Ocimum basilicum	*P. belbahrii*	Yadav et al. (2021)
Rust	*Mentha* spp.	*Puccinia menthae*	Kalra et al. (2008)
Powdery mildew	*Mentha* spp.	*Erysiphe cichoracearum*	Kalra et al. (2008)
Alternaria leaf spots	*Aloe barbadense*	*Alternaria alternata*	Ghosh and Banerjee (2014)
	Withania somnifera		Chauhan and Ravi (2020)
	Mentha spp.		Kalra et al. (2008)
	Cassia angustifolia		Chandel et al. (2014)
Colletotrichum leaf spot	*Withania somnifera*	*Colletotrichum gloeosporioides*	Sarkar and Dasgupta (2017)
Curvularia leaf spot	*Cymbopogon citratus*	*Curvularia andropogonis*	Sato and Ohkubo (1990)
Myrothecium leaf spot	*Withania somnifera*	*Myrothecium roridum*	Shivanna et al. (2014)
Leaf blight	*Chlorophytum borivilianum*	*Colletotrichum dematium*	Tekade et al. (2009)
	Tinospora cordifolia	*Macrophomina phaseolina*	Sharma (2016)
Lethal yellowing	*Cymbopogon citratus*	*Pythium aphanidermatum*	Alam et al. (1992)
Wart	*Andrographis paniculata*	*Synchytrium lepidagathis*	Mondal (2016)
Collar rot	*Cymbopogon*	*Fusarium moniliforme*	Alam et al. (1994)
Root rot	*Aloe barbadensis*	*Fusarium oxysporum*; *Phytophthora* spp.; *Pythium* spp.	Kawuri et al. (2012)
Wilts	*Ocimum sanctum*	*Fusarium oxysporum f.* sp. *basilicum*	Moreira et al. (2015)
	Withania somnifera	*Fusarium solani*	Gupta et al. (2004)
	Plantago ovata	*Fusarium solani*	Meena and Roy (2020)
Damping off	*Withania somnifera*	*Rhizoctonia solani*	Jetawat et al. (2015)
	Cassia angustifolia	*R. bataticola*	Chandel et al. (2014)

(continued)

Table 14.1 (continued)

Diseases	Host	Causal organism	References
Bacterial soft rot	*Aloe barbadense*	*Pectobacterium chrysanthemi*	Pervez et al. (2016)
Leaf spot	*Tinospora cordifolia*	*Xanthomonas campestris*	Achar et al. (2014)
Leaf spot	*Ocimum sanctum*	*Pseudomonas cichorii*	Moreira et al. (2015)

chlorosis, reduced leaf size, poor inflorescence, and stunted growth caused by a combined infection by *Andrographis yellow vein leaf curl betasatellite* and *Catharanthus yellow mosaic virus*, was reported from Lucknow, UP (Saeed et al. 2014). An incidence (15–20%) of yellow mosaic disease in *Withania somnifera* caused by the *Jatropha mosaic India virus* was reported from India (Baghel et al. 2010). Different varieties of mint (*Mentha* spp.) in the significant cultivating areas of India express yellowing, yellow vein, mosaic, crinkling, and cupping with 40–50% disease incidence, causing a tremendous loss in biomass and consequently high economic loss to the growers (Saeed et al. 2014).

Phytoplasmas are bacteria with pleomorphic characteristics, primarily confined to the phloem, and distinctive for their absence of cell walls. These microorganisms are predominantly spread by insect vectors and hold economic significance as plant pathogens. They impact a wide range of plants, including annual and perennial crops, shrubs, fruit trees, as well as ornamental, medicinal, and aromatic plants on a global scale. Infected plants are associated with diverse symptoms, including phyllody, witches' broom, virescence, rosetting, proliferation of axillary shoots, little leaf, phloem necrosis, and plant death. Phytoplasma diseases of medicinal and aromatic plants are majorly reported in European and Southeast Asian countries, affecting plants belonging to more than 70 families, predominantly Apiaceae and Asteraceae (Marcone et al. 2016). A total of 19 distinct phytoplasma ribosomal groups, along with numerous subgroups, have been identified so far, and diseases in medicinal plants are often linked to specific ribosomal groups (Marcone et al. 2016). Yellows and witches' broom disease of Ashwagandha were reported from the experimental fields of Central Institute of Medicinal and Aromatic Plants, Lucknow, India, in 1988 which in subsequent years spread to other parts of India, reaching an incidence of 70% (Zaim and Samad 1995). Similar symptoms were recorded in kalmegh in 2014, with incidence ranging from 7% to 10% (Rao et al. 2018). A little leaf of *Datura* exhibiting proliferation of branches with shortened internodes and reduced leaf size was reported by Raj et al. (2009). Apart from disturbing the host development, they also influence secondary metabolite production in terms of quantity and composition. For example, Hemmati and Nikooei (2019) in *Artemisia* reported an increase in the abundance of monoterpenes from 4.9% in the healthy plants to 45.2% in the infected plants and an equivalent reduction in monoterpenoid content in infected plants. In periwinkle plants, infections caused by phytoplasma resulted in an elevation of metabolites associated with the biosynthesis of terpenoid indole alkaloids, including vindoline, and phenylpropanoids like chlorogenic acid and polyphenols. A study by Choi et al. (2004) documented an increased accumulation of therapeutic compounds, namely, vinblastine and vincristine, in periwinkle following phytoplasma infection. Consequently, despite the adverse impacts on plant hosts, phytoplasma infections could be deemed advantageous from a pharmaceutical perspective due to the enhanced production of valuable medicinal compounds.

14.5 Pests: The Tiny Menaces

Insect pests can cause direct physical damage to medicinal plants by feeding on leaves, stems, flowers, or fruits, thus weakening the plants. In response to insect feeding, plants may produce defense compounds that may alter the balance of phytochemicals in medicinal plants. The concentrations of desired active ingredients may decrease, while the levels of other compounds, including toxins or repellents, may increase. Consequently, the overall medicinal potency and therapeutic efficacy of the plant may be compromised. Some insects can introduce contaminants into medicinal plants, including fungal spores, bacteria, or viruses, through feeding or physical contact, resulting in secondary infections. Over 700 species of insects and mites have been identified in connection with approximately 80 medicinal and aromatic plants, encompassing both cultivated and wild varieties. The majority of these organisms display polyphagous or oligophagous feeding habits, and their economic significance varies considerably, contingent upon factors such as the host plant, geographical location, and cultivation conditions. The most damaging species are sap-sucking pests (Hemiptera, Tetranychidae), defoliators, stem borers, and flower and fruit-feeding pests (Coleoptera and Lepidoptera) (Table 14.2) (Marimuthu et al. 2018).

Nematodes have emerged as significant pests of medicinal plants, among others, causing a substantial decline in active ingredient content, thus undermining market acceptance. Key pest species include root-knot nematodes, reniform nematodes, lesion nematodes, and pin nematodes, each exerting considerable pressure across diverse agroecosystems. Root-knot nematodes, in particular, inflict significant damage on numerous medicinal plants such as *Ocimum* spp., *Artemisia* sp., *Hyoscyamus* sp., *Mentha* spp., *Palmarosa* sp., *W. somnifera*, *Cannabis sativa*, *Portulaca grandiflora*, *Rosa alba*, *Catharanthus roseus*, *Datura inoxia*, *Calotropis gigantea*, *Foeniculum vulgare*, *Phyllanthus amarus*, and *Plantago* (Singh et al. 2016). Among these, *Meloidogyne incognita* and *M. javanica* are responsible for notable harm to *Ocimum* spp. and *Mentha* spp. Similarly, *Artemisia* sp. faces severe infestations from *M. arenaria*, while geranium and *Mentha arvensis* confront attacks from *M. hapla*. Lesion nematodes and pin nematodes represent formidable challenges as pests for medicinal plants.

Table 14.2 List of pests infecting medicinal plants

Host	Pests	Symptoms
Cassia angustifolia	*Aphis craccivora*	Crinkling and curling of leaves
	Catopsilia pyranthe; *Eurema hecabe*	Defoliation
	Etiella zinckenella	Pod borer
Withania somnifera	*Deilephila nerii*, *Myllocerus viridanus* *Myllocerus discolor* *Henosepilachna vigintioctopunctata* *Deilephila nerii* *Aphis gossypii* *Mealybug* *Ferrisia virgata* *Green plant bug* *Nezara viridula*	Defoliation
	Aphis gossypii *Ferrisia virgata* *Nezara viridula*	Crinkling and curling of leaves
	Tetranychus urticae	Yellowing and drying of leaves
Tinospora cordifolia	*Othreis maternal* *Myllocerus viridanus*	Feeding on leaves and notching of leaf margins
Plantago ovata	*Rhopalosiphum maidis* *Aphis gossypii*	Curling and crinkling of leaves

14.6 Phanerogamic Plant Parasites: The Unwanted Competitors

Besides the constraints mentioned above, another substantial obstacle to their commercial cultivation is competition from weeds. Weeds compete for essential resources such as nutrients, water, light, and space that ultimately suppress plant growth and reduce biomass yield and quality of produce. Across various crops, weeds are responsible for nearly 45% of losses, a trend that might be parallel in medicinal plant cultivation. For instance, weeds can cause a notable decrease, ranging from 7% to 97%, in the yield of medicinal plant seeds (such as fenugreek, fennel, and isabgol) and biomass (like basil, psyllium, and mint) (Carrubba 2017). The issue of weeds extends to include parasitic plants, which have evolved diverse strategies to leech resources from their hosts. Broomrape, a non-photosynthetic complete root parasite belonging to the *Orobanche* genus, infests a wide range of crop species, including broad-leaved legumes, vegetables, and ornamental and medicinal herbs, inducing negative impacts on the growth and yield of their hosts. This parasitic plant attaches itself to the roots, extracting nutrients and water, resulting in stunted growth, reduced vigor, and potential yield losses. Kalariya et al. (2023) recorded small growth, an 83% decline in leaf area, an 82% reduction in total biomass, a 68% reduction in root yield, and a 38% reduction in withanolide yield in weeded compared with those of non-weeded ashwagandha plants. Dodder, holoparasite in the family *Convolvulaceae*, grows on dicotyledonous plants and lacks leaves and roots. The *Cuscuta* genera are relatively large, with more than 200 species. They attach to host plants via haustoria and draw water, solutes, and organic molecules from the host conductive system, leading to hampered growth and yield. Saric-Krsmanovic et al. (2020) reported a negative impact of field dodder on two medicinal and aromatic plants, peppermint and chamomile, in terms of essential oil yield and oil composition.

14.7 Biotic Stress Management

14.7.1 Breeding for Resistance and Collection of Wild Germplasm

An essential technology for enhancing medicinal and aromatic plants is breeding. Breeding for higher uniformity among cultivars, tolerance to abiotic and biotic challenges, removal of undesirable chemicals, and enhanced output of beneficial compounds are essential goals (Carlen 2012). The cultivars must also be suitable for seed propagation as well as for methods of sustainable production, extraction, and purification. Nowadays, the breeding of medicinal and aromatic plants is in its early stages compared to traditional food crops, with the benefit that breeders can take use of a high level of natural variability within a species. One of the reasons that conventional breeding techniques have predominated up to now is the typically considerable biological diversity within a species. Other benefits include the affordability of these techniques and their ability to generate a profit even from modest initial sales.

Additionally, according to Canter et al. (2005) and Pank (2010), transgenic medicinal plants are not permitted on the European market. Depending on the species and the selection criteria, breeding a new cultivar takes 5–15 years. Employing morphological, phytochemical, and genetic markers in the initial phases of the reproductive cycle, enhancing the frequency of generations annually, and adopting rapid and cost-effective techniques for measuring desired traits are some instances of approaches to accelerate breeding processes. These methods should be considered to address stakeholder demands more promptly. In the breeding process, several stages are considered, comprising (1) an assessment of the inherent diversity within a species, (2) the establishment of selection criteria and breeding methodologies, and (3) the implementation of tools to enhance the efficiency of the selection process.

14.8 Variability Within a Species

14.8.1 Natural Variability

Genetic diversity is necessary for successful breeding. It is crucial to learn about the natural variability of the target qualities within the under-consideration species before initiating a breeding effort. If such information is not available, it is necessary to methodically gather seeds or vegetative propagation material from the wild, botanical gardens, gene banks, research facilities, and commercial businesses. The best technique is to examine these accessions in the same experimental fields with settings that will be used for future commercial production, including fertile conditions with irrigations, as well as similar soil and climatic conditions (Dudai 2012). In a Swiss study, *Tanacetum vulgare*, an *Asteraceae* perennial herbaceous flowering plant endemic to temperate Europe and Asia, had its morphological and phytochemical variability examined. The literature frequently refers to common tansy as a plant with anthelmintic benefits for animals (Waller et al. 2001). Its inclusion in industrial goods has also been documented (Valchev et al. 2009). However, the domestication and cultivation of this perennial plant have received relatively little research.

14.8.2 Creation of Variability

When the natural variability does not provide the desired trait expression or trait combination, additional variability might be produced. The most frequent methods for creating novel varieties involve crossing various genotypes of the same species or two closely related species. By introducing features or genes from one genotype to another, genotypes can be crossed to produce novel trait combinations. For instance, to achieve both features in particular offspring, a disease-resistant genotype may be mated with a high-yielding but vulnerable genotype. Other strategies include polyploidy induction, mutation induction, and biotechnological techniques like gene transfer (Bernath 2002). Chemicals or radiation can cause mutations. The mutants that are produced are examined and further chosen for the desired features. Examples of induced mutations that made intriguing outcomes include populations of *Mentha piperita* with higher menthol content (Zheljazkov et al. 1996) and *Cymbopogon flexuosus* with altered essential oil accumulation pattern (Kulkarni et al. 1992). Induced polyploidy is another way to increase the variety and the breeding foundation. Diploid plants make up a large percentage of therapeutic and aromatic species. By administering the chemical colchicine, which doubles the number of chromosomes, it is possible to artificially increase the number of sets of chromosomes in each cell. Induced polyploidy has been shown to successfully boost the productivity of secondary metabolites and essential oil in a variety of plant species, as demonstrated for *Acorus calamus* (Dhawan and Lavania 1996). Similarly to this, polyploidy has boosted the critical oil concentration of *Lavandula angustifolia* (Raev et al. 1996). Because of the extended duration of the breeding process and the substantial inherent variability within a species, induced mutation and polyploidy have not been commonly employed in the context of medicinal and aromatic plants.

14.9 Breeding Strategies

14.9.1 Species with Vegetative Propagation

For the majority of plant species, vegetative propagation can be utilized to produce a cultivar (clone). Following a screening procedure, the genotype that is best suited can be propagated, typically through cuttings or in vitro. This method has the advantages of a much shorter selection period and relatively homogeneous plants. The drawback is that, in contrast to seed-based propagation, vegetative propagation is significantly more expensive. According to Quennoz et al. (2006), this breeding strategy is primarily employed to multiply perennial species such as bush and tree species as well as species without seed development like Thymus × citriodorus (Pers.) Schreb. and Mentha × piperita in response to an industry demand for low quantities of plant material (Carron et al. 2009; Vouillamoz et al. 2013).

14.9.2 Apomictic Reproduction in Species

When seeds are formed through apomictic reproduction, the egg cells are not fertilized by the sperm cells. Because seed progenies share the same genetic code as the mother plants, breeders can employ the apomictic behavior to quickly genetically fixate the expression of desired traits (Pank and Kruger 2003). *Hypericum perforatum* is a significant therapeutic plant with this flower biology. In plant therapy, *H. perforatum* is advised due to its antiviral, vulnerary, and antidepressive qualities. However, *Colletotrichum gloeosporioides*, a wilt disease that causes anthracnose, severely affected this species (Debrunner et al. 2000). Therefore, tolerance or resistance to this anthracnose should be a key breeding goal. Within a Mediplant initiative, 21 wild accessions collected from Switzerland, Germany, Italy, and Canada underwent a comparative analysis against the commercially available cultivars Topas, Hypermedia, and Elixir. Notable variations were observed among the accessions in terms of anthracnose resistance, flowering period, plant morphologies, flower and drug yields, and phytochemical profiles. Among the accessions, P7 stood out due to its diminished susceptibility to anthracnose, leading to its selection for further consideration. Additionally, the yield and phytochemical profile (hypericin and flavonoids) met industrial standards and were comparable to those of the traditional cultivar Topas. Seeds may easily replicate the accession P7 because of apomictic reproduction, and it is now registered as the cultivar "Hyperivo 7" (Gaudin et al. 2003).

14.9.3 Self-Pollinated Species

In order to create seed production and a cultivar through self-pollination, the initial breeding stage for self-pollinated species is to choose individual plants within accessions from the wild that exhibit good performance (such as yield, quality, resistance, etc.). If no adapted genotype is discovered, the best single plants must be chosen from the progeny after a forced crossing between two adapted genotypes is conducted with one of the genotypes' anthers removed. In order to create a uniform line, it is necessary to produce progenies by self-pollination from each of the chosen plants throughout four to seven generations, this time selecting just the best individuals each time. As a result, a new cultivar is created, as with *Carum carvi* and *Ocimum basilicum* (Putievsky et al. 1994; Dudai et al. 2002).

14.9.4 Cross-pollinated Species

A population cultivar is made up of single, open-pollinated plants. Every plant in a population will have a distinct collection of genes, making them genetically different from each other. However, a population will collectively exhibit some traits that were passed down from generation to generation. Each population cultivar will be distinguished from other population cultivars of the same species by a specific set of traits. Positive and negative mass selections are two popular selection techniques. *Salvia officinalis* cultivar "Extracta" and *Thymus vulgaris* L. *cultivar* "Deutscher Winter" are two examples of population cultivars (Carron et al. 2005; Carlen et al. 2010). The cultivation of these varieties does not necessitate controlled pollination for seed production. Instead, population cultivars were derived by collecting seeds from the crop of the current year.

14.9.5 Cultivars Based on Several Genotypes

Multiple carefully chosen parents can generate cultivars, which can then be crossed under isolation via open pollination. A genotype's capacity to be a component of a cultivar with multiple parents (synthetic cultivar) is demonstrated by a polycross test of its general combining ability. According to research on forage grasses, the ideal number of parents is 4–6 (Piano et al. 2007). The parents are clones, or genotypes, that typically reproduce vegetatively. In contrast to population cultivars, synthetic variants have superior homogeneity, which makes them attractive. By combining 40 probable parents, a polycross test using *Carum carvi* L. *var. annuum* hort was created (Pank et al. 2007). The selection of parental lines was subsequently determined by evaluating the performance of their progeny to gain insights into the overall compatibility of the parental lines. The optimal parental lines were then employed to generate the desired synthetic varieties.

14.9.6 Tools to Improve the Breeding

Depending on the species and the breeding goals, a new cultivar requires 5–15 years from bioprospecting through cultivar registration. For enterprises creating and selling plant-based products, this is a very long time. Methods to speed up the breeding processes must be considered in order to respond to stakeholder demands more quickly. One of the options is to speed up generation succession through vernalization and greenhouse cultivation, which would enhance the number of generations produced annually (Pank and Schwarz 2005). Additional techniques involve rapid and cost-effective assessments, along with the early-stage utilization of morphological, phytochemical, and genetic markers in the reproductive cycle.

14.9.7 Molecular Marker-Assisted Selection

The goal of molecular breeding is to identify the genes of plants and their roles. Utilizing molecular markers enables plant breeders to screen sizable plant populations and aids in the selection of genotypes carrying the desired characteristic for use in breeding high-performing cultivars. Tharachand et al. (2012) and Sarwat et al. (2012) evaluated the role of molecular markers in the classification of medicinal plants. The markers used included single-nucleotide polymorphisms (SNPs), amplified fragment length polymorphisms (AFLPs), simple sequence repeats (SSRs or microsatellites), restriction fragment length polymorphisms (RFLPs), inter-simple sequence repeats (ISSRs), and randomly amplified polymorphic DNA (RAPD). The molecular foundation for marker-assisted breeding of *Artemisia annua* was developed in a recent study by Graham et al. (2010). There were found to be quantitative trait loci (QTL) associated with essential factors affecting artemisinin yield. A statistical technique called QTL analysis connects genotypic (information about genes) and phenotypic (information about specific traits) data to understand the genetic underpinnings of complex phenotypes. Positive artemisinin yield QTL in the parents will aid in the future production of new, high-yielding hybrids, allowing for a quicker response to the rising demand for artemisinin.

The characterization of medicinal plant materials can be done independently and objectively using molecular markers, which can also be used for breeding. Moreover, molecular markers play a pivotal role in efficiently overseeing germplasm collections, both in situ and ex situ. They facilitate the monitoring and evaluation of genetic diversity, the identification of duplicates within the collection, and the formulation of essential conservation strategies prioritizing what should be conserved. Future endeavors aimed at coordinating initiatives for the molecular characterization of medicinal plants and integrating existing molecular marker data are anticipated to yield significant productivity.

14.9.8 Germplasm Conservation

Preserving genetic diversity, particularly within species, is imperative for future conservation efforts, making germplasm preservation essential. The exploration and conservation of genetic resources in medicinal and aromatic plants are crucial for the well-being of current and future generations, given their untapped potential. Various policies, programs, and interventions have been proposed at international,

national, and state levels, as well as through initiatives like the Convention on Biological Diversity (CBD), established during the United Nations Conference on Environment and Development (UNCED) in 1992, to address these objectives. The increasing commitment of the global community to sustainable development served as the impetus for the creation of the Convention on Biological Diversity. It is a significant advancement for the preservation of biological variety, the sustainable use of its elements, and the just and equal distribution of the advantages brought about by the utilization of genetic resources. According to Janick and Whipkey (2007), germplasm preservation includes (1) the conservation of germplasm of various medicinal plant species through collection, acquisition, and exploration; (2) carrying out a variety of research related to germplasm, such as morphological, phytochemical, and genetic evaluation; and (3) promoting the use of germplasm collections and related information for research, crop improvement, and product development. As an illustration, the entire German parsley germplasm collection, which consists of 220 accessions from 35 different countries, includes both modern and historic cultivars as well as landraces (Lohwasser et al. 2010). Both parsley morphological types—leaf parsley and root parsley—were investigated in order to discover collection-wide diversity as well as the interactions between morphological, molecular, and phytochemical traits. There are two ways to approach the conservation of plant genetic resources: In situ conservation, on the other hand, is the creation and maintenance of natural reserves where species are permitted to continue existing in ideal ecosystems. Ex situ conservation, on the other hand, makes use of botanical gardens, field plantations, seed banks, and other facilities.

With regard to both the long-term preservation of these currently difficult-to-store germplasm categories and the mass production of endangered species, plant in vitro technology presents a viable solution. The key benefit of in vitro technology is the ability to preserve different types of cells and organized tissues, such as single cells, meristems, and embryos. A potent method for preserving germplasm is cryopreservation, which is the preservation of these tissues at temperatures below 80 °C and typically at liquid nitrogen temperature (196 °C). This genetic material needs to be preserved both in situ and ex situ for the future. Finding synergies, combining efforts, and proposing collaborative networks on ethnobotany and germplasm preservation, characterization, and evaluation, as done in Portugal, would be highly fruitful (Barata et al. 2011).

14.10 Biological Management

Plants produce allelochemicals or bioactive substances as by-products, end products, and even liberated metabolites from plants. The majority of them are terpenoids, steroids, alkaloids, and aromatic chemicals (organic cyanides, flavonoids, and phenolic acids). Despite these components, medicinal plants are still vulnerable to pests (insects, mites, nematodes) and diseases brought on by fungi, bacteria, and viruses, which reduce crop productivity and degrade the quality of plant parts used in medicines (such as roots, leaves, fruits, flowers, bark, and seeds). To lessen plant damage and the prevalence of pests and diseases, different chemical groups of pesticides are sprayed on medicinal crops. Toxic pesticides have been shown to have negative impacts on the ecosystem and helpful wildlife, though. Pesticides have been discovered in raw materials, infusions, decoctions, tinctures, and essential oils, all of which pose health risks (Zuin and Vilegas 2000). Nitrogen oxides in the air reacting with degrading hydrocarbons like linalool, myrcene, and ocimene, which are responsible for the floral aroma, can also have an impact on the scent of plant components (McFrederick et al. 2008). If the appropriate time following the final pesticide spray and before harvesting or collecting plant parts is not followed, chemical residues may persist in or on plant parts. In light of the strict requirements imposed by importing nations and the demand of local manufacturing enterprises for chemical-free raw materials, the quantity of pesticide residues is a critical criterion for therapeutic products with high value and export potential. Furthermore, even in the retail sector, organically cultivated plants can fetch higher prices.

14.11 The Use of Plant Parts and Plant-Derived Products Against Pests and Diseases

Among plants, shrubs and trees, the neem tree (*Azadirachta indica*) has traditionally been exploited for preparing neem leaf extract (NLE), neem seed extract (NSE), neem seed kernel extract (NSKE), neem cake (NC), and neem oil (NO) (Gahukar 1995), because neem contains a higher content of allelochemicals, azadirachtin (AZ) being the major bioactive isomer which is costly and difficult to synthesis. There have been numerous preparations researched for use against insects, including pest repellents, sterilants, growth retardants, oviposition deterrents, antifeedants, and fatal poisons. Other mechanisms of action that have been researched include root penetration obstruction, nematode development, spore germination inhibition, disease pathogen growth retardation, and spore retardation (Gahukar 1995; Akhtar 2000). Pure allelochemicals and other plant compounds can impart a single activity or a variety of effects. More than 50 neem-based commercial formulations (mainly emulsifiable concentrate) are currently available, thanks to advancements in formulation technology (Gahukar 1998). Neem is 1 of 29 plant species that are used to combat pests and diseases that affect medicinal plants.

Insect pests to control cotton aphid, *Aphis gossypii* Glover (Homoptera: *Aphididae*) on isabgol (*Plantago ovata*); in a field experiment, Patil et al. (2011) used three neem-based products (5% NSKE, 0.5% Godrej Achook, and 0.5% NO) and five chemical treatments to compare them to water and carbosulfan sprays. None of the therapies worked exceptionally well. When contrasted with the botanical treatment (1.5 aphids per plant) and the water control (1.8 aphids per plant) 10 days post-application, the application of 0.05% carbosulfan (25WP) significantly diminished the level of aphid infestation, resulting in 1.2 aphids per plant. As a result, botanical insecticides were shown to be less effective than chemical ones. This aphid species attacking neem and red amaranthus (*Amaranthus cruentus*) was successfully managed in a different field study using sprays of 5% water extract from 10 plants, which provided 40–60% pest repellency (Ramakrishna Mission 2008). Similar outcomes have been observed for the peach-potato aphid *Myzus persicae* (*Aphididae*), which attacks *Sida rhombifolia* and neem, as well as the plant bug *Urentius sentis* (Ramakrishna Mission 2008). Complete eradication of populations of the aphid *Schoutedenia emblica* was accomplished on the Indian gooseberry tree (*Emblica officinalis*) a week after spraying 0.045% NSKE containing 1500 ppm AZ (Sridhar 2011). Indian researchers tested the bioefficacy of 8 synthetic pesticides and 4 neem-based treatments (0.5% Margocide CK (20EC), 0.5% NO, 5% NSKE, and 5% NLE) against the fennel (*Foeniculum vulgare* Mill.) aphid *Hyadaphis coriander* (Das) (Jat et al. 2009). These workers reported a maximum reduction of up to 87% in the aphid population, an increase in seed yield (15.38 q ha^{-1}), and a more significant cost-benefit ratio of 1:49 from a field treated with 0.03% dimethoate (30EC), whereas the neem-derived products were least effective (64% reduction in pest infestation, 12 q ha^{-1} yield, and CB ratio of 1:2) (Jat et al. 2009). In order to impart resistance in ashwagandha plants, Ravikumar et al. (2008) used the combination of NC, FYM, and Azophos to confer resistance in ashwagandha plants. They observed a 69.8% decrease in plant damage, the lowest mealy bug feeding area, and the fewest plants bearing mealy bug eggs. Sprays containing a water extract of *Tagetes erecta* flowers or *Ocimum gratissimum* leaves completely prevented egg hatching and prevented the emergence of the swallow tail butterfly *Pachliopta aristolochiae* on shatawari (*Asparagus racemosus*) plants (Ramakrishna Mission 2008). It was observed that after 4 sprayings of neem derived solution 0.0.% EC under field conditions, the incidence of the gall wasp *Leptocybe invasa* (Hymenoptera: Eulophidae) was lowest (6.7% vs. 26.7% in control) with minimum oviposition damage (5.9% vs. 23.7%) and adult emergence was also lowest (adults seedling 1 vs. 13.7 adults in control) (Kavithakumari et al. 2009).

14.11.1 Mite Pests

Spraying a 5% water extract from *Mentha arvensis* in India resulted in the complete mortality of mites infesting medicinal plants, such as *Petrobia harti* (Tetranychidae) on Oxalis (*Oxalis corniculata*) and Tetranychus MacFarlane (Tetranychidae) on Ashwagandha. Additionally, in a neem plantation, the application of a 5% water extract from garlic (*Allium sativum* L.) bulbs demonstrated an 80% repellency against the two-spotted spider mite *Tetranychus urticae* (Ramakrishna Mission 2008).

14.11.2 Nematode Pests

By adding neem cake to the soil, *Meloidogyne incognita* (*Heteroderidae*), a root-knot nematode, was prevented from attacking Ashwagandha (Sharma and Pandey 2009). Ramakrishnan and Senthilkumar (2009) tested organic amendments as non-chemical alternatives for worm management on Ashwagandha and Indian senna (*Cassia angustifolia* and discovered that they were less efficient than humic acid and the bacterium *Pseudomonas fluorescens*). Under glasshouse and field conditions, populations of *M. incognita* juveniles in coleus plants were reduced by up to 60% resulting in 35.6% increase in the tuber yield after drenching with extracts of 6 plants, viz., *T. erecta*, *Crotalaria juncea*, *Vigna unguiculata*, *Brassica juncea*, *Allium cepa*, and *Ricinus communis* (Seenivasan 2011). Pot experiments conducted in Pakistan revealed a noteworthy enhancement in plant growth, coupled with a reduction in the quantity of galls on roots, the number of egg masses, and the reproductive factor of nematodes. This improvement was achieved through the utilization of neem, Datura, *Calotropis procera*, or *T. erecta* leaves. Toxic chemicals that were formed by the decomposition of organic materials in the soil led to the buildup of nematode mortality, and these criteria were dose-dependent (Hussain et al. 2011).

14.11.3 Disease Pathogens

Girish and Bhat (2008) reported complete inhibition in the mycelial growth of the fungus *Phomopsis azadirachtae* with foliar sprays of aqueous solutions of essential oils (at 3000 ppm) extracted from 5 plants, viz., *Eucalyptus citriodora*, *F. vulgare*, *Myristica fragrans*, *Capsicum annuum*, and *Curcuma domestica*. Pycnidia and colony width were both decreased by the spraying (Fathima et al. 2009). Neem leaves, or *Pongamia pinnata* seeds treated with NO (2%) or water extract (5%), totally prevented the growth of *Aspergillus flavus*, *Penicillium* spp., and *Mucor* spp., which would typically increase on neem seed and harm its quality (Srivastava et al. 1997). To combat leaf spots induced by the fungus *Cercospora rauwolfia* (Mycosphaerellaceae) on Sarpagandha (*Rauvolfia serpentina*) plants, a 2-year field study assessed 11 treatments, encompassing 6 biopesticides, 1 synthetic fungicide, and 4 botanicals. Notable results were obtained with the soil application of neem cake at a rate of 250 kg ha^{-1}, combined with *Eucalyptus* spp. leaf waste at 2.5 t ha^{-1} or *P. pinnata* cake, *Madhuca* indica cake at 10 g^{-1} plant, and foliar spray of NO (3%) or NSKE (5%), resulting in a 20% disease incidence 90 days after sowing. In contrast, the soil or foliar application of *P. fluorescens* exhibited greater efficacy, particularly in *Polyalthia longifolia* or *Catharanthus roseus* (Ramakrishna Mission 2008). Water extracts derived from the leaves of *C. odorata*, *P. longifolia*, or *C. citratus* have demonstrated the ability to inhibit *Curvularia lunata* spores by up to 100% and *Alternaria longipes* by up to 66.7%. *A. cepa* alone suppressed spore germination by up to 81% in *Colletotrichum gloeosporioides* (Penz.) (*Glomerellaceae*) and *P. longifolia* alone inhibited spore germination by up to 87.5% in *Alternaria dianthicola* (Ramakrishna Mission 2008). Neem or *Ailanthus excels*, *C. procera* water extract combined with cow urine wholly inhibited mycelial growth and spore germination in *Lasiodiplodia theobromae* and *F. oxysporum*, the causes of plant wilting and rotting in the lab (TFRI 2010).

14.11.4 Gene Editing for Medicinal and Aromatic Plants

An area that is ready for growth is direct DNA sequence editing to modify gene expression in medicinal plants. The possibility of changing a trait exists as long as it can be linked to one or a small number of genes. To develop as crops, essential agronomic qualities relating to uniformity, stability, growth and development, and tolerance to biotic and abiotic challenges must also be enhanced. Although the content of active chemicals is the primary aim for trait manipulation in medicinal plants, this is not the only factor to consider (Bajaj 1998; Dubey and Guerra 2002; Bernath 2002). The use of biotechnological methods with medicinal plants does not begin at a low point. There is already a lot of interest in modifying plant biosynthetic pathways to create insecticides, food ingredients, or medication precursors. One such method is to use trichomes, specialized surface organs that synthesize, store, and emit secondary metabolites (Stevenson 2004). Biosynthetic pathways have been modified in *Mentha* spp. (mints) to change the production of essential oils in the trichomes and to improve the plant's tolerance to fungus and abiotic stressors (Sevon and Oksman-Caldentey 2002). High-value phytochemicals have long been produced experimentally and commercially using tissue culture, an in vitro approach for cultivating plant organs, explants, tissues, cells, or protoplasts. To increase the productivity of such systems, the genetic transformation of cultures utilizing bacterial vectors to deliver genes into the cultured plant DNA has been frequently used. For instance, in hormone-free culture conditions, hairy root cultures, which are altered by infecting cells with the bacterium *Agrobacterium rhizogenes*, often retain stable and high output (Park and Facchini 2000). Tissue culture and regeneration are crucial components of a system for rationally changing the genetic makeup of medicinal plants, but they can also significantly improve other facets of plant breeding (Rout et al. 2000). The range of beneficial variation that is available to breeders is substantially increased by tissue culture, which frequently encourages genetic disruptions that lead to somaclonal variation (Rout et al. 2000; Kukreja and Dhawan 2000; Sevon and Oksman-Caldentey 2002). Tissue culture offers a valuable contribution to the conservation of biodiversity (Augustine and D'Souza 1997; Purohit et al. 1994). Somatic embryogenesis, a technique involving the micropropagation of entire plants from cultured cells, holds considerable potential in creating breeding material from wild populations (Kintzios and Michaelakis 1999). Moreover, it plays a pivotal role in the large-scale production of material for selection or genetic engineering purposes (Prakash et al. 2002).

A well-known goal of genetic engineering is to increase the synthesis of active phytochemical components; however, this task is challenging. Particularly, relatively few genes encoding necessary enzymatic or regulatory steps have been identified, and the metabolic pathways by which active chemicals are biosynthesized are largely poorly known. Nonetheless, there are instances of pathway engineering that resulted in advancements with the potential to be helpful in the breeding of therapeutic plants (Ferreira and Duke 1997; Charlwood and Pletsch 2002). The production of the sedative compound scopolamine in hairy root cultures of *Hyoscyamus niger* (black henbane) was increased ninefold by simultaneously overexpressing two genes encoding the rate-limiting upstream and downstream biosynthetic enzymes, according to a recent article (Zhang et al. 2004) illustrating the difficulties and opportunities of this approach. Hyoscyamine 6 b-hydroxylase from *Hyoscyamus* was used to increase the generation of scopolamine in *A. belladonna* from its naturally occurring chemical precursor. Engineering the production of alkaloids in *P. somniferum* has made some initial progress (Facchini et al. 2000). Transgenic *Artemisia* plants overexpressing farnesyl diphosphate synthase, the enzyme that comes before the first committed biosynthetic step, have been shown to produce the putative antimalarial and anticancer drug artemisinin three times more than usual (Chen et al. 2000). Exploiting transcription factors that turn entire secondary pathways on or off shows significant promise as a metabolic engineering tactic as an alternative to focusing on a single rate-limiting enzyme process (Robbins et al. 2003). Expanding the scope and accuracy of modifications by transgenesis will indeed produce potentially superior material for the breeder by applying new genomic methodologies and effective gene isolation techniques to challenging secondary pathways in medicinal plant metabolism.

14.12 Conclusion

Despite the significant potential of medicinal and aromatic plants in contributing to human health, there has been a notable lack of comprehensive research focused on pests and diseases affecting these valuable botanical resources. Unlike staple crops, these plants often receive less attention in terms of pest and disease management research, potentially putting their sustainability and the availability of therapeutic compounds at risk. In conclusion, safeguarding the future of medicinal plants as a source of therapeutic benefits necessitates a holistic approach, implementing integrated pest and disease management strategies, regular crop surveillance, crop diversification, and strengthening research by using advanced molecular techniques for the development of new resistant varieties. By prioritizing research and sustainable practices, we can ensure the continued availability of these vital natural resources for generations to come, bolstering both human health and ecosystem resilience.

References

Achar KGS, Parashurama TR, Shivanna MB (2014) A new record of leaf spot caused by Xanthomonas campestris in Tinospora cordifolia in India. Int J Curr Microbiol App Sci 3(1):269–273

Akhtar M (2000) The nematicidal potential of the neem tree, *Azadirachta indica*. Integr Pest Manag Rev 5:57–66

Alam M, Sattar A, Janardhanan KK, Husain A (1992) Lethal yellowing of java citronella (Cymbopogon winterianus) caused by Pythium aphanidermatum. Plant Dis 76:1074–1076

Alam M, Chourasia HK, Sattar A, Janardhanan KK (1994) Collar rot and wilt: a new disease of Java citronella (Cymbopogon winterianus) caused by Fusarium moniliforme Sheldon. Plant Pathol 43(6):1057–1061

Augustine AC, D'Souza L (1997) Regeneration of an anticarcinogenic herb, Curculigo orchioides (Gaertn.). In Vitro Cell Dev Biol Plant 33:111–113
Baghel G, Jahan T, Afreen B, Naqvi QA, Snehi SK, Raj SK (2010) Detection of a begomovirus associated with yellow mosaic disease of Ashwagandha (*Withania somnifera* L.) and its impact on biomass yield. Med Plants 2:219–223
Bajaj YPS (1998) Biotechnology for the improvement of medicinal plants. Acta Hortic 457:37–45
Barata AM, Rocha FA, Lopes VM, Morgado J, Maia J, Bettencourt E, Dias S, Delgado F, Costa M, Farinha N, Povoa O, Salgueiro L, Figueiredo AC (2011) Networking on conservation and use of medicinal, aromatic and culinary plants genetic resources in Portugal. Acta Hortic 925:21–35
Bernath J (2002) Strategies and recent achievements in selection of medicinal and aromatic plants. Acta Hortic 576:115–128
Beute MK (1970) Effect of virus infections on susceptibility to certain fungus diseases and yield of gladiolus. Phytopathology 60:1809–1813
Blum H, Meyer U, Schmidt R (2010) Inquiry about the occurrence of pests and diseases on medicinal and aromatic plants in Germany. Zeit Arznei Gewurzpfl 15:182–187
Bowers MD, Stamp NE (1993) Effects of plant-age, genotype, and herbivory on Plantago performance and chemistry. Ecology 74:1778–1791
Canter PH, Thomas H, Ernst E (2005) Bringing medicinal plants into cultivation: opportunities and challenges for biotechnology. Trends Biotechnol 23:180–185
Carlen C (2012) Breeding and cultivation of medicinal plants. In: Bagetta G, Cosentino M, Corasaniti MT, Sakurada S (eds) Herbal medicines for human health. Taylor & Francis, London, pp 79–91
Carlen C, Schaller M, Carron CA, Vouillamoz JF, Baroffio CA (2010) The new Thymus vulgaris L. hybrid cultivar 'Varico 3' compared to five established cultivars from Germany, France and Switzerland. Acta Hortic 860:161–166
Carron CA, Previdoli S, Cottagnoud A, Rey C, Carlen C (2005) *Salvia officinalis*: productivity and quality of the new cultivar Regula. Revue suisse Vitic Arboric Hortic 37:235–239
Carron CA, Baroffio CA, Vouillamoz JF (2009) Lemon thyme: comparison of two clones and definition of the optimum harvest stage. Revue suisse Vitic Arboric Hortic 41(6):357–361
Carrubba A (2017) Weed and weeding effects on medicinal herbs. In: Medicinal plants and environmental challenges. Springer, Cham, pp 295–327
Chandel S, Dubey K, Kaushal P (2014) Major diseases of medicinal and aromatic plants recorded in Himachal Pradesh-India. J Plant Dis Sci 9(2):145–153
Charlwood BV, Pletsch M (2002) Manipulation of natural product accumulation in plants through genetic engineering. Int J Geogr Inf Syst 9:139–151
Chauhan P, Ravi S (2020) Optimization of media and appraisement of fungicides and bioagents on leaf spot of Ashwagandha, Withania somnifera (L.) Dunal caused by Alternaria alternata. J Pharmacogn Phytochem 9(5):1049–1052
Chen DH et al (2000) Expression of a chimeric farnesyl diphosphate synthase gene in Artemisia annua L. transgenic plants via Agrobacterium tumefaciens-mediated transformation. Plant Sci 155:179–185
Choi YH, Tapias EC, Kim HK, Lefeber AW, Erkelens C, Verhoeven JT, Brzin J, Zel J, Verpoorte R (2004) Metabolic discrimination of Catharanthus roseus leaves infected by phytoplasma using 1H-NMR spectroscopy and multivariate data analysis. Plant Physiol 135(4):2398–2410
Cipollini DF (1997) Wind-induced mechanical stimulation increases pest resistance in common bean. Oecologia 111:84–90
Debrunner N, Rauber AL, Schwarz A, Michel V (2000) First report of St John's-wort anthracnose caused by *Colletotrichum gloeosporioides* in Switzerland. Plant Dis 84:203
Dhawan OP, Lavania UC (1996) Enhancing the productivity of secondary metabolites via induced polyploidy: a review. Euphytica 87(2):81–89
Dubey T, Guerra DJ (2002) Use of biotechnology for growing medicinal plants. Rec Prog Med Plants 5:47–61
Dudai N (2012) Domestication and breeding of wild medicinal and aromatic plants—30 years of experience in Israel. Acta Hortic 955:175–183

Dudai N, Chaimovitsh D, Reuveni R, Ravid R, Larkov O, Putievsky E (2002) Breeding of sweet basil (*Ocimum basilicum*) resistant to Fusarium wilt caused by *Fusarium oxysporum* f. sp. *basilicum*. In: Johnson CB, Franz C (eds) Breeding research of aromatic and medicinal plants. The Haworth Press, New York, pp 45–51

Facchini PJ et al (2000) Toward the metabolic engineering of benzylisoquinoline alkaloid biosynthesis in opium poppy and related species. Rec Res Dev Phytochem 4:31–47

Fathima SK, Bhat S, Girish K (2009) Efficacy of some essential oils against *Phomopsis azadirachtae*: the incitant of die-back of neem. J Biopest 2:157–160

Ferreira JFS, Duke SO (1997) Approaches for maximizing biosynthesis of medicinal plant secondary metabolites. AgBiotech News Inf 9:309N–315N

Gahukar RT (1995) Neem in plant protection, 1st edn. Agri-Horticultural Publishing House, Nagpur. 95 pp

Gahukar RT (1998) Commercial and industrial aspects of neem-based pesticides. Pestology 22:5–41

Garg VK (2011) Status of insect pests of Indian gooseberrym *Emblica officinalis,* Gaertn. J Insect Sci 24:304–306

Gaudin M, Simonnet X, Debrunner N (2003) *Colletotricum gloeosporioides* as a cause of St. John's wort (*Hypericum perforatum*) dieback in Switzerland and breeding for tolerant variety. In: Ernst E (ed) Hypericum—the genus Hypericum. Taylor and Francis, London, pp 23–42

Ghosh SK, Banerjee S (2014) First report of Alternaria brassicae leaf spot disease of Aloe vera and its disease intensity in West Bengal. Eur J Biotechnol Biosci 2(1):37–43

Girish K, Bhat SS (2008) *Phomopsis azadirachtae*: the die-back of neem pathogen. J Biol 4:112–119

Gohel N, Prajapati BK (2022) Diseases of Isabgol (Plantago Ovata Forsk.) and their management. In: Diseases of horticultural crops: diagnosis and management. Taylor & Francis Group, London. https://doi.org/10.1201/9781003160472-9

Graham JS, Hall G, Pearce G, Ryan CA (1986) Regulation of synthesis of proteinase inhibitor-I and inhibitor-II messenger-RNAs in leaves of wounded tomato plants. Planta 169:399–405

Graham IA, Besser K, Blumer S, Branigan CA, Czechowski T, Elias L, Guterman I, Harvey D, Isaac PD, Khan AW, Larson TR, Li Y, Pawson T, Penfield T, Rae AM, Rathbone DA, Reid S, Ross J, Smallwood MF, Segura V, Townsend T, Vyas D, Winzer T, Bowles D (2010) The genetic map of *Artemisia annua* L. identifies loci affecting yield of the antimalarial drug artemisinin. Science 327:328–331

Gupta ML, Misra HO, Kalra A, Khanuja SPS (2004) Root rot and wilt: a new disease of Ashwagandha (Withania somnifera) caused by Fusarium solani. J Med Aromat Plant Sci 26(2):285–287

Harborne JB (1989) Higher plant-lower plant interactions: phytoalexins and phytotoxins. In: Harborne JB (ed) Introduction to ecological biochemistry. Academic, London, pp 302–340

Heil M, Bostock RM (2002) Induced systemic resistance (I.S.R.) against pathogens in the context of induced plant defences. Ann Bot 89:503–512

Hemmati C, Nikooei M (2019) Phytoplasma infection could affect chemical composition of *Artemisia sieberi*. Plant Pathol J 35(3):274–279

Hilder VA, Gatehouse AMR, Sheerman SE, Barker RF, Boulter D (1987) A novel mechanism of insect resistance engineered into tobacco. Nature 330:160–163

Holopainen JK, Gershenzon J (2010) Multiple stress factors and the emission of plant VOCs. Trends Plant Sci 15:176–184

Hussain MA, Mukhtar T, Kayani MZ (2011) Efficacy evaluation of *Azadirachta indica*, *Calotropis procera*, *Datura stramonium* and *Tagetes erecta* against root-knot nematode, *Meloidogyne incognita*. Pak J Bot 43(SI):197–204

Janick J, Whipkey A (eds) (2007) Issues in new crops and new uses. ASHS Press, Alexandria, p 366

Jat BL, Choudhary RK, Kumawat KC (2009) Bioefficacy and economics of newer insecticides and neem products against aphid, *Hyadaphis coriandri* (Das) on fennel. Indian J Entomol 71:133–136

Jetawat RPS, Mathur K, Singh K, Singh A (2015) Effect of fungicides against Fusarium solani and Rhizoctonia solani infecting Ashwagandha. Ecoscan 7:75–78

Kalariya K, Shahi D, Saran PL, Prasnna R, Gajbhiye N, Choyal P, Roy S (2023) Effects of nodding broomrape parasitism on growth, physio-biochemical changes and yield loss of Withania somnifera (Linn.) Dunal plant. Vegetos 3. https://doi.org/10.1007/s42535-023-00628-y
Kalra A, Singh HB, Pandey R, Samad A, Patra NK, Kumar S (2005) Diseases in mint: causal organisms, distribution, and control measures. J Herbs Spices Med Plants 11:71–91
Kalra A, Singh HB, Pandey R, Samad A, Patra NK, Kumar S (2008) Diseases in mint: causal organisms, distribution, and control measures. J Herbs Spices Med Plants 11:71–91
Kavithakumari N, Vastrad AS, Basavana Goud K (2009) Evaluation of insecticides for the management of eucalyptus gall wasp, Leptocybe invasa Fisher & La Salle. Indian J Entomol 71:267–269
Kawuri R, Suprapta DN, Nitta Y, Homma T (2012) Destructive leaf rot disease caused by Fusarium oxysporum on Aloe barbadensis Miller in Bali. Agric Sci Res J 2(6):295–301
Kintzios S, Michaelakis A (1999) Induction of somatic embryogenesis and in vitro flowering from inflorescences of chamomile (Chamomilla recutita L.). Plant Cell Rep 8:684–690
Korth KL, Dixon RA (1997) Evidence for chewing insect-specific molecular events distinct from a general wound response in leaves. Plant Physiol 115:1299–1305
Kukreja AK, Dhawan OP (2000) Somaclonal breeding for improvement of herb and essential oil yield in mints: a case study. J Med Aromat Plant Sci 22:431–442
Kulkarni RN, Mallavarapu GR, Ramesh S (1992) The oil content and composition of new variants of Cymbopogon flexuosus. J Essent Oil Res 4:511–514
Lohwasser U, Declercq M, Beorner A, Struckmeyer T, Budahn H, Kruger H, Ulrich D, Marthe F (2010) The German parsley germplasm collection—interaction of morphological, molecular and phytochemical characters. Acta Hortic 860:235–240
Malcolm SB, Zalucki MP (1996) Milkweed latex and cardenolide induction may resolve the lethal plant defence paradox. Entomol Exp Appl 80:193–196
Marcone C, Bellardi MG, Bertaccini A (2016) Phytoplasma diseases of medicinal and aromatic plants. J Plant Pathol 98:379–404
Marimuthu T, Suganthy M, Nakkeeran S (2018) Common pests and diseases of medicinal plants and strategies to manage them. In: New age herbals. Springer, Cham
Mccloud ES, Tallamy DW, Halaweish FT (1995) Squash beetle trenching behavior—avoidance of cucurbitacin induction or mucilaginous plant sap. Ecol Entomol 20:51–59
McFrederick QS, Kathilankal JC, Fuentes JD (2008) Air pollution modifies floral scent trails. Atmos Environ 42:2336–23348
Meena RP, Roy S (2020) Morphological and molecular characterization of Fusarium sp. causing wilt disease of isabgol (Plantago ovata Forsk.) and its management strategies. J Appl Res Med Aromat Plants 16:100244
Mitra B, Biswas B (2002) Insects of aswagandha in South 24 Paraganas, West Bengal. Insect Environ 8:122
Mondal G (2016) Occurrence of Synchytrium inducing gall on punarnava (Boerhavia diffusa) from West Bengal, India. Presented in 6th international conference on "plant, pathogens and people" with the mission "challenges in plant pathology to benefit humankind" held at NASC Complex, New Delhi during February 23–27, 2016, organized by I.P.S., Division of Plant Pathology, IARI, New Delhi 110012, India. Abstract, p 236
Moreira ZPM, dos Santos PO, de Oliveira TAS, de Souza JT (2015) Occurrence of basil leaf spot caused by Pseudomonas cichorii in Bahia State, Brazil. Summa Phytopathol 41(1):73
Neuvonen S, Haukioja E, Molarius A (1987) Delayed inducible resistance against a leaf-chewing insect in 4 deciduous tree species. Oecologia 74:363–369
Opitz S, Kunert G, Gershenzon J (2008) Increased terpenoid accumulation in cotton (Gossypium hirsutum) foliage is a general wound response. J Chem Ecol 34:508–522
Osier TL, Lindroth RL (2001) Effects of genotype, nutrient availability, and defoliation on aspen phytochemistry and insect performance. J Chem Ecol 27:1289–1313
Pandey R (2017) Diseases of medicinal and aromatic plants: insights in nematode biomanagement. Indian Phytopathol 70:12–21. https://doi.org/10.24838/ip.2017.v70.i1.48984

Pank F (2010) Conventional breeding of medicinal and aromatic plants—fundamentals and examples. Acta Hortic 860:135–146
Pank F, Kruger H (2003) Sources of variability of thyme populations (*Thymus vulgaris* L.) and conclusions for breeding. Z Arznei Gewurzpfl 8:117–124
Pank F, Schwarz S (2005) Cultivation of annual caraway (*Carum carvi* L. var. *annuum hort.*) in the greenhouse during winter to accelerate the breeding progress. Z Arznei Gewurzpfl 10:194–197
Pank F, Kruger H, Quilitzsch R (2007) Results of a polycross-test with annual caraway (*Carum carvi* L. var *annuum* hort). Z Arznei Gewurzpfl 12(3):127–135
Park SU, Facchini PJ (2000) Agrobacterium rhizogenes mediated transformation of opium poppy, Papaver somniferum L., and California poppy, Eschscholzia californica Cham., root cultures. J Exp Bot 51:1005–1016
Patil SJ, Patel BR, Chaudhari RK (2011) Efficacy of synthetic and botanical insecticides against aphid, *Aphis gossypii* Glover in isabgol crop. Pestology 35:32–34
Pervez Z, Alam MS, Islam MS (2016) First report of bacterial soft rot of Aloe vera (Aloe barbadensis) caused by Pectobacterium chrysanthemi in Bangladesh. J Plant Pathol Microbiol 7:12
Piano E, Annicchiarico P, Romani M, Pecetti L (2007) Effect of the number of parents and their combining ability on the performance of synthetic varieties in tall fescue. Aust J Agric Res 58(11):1100–1105
Prakash L et al (2002) Production of four commercially cultivated Echinacea species by different methods of in vitro regeneration. J Hortic Sci Biotechnol 77:158–163
Purohit SD et al (1994) Micropropagation of Safed musli (Chlorophytum borivilianum), a rare Indian medicinal herb. Plant Cell Tissue Organ Cult 3:93–96
Putievsky E, Dudai N, Ravid U, Katzir I (1994) A new cultivar of caraway (*Carum carvi* L.) and its essential oil. Int J Geogr Inf Syst 2(2):81–84
Quennoz M, Simonnet X, Verge'res C, Hausamann H (2006) Sea buckthorn (Hippophae rhamnoides L.) is a species for the cosmetics industry. Revue suisse Vitic Arboric Hortic 38(4):215–217
Raj SK, Snehi SK, Kumar S, Khan MS (2009) First finding of 'Candidatus Phytoplasma trifolii'(16SrVI group) associated with little leaf disease of Datura inoxia in India. Plant Pathol 58(4):791. https://doi.org/10.1111/j.1365-3059.2009.02074.x
Raev RT, Jordanov R, Zheljazkov V, Rabotyagov VD (1996) Induced polyploidity in lavender. Acta Hortic 426:561–572
Rao GP, Marcone C, Bellardi MG, Madhupriya (2018) Phytoplasma diseases of medicinal crops. In: Rao G, Bertaccini A, Fiore N, Liefting L (eds) Phytoplasmas: plant pathogenic bacteria - I. Springer, Singapore, pp 236–266. https://doi.org/10.1007/978-981-13-0119-3_8
Rai PK, Tetrawal ML (2010) Biochemical changes in senna (Cassia angustifolia Vahl.) leaves infected with Alternaria alternata. Prog Agric 10:168–169
Ramakrishna Mission (2008) Assessment of bioactivity of some locally available medicinal and aromatic plants. In: Management of pests and diseases of medicinal plants in West Bengal. Medicinal Plants Institute, Kolkata
Ramakrishnan S, Senthilkumar T (2009) Non-chemical management of root-knot nematode in Ashwagandha (*Withania somnifera* Dunal) and senna (*Cassia angustifolia* Vahl.). Indian J Nematol 39:170–174
Ramappa Parashurama T, Shivanna MB (2013) Fungal foliar diseases of Rauwolfia serpentina in wild, its seasonal occurrence, seed transmission and disease management. Arch Phytopath Plant Protec 46(13):1609–1621
Ravikumar A, Rajendran R, Chinniah C, Irulandi S, Pandi RJ (2008) Induction of resistance through organic amendments for the management of spotted leaf beetle, *Epilachna vigintiocto-punctata* Fab. on Aswagandha. J Biopest 1:23–27
Reymond P, Weber H, Damond M, Farmer EE (2000) Differential gene expression in response to mechanical wounding and insect feeding in Arabidopsis. Plant Cell 12:707–719
Robbins MP et al (2003) Sn, a maize bHLH gene, modulates anthocyanin and condensed tannin pathways in Lotus corniculatus 2003. J Exp Bot 54:239–248
Rout GR et al (2000) In vitro manipulation and propagation of medicinal plants. Biotechnol Adv 18:91–120

Saeed ST, Samad A (2017) Emerging threats of begomoviruses to the cultivation of medicinal and aromatic crops and their management strategies. Virus Disease 28:1–17. https://doi.org/10.1007/s13337-016-0358-0
Saeed ST, Khan A, Kumar B, Ajayakumar PV, Samad A (2014) First report of *Chilli leaf curl India virus* infecting *Mentha spicata* (Neera) in India. Plant Dis 98:164
Salisbury FB, Ross CW (1991) Plant physiology. Wadsworth Publishing, Belmont
Saric-Krsmanovic M, Dragumilo A, Umiljendić J, Radivojevic L, Šantrić L, Đurović-Pejčev R (2020) Infestation of field dodder (Cuscuta campestris Yunck.) promotes changes in host dry weight and essential oil production in two aromatic plants, peppermint and chamomile. Plants 9:1286. https://doi.org/10.3390/plants9101286
Sarkar S, Dasgupta B (2017) First report of leaf spot of Aswagandha (Withania somnifera Dunal) caused by Colletotrichum gloeosporioides from West Bengal, India. J Mycopathol Res 55(3):257–259
Sarwat M, Nabi G, Das S, Srivastava PS (2012) Molecular markers in medicinal plant biotechnology: past and present. Crit Rev Biotechnol 32(1):74–92
Sato T, Ohkubo H (1990) Newly discovered leaf blight of citronella grass caused by Curvularia andropogonis in The Philippines. Jpn Agric Res Q 23(3):170–175
Schultz JC, Baldwin IT (1982) Oak leaf quality declines in response to defoliation by gypsy-moth larvae. Science 217:149–150
Seenivasan N (2011) Bioefficacy of anti-nemic plants against root-knot nematode in medicinal coleus. J Ecofriendly Agric 6:92–96
Sevon N, Oksman-Caldentey KM (2002) Agrobacterium rhizogenes-mediated transformation: root cultures as a source of alkaloids. Planta Med 68:859–868
Sharma V (2016) A new leaf spot disease of an Indian medicinal plant Tinospora cordifolia (Willd.) Miers. Int J Adv Life Sci 9(3):381–384
Sharma P, Pandey R (2009) Biological control of root-knot nematode, *Meloidogyne incognita* in the medicinal plant, *Withania somnifera* and the effect of biocontrol agents on plant growth. Afr J Agric Res 4:564–567
Shivanna MB, Parashurama TR, Achar KS, Vasanthakumari MM (2014) Fungal foliar diseases in Withania somnifera and its effect on secondary metabolites. Plant Biosyst 148:907–916
Singh A, Gupta R, Saikia SK, Pant A, Pandey R (2016) Diseases of medicinal and aromatic plants, their biological impact and management. Plant Genet Res 14(4):370–383
Sinha R, Chattopadhyay S (2011) Changes in the leaf proteome profile of Mentha arvensis in response to Alternaria alternata infection. J Proteomics 74(3):327–336. https://doi.org/10.1016/j.jprot.2010.11.009
Sofowora A, Ogunbodede E, Onayade A (2013) The role and place of medicinal plants in the strategies for disease prevention. Afr J Tradit Complement Altern Med 10(5):210–229
Sridhar Y (2011) Field evaluation of different insecticides against aonla aphid, *Schoutedenia emblica*. Indian J Plant Prot 39:35–37
Srivastava KK, Gupta PK, Tripathi YC, Sarvate R (1997) Antifungal activity of plant products on spermoplane fungi of *Azadirachta indica* (neem) seeds. Indian For 23:157–161
Stevenson R (2004) Chemical harvest. Chemistry & industry, 6th Sept, pp 16–18
Tara JS, Sudan M, Sharma B (2011) A report on the occurrence of insect pests on *Zanthoxylum armatum* D.C. (Family: *Rutaceae*): an essential medicinal plant in Jammu region. Bioscan 6:223–228
Tekade A, Koche M, Raut B (2009) Influence of weather factor on fungal diseases of Musli. J Plant Dis Sci 4(2):173–175
TFRI (2010) Tropical Forest Research Institute, Jabalpur; report no. 129/TFRI-2008/Patho-1(13)
Thakur S, Harsh NS (2014) In vitro potential of volatile metabolites of phylloplane fungi of Piper longum as biocontrol agent against plant pathogen. Int J Sci Nat 5(1):33–36
Tharachand C, Immanuel Selvaraj C, Mythili MN (2012) Molecular markers in characterization of medicinal plants: an overview. Res Plant Biol 2(2):1–12

Valchev G, Popova-Ralcheva S, Bonovska M, Zaprianova I, Gudev D (2009) Effect of dietary supplements of herb extracts on performance in growing pigs. Biotechnol Anim Hus 25(5/6):859–870

Vaou N, Stavropoulou E, Voidarou C, Tsigalou C, Bezirtzoglou E (2021) Towards advances in medicinal plant antimicrobial activity: a review study on challenges and future perspectives. Microorganisms 9(10):2041

Vazquez-Flota F, Carrillo-Pech M, Minero-Garcia Y, Miranda-Ham MD (2004) Alkaloid metabolism in wounded Catharanthus roseus seedlings. Plant Physiol Biochem 42:623–628

Vouillamoz JF, D'Anna E, Carron CA, Baroffio C (2013) *Mentha piperita* '541': D.N.A. certification of peppermint in Switzerland by RAPD. Revue suisse Vitic Arboric Hortic 45(5):284–288

Waller P, Bernes G, Thamsborg S, Sukura A, Richter S, Ingebrigtsen K, Hoglund J (2001) Plants as de-worming agents of livestock in the Nordic countries: historical perspective, popular beliefs and prospects for the future. Acta Vet Scand 42:31–44

Waterman PG, Mole S (1989) Extrinsic factors influencing production of secondary metabolites in plants. In: Bernays EA (ed) Insect–plant interactions. CRC Press, Boca Raton, pp 107–134

Wurst S, Van Dam NM, Monroy F, Biere A, Van der Putten WH (2008) Intraspecific variation in plant defense alters the effects of root herbivores on leaf chemistry and aboveground herbivore damage. J Chem Ecol 34:1360–1367

Yadav SS, Suryavanshi P, Nishad I, Sinha S (2021) First report of downy mildew on sweet basil caused by Peronospora belbahrii in India. Plant Dis 106:318

Yuan JS, Kollner TG, Wiggins G, Grant J, Degenhardt J, Chen F (2008) Molecular and genomic basis of volatile-mediated indirect defense against insects in rice. Plant J 55:491–503

Zahra W et al (2020) Economic importance of medicinal plants in Asian countries. In: Keswani C (ed) Bioeconomy for sustainable development. Springer, Singapore

Zaim M, Samad A (1995) Association of phytoplasmas with a witches-broom disease of Withania somnifera (L.) Dunal in India. Plant Sci 109(2):225–229. https://doi.org/10.1016/0168-9452(95)04158-Q

Zeilinger S, Gupta VK, Dahms TE, Silva RN, Singh HB, Upadhyay RS, Gomes EV, Tsui CK, Nayak SC (2016) Friends or foes? Emerging insights from fungal interactions with plants. FEMS Microbiol Rev 40(2):182–207

Zhang L et al (2004) Engineering tropane biosynthetic pathway in Hyoscyamus Niger hairy root cultures. Proc Natl Acad Sci U S A 101:6786–6791

Zheljazkov V, Margina A, Stoeva T (1996) Effect of gamma irradiation on some quantitative and qualitative characteristics in mint and cormint. Acta Hortic 426:381–388

Zheng ZU, Wu M (2004) Cadmium treatment enhances the production of alkaloid secondary metabolites in Catharanthus roseus. Plant Sci 166:507–514

Zuin VG, Vilegas JHY (2000) Pesticide residues in medicinal plants and phytomedicines. Phytother Res 14:73–88

15 Metabolomics in Medicinal Plants: Unraveling the Therapeutic Potential of Natural Compounds

Naveen Reddy Penumallu, Sageer Abass, Rabea Parveen, and Sayeed Ahmed

Abstract

The importance of metabolomics in furthering the study of medicinal plants is discussed in this chapter. It clarifies the differences between primary and secondary metabolites; examines important metabolomics techniques such as LC-MS, GC-MS, and NMR; and highlights their uses in the identification of bioactive compounds, quality assurance, authentication, and metabolic pathway research. The essay discusses current issues as well as prospective developments in the future, highlighting how metabolomics has the ability to revolutionize the field of medicinal plant research and customized herbal medicine and ultimately provide new drugs and comprehensive healthcare solutions.

N. R. Penumallu
Centre of Excellence in Unani Medicine (Pharmacognosy and Pharmacology), School of Pharmaceutical Education and Research, Jamia Hamdard, New Delhi, India

Department of Pharmacology, School of Pharmaceutical Education and Research, Jamia Hamdard, New Delhi, India

S. Abass
Centre of Excellence in Unani Medicine (Pharmacognosy and Pharmacology), School of Pharmaceutical Education and Research, Jamia Hamdard, New Delhi, India

R. Parveen
Centre of Excellence in Unani Medicine (Pharmacognosy and Pharmacology), School of Pharmaceutical Education and Research, Jamia Hamdard, New Delhi, India

Department of Pharmaceutics, School of Pharmaceutical Education and Research, Jamia Hamdard, New Delhi, India

S. Ahmed (✉)
Centre of Excellence in Unani Medicine (Pharmacognosy and Pharmacology), School of Pharmaceutical Education and Research, Jamia Hamdard, New Delhi, India

Department of Pharmacognosy, School of Pharmaceutical Education and Research, Jamia Hamdard, New Delhi, India

© The Author(s), under exclusive license to Springer Nature Singapore Pte Ltd. 2024
M. Nandave et al. (eds.), *Ethnopharmacology and OMICS Advances in Medicinal Plants Volume 1*, https://doi.org/10.1007/978-981-97-2367-6_15

Keywords
Metabolomics · LC-MS · GC-MS · NMR · Metabolites

15.1 Introduction

Metabolomics is a scientific technique that focuses on analyzing all low-molecular-weight metabolites in a biological system. Metabolomics research in the context of medicinal plants aims to give a detailed and exact assessment of the numerous chemical components contained in these plants (Salem et al. 2020). This method of analysis is critical for comprehending the complex biochemical processes and metabolic pathways that occur within medicinal plants. It is a relatively new branch of study in medicinal plants that aims to explore the whole range of tiny molecules. The therapeutic capabilities of medicinal plants are highly regarded, and the bioactive molecules responsible for their medicinal benefits are frequently low-molecular-weight metabolites (Nasim et al. 2022). Understanding these plants' whole metabolome is critical for unlocking their full therapeutic potential. Metabolomics is focused on providing a holistic and accurate examination of these metabolites which aims to identify and quantify a wide range of compounds, including primary and secondary metabolites, such as sugars, amino acids, organic acids, alkaloids, and polyphenols. The variety of metabolites is likely far more than the number of genes, proteins, or mRNAs, increasing the sample's complexity. By doing so, metabolomics enables researchers to gain a more detailed picture of the complex metabolic network within medicinal plants (Patel et al. 2021). Plants are a large reservoir of natural chemicals with a wide range of medicinal properties. These plant resources are constantly being researched in order to identify novel medications (Nasim et al. 2022). Traditional healing practices have depended on natural medicines to treat a wide range of disorders throughout history. In the present period, these natural sources provide a significant share of pharmaceutical medications. These natural materials are made up of a plethora of physiologically active components. These active ingredients have therapeutic actions against a variety of disease-causing substances. A multitude of secondary metabolites with different structures and pharmacological properties have been identified from plant specimens over time (Dias et al. 2012). The knowledge passed down from traditional medicine has paved the way for further research into medicinal plants as a source for pharmaceutical products (Mushtaq et al. 2018). More than 85% of the world's population depends on the traditional medicine system for combating various diseases (Abass et al. 2022). Medicinal plants have been used in human treatment for ages, with traditional medical systems such as Ayurveda, traditional Chinese treatment, and indigenous practices relying on plant-based medicines. Natural sources of medicinal chemicals have influenced contemporary drug development efforts and continue to play an important role in world health care (Yuan et al. 2016). Currently, the major use of plant metabolomics has been focused on activities such as source confirmation, authenticity verification, and quality control of medicinal plants (Lee et al. 2017).

In order to get a more thorough metabolite coverage in plant metabolomics, a variety of analytical methodologies must be used due to the large diversity of chemical classes and properties, as well as the significant variation of metabolite concentrations inside plants. Nuclear magnetic resonance (NMR), liquid chromatography-mass spectrometry (LC-MS), and gas chromatography-mass spectrometry (GC-MS) are significant analytical techniques that are frequently used in metabolomics studies; all these devices are capable of measuring a large number of metabolites with high throughput and consistency (Nagana Gowda and Djukovic 2014). These tools are important detection techniques of plant metabolomics, and the chapter highlighted their application.

15.2 Methodologies in Metabolomics

15.2.1 LC-MS (Liquid Chromatography-Mass Spectrometry)

In the study of plant metabolomics, liquid chromatography-mass spectrometry (LC-MS) is a particularly important analytical tool. It assists in the complete investigation of plants' broad variety of metabolites, including primary and secondary metabolites, and LC-MS is suitable for metabolites with low volatility and poor thermal stability (Patel et al. 2021). The exceptional sensitivity of LC-MS allows for the identification and measurement of metabolites at trace levels. This nontargeted technique is very helpful for detecting new or unexpected molecules, making it critical for discovering novel bioactive substances in plants. In addition to structural elucidation, LC-MS may be used in conjunction with bioinformatics tools for metabolic profiling and pathway analysis. It has uses in plant quality monitoring, plant research, and biotechnology, leading to a better knowledge of plant metabolism and its potential for use in agriculture, food, and medicine (Liu et al. 2021). The leaves and bark of *Myristica dactyloides* were subjected to a metabolomics analysis using high-resolution liquid chromatography-mass spectrometry (HR-LCMS) and ultrahigh-performance liquid chromatography electrospray ionization quadrupole time-of-flight mass spectrometry (UHPLC-ESI-QTOF-MS) techniques. Numerous metabolites, including monotropein, malabaricone B, malabaricone C, fragransol B, guaiacin, myricanone, and epicatechin, were discovered as a result of this investigation (Marulasiddaswamy et al. 2021). One of the studies revealed that *Ephedra alata* extract is rich in potent flavonoid glycosidic compounds as revealed UHPLC-MS analysis (Al-Rimawi et al. 2017). Phytochemical constituents of stems of *J. macrantha* ethyl acetate fraction determined by LC-ESI-MS/MS found 42 metabolites most of having the therapeutic potential (Tinco-Jayo et al. 2022). The whole alcoholic extract of *Y. aloifolia variegata* L. leaves was analyzed by LC-MS/MS using different voltages, negative and positive-ion mode analyses at low fragmentation energy permitted the preliminary identification of 41 and 34 chemicals, respectively. The LC-ESI-MS/MS analysis in positive mode found to be more effective in identifying saponins (El Sayed et al. 2020). Another studies investigated that different classes of alkaloids in plant extracts through the liquid chromatography with electrospray ionization-tandem mass spectrometry (LC–ESI-MS/MS) approach a total of 56 alkaloids were identified based on the reference standard retention times, HR-MS spectra, and MS/MS spectra (Zareena et al. 2021). In one of the research, proteinogenic amino acids in plant extracts were evaluated using liquid chromatography-electrospray ionization-tandem mass spectrometry (Thiele et al. 2019). In the extract of *Zanthoxylum zanthoxyloides*, different kinds of flavonoids like atiscin and quercitrin were characterized with the LC-MS analysis (Tine et al. 2017). Another study discovered the presence of chlorogenic acid, kaempferol 3-O-rutinoside, acacetin 7-O-rutinoside, and genistein in *Potentilla anserina* L. using LC-MS and GC-MS (Mari et al. 2013). Some secondary metabolites like gallic acid, rutin, and quercetin were quantitatively estimated in some plant extracts through the LC-MS/MS analysis (Saini et al. 2023).

15.2.2 GC-MS (Gas Chromatography-Mass Spectrometry)

GC-MS is an excellent approach for identifying and quantifying low-molecular-weight metabolites, which generally weigh approximately 500 Da. These compounds include amino acids, fatty acids, hydroxyl acids, alcohols, sugars, sterols, and amines, and their identification is mostly based on chemical derivatization to increase volatility, allowing gas chromatography examination (Fiehn 2017). With over 50 years of known procedures for study of metabolites such as sugars, amino acids, sterols, hydroxyl acids, fatty acids, aromatics, and many other intermediates of primary metabolism, GC-MS is the most standardized approach in metabolomics (Fiehn 2016). Various research on medicinal plants have been conducted, and the composition of various metabolites and their differences was examined using GC-MS analysis. In one investigation, it was discovered that cinnamon oil contains 27 phenolic compounds. Benzyl benzoate, trans-cinnamyl acetate, and cinnamaldehyde were the most prevalent essential oils discovered (Mutlu et al. 2023). Another study showed the presence of 23 phytoconstituents which contribute the activities like antimicrobial, antioxidant, anticancer, and other activities in *Hibiscus asper* leaves (Olivia et al. 2021). In one of the research, the ethanolic extract of *Evolvulus alsinoides* was shown to contain a number of chemicals, including piperine, octadecenoic acids, and squalene, according to the results of the GC-MS analysis (Gomathi et al. 2015). In other research, the GC-MS analysis of ethyl acetate and methanol extracts from *A. nilgiricum* leaves and methanol extract from rhizomes found the presence of 25 phytochemical components (Konappa et al. 2020). One of the investigations revealed the presence of 21 compounds in the methanolic extract of *Lactuca runcinata* (Kanthal et al. 2014). Methyl octadeca-17-enoate Methyl palmitate (-)-1,2-Didehydroaspidospermidine and strictamine were the main metabolites found in the ethanolic extract of *Rhazya stricta* (Baeshen et al. 2023). Phytocompounds like thymol, Z-sabinene hydrate, γ-terpinene, p-cymene, linalyl acetate, sabinene, and carvacrol methyl ether in *Origanum vulgare* L. essential oil were determined in the studies (Simirgiotis et al. 2020). One study used GC-MS analysis to identify many antidiabetic components in the extract of *Pterocarpus marsupium*, including myristic acid, succinic acid, isoeugenol, and pentadecanoic acid (Dar et al. 2022). A total of 17 metabolites were identified in *Pistacia integerrima* extract through GC-MS analysis with most of the compounds exhibiting antibacterial properties (Abass et al. 2023). Another research work showed that GC-MS analysis revealed several volatile compounds in the rose flower oil (Rusanov et al. 2011). Rose Petals in Unani Formulations revealed different metabolites via GC-MS analysis (Parveen et al. 2020).

15.2.3 NMR (Nuclear Magnetic Resonance)

A crucial technique in the study of tiny molecules in biological processes known as metabolomics is nuclear magnetic resonance (NMR) spectroscopy. NMR is noninvasive and offers quantitative information and structural understanding of a variety of metabolites. It is utilized to identify biomarkers, diagnose illnesses, and comprehend metabolic alterations. Understanding the intricacies of metabolism in many biological systems requires the use of NMR because of its capacity to monitor metabolic events, track particular metabolites, and offer structural data (Moco 2022). NMR spectroscopy involves putting a substance in a magnetic field. The compound's isotopes, such as ^{1}H, ^{13}C, ^{14}N, ^{15}N, and ^{17}O, absorb the radiation and resonate at a frequency that depends on where they are in the tiny molecule (Wang et al. 2010). One study used low-field NMR analysis to assess the impact of different pruning and irrigation managements on the oil content and lipid profile of *Jatropha curcas* L. seeds (Santos et al. 2017). One study used high-resolution magic angle spinning (HR-MAS) NMR to profile the metabolic profiles of intact floral primordia of *Ribes nigrum* isolated from buds with varying dormancy status. The technique proved useful in monitoring different groups of metabolites, such as carbohydrates and amino acids (Pagter et al. 2017). In one of the investigations, HRMAS NMR analysis was used to assess the metabolic studies of tomato fruit and tomato tissues (Pérez et al. 2010). A different investigation identified 17 metabolites, such as caffeic acid and berberine, a notable alkaloid derived from *Berberis laurina*, through the application of NMR analysis. The alterations in metabolic patterns in the leaves over time were observed to be linked to seasonal fluctuations and environmental variations (Ali et al. 2020). In transgenic plant *Zea mays*, some key metabolites like lactic acid, lysine, arginine, trehalose, galactose, etc. were analyzed through the NMR analysis (Piccioni et al. 2009). In *Withania somnifera* (L.), 41 metabolites were discovered from both the leaf and root tissues of the chemotypes. HR-MAS NMR spectroscopy was used to perform metabolic profiling on fresh leaf and root tissue specimens from its different chemotypes (Bharti et al. 2011). Figure 15.1 shows the workflow of metabolomics analysis of plant extracts.

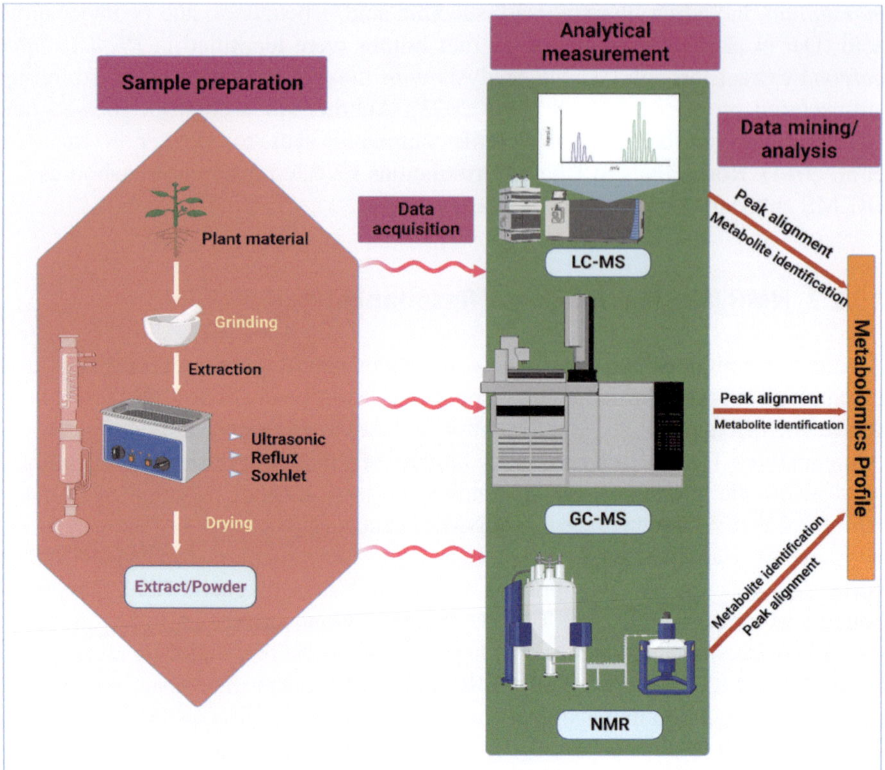

Fig. 15.1 Metabolomics analysis workflow

15.3 Metabolite Classes in Medicinal Plants

The phytochemical elements of medicinal plants are primarily responsible for their diverse pharmacological effects. Plant phytochemical elements are classified into two types depending on their participation in basic metabolic processes: primary and secondary metabolites. Primary metabolites are organic substances that are required for the growth, development, and maintenance of living organisms. These molecules play crucial roles in biochemical processes such as energy generation, cell structure, and fundamental cellular functioning. Carbohydrates, lipids, proteins, and nucleic acids are examples of primary metabolites (Irchhaiya et al. 2015). These substances are essential for life and serve as the foundation for more complex biological entities. These molecules are highly conserved across species and are essential for all living creatures to operate. Understanding basic metabolites is critical in sciences like biochemistry, molecular biology, and biotechnology, because it provides insights into the underlying mechanisms that support life (Rolin 2006). Metabolomics, a cutting-edge science in plant biology, analyzes the full collection of small molecules within a plant using advanced analytical techniques such as

LC-MS, GC-MS, and NMR spectroscopy. This method allows researchers to investigate the stress responses of primary metabolites in plants when they are subjected to varied environmental stressors. Integrating metabolomics with these analytical technologies allows researchers to decipher the complicated dynamics of primary metabolites and acquire insights into how plants adapt to and deal with environmental stresses (Jorge et al. 2016). Another class of metabolites are secondary which are organic molecules that are synthesized by plants, microbes, and certain animals. These chemicals are not directly engaged in the key processes of growth and development, but they play critical roles in interactions with the environment, defense mechanisms, and adaptability to varied ecological niches. Secondary metabolites include a diverse spectrum of substances such as alkaloids, terpenoids, flavonoids, and phenolic compounds that contribute to the distinct features and bioactivity of many plant species (Twaij and Hasan 2022). Metabolomics plays a critical role in secondary metabolite studies. It contributes to understanding the chemical variety of secondary metabolites inside organisms, as well as their biosynthesis processes and ecological and physiological roles (Breitling et al. 2013). LC-MS analysis was used in one study to examine the metabolic profiles for carbs, amino acids, lipids, and flavones during grain growth in *Indica rice* varieties (Yibo et al. 2022). In a separate research, an unrivalled set of complementary analytical methods was used to investigate the spatial and developmental changes that occur throughout the ripening of melon fruit. These approaches included ^{1}H-NMR, GC-MS, and LC-MS screenings for metabolomic profiling, as well as inductively coupled plasma mass spectrometry (ICP-MS) assessments of macro- and microelements (Moing et al. 2011). One of the research projects employed LC–MS techniques to analyze and characterize the structural properties and profiles of phenolic secondary metabolites in Mexican lupine species (Wojakowska et al. 2013). The chemical composition of primary and secondary metabolites in various portions of fruit of *Garcinia mangostana* Linn. (pericarp, aril, and seed) was examined concurrently using GC-MS and LC-MS methods at four unique ripening phases. The GC-MS and LC-MS studies effectively detected a total of 57 and 98 metabolites, respectively (Mamat et al. 2020). In one of the investigations, the use of GC-MS and LC-MS for metabolomics profiling in *Panax ginseng* and *Panax quinquefolius* revealed a different pattern of primary metabolite and ginsenoside accumulation that is specific to both the tissue type and the plant species (Liu et al. 2017). Another study successfully identified potent antimalarial compounds in the methanolic extract of *A. paniculata*. These chemicals were identified and characterized using LC-MS, with retention duration and the mass spectrometry spectrum being important considerations. The spectral data was used to calculate the distribution of diterpenes and flavones (Dwivedi et al. 2021). One more study reported that a combination of NMR and LC-MS analysis revealed alterations in the chemical makeup of *Salvia miltiorrhiza* Bunge when exposed to a shortage of water (Dai et al. 2010).

Secondary metabolites, as previously discussed, play an important part in a variety of ecological and biological activities and have substantial therapeutic potential in humans. Since plant-based materials make up the majority of medications, secondary metabolites improve human immunity. Plant secondary compounds may have therapeutic benefits for people (Zeeshan Bhatti et al. 2022). One of the review covers a range of plant secondary metabolites mainly alkaloids that have shown substantial efficacy in preclinical trials for treating autoimmune disorders (Khan et al. 2021). A different research investigation revealed the presence of 11 alkaloids with potential therapeutic benefits against the severe acute respiratory syndrome coronavirus-2 (SARS-CoV-2) in the scientific literature (Fielding et al. 2020). Another study discovered that the data currently available provides confirmation from both preclinical and clinical investigations that indole alkaloids have the potential to be effective in treating respiratory conditions (Mitra et al. 2021). Flavonoids have been found as a potential neuroprotective agent in Alzheimer's disease (Minocha et al. 2022). A different study examines the potential of flavonoids as natural treatment options for cancer (Slika et al. 2022). In one study, essential oils were assessed for their potential in preventing or treating COVID-19 (Salem and Ezzat 2021).

15.3.1 Importance of Primary and Secondary Metabolites

Metabolites, integral to plant growth and development, are categorized into primary and secondary metabolites, each with distinct functions (Table 15.1). Primary metabolites, such as sugars (e.g., glucose, fructose, sucrose, maltose), fatty acids (e.g., palmitic acid, stearic acid, oleic acid, linoleic acid), amino acids (e.g., alanine, glycine, valine, leucine, isoleucine), organic acids (e.g., citric acid, malic acid, oxalic acid, fumaric acid), and nucleotides (e.g., ATP, GTP, CTP), are fundamental compounds that facilitate energy production, protein synthesis, membrane structure, and essential signaling pathways. Additionally, they serve as osmolytes and osmoprotectants, playing key roles in responding to biotic and abiotic stresses. In contrast, secondary metabolites represent a diverse group of compounds not directly involved in core growth and reproduction but instead fulfill specialized functions, including defense, pollinator attraction, and inter-organism communication. These secondary metabolites span alkaloids, terpenoids, phenolic compounds, glycosides, polyketides, coumarins, quinones, volatile organic compounds (VOCs), cyanogenic compounds, and betalains, showcasing the rich diversity of plant chemistry and its ecological significance.

15.4 Metabolomics Approaches in Medicinal Plant Research

Metabolomics is the study of metabolites within a sample, and it encompasses two principal methodologies: targeted and untargeted approaches. Targeted metabolomics is centered on quantifying a specific, predefined set of known metabolites, while untargeted metabolomics seeks to comprehensively analyze the entire metabolome of an organism, encompassing both recognized compounds and unidentified chemical entities, offering a more expansive exploration of the metabolic profile

Table 15.1 Metabolite profile and applications

Metabolite type	Examples	Plant source	Therapeutic use/ pharmacological activity	Methods of analysis by metabolomics	Applications/clinical and preclinical data
Primary metabolites					
Sugars	Glucose, fructose, sucrose, maltose	Various plants	Primary source of energy for cells	NMR, GC-MS (Lara-Cruz and Jaramillo-Botero 2022)	• Signaling molecules that trigger defense responses • Energy and carbon sources • Regulators that control plant growth and development • Stress adaptation • Defense against biotic factors such as herbivores, insects, and pathogens • Enhancing resistance against abiotic factors (Jeandet et al. 2022; Kurzyna-Szklarek et al. 2022; Plant Primary Metabolites Analysis Service - Creative Proteomics n.d.; Salam et al. 2023; Zaynab et al. 2019)
Fatty acids	Palmitic acid, stearic acid, oleic acid, linoleic acid	Plant and animal lipids	Structural components, energy storage	GC-MS, HPLC, FTIR, GC-IR, AFS, EM (Rahim et al. 2023)	• Growth and development • Respiration and photosynthesis • Hormone and protein synthesis • Cell reproduction • Signaling to trigger defense responses • The biogenesis of cell membranes (Kim 2020)

(continued)

Table 15.1 (continued)

Metabolite type	Examples	Plant source	Therapeutic use/ pharmacological activity	Methods of analysis by metabolomics	Applications/clinical and preclinical data
Amino acids	Alanine, glycine, valine, leucine, isoleucine	Various plants, proteins	Protein synthesis, neurotransmitters	NMR spectroscopy, LC-MS, GC-MS, HPLC	• Triggering defense responses • Controlling bacterial, viral, and fungal pathogens and insects • Enhancing the texture and nutritive quality of food (Chaturvedi and Gupta 2021)
Organic acids	Citric acid, malic acid, oxalic acid, fumaric acid	Various plants	Intermediates in metabolic pathways	NMR spectroscopy, LC-MS, GC-MS, HPLC	• Trigger defense responses • Control bacterial, viral, and fungal pathogens and insects • Flavor food and drinks • Prevent oxidation and rancidity of fats and oils (Elateeq et al. 2023; Glazer and Nikaido 2007)
Nucleotides	Adenosine triphosphate (ATP), guanosine triphosphate (GTP), cytidine triphosphate (CTP)	All living organisms	Cellular energy currency and signaling	NMR spectroscopy, LC-MS, GC-MS, HPLC	• Building blocks of nucleic acids • Storage of chemical energy • Carriers of activated metabolites for biosynthesis • Structural moieties of coenzymes • Metabolic regulators • Essential chemical links in the cell's response to hormones and other extracellular stimuli • Providing genetic information that is produced in proteins (Lopatkin and Yang 2021)
Secondary metabolites					

Metabolite type	Examples	Plant source	Therapeutic use/ pharmacological activity	Methods of analysis by metabolomics	Applications/clinical and preclinical data
Alkaloids	Caffeine, nicotine, morphine, quinine, vinblastine	Various plants	Various pharmacological activities, e.g., pain relief, antimalarial, anticancer	LC-MS, GC-MS, NMR spectroscopy, HPLC	• Anti-inflammatory, anticancer, analgesic, and pain relief activities • Cardioprotective activities • Defense systems against pathogens and animals (Kurek 2019; Patel et al. 2012)
Terpenoids	Carotenoids, steroids, essential oils	Various plants	Antioxidants, hormones, aroma compounds	GC-MS, LC-MS, NMR spectroscopy, HPLC	• Antitumor effects • Anti-inflammatory effects • Antibacterial effects • Antiviral effects • Antimalarial effects • Promoting transdermal absorption • Preventing and treating cardiovascular diseases • Hypoglycemic activities (Mani et al. 2021; Yang et al. 2020)

(continued)

Table 15.1 (continued)

Metabolite type	Examples	Plant source	Therapeutic use/pharmacological activity	Methods of analysis by metabolomics	Applications/clinical and preclinical data
Phenolic compounds	Flavonoids, tannins, lignans, phenolic acids	Various plants	Antioxidants, anti-inflammatory, anticancer	LC-MS, GC-MS, NMR spectroscopy, HPLC	• Antioxidant • Antiseptic • Protective against cardiovascular diseases • Anti-obesity activity • Antidiabetic activity • Antiaging activity • Protective action on Alzheimer's disease • Anticancer activity • Anti-inflammatory • Antitumor • Antiviral • Anti-allergic (Hu et al. 2022a; Lin et al. 2016; Luna-Guevara et al. 2018)
Glycosides	Cardiac glycosides, cyanogenic glycosides, saponins	Various plants	Cardiac medications, toxic when hydrolyzed, plant defense	LC-MS, GC-MS, HPLC, NMR spectroscopy	• Analgesic • Anti-inflammatory • Cardiotonic • Antibacterial • Antifungal • Antiviral • Anticancer (Minic 2008; Shah et al. 2014; Yeoman et al. 2010)

Metabolite type	Examples	Plant source	Therapeutic use/ pharmacological activity	Methods of analysis by metabolomics	Applications/clinical and preclinical data
Polyketides	Erythromycin, tetracycline, antibiotics (produced by bacteria and fungi)	Various sources including bacteria and fungi	Antibiotics and other bioactive compounds	LC-MS, GC-MS, NMR spectroscopy, HPLC	• Antimicrobial properties • Anticancer properties • Immunosuppressive properties • Antioxidant properties • Anti-inflammatory properties (El-Bondkly 2014; Ridley and Khosla 2009; Risdian et al. 2019; Xu et al. 2021)
Coumarins	Coumarin, scopoletin, aromatic compounds	Various plants	Anticoagulants, antioxidant properties	LC-MS, GC-MS, HPLC, NMR spectroscopy	• Anti-inflammatory • Anticoagulant, antibacterial, antifungal, antiviral, anticancer • Antihypertensive, antituberculous, anticonvulsant • Antiadipogenic, antihyperglycemic, antioxidant, neuroprotective (Robe et al. 2021; Sharifi-Rad et al. 2021; Stringlis et al. 2019)
Quinones	Ubiquinone, plumbagin, naphthoquinones	Various plants	Electron transport, antimicrobial, plant defense	LC-MS, GC-MS, NMR spectroscopy, HPLC	• Quinine is used alone or with other medications to treat malaria • Quinolones act by inhibiting the activity of two essential bacterial type II topoisomerases (Christiansen et al. 2021; Correia et al. 2017; Madeo et al. 2013; Niaz et al. 2020)

(continued)

Table 15.1 (continued)

Metabolite type	Examples	Plant source	Therapeutic use/pharmacological activity	Methods of analysis by metabolomics	Applications/clinical and preclinical data
Volatile organic compounds (VOCs)	Small, volatile molecules involved in plant communication	Various plants	Scent, signaling between plants and organisms	GC-MS, HPLC, PTR-MS, TLC (Liu et al. 2023; Sadgrove et al. 2022)	• Irritation to skin and eye • Sensitization • Central nervous system effects • Carcinogenicity • Liver and kidney effects (Almeida et al. 2023; David and Niculescu 2021; Rios-Navarro et al. 2021; Schulz-Bohm et al. 2017)
Cyanogenic glycosides	Compounds that release toxic cyanide when hydrolyzed (e.g., in cassava and cherry laurel)	Various plants	Toxic when hydrolyzed, plant defense	LC-MS, GC-MS, HPLC-MS/MS UV-Vis spectroscopy (Cai et al. 2022; Toomey et al. 2012)	• Serve as storage forms for reduced nitrogen in plants • In humans, the acute lethal oral dose of cyanide is between 0.5 and 3.5 mg/kg body weight. The toxic threshold value for cyanide in blood is between 0.5 mg/L and 1.0 mg/L. The lethal threshold value ranges between 2.5 mg/L • Cyanide is detoxified by the enzyme rhodanese, forming thiocyanate, which is excreted in the urine (Lechtenberg 2021; Schrenk et al. 2019; Yulvianti and Zidorn 2021; Zuk et al. 2020)

Metabolite type	Examples	Plant source	Therapeutic use/ pharmacological activity	Methods of analysis by metabolomics	Applications/clinical and preclinical data
Betalains	Pigments found in beets, cacti (betacyanins and betaxanthins)	Plants such as beets and cacti	Coloration, potential antioxidant properties	LC-MS, GC-MS, HPLC-MS/MS, NMR UV-Vis spectroscopy (Aztatzi-Rugerio et al. 2019; Coy-Barrera 2020; Sawicki et al. 2017; Trych et al. 2022)	• Anti-inflammatory • Anticancer • Anti-hepatitis • Blood pressure and lipid lowering • Antidiabetic • Anti-obesity • Support the body's natural detox process • Reduce cancer risk • Help relax the mind and provide a sense of well-being • Support emotional health (Georgiev et al. 2010; Hadipour et al. 2020; Li et al. 2019; Madadi et al. 2020)

(continued)

Table 15.1 (continued)

Metabolite type	Examples	Plant source	Therapeutic use/ pharmacological activity	Methods of analysis by metabolomics	Applications/clinical and preclinical data
Glucosinolates	Sulforaphane, allyl isothiocyanate, indole-3-carbinol	Cruciferous vegetables (e.g., broccoli, cabbage)	Potential anticancer, antioxidant properties	UPLC, LC-DAD, LC-MS, HPLC, NMR (Kim et al. 2017; Meza et al. 2022)	• Anticancer properties • Anti-inflammatory properties • Antidiabetic properties • Lipid-lowering properties • Antimicrobial properties • Chemoprotective properties against chemical carcinogens • Regulatory functions in inflammation, stress response, and phase I metabolism • Antioxidant activities (Akram et al. 2021; Bischoff 2016; Chowdhury and Chowdhury 2022; Singh 2017; Zukalová and Vašák 2002)

Metabolite type	Examples	Plant source	Therapeutic use/ pharmacological activity	Methods of analysis by metabolomics	Applications/clinical and preclinical data
Saponins	Ginsenosides, glycyrrhizic acid, quillaja saponins	Various plants	Potential anti-inflammatory, immune-modulating	LC-MS, HPLC, GC-MS (Lim et al. 2020; Oleszek 2002; Savarino et al. 2023; Wang et al. 2022)	• Decreasing blood lipids • Lowering cancer risks • Lowering blood glucose response • Inhibiting dental caries and platelet aggregation • Treating hypercalciuria in humans • Acting as an antidote against acute lead poisoning • Immunostimulating • Hypocholesterolemic • Anti-inflammatory • Antimicrobial • Anti-protozoan • Molluscicidal • Antioxidant (Bouzier et al. 2023; Guzmán et al. 2020; Kareem et al. 2022; Moses et al. 2014)
Isoflavones	Genistein, daidzein, biochanin A	Legumes (e.g., soybeans)	Potential estrogenic effects, antioxidant properties	HPLC, NMR, ICP-AES, LC/MS/MS (Jung et al. 2020; Otieno et al. 2007; Soni et al. 2023)	• Protection against age-related diseases • Anti-inflammation • Anticancer • Anti-obesity • Antidiabetes • Gut biota regulation • Menopausal relief (Hsiao et al. 2020; Kim 2021; Sajid et al. 2021; Wong et al. 2008)

(Azad and Shulaev 2019; Roberts et al. 2012; Schrimpe-Rutledge et al. 2016; Waris et al. 2022).

15.4.1 Untargeted Metabolomics

Untargeted metabolomics employs liquid or gas chromatography (LC/GC) coupled with mass spectrometry (MS) and nuclear magnetic resonance (NMR) to enable the simultaneous detection of a broad spectrum of metabolites in a given sample. In the context of medicinal plant research, untargeted metabolomics has proven invaluable in various applications. For instance, it has been used to explore phytochemical profiles and assess antimicrobial and antioxidant activities in the *Zanthoxylum* genus. Additionally, untargeted metabolomics has been instrumental in characterizing the metabolomic profiles of several medicinal plants recognized for their sedative effects, including *Valeriana officinalis* L., *Melissa officinalis* L., *Hypericum perforatum* L., and *Passiflora incarnata* L. Furthermore, this approach has been applied to discriminate differences in steaming *Panax notoginseng*, showcasing its utility in deciphering the complex metabolic changes associated with various medicinal plant processes and activities (Khoomrung et al. 2017; Marchev et al. 2021; Xiao et al. 2022).

15.4.2 Targeted Metabolomics

Targeted metabolomics focuses on specific metabolite classes or pathways of interest. This approach is particularly useful when investigating known bioactive compounds or understanding specific metabolic responses. A targeted metabolomics method was developed utilizing the power of UHPLC-triple quadrupole MS, enabling the precise profiling of more than 400 biologically significant metabolites, including amino acids, sugars, nucleotides, dipeptides, coenzymes, and fatty acids. This comprehensive approach covers a wide array of 92 metabolic pathways, such as the Krebs cycle, glycolysis, amino acid metabolism, ammonia recycling, and one-carbon metabolism. The use of UHPLC-triple quadrupole MS allows for the specific and high-throughput analysis of these metabolites, providing valuable insights into the intricate biochemical processes associated with these metabolic pathways (Cao et al. 2020).

15.4.3 Comparative Metabolomics

Comparative metabolomics compares metabolite profiles under different conditions, such as different plant growth stages, environmental stresses, or treatments. This approach provides insights into how metabolites are influenced by external factors. Comparative metabolomics offers a valuable approach to uncover fresh perspectives, encompassing insights into variations in the accumulation of taxoids and active ingredients among *Taxus* species, the potential antitumor properties of *Brassica rapa* L., the genetic changes induced by diverse ecological factors, and distinctions in metabolite profiles between differently colored varieties of Danshen *Salvia miltiorrhiza* (Li et al. 2022; Wei et al. 2020; Xiao et al. 2022; Zhou et al. 2019).

15.5 Applications of Metabolomics in Medicinal Plants

15.5.1 Discovery of Bioactive Compounds

Metabolomics serves as a powerful tool not only for the discovery of novel active compounds in medicinal plants but also for a wide range of applications, such as studying changes in metabolites in response to varying environmental conditions, investigating variations in metabolites among different plant parts or across different time periods, determining the vegetation period of plants, detecting adulteration, assessing the correlation between biological activity and metabolite levels, and ensuring the quality control of medicinal plants. Additionally, metabolomics finds applications in diverse fields beyond medicine, including food science, pharmaceuticals, and nutrition, where it plays a crucial role in elucidating metabolic processes, identifying biomarkers, and enhancing our understanding of complex biological systems (Hu et al. 2022b; Mishra and Priyanka 2022; Waris et al. 2022).

15.5.2 Quality Control and Authentication

Metabolomics stands as a potent instrument in the quality control and authentication of medicinal plants. It plays a multifaceted role, encompassing the prediction of the biological activity of medicinal plants, assistance in bioactivity-guided fractionation and lead discovery, the identification of novel active compounds, determination of the vegetation period, detection of adulteration, assessment of the correlation between bioactivity and metabolite levels, continuous monitoring of the quality of medicinal plants, and evaluation of natural product-derived medicines. Furthermore, it aids in pinpointing the source, authenticity, and processing methods of traditional Chinese medicines (CMM) while also contributing to drug discovery, systems biology, gene-function analysis, and diverse diagnostic techniques. This broad utility is particularly crucial for ensuring the consistency and standardization of natural product-derived medicines, given their complex composition consisting of hundreds of natural compounds (Lee et al. 2017; Mukherjee et al. 2016; Xiao et al. 2022).

15.5.3 Study of Metabolic Pathways and Biosynthetic Networks

Metabolomics is a versatile tool for the study of medicinal plants, enabling researchers to unravel the synthesis and accumulation of bioactive components in these plants. Its applications span diverse areas, including improving breeding programs, identifying metabolites, enhancing disease resistance, assessing medicinal value, profiling metabolites in various crops and medicinal plants, differentiating taxonomic or biochemical variations in plant species across different ecotypes, investigating metabolic responses to biotic and abiotic stimuli, tracking changes in metabolites in response to shifting environmental conditions, analyzing variations in metabolite types and contents within different plant parts or time periods, enhancing abiotic stress tolerance, bolstering pathogen resistance, facilitating metabolic-assisted crop breeding, evaluating the influence of ecotypes on plant growth and development, discovering biomarkers, aiding in genotyping, and elucidating gene functions. Metabolomics serves as a comprehensive approach for advancing our understanding of the intricate biochemistry of medicinal plants and their potential applications in various fields (Sharma and Yadav 2022; Yuan et al. 2022).

15.6 Challenges and Future Directions

In the field of metabolomics, challenges and promising future directions coexist. Accurate metabolite identification and annotation are intricate due to the vast compound diversity, necessitating collaboration between metabolomics and natural products chemistry experts. Standardized protocols for sample preparation, data acquisition, and analysis are vital for enhancing study reliability and facilitating cross-study comparisons. The field benefits from evolving technologies, including high-resolution mass spectrometry and advanced data processing algorithms, promising greater precision and depth in metabolomics analyses. Furthermore, metabolomics holds the potential for personalized medicine, accommodating individual variations in metabolite profiles to offer tailored herbal therapies. The future of metabolite identification relies on metabolome data repositories and advanced data analysis tools, streamlining data sharing, reanalysis, and automated re-annotation. Challenges encompass the lack of standardized data production methods, limited annotations for a fraction of features and identified metabolites, and the pursuit of a comprehensive single-method approach for measuring a biological entity's entire metabolome, as well as the precise identification and quantification of specific key metabolites. The most reliable identification method involves comparing a metabolite's characteristics with authenticated standards, often facilitated by popular software packages such as Cytoscape (41%) and MetaboAnalyst (55%). These developments and challenges collectively shape the ever-evolving landscape of metabolomics and its vital role in unraveling the complexities of biological systems (Castelli et al. 2022; Chaleckis et al. 2019; Pinu et al. 2019; Weber et al. 2017; Wishart 2011; Xiao et al. 2012).

15.7 Conclusion

In conclusion, metabolomics has emerged as a powerful tool in advancing medicinal plant research. It offers insights into plant biochemistry, enabling the discovery of bioactive compounds, validating traditional knowledge, and contributing to personalized herbal medicine. The dynamic synergy between metabolomics and medicinal plants holds the potential for novel drug discovery and transformative integrative healthcare approaches.

References

Abass S, Parveen R, Irfan M, Jan B, Husain SA, Ahmad S (2022) Synergy based extracts of medicinal plants: future antimicrobials to combat multidrug resistance. Curr Pharm Biotechnol 23:1527–1540

Abass S, Dar MI, Zahiruddin S, Beg MA, Nagar P, Jan B, Husain SA, Ahmad S (2023) Phytochemical and antibacterial analysis of *Pistacia integerrima*: an integrated in vitro and in silico approach. Process Biochem 132:236–247

Akram M, Jabeen F, Riaz M, Khan FS, Okushanova E, Imran M, Shariati MA, Riaz T, Egbuna C, Ezeofor NJ (2021) Health benefits of glucosinolate isolated from cruciferous and other vegetables. In: Preparation of phytopharmaceuticals for the management of disorders: the development of nutraceuticals and traditional medicine. Elsevier, Amsterdam, pp 361–371

Ali S, Badshah G, Da Ros Montes C, Campos FR, Nagata N, Khan A, Santos MDFC, Barison A (2020) High-resolution magic angle spinning (HR-MAS) NMR-based fingerprints determination in the medicinal plant berberis laurina. Molecules 25(16):3647

Almeida OAC, de Araujo NO, Mulato ATN, Persinoti GF, Sforça ML, Calderan-Rodrigues MJ, Oliveira JV (2023) Bacterial volatile organic compounds (VOCs) promote growth and induce metabolic changes in rice. Front Plant Sci 13:1056082

Al-Rimawi F, Abu-Lafi S, Abbadi J, Alamarneh AAA, Sawahreh RA, Odeh I (2017) Analysis of phenolic and flavonoids of wild ephedra Alata plant extracts by LC/PDA and LC/MS and their antioxidant activity. Afr J Tradit Complement Altern Med 14(2):130–141

Azad RK, Shulaev V (2019) Metabolomics technology and bioinformatics for precision medicine. Brief Bioinform 20:1957

Aztatzi-Rugerio L, Granados-Balbuena SY, Zainos-Cuapio Y, Ocaranza-Sánchez E, Rojas-López M (2019) Analysis of the degradation of betanin obtained from beetroot using Fourier transform infrared spectroscopy. J Food Sci Technol 56:3677

Baeshen NA, Almulaiky YQ, Afifi M, Al-Farga A, Ali HA, Baeshen NN, Abomughaid MM, Abdelazim AM, Baeshen MN (2023) GC-MS analysis of bioactive compounds extracted from plant Rhazya stricta using various solvents. Plants 12(4):960

Bharti SK, Bhatia A, Tewari SK, Sidhu OP, Roy R (2011) Application of HR-MAS NMR spectroscopy for studying chemotype variations of Withania somnifera (L.) Dunal. Magn Reson Chem 49(10):659–667

Bischoff KL (2016) Glucosinolates. In: Nutraceuticals: efficacy, safety and toxicity. Academic, New York, pp 551–554

Bouzier A, Rojas J, Kerthy S, Ibinga K, Lamarti A, Martin P, Morillo M (2023) The impact of saponins on health-review. Biointerf Res Appl Chem 13:362

Breitling R, Ceniceros A, Jankevics A, Takano E (2013) Metabolomics for secondary metabolite research. Metabolites 3(4):1076–1083

Cai BD, Wu JY, Bai YL, Feng YQ (2022) Highly sensitive analysis of cyanogenic glycosides in cold-pressed flaxseed oil by employing cigarette filter fiber-based SPE coupled with ultra-performance liquid chromatography-tandem mass spectrometry. Food Chem 377:131962

Cao G, Song Z, Hong Y, Yang Z, Song Y, Chen Z, Chen Z, Cai Z (2020) Large-scale targeted metabolomics method for metabolite profiling of human samples. Anal Chim Acta 1125:144–151

Castelli FA, Rosati G, Moguet C, Fuentes C, Marrugo-Ramírez J, Lefebvre T, Volland H, Merkoçi A, Simon S, Fenaille F, Junot C (2022) Metabolomics for personalized medicine: the input of analytical chemistry from biomarker discovery to point-of-care tests. Anal Bioanal Chem 414:759

Chaleckis R, Meister I, Zhang P, Wheelock CE (2019) Challenges, progress and promises of metabolite annotation for LC-MS-based metabolomics. Curr Opin Biotechnol 55:44–50

Chaturvedi S, Gupta P (2021) Plant secondary metabolites for preferential targeting among various stressors of metabolic syndrome. Stud Nat Prod Chem 71:221–261

Chowdhury P, Chowdhury P (2022) Glucosinolates and its role in mitigating abiotic and biotic stress in *Brassicaceae*. In: Plant stress physiology - perspectives in agriculture. IntechOpen, London

Christiansen JV, Isbrandt T, Petersen C, Sondergaard TE, Nielsen MR, Pedersen TB, Sørensen JL, Larsen TO, Frisvad JC (2021) Fungal quinones: diversity, producers, and applications of quinones from Aspergillus, Penicillium, Talaromyces, Fusarium, and Arthrinium. Appl Microbiol Biotechnol 105:8157–8193

Correia S, Poeta P, Ebraud MH, Luis Capelo J, Igrejas G (2017) Mechanisms of quinolone action and resistance: where do we stand? J Med Microbiol 66(5):551–559

Coy-Barrera E (2020) Analysis of betalains (betacyanins and betaxanthins). In: Recent advances in natural products analysis. Elsevier, Amsterdam, pp 593–619

Dai H, Xiao C, Liu H, Tang H (2010) Combined NMR and LC-MS analysis reveals the metabonomic changes in salvia miltiorrhiza bunge induced by water depletion. J Proteome Res 9:1460–1475

Dar MI, Rafat S, Dev K, Abass S, Khan MU, Abualsunun WA, Murshid SS, Ahmad S, Qureshi MI (2022) Heartwood extract of Pterocarpus marsupium Roxb. offers defense against oxyradicals and improves glucose uptake in HepG2 cells. Metabolites 12:947

David E, Niculescu VC (2021) Volatile organic compounds (VOCs) as environmental pollutants: occurrence and mitigation using nanomaterials. Int J Environ Res Public Health 18:13147

Dias DA, Urban S, Roessner U (2012) A historical overview of natural products in drug discovery. Metabolites 2(2):303–336

Dwivedi MK, Sonter S, Mishra S, Singh P, Singh PK (2021) Secondary metabolite profiling and characterization of diterpenes and flavones from the methanolic extract of Andrographis paniculata using HPLC-LC-MS/MS. Future J Pharm Sci 7:184

El Sayed AM, Basam SM, El-Naggar EM, Bellah A, Marzouk HS, El-Hawary S (2020) LC–MS/MS and GC–MS profiling as well as the antimicrobial effect of leaves of selected Yucca species introduced to Egypt. Sci Rep 10:17778

Elateeq Z-H, Khan AA, Khan Y, Khan J, Ali A, Plant S, Hanada K, Barna B, Salam U, Ullah S, Tang Z-H, Elateeq AA, Khan Y, Khan J, Khan A, Ali S (2023) Plant metabolomics: an overview of the role of primary and secondary metabolites against different environmental stress factors. Life 2023(13):706

El-Bondkly AMA (2014) Sequence analysis of industrially important genes from trichoderma. In: Biotechnology and biology of trichoderma. Elsevier, Amsterdam, pp 377–392

Fiehn O (2016) Metabolomics by gas chromatography-mass spectrometry: combined targeted and untargeted profiling. Curr Protoc Mol Biol 114:1–32

Fiehn O (2017) Metabolomics by gas chromatography-mass spectrometry: the combination of targeted and untargeted profiling. Curr Protoc Mol Biol 114:30.4.1–30.4.32

Fielding BC, Filho C, Ismail NSM, de Sousa DP (2020) Alkaloids: therapeutic potential against human coronaviruses. Molecules 25(23):5496

Georgiev VG, Weber J, Kneschke EM, Denev PN, Bley T, Pavlov AI (2010) Antioxidant activity and phenolic content of betalain extracts from intact plants and hairy root cultures of the red beetroot Beta vulgaris cv. Detroit dark red. Plant Foods Hum Nutr 65:105–111

Glazer AN, Nikaido H (2007) Primary metabolites: organic acids and amino acids. In: Microbial biotechnology. Cambridge University Press, Cambridge, pp 299–323

Gomathi D, Kalaiselvi M, Ravikumar G, Devaki K, Uma C (2015) GC-MS analysis of bioactive compounds from the whole plant ethanolic extract of Evolvulus alsinoides (L.) L. J Food Sci Technol 52(2):1212–1217
Guzmán DC, Olguín HJ, Corona QV, Herrera MO, Brizuela NO, Mejía GB (2020) Consumption of cooked common beans or saponins could reduce the risk of diabetic complications. Diabetes Metab Syndr Obes 13:3481
Hadipour E, Taleghani A, Tayarani-Najaran N, Tayarani-Najaran Z (2020) Biological effects of red beetroot and betalains: a review. Phytother Res 34:1847–1867
Hsiao YH, Ho CT, Pan MH (2020) Bioavailability and health benefits of major isoflavone aglycones and their metabolites. J Funct Foods 74:104164
Hu S, Liu C, Liu X (2022a) Innovative application of metabolomics on bioactive ingredients of foods. Foods 11:2974
Hu W, Sarengaowa W, Guan Y, Feng K (2022b) Biosynthesis of phenolic compounds and antioxidant activity in fresh-cut fruits and vegetables. Front Microbiol 13:906069
Irchhaiya R, Kumar A, Yadav A, Gupta N, Kumar S, Gupta N, Kumar S, Yadav V, Prakash A, Gurjar H (2015) Metabolites in plants and its classification. World J Pharm Pharm Sci 24(4):3266
Jeandet P, Formela-Luboińska M, Labudda M, Morkunas I (2022) The role of sugars in plant responses to stress and their regulatory function during development. Int J Mol Sci 23(9):5161
Jorge TF, Rodrigues JA, Caldana C, Schmidt R, van Dongen JT, Thomas-Oates J, António C (2016) Mass spectrometry-based plant metabolomics: metabolite responses to abiotic stress. Mass Spectrom Rev 35:620–649
Jung YS, Rha CS, Baik MY, Baek NI, Kim DO (2020) A brief history and spectroscopic analysis of soy isoflavones. Food Sci Biotechnol 29:1605
Kanthal LK, Dey A, Satyavathi K, Bhojaraju P (2014) GC-MS analysis of bio-active compounds in methanolic extract of Lactuca runcinata DC. Pharm Res 6(1):58–61
Kareem O, Ali T, Dar LA, Mir SA, Rashid R, Nazli N, Gulzar T, Bader GN (2022) Positive health benefits of saponins from edible legumes: phytochemistry and pharmacology. In: Edible plants in health and diseases: volume II: phytochemical and pharmacological properties, vol 2. Springer, Cham, pp 279–298
Khan H, Ullah H, Khattak S, Aschner M, Aguilar CN, Halimi SMA, Cauli O, Shah SMM (2021) Therapeutic potential of alkaloids in autoimmune diseases: promising candidates for clinical trials. Phytother Res 35(1):50–62
Khoomrung S, Wanichthanarak K, Nookaew I, Thamsermsang O, Seubnooch P, Laohapand T, Akarasereenont P (2017) Metabolomics and integrative omics for the development of Thai traditional medicine. Front Pharmacol 8:265080
Kim HU (2020) Lipid metabolism in plants. Plants 9:1–4
Kim IS (2021) Current perspectives on the beneficial effects of soybean isoflavones and their metabolites for humans. Antioxidants 10:1064
Kim HJ, Lee MJ, Jeong MH, Kim JE (2017) Identification and quantification of glucosinolates in kimchi by liquid chromatography-electrospray tandem mass spectrometry. Int J Anal Chem 2017:6753481
Konappa N, Udayashankar AC, Krishnamurthy S, Pradeep CK, Chowdappa S, Jogaiah S (2020) GC–MS analysis of phytoconstituents from Amomum nilgiricum and molecular docking interactions of bioactive serverogenin acetate with target proteins. Sci Rep 10(1):16438
Kurek J (2019) Alkaloids - their importance in nature and for human life. IntechOpen, London
Kurzyna-Szklarek M, Cybulska J, Zdunek A (2022) Analysis of the chemical composition of natural carbohydrates – an overview of methods. Food Chem 394:133466
Lara-Cruz GA, Jaramillo-Botero A (2022) Molecular level sucrose quantification: a critical review. Sensors 22:09511
Lechtenberg M (2021) Cyanogenic glycosides and biogenetically related compounds in higher plants and animals. eLS 2:1–18
Lee KM, Jeon JY, Lee BJ, Lee H, Choi HK (2017) Application of metabolomics to quality control of natural product derived medicines. Biomol Ther 25:559

Li G, Meng X, Zhu M, Li Z (2019) Research progress of betalain in response to adverse stresses and evolutionary relationship compared with anthocyanin. Molecules 2019(24):3078

Li Z, Wang H, Feng L, Song L, Lu Y, Li H, Li Y, Tian G, Yang Y, Li H, Mei X, Peng L (2022) Comparative metabolomics provides novel insights into correlation between dominant habitat factors and constituents of Stellaria Radix (Stellaria dichotoma L. var. lanceolata Bge.). Front Plant Sci 13:1035712

Lim JG, Park HM, Yoon KS (2020) Analysis of saponin composition and comparison of the antioxidant activity of various parts of the quinoa plant (Chenopodium quinoa Willd.). Food Sci Nutr 8:694

Lin D, Xiao M, Zhao J, Li Z, Xing B, Li X, Kong M, Li L, Zhang Q, Liu Y, Chen H, Qin W, Wu H, Chen S (2016) An overview of plant phenolic compounds and their importance in human nutrition and management of type 2 diabetes. Molecules 21(10):1374

Liu J, Liu Y, Wang Y, Abozeid A, Zu YG, Tang ZH (2017) The integration of GC–MS and LC–MS to assay the metabolomics profiling in Panax ginseng and Panax quinquefolius reveals a tissue- and species-specific connectivity of primary metabolites and ginsenosides accumulation. J Pharm Biomed Anal 135:176–185

Liu R, Bao ZX, Zhao PJ, Li GH (2021) Advances in the study of metabolomics and metabolites in some species interactions. Molecules 26(11):3311

Liu X, Wang X, Cheng Y, Wu Y, Yan Y, Li Z (2023) Variations in volatile organic compounds in Zhenyuan Daocai samples at different storage durations evaluated using E-nose, E-tongue, gas chromatography, and spectrometry. LWT 173:114186

Lopatkin AJ, Yang JH (2021) Digital insights into nucleotide metabolism and antibiotic treatment failure. Front Digit Health 3:583468

Luna-Guevara ML, Luna-Guevara JJ, Hernández-Carranza P, Ruíz-Espinosa H, Ochoa-Velasco CE (2018) Phenolic compounds: a good choice against chronic degenerative diseases. Stud Nat Prod Chem 59:79–108

Madadi E, Mazloum-Ravasan S, Yu JS, Ha JW, Hamishehkar H, Kim KH (2020) Therapeutic application of betalains: a review. Plants 9:1–27

Madeo J, Zubair A, Marianne F (2013) A review on the role of quinones in renal disorders. Springerplus 2:1–8

Mamat SF, Azizan KA, Baharum SN, Noor NM, Aizat WM (2020) GC-MS and LC-MS analyses reveal the distribution of primary and secondary metabolites in mangosteen (Garcinia mangostana Linn.) fruit during ripening. Sci Hortic 262:109004

Mani V, Park S, Kim JA, Lee SI, Lee K (2021) Metabolic perturbation and synthetic biology strategies for plant terpenoid production—an updated overview. Plants 10(10):2179

Marchev AS, Vasileva LV, Amirova KM, Savova MS, Balcheva-Sivenova ZP, Georgiev MI (2021) Metabolomics and health: from nutritional crops and plant-based pharmaceuticals to profiling of human biofluids. Cell Mol Life Sci 78:6487–6503

Mari A, Lyon D, Fragner L, Montoro P, Piacente S, Wienkoop S, Egelhofer V, Weckwerth W (2013) Phytochemical composition of Potentilla anserina L. analyzed by an integrative GC-MS and LC-MS metabolomics platform. Metabolomics 9(3):599–607

Marulasiddaswamy KM, Nuthan BR, Sunilkumar CR, Bajpe SN, Kumara KKS, Sekhar S, Kini KR (2021) HR-LC-MS based profiling of phytochemicals from methanol extracts of leaves and bark of myristica dactyloides gaertn. From Western Ghats of Karnataka, India. J Appl Biol Biotechnol 9(5):124–135

Meza S, Zhou Y, Chastain J, Yang Y, Cheng HH, Iassonova D, Rivest J, You H (2022) Eco-efficient quantification of glucosinolates in camelina seed, oil, and defatted meal: optimization, development, and validation of a UPLC-DAD method. Antioxidants 11:2441

Minic Z (2008) Physiological roles of plant glycoside hydrolases. Planta 227:723–740

Minocha T, Birla H, Obaid AA, Rai V, Sushma P, Shivamallu C, Moustafa M, Al-Shehri M, Al-Emam A, Tikhonova MA, Yadav SK, Poeggeler B, Singh D, Singh SK (2022) Flavonoids as promising neuroprotectants and their therapeutic potential against Alzheimer's disease. Oxidative Med Cell Longev 2022:6038996

Mishra S, Priyanka S (2022) Metabolomic insights into endophyte-derived bioactive compounds. Front Microbiol 13:835931

Mitra S, Prova SR, Sultana SA, Das R, Nainu F, Emran TB, Tareq AM, Uddin MS, Alqahtani AM, Dhama K, Simal-Gandara J (2021) Therapeutic potential of indole alkaloids in respiratory diseases: a comprehensive review. Phytomedicine 90:153649

Moco S (2022) Studying metabolism by NMR-based metabolomics. Front Mol Biosci 9:882487

Moing A, Aharoni A, Biais B, Rogachev I, Meir S, Brodsky L, Allwood JW, Erban A, Dunn WB, Kay L, de Koning S, de Vos RCH, Jonker H, Mumm R, Deborde C, Maucourt M, Bernillon S, Gibon Y, Hansen TH, Husted S, Goodacre R, Kopka J, Schjoerring JK, Rolin D, Hall RD (2011) Extensive metabolic cross-talk in melon fruit revealed by spatial and developmental combinatorial metabolomics. New Phytol 190(3):683–696

Moses T, Papadopoulou KK, Osbourn A (2014) Metabolic and functional diversity of saponins, biosynthetic intermediates and semi-synthetic derivatives. Crit Rev Biochem Mol Biol 49:439

Mukherjee PK, Harwansh RK, Bahadur S, Biswas S, Kuchibhatla LN, Tetali SD, Raghavendra AS (2016) Metabolomics of medicinal plants-a versatile tool for standardization of herbal products and quality evaluation of Ayurvedic formulations Ayurvedic formulations, their uses and limitations. Curr Sci 111:1624

Mushtaq S, Abbasi BH, Uzair B, Abbasi R (2018) Natural products as reservoirs of novel therapeutic agents. EXCLI J 17:420–451

Mutlu M, Bingol Z, Uc EM, Köksal E, Goren AC, Alwasel SH, Gulcin İ (2023) Comprehensive metabolite profiling of cinnamon (Cinnamomum zeylanicum) leaf oil using LC-HR/MS, GC/MS, and GC-FID: determination of antiglaucoma, antioxidant, anticholinergic, and antidiabetic profiles. Life 13(1):136

Nagana Gowda GA, Djukovic D (2014) Overview of mass spectrometry-based metabolomics: opportunities and challenges. Methods Mol Biol 1198:3–12

Nasim N, Sandeep IS, Mohanty S (2022) Plant-derived natural products for drug discovery: current approaches and prospects. Nucleus 65(3):399–411

Niaz K, Khan F, Ajmal Shah M (2020) Analysis of quinonoids. In: Recent advances in natural products analysis. Elsevier, Amsterdam, pp 749–766

Oleszek WA (2002) Chromatographic determination of plant saponins. J Chromatogr A 967:147–162

Olivia NU, Goodness UC, Obinna OM (2021) Phytochemical profiling and GC-MS analysis of aqueous methanol fraction of Hibiscus asper leaves. Futur J Pharm Sci 7:59

Otieno DO, Rose H, Shah NP (2007) Profiling and quantification of isoflavones in soymilk from soy protein isolate using extracted ion chromatography and positive ion fragmentation techniques. Food Chem 105:1642–1651

Pagter M, Yde CC, Kjær KH (2017) Metabolic fingerprinting of dormant and active flower primordia of Ribes nigrum using high-resolution magic angle spinning NMR. J Agric Food Chem 65:10123–10130

Parveen R, Zahiruddin S, Charegaonkar A, Khale A, Mallick S (2020) Chromatographic profiling of rose petals in Unani formulations (Gulkand, Arq-e-Gulab, and Rose Sharbat) using HPTLC and GC–MS. J AOAC Int 103:684–691

Patel K, Gadewar M, Tripathi R, Prasad SK, Patel DK (2012) A review on medicinal importance, pharmacological activity and bioanalytical aspects of beta-carboline alkaloid "Harmine". Asian Pac J Trop Biomed 2:660

Patel MK, Pandey S, Kumar M, Haque MI, Pal S, Yadav NS (2021) Plants metabolome study: emerging tools and techniques. Plants 10(11):2409

Pérez EMS, Iglesias MJ, Ortiz FL, Pérez IS, Galera MM (2010) Study of the suitability of HRMAS NMR for metabolic profiling of tomatoes: application to tissue differentiation and fruit ripening. Food Chem 122:877–887

Piccioni F, Capitam D, Zolla L, Mannina L (2009) NMR metabolie profiling of transgenic maize with the Cry1A(b) gene. J Agric Food Chem 57(14):6041–6049

Pinu FR, Goldansaz SA, Jaine J (2019) Translational metabolomics: current challenges and future opportunities. Metabolites 9:108

Plant Primary Metabolites Analysis Service - Creative Proteomics (n.d.). https://www.creative-proteomics.com/services/plant-primary-metabolites-analysis-service.htm. Accessed 29 October 2023
Rahim MA, Ayub H, Sehrish A, Ambreen S, Khan FA, Itrat N, Nazir A, Shoukat A, Shoukat A, Ejaz A, Özogul F, Bartkiene E, Rocha JM (2023) Essential components from plant source oils: a review on extraction, detection, identification, and quantification. Molecules 28:6881
Ridley CP, Khosla C (2009) Polyketides. In: Encyclopedia of microbiology, 3rd edn. Elsevier, Amsterdam, pp 472–481
Rios-Navarro A, Gonzalez M, Carazzone C, Celis Ramírez AM (2021) Learning about microbial language: possible interactions mediated by microbial volatile organic compounds (VOCs) and relevance to understanding Malassezia spp. metabolism. Metabolomics 17:1–11
Risdian C, Mozef T, Wink J (2019) Biosynthesis of polyketides in streptomyces. Microorganisms 7:124
Robe K, Izquierdo E, Vignols F, Rouached H, Dubos C (2021) The coumarins: secondary metabolites playing a primary role in plant nutrition and health. Trends Plant Sci 26:248–259
Roberts LD, Souza AL, Gerszten RE, Clish CB (2012) Targeted metabolomics. Curr Protoc Mol Biol 30:2
Rolin D (2006) Control of primary metabolism in plants. Annual Plant Reviews 22. Ann Bot 98(6):1331–1332
Rusanov KE, Kovacheva NM, Atanassov II (2011) Comparative GC/MS analysis of rose flower and distilled oil volatiles of the oil bearing rose Rosa damascena. Biotechnol Biotechnol Equip 25:15
Sadgrove NJ, Padilla-González GF, Phumthum M (2022) Fundamental chemistry of essential oils and volatile organic compounds, methods of analysis and authentication. Plants 11:789
Saini AK, Sawant L, Zahiruddin S, Shrivastva D, Mitra R, Rai RK, Ahmad S (2023) LC-MS/MS-based targeted metabolomic profiling of aqueous and hydro-alcoholic extracts of Pistacia integerrima Linn., Quercus infectoria olivier and Terminalia chebula Retz. Pharmacogn Mag 19:097312962211448
Sajid M, Stone SR, Kaur P (2021) Recent advances in heterologous synthesis paving way for future green-modular bioindustries: a review with special reference to isoflavonoids. Front Bioeng Biotechnol 9:673270
Salam U, Ullah S, Tang ZH, Elateeq AA, Khan Y, Khan J, Khan A, Ali S (2023) Plant metabolomics: an overview of the role of primary and secondary metabolites against different environmental stress factors. Life 13:706
Salem MA, Ezzat SM (2021) The use of aromatic plants and their therapeutic potential as antiviral agents: a hope for finding anti-COVID 19 essential oils. J Essent Oil Res 33:105–113
Salem MA, De Souza LP, Serag A, Fernie AR, Farag MA, Ezzat SM, Alseekh S (2020) Metabolomics in the context of plant natural products research: from sample preparation to metabolite analysis. Metabolites 10(1):37
Santos ONA, Folegatti MV, Dutra LM, Andrade IP, Fanaya ED, Lena BP, Barison A, Santos AD (2017) Tracking lipid profiles of Jatropha curcas L. seeds under different pruning types and water managements by low-field and HR-MAS NMR spectroscopy. Ind Crop Prod 109:918–922
Savarino P, Demeyer M, Decroo C, Colson E, Gerbaux P (2023) Mass spectrometry analysis of saponins. Mass Spectrom Rev 42:954–983
Sawicki T, Juśkiewicz J, Wiczkowski W (2017) Using the SPE and micro-HPLC-MS/MS method for the analysis of betalains in Rat plasma after red beet administration. Molecules 22:2137
Schrenk D, Bignami M, Bodin L, Chipman JK, del Mazo J, Grasl-Kraupp B, Hogstrand C, Hoogenboom L, Leblanc JC, Nebbia CS, Nielsen E, Ntzani E, Petersen A, Sand S, Vleminckx C, Wallace H, Benford D, Brimer L, Mancini FR, Metzler M, Viviani B, Altieri A, Arcella D, Steinkellner H, Schwerdtle T (2019) Evaluation of the health risks related to the presence of cyanogenic glycosides in foods other than raw apricot kernels. EFSA J 17:e05662
Schrimpe-Rutledge AC, Codreanu SG, Sherrod SD, McLean JA (2016) Untargeted metabolomics strategies – challenges and emerging directions. J Am Soc Mass Spectrom 27:1897

Schulz-Bohm K, Martín-Sánchez L, Garbeva P (2017) Microbial volatiles: small molecules with an important role in intra- and inter-kingdom interactions. Front Microbiol 8:289291

Shah A, Varma C, Patankar S, Kadam V (2014) Plant glycosides and aglycones displaying antiproliferative and antitumour activities – a review. Curr Bioact Compd 9:288–305

Sharifi-Rad J, Cruz-Martins N, López-Jornet P, Lopez EPF, Harun N, Yeskaliyeva B, Beyatli A, Sytar O, Shaheen S, Sharopov F, Taheri Y, Docea AO, Calina D, Cho WC (2021) Natural coumarins: exploring the pharmacological complexity and underlying molecular mechanisms. Oxidative Med Cell Longev 2021:6492346

Sharma B, Yadav DK (2022) Metabolomics and network pharmacology in the exploration of the multi-targeted therapeutic approach of traditional medicinal plants. Plants 11:3243

Simirgiotis MJ, Burton D, Parra F, López J, Muñoz P, Escobar H, Parra C (2020) Antioxidant and antibacterial capacities of Origanum vulgare L. Essential oil from the arid Andean region of Chile and its chemical characterization by GC-MS. Metabolites 10(10):414

Singh A (2017) Glucosinolates and plant defense. Springer, Cham, pp 237–246

Slika H, Mansour H, Wehbe N, Nasser SA, Iratni R, Nasrallah G, Shaito A, Ghaddar T, Kobeissy F, Eid AH (2022) Therapeutic potential of flavonoids in cancer: ROS-mediated mechanisms. Biomed Pharmacother 146:112442

Soni K, Frew R, Kebede B (2023) A review of conventional and rapid analytical techniques coupled with multivariate analysis for origin traceability of soybean. Crit Rev Food Sci Nutr 3:1–20

Stringlis IA, De Jonge R, Pieterse CMJ (2019) The age of coumarins in plant-microbe interactions. Plant Cell Physiol 60:1405–1419

Thiele B, Hupert M, Santiago-Schübel B, Oldiges M, Hofmann D (2019) Direct analysis of underivatized amino acids in plant extracts by LC-MS/MS (improved method). Methods Mol Biol 2030:403–414

Tinco-Jayo JA, Aguilar-Felices EJ, Enciso-Roca EC, Arroyo-Acevedo JL, Herrera-Calderon O (2022) Phytochemical screening by LC-ESI-MS/MS and effect of the ethyl acetate fraction from leaves and stems of Jatropha macrantha Müll Arg. on ketamine-induced erectile dysfunction in rats. Molecules 27(1):115

Tine Y, Yang Y, Renucci F, Costa J, Wélé A, Paolini J (2017) LC-MS/MS analysis of flavonoid compounds from Zanthoxylum zanthoxyloides extracts and their antioxidant activities. Nat Prod Commun 12(12):1865–1868

Toomey VM, Nickum EA, Flurer CL (2012) Cyanide and amygdalin as indicators of the presence of bitter almonds in imported raw almonds. J Forensic Sci 57:1313–1317

Trych U, Buniowska-Olejnik M, Marszałek K (2022) Bioaccessibility of betalains in beetroot (Beta vulgaris L.) juice under different high-pressure techniques. Molecules 27:7093

Twaij BM, Hasan MN (2022) Bioactive secondary metabolites from plant sources: types, synthesis, and their therapeutic uses. Int J Plant Biol 13(1):4–14

Wang JH, Byun J, Pennathur S (2010) Analytical approaches to metabolomics and applications to systems biology. Semin Nephrol 30(5):500–511

Wang Y, Ma Y, Tao L, Zhang X, Hao F, Zhao S, Han L, Bai C (2022) Recent advances in separation and analysis of saponins in natural products. Separations 9:163

Waris M, Koçak E, Gonulalan EM, Demirezer LO, Kır S, Nemutlu E (2022) Metabolomics analysis insight into medicinal plant science. TrAC Trends Anal Chem 157:116795

Weber RJM, Lawson TN, Salek RM, Ebbels TMD, Glen RC, Goodacre R, Griffin JL, Haug K, Koulman A, Moreno P, Ralser M, Steinbeck C, Dunn WB, Viant MR (2017) Computational tools and workflows in metabolomics: an international survey highlights the opportunity for harmonisation through galaxy. Metabolomics 13:1–5

Wei J, Chen F, Liu Y, Abudoukerimu A, Zheng Q, Zhang X, Sun Y, Yimiti D (2020) Comparative metabolomics revealed the potential antitumor characteristics of four endophytic fungi of Brassica rapa L. ACS Omega 5:5939–5950

Wishart DS (2011) Advances in metabolite identification. Bioanalysis 3:1769–1782

Wojakowska A, Piasecka A, García-López PM, Zamora-Natera F, Krajewski P, Marczak Ł, Kachlicki P, Stobiecki M (2013) Structural analysis and profiling of phenolic secondary metabolites of Mexican lupine species using LC-MS techniques. Phytochemistry 92:71–86

Wong MCY, Emery PW, Preedy VR, Wiseman H (2008) Health benefits of isoflavones in functional foods? Proteomic and metabonomic advances. Inflammopharmacology 16:235–239

Xiao JF, Zhou B, Ressom HW (2012) Metabolite identification and quantitation in LC-MS/MS-based metabolomics. Trends Anal Chem 32:1

Xiao Q, Mu X, Liu J, Li B, Liu H, Zhang B, Xiao P (2022) Plant metabolomics: a new strategy and tool for quality evaluation of Chinese medicinal materials. Chin Med 17:1–19

Xu X, Qu R, Wu W, Jiang C, Shao D, Shi J (2021) Applications of microbial co-cultures in polyketides production. J Appl Microbiol 130:1023–1034

Yang W, Chen X, Li Y, Guo S, Wang Z, Yu X (2020) Advances in pharmacological activities of terpenoids. Nat Prod Commun 15:3

Yeoman CJ, Han Y, Dodd D, Schroeder CM, Mackie RI, Cann IKO (2010) Thermostable enzymes as biocatalysts in the biofuel industry. Adv Appl Microbiol 70:1–55

Yibo C, Zhidong W, Chongrong W, Hong L, Daoqiang H, Degui Z, Lei Z, Yangyang P, Rong G, Shaochuan Z (2022) Comparisons of metabolic profiles for carbohydrates, amino acids, lipids, fragrance and flavones during grain development in indica rice cultivars. Rice Sci 29:155–165

Yuan H, Ma Q, Ye L, Piao G (2016) The traditional medicine and modern medicine from natural products. Molecules 21(5):559

Yuan Y, Zuo J, Zhang H, Li R, Yu M, Liu S (2022) Integration of transcriptome and metabolome provides new insights to flavonoids biosynthesis in Dendrobium huoshanense. Front Plant Sci 13:850090

Yulvianti M, Zidorn C (2021) Chemical diversity of plant cyanogenic glycosides: an overview of reported natural products. Molecules 26(3):719

Zareena B, Khadim A, Jeelani SUY, Hussain S, Ali A, Musharraf SG (2021) High-throughput detection of an alkaloidal plant metabolome in plant extracts using LC-ESI-QTOF-MS. J Proteome Res 20(8):3826–3839

Zaynab M, Fatima M, Sharif Y, Zafar MH, Ali H, Khan KA (2019) Role of primary metabolites in plant defense against pathogens. Microb Pathog 137:103728

Zeeshan Bhatti M, Ismail H, Khan Kayani W (2022) Plant secondary metabolites: therapeutic potential and pharmacological properties. In: Secondary metabolites - trends and reviews. IntechOpen, London

Zhou T, Luo X, Zhang C, Xu X, Yu C, Jiang Z, Zhang L, Yuan H, Zheng B, Pi E, Shen C (2019) Comparative metabolomic analysis reveals the variations in taxoids and flavonoids among three Taxus species. BMC Plant Biol 19:1–12

Zuk M, Pelc K, Szperlik J, Sawula A, Szopa J (2020) Metabolism of the cyanogenic glucosides in developing flax: metabolic analysis, and expression pattern of genes. Metabolites 10:1–13

Zukalová H, Vašák J (2002) The role and effects of glucosinolates species – a review. Rostlinna Vyroba 48:175–180

Exploring the Therapeutic Potential of Secondary Metabolites in Medicinal Plants

16

Monalisha Samal, Abdur Rahman, and Sayeed Ahmad

Abstract

The study of plant secondary metabolites has advanced throughout the previous few decades. Secondary metabolites are organic compounds produced by plants that are not directly involved in the primary metabolic processes of growth and development. Instead, these compounds often serve various ecological functions, such as defense against herbivores, attraction of pollinators, and competition with other plants. Many secondary metabolites have also been found to possess medicinal properties, making them valuable resources in traditional and modern system of medicine. Some of the well-studied secondary metabolites for their potential pharmacological activities and health benefits include alkaloids, terpenoids, phenolics, glycosides saponins, anthocyanins, lignans, coumarins, etc. They have special applications as flavors, dyes, insecticides, and colorants, making them economically valuable. It is important to note that while many medicinal plants have shown promising effects in traditional medicine and early research. Further studies, including clinical trials, are often needed to establish their safety and efficacy for modern medicinal use. This chapter consists of an introductory overview of secondary metabolites, methods for screening analysis using modern chromatographic and spectroscopic techniques, therapeutic applications, bioavailability and pharmacokinetics, synergistic combination and therapies, and safety and toxicity combinations.

Keywords

Secondary metabolites · Chromatographic and spectroscopic techniques · Bioavailability · Synergy

---

M. Samal · A. Rahman · S. Ahmad (✉)
Centre of Excellence in Unani Medicine, Bioactive Natural Product Laboratory, Department of Pharmacognosy and Phytochemistry, School of Pharmaceutical Education and Research, Jamia Hamdard, New Delhi, India

© The Author(s), under exclusive license to Springer Nature Singapore Pte Ltd. 2024
M. Nandave et al. (eds.), *Ethnopharmacology and OMICS Advances in Medicinal Plants Volume 1*, https://doi.org/10.1007/978-981-97-2367-6_16

16.1 Introduction

Traditional medicines have played a significant role in human health care and healing for thousands of years (Shi et al. 2021). They are important in many cultures all over the world, because they provide all-natural treatments for a variety of diseases and health issues. Traditional medical practices and the usage of therapeutic herbs precede written history (Rasool et al. 2020). Traditional remedies generally use a holistic approach to health, considering not only the physical, mental, emotional, and spiritual aspects of well-being (Weragoda 1980). The harmony and balance of the body are highly valued by them. The World Health Organization values the use of conventional medicines and supports their safe and efficient integration into healthcare systems (Samal and Dehury 2019).

Compounds known as secondary metabolites are not directly engaged in the three main metabolic processes of growth, development, and reproduction in medicinal plants (Divekar et al. 2022). These bioactive substances help plants adapt and perform a variety of ecological tasks. Numerous secondary metabolites have pharmacological properties that make them useful for maintaining human health (Jusuf et al. 2022). These substances are frequently the active components of contemporary pharmaceutical medications and conventional herbal treatments. For example, alkaloids like morphine from the opium poppy and quinine from cinchona bark are powerful analgesics and antimalarial agents, respectively (Udeani et al. 2021). Secondary metabolites often function as a plant's natural defense mechanism against herbivores, pathogens, and pests. These compounds can be toxic or repellent to potential threats, helping the plant survive and reproduce (Erb and Kliebenstein 2020). Medicinal plants use these compounds to protect themselves from diseases, and some of these protective properties can be beneficial to humans (Divekar et al. 2022; Mazid et al. 2011). Secondary metabolites represent a vast and diverse group of compounds, including alkaloids, terpenoids, flavonoids, phenolic compounds, and more. This diversity provides a wide range of chemical structures and potential therapeutic applications (Armijos et al. 2021; Abass et al. 2022, 2023). Secondary metabolites can have a range of pharmacological activities, including antimicrobial, anti-inflammatory, antioxidant, anticancer, and immunomodulatory effects (Yeshi et al. 2022). These properties make them valuable sources of new drug leads and therapeutic agents. Ayurveda and Traditional Chinese Medicine are two traditional medical systems that have long used medicinal plants and their secondary metabolites as a form of treatment. These systems possess a plethora of information regarding the use of plant-based treatments for a variety of illnesses (Shi et al. 2021; Ma et al. 2021). Numerous significant medications have been made from secondary metabolites of medicinal plants. Examples are the pain reliever aspirin, which comes from the willow tree, and the anticancer medication paclitaxel, which comes from the Pacific yew tree. The preservation of these plant species and any possible health advantages they may have for people depends on their conservation and appropriate management. Studying secondary metabolites advances our knowledge of plant physiology and biochemistry. Additionally, it aids in the discovery of novel molecules with medicinal potential. Therefore, secondary metabolites in medicinal plants are crucial because of their variety of pharmacological characteristics and potential therapeutic advantages. They have been essential to conventional medicine and are still a great source of ideas for developing new drugs and improving health care. The complicated interactions between plants and their surroundings are further highlighted by their ecological roles in plant defense and adaptation.

16.2 Types and Functions of Secondary Metabolites

Secondary metabolites can be classified into several major groups based on their chemical structures and biosynthetic origins (Samal et al. 2023). These classifications help categorize and study the wide variety of secondary metabolites found in plants, microorganisms, and other organisms. The primary classes of secondary metabolites include:

Terpenoids (Isoprenoids) Terpenoids are derived from isoprene units and encompass a vast group of secondary metabolites. They include compounds like essential oils, steroids, and carotenoids. Terpenoids have diverse roles, such as plant defense, pigmentation, and aroma production. More than 40,000 different chemicals make up terpenes. In the secondary metabolism of insects, terpenes also serve as protective agents, poisons, and deterrents (Ludwiczuk et al. 2017).

Phenolic Compounds Phenolic compounds are characterized by the presence of phenol rings. Phenolic compounds with a single carboxylic acid group are referred to as "phenolic acids." They can be found naturally in a variety of plant-based foods, with seeds, fruit skins, and vegetable leaves often containing the largest concentrations. Due to their potent antioxidant effects and related health advantages, these plant phenolic acids are a crucial component of the human diet. According to epidemiological research, eating a diet high in antioxidant-rich fruits and vegetables is linked to a lower prevalence of oxidative disorders like cancer, diabetes, and cardiovascular diseases. These substances also have indirect antioxidant action, because they can activate defense mechanisms that benefit signalling pathways. The actions of the glucose and insulin receptors are significantly influenced by phenolic acids. Through the PI3K/Akt and AMP-activated kinase pathways, they can raise the levels of the GLUT2 glucose transporter in insulin-producing pancreatic cells and encourage GLUT4 transportation. For instance, ferulic and chlorogenic acids have demonstrated exact mechanisms of transporter activation and have been found to have antidiabetic effects. Phenolic acids are the most common phenolic component detected in wastewater. They can be used as food preservatives and have antimicrobial characteristics. The prevention and therapy of cancer are both greatly aided by phenolic acids and their derivatives. Plant phenolics could make a significant contribution in this area. Notably, numerous clinical trials in this field are continuing, and more than half of the anticancer medications approved internationally between the 1940s and 2006 were made from natural compounds or their derivatives. These substances prevent the synthesis of genotoxic substances, prevent the creation of DNA adducts, and reduce the action of mutagens (Kumar et al. 2020).

Flavonoids Flavonoids are a subgroup of phenolic compounds with a distinctive structure. They are known for their antioxidant, anti-inflammatory, and anticancer properties. Examples include quercetin, kaempferol, and anthocyanins. Flavonoids, a class of naturally occurring polyphenolic compounds, are among the most common compounds found in vegetables, fruits, and plant-based beverages. They are considered dietary supplements that promote overall health and help prevent various diseases. Flavonoids have now become an integral component of a wide range of products, including nutraceuticals, pharmaceuticals, and medical interventions.

In addition to their well-known antioxidant properties, flavonoids exhibit a diverse array of biological activities that contribute to human well-being. These activities include anti-inflammatory, antiulcer, antiviral, anticancer, antidiabetic, and cytotoxic effects, to name just a few. Various studies have highlighted the dietary benefits of flavonoids in enhancing antioxidant activity and protecting cell membranes from damage caused by lipid peroxidation. Consequently, flavonoids play a crucial role as antioxidants in diseases related to oxidative stress (Vazhappilly et al. 2019).

Flavonoids have also been associated with inflammatory disorders such as leukemia, asthma, sepsis, atherosclerosis, sclerosis, allergic rhinitis, psoriasis, rheumatoid arthritis, ileitis/colitis, and more. These compounds are involved in the recruitment of inflammatory cells and the release of reactive nitrogen species (RNS), reactive oxygen species (ROS) (Agati et al. 2013), and proinflammatory cytokines, which are important for eliminating pathogens and healing damaged tissues. While inflammation is typically a self-limiting process, unresolved or prolonged inflammation can lead to various chronic diseases.

Furthermore, flavonoids inhibit phosphodiesterases that play a role in cell activation, and they have been found to possess antibacterial properties in plant extracts. These compounds have demonstrated strong antiviral actions, dating back to the 1940s, by blocking enzymes involved in the virus life cycle. For instance, flavonoids like hesperetin, quercetin, and naringin exhibit anti-dengue activity.

In the context of cancer, flavonoids have a significant impact on the immune responses associated with the initiation and progression of the disease. They can influence various biological signals in cancer development, affecting processes like angiogenesis, apoptosis, cell proliferation, and cell differentiation. Overall, flavonoids primarily affect the initiation and promotion stages of carcinogenesis and influence tumor growth and hormonal activity.

Tannins Tannins are a type of polyphenolic compound that can bind and precipitate proteins. They are commonly found in plants and have astringent properties. Tannins are important for plant defense against herbivores. Chestnut tannin, a prominent member of the hydrolyzable tannin, is known for its possible antibacterial characteristics and capacity to lessen mycotoxins. Tannins, such as ellagitannins and gallotannins, have a variety of uses, such as the management of cytotoxin generation, the treatment of bacterial infections, the prevention of rhinitis, the reduction of antihistamine effects, and the inhibition of HIV replication in human cells. The prevention of *Staphylococcus aureus* and other gram-positive bacterial infections, leukemia, obesity, arteriosclerosis, thrombosis, triglyceride reduction, and other illnesses have also been reported to benefit from tannin-derived chemicals. Additionally, type II diabetes, obesity, hypertension, and hypercholesterolemia have all been the subject of patent applications for the noncommercial use of tannins. *Acacia mearnsii* and *Acacia nilotica* consists of condensed tannins are known for their antipyretic effects, with the former also having antidiarrheal activities. Combined condensed and hydrolyzable tannins found in sumac are thought to have anti-inflammatory, antibacterial, and immunomodulatory properties (Saarinen et al. 2007).

Lignans The word "lignan" describes a group of phenylpropanoids known as dimeric phenylpropanoids, which are made up of two C6-C3 phenylpropanoids connected by a C8 phenylpropanol. Lignans are a subclass of nonflavonoid polyphenols and are found in more than 60 distinct types of vascular plants. When incorporated in the diet, these substances contain substantial functional significance and have been linked to a lower risk of cardiovascular disease. In addition to barley, flaxseed, wheat bran, almonds, legumes, sesame seeds, fruits, and vegetables, lignans can also be found in a wide range of other foods. Clinical studies have also demonstrated that adding 30–50 g of flaxseed to your diet every day for 4–12 weeks will lower your LDL cholesterol by 8–14%. In addition, a study looking at how dietary lignans affect breast cancer risk discovered that women who ingested dietary lignans had a 17% decreased risk of breast cancer compared to those in the lowest quartile. The study found that endometrial cancer risk is lower in women who consume a considerable amount of dietary lignans. Both enterodiol and enterolactone, which are generated from lignans, have shown to reduce the incidence of malignancies linked to hormones. Lignans have a number of health advantages, such as their hypotensive (blood pressure lowering), anticarcinogenic, cardiac-protective, cholesterol-reducing, and digestive qualities (they can extend the time food remains in the stomach). Silymarin, a compound that has been used for years to treat conditions affecting the liver, spleen, and gallbladder, has hepatoprotective, antioxidant, anti-inflammatory, anticarcinogenic, and antidiabetic properties (Saarinen et al. 2007).

Glucosinolates Glucosinolates are sulfur-containing compounds primarily found in cruciferous vegetables like broccoli and cabbage. They are involved in plant defense against herbivores and have potential anticancer properties (Variyar et al. 2014).

Polyketides Polyketides are complex secondary metabolites produced by multienzyme systems. They have diverse structures and functions and include compounds like antibiotics (e.g., erythromycin), pigments (e.g., anthraquinones), and mycotoxins (Pankewitz and Hilker 2008).

Stilbenoids Stilbenes, a group of phenylpropanoids, are distinguished by their 1,2-diphenylethylene structure. Resveratrol, found in red grapes and wine, is a well-known stilbenoid with potential health benefits, and the majority of plant stilbenes are built on transresveratrol as nucleus. It has been demonstrated that stilbenes exhibit anti-inflammatory properties, antifungal, antiviral, antibacterial, and antiviral activity and act as agonists for estrogen receptors and apoptosis. Notably, resveratrol is the stilbene that has been the subject of the most research and is well known for having potent anticancer, anti-inflammatory, and antioxidant effects. Additionally, pterostilbene has been discovered to possess properties that are advantageous for controlling diabetes (Meier 1992).

Alkaloids Heterocyclic nitrogen compounds are alkaloids. Tryptophan, tyrosine, and lysine are only a few examples of the amino acids that make up the primary metabolites from which they are generated. Alkaloid production pathways can be lengthy, and many alkaloids have chemical structures that are equally complex. In the past 3000 years, alkaloids have been used medicinally by humans as purgatives, antitussives, and sedatives for snakebite, fever, and insanity. Papaverine, a benzylisoquinoline alkaloid, has been demonstrated to have an inhibitory effect on a number of viruses and indoquinoline alkaloids from *Cryptolepis sanguinolenta* showed efficacy against several gram-negative bacteria and yeast. Quinine is well known for its ability to combat the malaria parasite. There is evidence in the literature that plant alkaloids have significant biological activity (Waller et al. 1978).

Saponins The presence of a triterpenoid or steroidal aglycone linked to one or more sugar chains distinguishes saponins, as a varied group of compounds found in plants. They have a variety of biological and pharmacological effects and can be found in various plant species, such as ginseng, legumes, and several kinds of beans. Their distinctive shape is made up of a hydrophilic ("water-attracting") "head" and a hydrophobic ("water-repelling") "tail." Saponins are named after "sapo," the Latin word for soap, because they are known for their capacity to create stable frothy or foamy solutions when agitated in water. Since the 1950s, their immune-stimulating qualities have been used as adjuvants in vaccine formulations. Because of their potential for both cancer prevention and treatment, dammarane sapogenins produced from ginseng have generated interest, resulting in the development of anticancer medicines at various stages (Hussain et al. 2019).

Glycosides Glycosides consist of a sugar molecule (glycone) linked to a non-sugar moiety (aglycone). They are often found in plants and can have various medicinal properties. Examples include cardiac glycosides (e.g., digoxin) and saponins (Kren and Martínková 2001).

These classifications are not exhaustive, as secondary metabolites represent an incredibly diverse and continually expanding group of compounds. Many secondary metabolites serve vital ecological and physiological functions in organisms and have important applications in pharmacology, agriculture, and various industries. Here are some of the key functions of secondary metabolites:

Defense Against Herbivores and Predators: Secondary metabolites often act as chemical defenses against herbivores and predators. They can be toxic, repellent, or unpalatable, deterring animals from consuming the organism.

Protection Against Pathogens: Some secondary metabolites have antimicrobial properties and protect the organism from pathogenic bacteria, fungi, and other microorganisms. They play a role in preventing infections and diseases.

Attracting Pollinators and Dispersers: Secondary metabolites, particularly pigments and fragrant compounds, attract pollinators like bees and butterflies. They also aid in the dispersal of seeds and fruits by attracting animals to consume or transport them.

Communication and Signaling: Some secondary metabolites act as signaling molecules within and between organisms. For example, plant secondary metabolites may signal neighboring plants about herbivore attacks or pathogen threats.

Regulation of Growth and Development: Secondary metabolites often play roles in regulating the growth and development of the producing organism. For example, phytohormones like gibberellins and auxins are secondary metabolites involved in plant growth.

Adaptation to Environmental Stress: Secondary metabolites can help organisms adapt to environmental stressors, such as extreme temperatures, drought, or high salinity. They may enhance the organism's resilience and survival under adverse conditions.

Medicinal and Pharmacological Uses: Many secondary metabolites have medicinal properties and are used in traditional and modern medicine. They can have therapeutic effects on human health, such as pain relief, treatment of infections, and cancer therapy.

Aroma and Flavor: Aromatic secondary metabolites, such as terpenoids and essential oils, contribute to the aroma and flavor of plants and food products. They are used in the fragrance and food industries (Diez-Simon et al. 2019).

Pigmentation: Secondary metabolites like anthocyanins and carotenoids are responsible for the pigmentation of flowers, fruits, and leaves. Pigments play roles in pollinator attraction and fruit ripening (Fistarol and Itin 2010).

These functions of secondary metabolites highlight their ecological, physiological, and economic significance. They play crucial roles in the survival, adaptation, and interactions of organisms with their environments and other living entities.

16.3 Methods for Secondary Metabolite Analysis

Analysing secondary metabolites is essential for understanding their presence, concentration, and potential applications (Roopashree and Naik 2019). Various methods are available for secondary metabolite analysis, each tailored to specific compound classes. Here are some common methods used for secondary metabolite analysis:

Chromatography Techniques High-performance liquid chromatography (HPLC) is widely used to separate, quantify, and analyze secondary metabolites. It offers high sensitivity and accuracy. Different detectors such as UV, fluorescence, and mass spectrometry (MS) can be coupled with HPLC for compound identification. Gas chromatography (GC) is suitable for volatile and thermally stable compounds. It is often used for the analysis of terpenoids, fatty acids, and some alkaloids. GC can be coupled with detectors like flame ionization detector (FID) or mass spectrometry. Thin-layer chromatography (TLC) is a cost-effective method for qualitative analysis and compound separation. It is useful for identifying compounds based on their migration distances and visualizing them with stains or chemical reagents (Cortes 2020).

Mass Spectrometry (MS) Liquid chromatography-mass spectrometry (LC-MS) combines the separation capabilities of HPLC with mass spectrometry detection. It is highly versatile and widely used for compound identification and quantification. Gas chromatography-mass spectrometry (GC-MS) is used for volatile and semi-volatile compound analysis. It is particularly suitable for compounds that can be vaporized without decomposition. Matrix-assisted laser desorption/ionization mass spectrometry (MALDI-MS) is employed for the analysis of larger biomolecules and some secondary metabolites. It is useful for high-throughput analyses. Nuclear magnetic resonance (NMR) spectroscopy is a powerful technique for elucidating the chemical structure of secondary metabolites. It provides information about the connectivity of atoms and the arrangement of functional groups within a molecule. UV-visible spectroscopy is employed for the quantification of compounds that absorb in the UV or visible range. It is often used for analyzing pigments, flavonoids, and phenolic compounds. Fluorescence spectroscopy measures the fluorescence emissions of compounds when excited by specific wavelengths of light. It is suitable for studying compounds with intrinsic fluorescence properties, such as certain flavonoids (Mann et al. 2001).

Enzymatic Assays Enzyme assays are used to measure the enzymatic activity of secondary metabolites, particularly in the context of their interactions with target enzymes. This is often used in studies related to antidiabetic or antimicrobial compounds (Fried 1975).

Bioassays Bioassays involve using living organisms or cells to test the biological activity of secondary metabolites, such as their antimicrobial, anticancer, or antioxidant properties. These assays provide insights into the compound's physiological effects (Kherlopian et al. 2008).

Imaging Techniques Techniques like fluorescence microscopy or confocal microscopy can be used to visualize the distribution and localization of secondary metabolites within cells and tissues (Kherlopian et al. 2008).

16.3.1 Bioassays and Biological Screening

Bioassays are critical tools for the biological screening of secondary metabolites, helping to assess their effects on various living organisms and biological systems. These assays provide insights into the potential bioactivity and therapeutic applications of secondary metabolites. Some common bioassays used for the biological screening of secondary metabolites are shown in Fig. 16.1:

16.3.1.1 Antioxidant Activity Assays

- *DPPH (1,1-diphenyl-2-picrylhydrazyl) assay* measures the ability of secondary metabolites to scavenge free radicals and evaluate their antioxidant potential.
- *ABTS (2,2′-azino-bis (3-ethylbenzothiazoline-6-sulfonic acid)) assay* measures the ability of compounds to quench the ABTS radical cation, another method to assess antioxidant activity (Moon and Shibamoto 2009).

16.3.1.2 Anti-inflammatory Assays

- *Cytokine and inflammatory mediator inhibition assays* measure the ability of secondary metabolites to reduce the production of pro-inflammatory cytokines and mediators in cell or tissue cultures.
- *Cyclooxygenase (COX) and lipoxygenase (LOX) inhibition assays* evaluate the potential of compounds to inhibit enzymes involved in the inflammatory pathway (Ghosh et al. 2008).

16.3.1.3 Antimicrobial Assays

- *Agar well diffusion assay* determines the inhibitory activity of secondary metabolites against bacteria or fungi by measuring the diameter of the inhibition zone around the well.

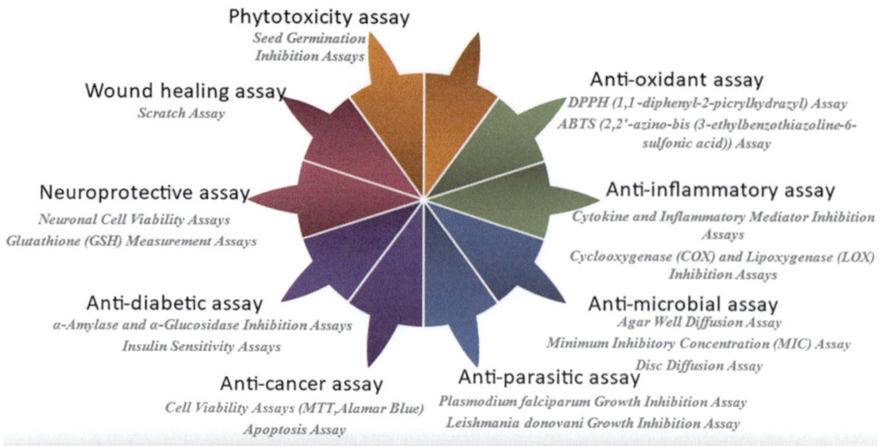

Fig. 16.1 General bioassays used for the biological screening of secondary metabolites

- *Minimum inhibitory concentration (MIC) assay* measures the lowest concentration of a compound required to inhibit the visible growth of microorganisms (Usman and Osuji 2007).
- *Disc diffusion assay*, similar to agar well diffusion, assesses the antimicrobial activity of compounds by measuring the zone of inhibition around a paper disc soaked with the secondary metabolite (Ghosh et al. 2008).

16.3.1.4 Antiparasitic Assays

- *Plasmodium falciparum growth inhibition assay* evaluates the antimalarial potential of secondary metabolites.
- *Leishmania donovani growth inhibition assay* measures the ability of compounds to inhibit the growth of *Leishmania* parasites (Jimoh et al. 2020).

16.3.1.5 Anticancer Assays

- *Cell viability assays (MTT, Alamar Blue)* assess the effect of secondary metabolites on the viability and proliferation of cancer cells.
- *Apoptosis assay* measures the induction of apoptosis or programmed cell death in cancer cells by secondary metabolites (McCauley et al. 2013).

16.3.1.6 Antidiabetic Assays

- *α-amylase and α-glucosidase inhibition assays* evaluate the potential of compounds to inhibit enzymes involved in carbohydrate digestion and postprandial glucose control.
- *Insulin sensitivity assays* assess the effect of compounds on insulin signaling pathways in cell cultures (Phoboo et al. 2015).

16.3.1.7 Neuroprotective Assays

- *Neuronal cell viability assays* evaluate the protective effects of secondary metabolites on neuronal cells exposed to oxidative stress or neurotoxic agents (Ying et al. 2000).
- *Glutathione (GSH) measurement assays* assess the impact of compounds on cellular GSH levels, an indicator of neuroprotection (Rahman et al. 2006).

16.3.1.8 Phytotoxicity Assays

- *Seed germination inhibition assays* determine the potential phytotoxicity of secondary metabolites on plant growth and germination (Kirk et al. 2002).

16.3.1.9 Wound Healing Assays

- *In vitro scratch assays* measure the ability of compounds to promote the migration and closure of cell monolayers, simulating wound healing (Maugeri et al. 2023).

These bioassays provide valuable insights into the pharmacological and biological activities of secondary metabolites, enabling researchers to assess their potential for various therapeutic applications. The choice of assay depends on the specific research objectives and the type of secondary metabolites being investigated.

16.4 Major Metabolic Disorders

16.4.1 Anticancer Activity

Despite major advancements in therapy, cancer continues to be one of the most serious health issues in the world. To create treatments that are more efficient and have fewer adverse effects, researchers are constantly looking for novel approaches. With compounds that may have anticancer effects that can suppress cancer cell development, trigger cancer cell death, and battle multidrug resistance in some malignancies, plant secondary metabolites have become a key topic of research in the fight against cancer. Flavonoids, such as 6,7,30-trimethoxy-3,5,40-trihydroxy-flavone and 5,40-dihydroxy-3,6,30-trihydroxy-flavone 7-O-d-glucoside isolated from *Chrysosplenium nudicaule*, spearmary, have been shown to have cytotoxic and antitumor activities in human leukemia and gastric cancer cell lines, suggesting their potential as anticancer agents. Leukemia cells have shown that agathisflavone can cause apoptosis and have an antiproliferative impact. Citrus flavonoids are shown to have a strong inhibitory effect on the growth of leukemia cells. By preventing cell growth, cell cycle arrest, and cell cycle termination, the flavonoid quercetin has demonstrated antiproliferative effects. Studies on the antiproliferative properties of kaempferol and quercetin (Maugeri et al. 2023) against a variety of cancer cell lines, including those from the human colon, breast, and ovaries, have been conducted. Human leukemic T cells and human gastric cancer cells have demonstrated that quercetin inhibits the G1 phase of the cell cycle. Quercetin's impact on cell growth and proliferation was biphasic in a human oral squamous carcinoma cell line (SCC-25). Consistent results from quercetin in vivo study point to the substance's potential as a chemopreventive agent for skin cancer. On the other hand, the human lung cancer cell line A549 has shown that kaempferol (Tu et al. 2016) reduces cell survival and DNA synthesis in a dose- and time-dependent manner. Additionally, it drastically decreased the viability of MCF-7 breast cancer cells.

Human oral, breast, colon, and prostate cancer cell lines have been demonstrated to be inhibited by strawberry polyphenols like anthocyanins, kaempferol, quercetin, and ellagic acid esters. The major polyphenol in green tea, epigallocatechin-3-gallate (EGCG), has exhibited anticancer benefits in a variety of cancer types. According to studies, EGCG may control many signaling pathways to increase the formation of reactive oxygen species and decrease angiogenesis in cancer cells. One of *Plumbago zeylanica*'s main ingredients, naphthoquinone, has shown in vitro and in vivo anticancer activity against a number of cancers, including breast, pancreatic, lung, prostate, melanoma, and leukemia. Hela cells and pituitary adenoma cells have been shown to proliferate less rapidly when exposed to cardanol, anacardic acid, and methyl cardol. Anacardic acid induces polymerase breakdown, cell cycle arrest, and apoptotic regulation. Using xenograft animal models, in vivo investigations have demonstrated the anticancer efficacy of plant-derived phenolic compounds against skin, colon, lung, breast, liver, prostate, stomach, esophagus, small intestine, pancreas, and mammary gland tumors. In a different study, the main anthocyanin in blackberries, cyanidin-3-glucoside (C3G), was tested for its ability to prevent the development of skin papillomas in an animal model that had been generated by the chemicals 7,12-dimethylbenzanthracene (DMBA) and 12-O-tetradecanoylphorbol-13-acetate (TPA). Similar to this, natural anthraquinones with anticancer effects include rhein and emodin. Through apoptosis and cell cycle arrest, tetrahydroanthraquinones, a subclass of anthraquinones, have been reported to limit cell proliferation, invasion, metastasis, and angiogenesis. Tetrahydroanthraquinone altersolanol A (a compound having anti-invasive and pro-apoptotic characteristics) has demonstrated anticancer activity against stomach, colon, and bladder cancer. Both in vitro and in vivo studies have shown that altersolanol A inhibits angiogenesis. Altersolanol N showed lethal effects against the murine cancer cell line L5178Y, while Altersolanol F decreased the viability of colorectal and cervical cancer cells (Solárová et al. 2020).

16.4.2 Antimicrobial Activity

16.4.2.1 Antibacterial Activity of Secondary Metabolites

On a global level, bacterial infections provide a serious public health concern. Antibiotic resistance has become a critical global issue, and the efficacy of numerous existing medications is in danger due to the multidrug-resistant bacteria's rising prevalence. The attention has shifted to plant-derived secondary metabolites, which are abundant sources of natural chemicals with the ability to meet this demand.

The antibacterial properties of gallic acid (Rattanata et al. 2016) and its derivatives have showed promise in the treatment of bacterial illnesses. Particularly, methyl gallate and gallic acid have shown noteworthy antibacterial efficacy against *Salmonella*. Additionally, it has been discovered that a number of phenolic substances, such as stilbenes, tannins, and isoflavones, limit the growth of the bacteria *Bacillus* and *E. coli*. *S. aureus* and *S. pyogenes* have been resistant to the antibacterial effects of anacardic acid. Alkaloids have also been used as the building blocks for the creation of additional antibiotics including linezolid and trimethoprim. For instance, *Shigella*, *Campylobacter jejuni*, and *E. coli* strains have been inhibited by the alkaloid cocsoline produced from *Epinetrum villosum*, demonstrating broad-spectrum antibacterial action. Broad-spectrum steroidal antibiotic squalamine, a polyamine alkaloid isolated from the tissue of the squalus shark (Acanthosis), has demonstrated potent bactericidal activities against both gram-positive and gram-negative bacterial strains. Numerous *Solanum dulcamara* L. alkaloids, including as solanine, solasodine, and B-solamarine, have been found to have antibacterial action against *S. aureus*. Additionally, methicillin-resistant *S. aureus* (MRSA) and *S. aureus* have been shown to be susceptible to the antibacterial effects of bis-indole alkaloids from marine invertebrates. In particular, against *S. pyogenes* and *S. aureus*, the Goldenseal alkaloids berberine and hydrastine have shown substantial antibacterial action. *Shigella strains*, *Campylobacter jejuni*, and *E. coli* are all susceptible to the antibacterial effects of the cocsoline alkaloid found in *Epinetrum villosum* (Exell). *Trichoderma quinone* has antibacterial activity against MRSA, while coniothranthraquinone 1 (Khamthong et al. 2012) has antibacterial activity against *S. aureus*. A number of bacteria, such as S. *aureus*, *E. coli*, *Pseudomonas aeruginosa*, *Sarcina ventriculi*, and *Bacillus subtilis*, are resistant to deoxybostrycin and bostrycin, two compounds with antibacterial activity (Sokmen et al. 1999).

16.4.2.2 Antifungal Activity of Secondary Metabolites

Antifungal drug resistance has recently emerged, increasing the morbidity and death linked to fungal infections. There is a persistent worry that antifungal medicines may also be damaging to host cells, since human and fungal molecular pathways are comparable. Patients with impaired immune systems, such as transplant recipients, cancer patients, and those with diabetes, may not respond well to conventional antifungal medicines and require new antifungal therapy, which is especially problematic for them. Additionally, many fungi have developed drug resistance, reducing the efficiency of the antifungal medications now in use. These adverse effects include nausea, vomiting, diarrhea, and itching.

Numerous alkaloids have proven to have strong antifungal properties. For instance, protoberberine (Volleková et al. 2003) alkaloid jatrorrhizine from *Mahonia aquifolium*'s had notable inhibitory effects. Another illustration is the bisbenzylisoquinoline alkaloid (+)-Cocsoline, which is produced from the *Epinetrum villosum* and has been shown to have antifungal properties. Alkaloids from *Corydalis longipes*, such as N-ethylhydrasteine hydroxylactam and 1-methoxyberberine chloride, have also demonstrated notable inhibitory effects. Dicentrine, glaucine, protopine, and alpha-allocryptopine, among other alkaloids discovered in *Glaucium oxylobum*

(Morteza-Semnani et al. 2003), demonstrated antifungal efficacy against a number of fungus, including *Microsporum gypseum*, *Microsporum canis*, *T. mentagrophytes*, and *Epidermophyton floccosum. Zanthoxylum chiloperone* var. angustifolium canthin-6-one and 5-methoxy-canthin-6-one have antifungal activity against *Candida albicans*, *Aspergillus fumigatus*, and *T. mentagrophytes*.

Two quinolinone alkaloids produced from *Melochia odorata*, frangulanine, a cyclic peptide alkaloid, and waltherione A, have shown antifungal action against a variety of pathogenic fungi. Additionally, anodic alkali aninolinate exhibits antifungal effects. In the root of Dictamnus dasycarpus, two fructoxin alkaloids with antimycotic properties were found. *Cleistopholis patens*' 3-Methoxisampangin significantly inhibited *C. albicans*, *A. fumigatus*, and *C. neoformans*. Against *Aspergillus* and *Candida* species, a novel alkaloid known as 2-(3,4-dimethyl-2,5-dihydro-1H-pyrrol-2-yl)-1-methylethyl pentanoate (Dabur et al. 2005), which was derived from the plant *Datura metel*, has shown in vitro and in vivo action.

16.5 Bioavailability and Pharmacokinetics

Several factors influence the bioavailability of secondary metabolites, affecting their absorption and overall pharmacological activity. These include chemical structure of secondary metabolites, molecular weight, polarity, and lipophilicity, which can influence their bioavailability, and the solubility of secondary metabolites in water and lipids significantly affects their absorption in the gastrointestinal tract and formulation of secondary metabolites into dosage forms, such as solid dispersions, liposomes (Valença et al. 2023), or nanoparticles; can enhance their bioavailability, the extent to which secondary metabolites are metabolized in the liver or other tissues; can influence their systemic availability, efflux transporters, such as P-glycoprotein (El-Readi et al. 2010); and can limit the absorption of secondary metabolites by actively pumping them out of enterocytes, and secondary metabolite absorption can be influenced by food components, as some compounds may require dietary fats or interact with specific nutrients, and the pH of the stomach and intestines can affect the solubility and stability of secondary metabolites, the extent to which secondary metabolites undergo first-pass metabolism in the liver which can significantly affect their bioavailability. The particle size of secondary metabolite formulations can impact their dissolution and absorption, and drug-drug interactions can affect the bioavailability of secondary metabolites, either by inhibiting or inducing metabolic enzymes. Understanding these factors and their impact on the bioavailability of secondary metabolites is crucial for optimizing their therapeutic use and ensuring consistent and predictable drug effects. These studies contribute to the development of safe and effective drug formulations (Shinyuy et al. 2023).

Pharmacokinetics is the study of how the body processes drugs, including the absorption, distribution, metabolism, and excretion (ADME) of these compounds. When it comes to secondary metabolites, there are unique challenges and considerations in pharmacokinetic studies.

16.6 Challenges in Pharmacokinetic Studies of Secondary Metabolites

There are several challenges when pharmacokinetic studies are undertaken for secondary metabolites as the following:

- *Complex Chemical Structures*: Secondary metabolites often have complex and diverse chemical structures. This complexity can make their isolation, quantification, and structural identification challenging.
- *Variable Bioavailability*: The bioavailability of secondary metabolites can vary widely. Factors such as chemical structure, solubility, and formulation play significant roles in determining whether a compound can be effectively absorbed into the bloodstream.
- *Low Concentrations*: Many secondary metabolites are found in nature in low concentrations, which can make it difficult to obtain sufficient quantities for pharmacokinetic studies and clinical trials.
- *Metabolism*: Secondary metabolites are often subject to extensive metabolism in the body. The metabolic pathways of these compounds can be complex, leading to the formation of various metabolites with potentially different pharmacological activities.
- *Tissue Distribution*: The distribution of secondary metabolites to target tissues and organs may not be uniform. Some compounds may have a preference for specific tissues, making it challenging to achieve therapeutic levels at the intended site of action.
- *Transporter Interactions*: Secondary metabolites may interact with transport proteins in the body, affecting their distribution and elimination. This can lead to drug-drug interactions or variations in pharmacokinetics.
- *Excretion*: The excretion of secondary metabolites can occur via multiple routes, including renal excretion, biliary excretion, and enterohepatic circulation. Understanding the elimination pathways is crucial for dose optimization.
- *Lack of Standardization*: The natural sources of secondary metabolites can vary in chemical composition due to factors like geographic location, climate, and soil conditions. This variability makes standardization of dosage forms challenging (Nascimento and Fett-Neto 2010).

16.7 Conclusion

Nowadays, with the immense of the exploitation of secondary metabolites like alkaloids, glycosides, triterpenoids, etc. day by day, it is the matter of extreme concern to isolate, standardize, and authenticate by various assays of secondary metabolites.

References

Abass S, Parveen R, Irfan M, Jan B, Husain SA, Ahmad S (2022) Synergy based extracts of medicinal plants: future antimicrobials to combat multidrug resistance. Curr Pharm Biotechnol 23:1–14. https://doi.org/10.2174/1389201023666220126115656

Abass S, Dar MI, Zahiruddin S, Beg MA, Nagar P, Jan B, Husain SA, Ahmad S (2023) Phytochemical and antibacterial analysis of Pistacia integerrima: an integrated in vitro and in silico approach. Process Biochem 132:236–247

Agati G, Brunetti C, Di Ferdinando M, Ferrini F, Pollastri S, Tattini M (2013) Functional roles of flavonoids in photoprotection: new evidence, lessons from the past. Plant Physiol Biochem 72:35–45

Armijos C, Ramírez J, Salinas M, Vidari G, Suárez AI (2021) Pharmacology and phytochemistry of Ecuadorian medicinal plants: an update and perspectives. Pharmaceuticals 14(11):1145. https://doi.org/10.3390/ph14111145

Cortes HJ (2020) Multidimensional chromatography: techniques and applications. CRC Press, Boca Raton

Dabur R, Chhillar AK, Yadav V, Kamal PK, Gupta J, Sharma GL (2005) In vitro antifungal activity of 2-(3,4-dimethyl-2,5-dihydro-1H-pyrrol-2-yl)-1-methylethyl pentanoate, a dihydropyrrole derivative. J Med Microbiol 54(6):549–552

Diez-Simon C, Mumm R, Hall RD (2019) Mass spectrometry-based metabolomics of volatiles as a new tool for understanding aroma and flavour chemistry in processed food products. Metabolomics 15:1–20

Divekar PA, Narayana S, Divekar BA, Kumar R, Gadratagi BG, Ray A, Singh AK, Rani V, Singh V, Singh AK, Kumar A, Singh RP, Meena RS, Behera TK (2022) Plant secondary metabolites as defense tools against herbivores for sustainable crop protection. Int J Mol Sci 23(5):2690. https://doi.org/10.3390/ijms23052690

El-Readi MZ, Hamdan D, Farrag N, El-Shazly A, Wink M (2010) Inhibition of P-glycoprotein activity by limonin and other secondary metabolites from Citrus species in human colon and leukaemia cell lines. Eur J Pharmacol 626(2–3):139–145

Erb M, Kliebenstein DJ (2020) Plant secondary metabolites as defenses, regulators, and primary metabolites: the blurred functional trichotomy1. Plant Physiol. https://doi.org/10.1104/PP.20.00433

Fistarol SK, Itin PH (2010) Disorders of pigmentation. JDDG 8(3):187–202

Fried R (1975) Enzymatic and non-enzymatic assay of superoxide dismutase. Biochimie 57(5):657–660

Ghosh S, Subudhi E, Nayak S (2008) Antimicrobial assay of Stevia rebaudiana Bertoni leaf extracts against 10 pathogens. Int J Integr Biol 2(1):27–31

Hussain M, Debnath B, Qasim M, Bamisile BS, Islam W, Hameed MS, Wang L, Qiu D (2019) Role of saponins in plant defense against specialist herbivores. Molecules 24(11):2067

Jimoh MA, Idris OA, Jimoh MO (2020) Cytotoxicity, phytochemical, antiparasitic screening, and antioxidant activities of Mucuna pruriens (Fabaceae). Plants 9(9):1249

Jusuf H, Elveny M, Azizova F, Shichiyakh RA, Kulikov D, Al-Taee MM, Atiyah KK, Jalil AT, Aravindhan S (2022) Plant secondary metabolites as defenses: a review. J Water Land Dev. https://doi.org/10.24425/jwld.2022.142323

Khamthong N, Rukachaisirikul V, Tadpetch K, Kaewpet M, Phongpaichit S, Preedanon S, Sakayaroj J (2012) Tetrahydroanthraquinone and xanthone derivatives from the marine-derived fungus Trichoderma aureoviride PSU-F95. Arch Pharm Res 35:461–468

Kherlopian AR, Song T, Duan Q, Neimark MA, Po MJ, Gohagan JK, Laine AF (2008) A review of imaging techniques for systems biology. BMC Syst Biol 2(1):1–18

Kirk JL, Klirnomos JN, Lee H, Trevors JT (2002) Phytotoxicity assay to assess plant species for phytoremediation of petroleum-contaminated soil. Biorem J 6(1):57–63

Kren V, Martínková L (2001) Glycosides in medicine: "the role of glycosidic residue in biological activity". Curr Med Chem 8(11):1303–1328

Kumar S, Abedin MM, Singh AK, Das S (2020) Role of phenolic compounds in plant-defensive mechanisms. Plant Phenol Sustainable Agric 1:517–532
Ludwiczuk A, Skalicka-Woźniak K, Georgiev MI (2017) Terpenoids. In: Pharmacognosy. Elsevier, Amsterdam, pp 233–266
Ma D, Wang S, Shi Y, Ni S, Tang M, Xu A (2021) The development of traditional Chinese medicine. J Trad Chin Med Sci. https://doi.org/10.1016/j.jtcms.2021.11.002
Mann M, Hendrickson RC, Pandey A (2001) Analysis of proteins and proteomes by mass spectrometry. Annu Rev Biochem 70(1):437–473
Maugeri A, Calderaro A, Patanè GT, Navarra M, Barreca D, Cirmi S, Felice MR (2023) Targets involved in the anti-cancer activity of quercetin in breast, colorectal and liver neoplasms. Int J Mol Sci 24(3):2952
Mazid M, Khan TA, Mohammad F (2011) Role of secondary metabolites in defense mechanisms of plants. Biol Med 3:232–249
McCauley J, Zivanovic A, Skropeta D (2013) Bioassays for anticancer activities. In: Metabolomics tools for natural product discovery: methods and protocols. Springer, Cham, pp 191–205
Meier H (1992) The photochemistry of stilbenoid compounds and their role in materials technology. Angew Chem Int Ed 31(11):1399–1420
Moon J-K, Shibamoto T (2009) Antioxidant assays for plant and food components. J Agric Food Chem 57(5):1655–1666
Morteza-Semnani K, Amin G, Shidfar MR, Hadizadeh H, Shafiee A (2003) Antifungal activity of the methanolic extract and alkaloids of Glaucium oxylobum. Fitoterapia 74(5):493–496
Nascimento NC, Fett-Neto AG (2010) Plant secondary metabolism and challenges in modifying its operation: an overview. In: Plant secondary metabolism engineering: methods and applications. Springer, Cham, pp 1–13
Pankewitz F, Hilker M (2008) Polyketides in insects: ecological role of these widespread chemicals and evolutionary aspects of their biogenesis. Biol Rev 83(2):209–226
Phoboo S, Shetty K, ElObeid T (2015) In vitro assays of anti-diabetic and anti-hypertensive potential of some traditional edible plants of Qatar. J Med Active Plants 4(3):22–29
Rahman I, Kode A, Biswas SK (2006) Assay for quantitative determination of glutathione and glutathione disulfide levels using enzymatic recycling method. Nat Protoc 1(6):3159–3165
Rasool A, Bhat KM, Sheikh AA, Jan A, Hassan S (2020) Medicinal plants: role, distribution and future. J Pharmacogn Phytochem 9:2111–2114
Rattanata N, Klaynongsruang S, Leelayuwat C, Limpaiboon T, Lulitanond A, Boonsiri P, Chio-Srichan S, Soontaranon S, Rugmai S, Daduang J (2016) Gallic acid conjugated with gold nanoparticles: antibacterial activity and mechanism of action on foodborne pathogens. Int J Nanomedicine 11:3347–3356
Roopashree KM, Naik D (2019) Advanced method of secondary metabolite extraction and quality analysis. J Pharmacogn Phytochem 8(3):1829–1842
Saarinen NM, Wärri A, Airio M, Smeds A, Mäkelä S (2007) Role of dietary lignans in the reduction of breast cancer risk. Mol Nutr Food Res 51(7):857–866
Samal J, Dehury RK (2019) Utilization, preference, perception and characteristics of people adopting traditional and AYUSH systems of medicine in India: a systematic review. J Complementary Integr Med. https://doi.org/10.1515/jcim-2018-0020
Samal M, Abass S, Parveen R, Ahmad S, Iqbal M (2023) Effect of abiotic stress on production of secondary metabolites in plants. Plants Med Aromat. https://doi.org/10.1201/9781003226925-11
Shi Y, Zhang C, Li X (2021) Traditional medicine in India. J Trad Chin Med Sci. https://doi.org/10.1016/j.jtcms.2020.06.007
Shinyuy LM, Loe GE, Jansen O, Mamede L, Ledoux A, Noukimi SF, Abenwie SN, Ghogomu SM, Souopgui J, Robert A (2023) Secondary metabolites isolated from Artemisia afra and Artemisia annua and their anti-malarial, anti-inflammatory and immunomodulating properties—pharmacokinetics and pharmacodynamics: a review. Metabolites 13(5):613
Sokmen A, Jones BM, Erturk M (1999) The in vitro antibacterial activity of Turkish medicinal plants. J Ethnopharmacol 67(1):79–86

Solárová Z, Liskova A, Samec M, Kubatka P, Büsselberg D, Solár P (2020) Anticancer potential of lichens' secondary metabolites. Biomolecules 10(1):87

Tu L-Y, Pi J, Jin H, Cai J-Y, Deng S-P (2016) Synthesis, characterization and anticancer activity of kaempferol-zinc (II) complex. Bioorg Med Chem Lett 26(11):2730–2734

Udeani TK, Ugwu LO, Nnadi CO, Okwuosa CN (2021) Evaluation of anti-microbial activity and proximate composition of alkaloids from vitex doniana seed. Dhaka Univ J Pharm Sci. https://doi.org/10.3329/dujps.v20i1.54035

Usman H, Osuji JC (2007) Phytochemical and in vitro antimicrobial assay of the leaf extract of Newbouldia laevis. Afr J Tradit Complement Altern Med 4(4):476–480

Valença CAS, Barbosa AAT, Dolabella SS, Severino P, Matos C, Krambeck K, Souto EB, Jain S (2023) Antimicrobial bacterial metabolites: properties, applications and loading in liposomes for site-specific delivery. Curr Pharm Des 29(28):2191–2203

Variyar PS, Banerjee A, Akkarakaran JJ, Suprasanna P (2014) Role of glucosinolates in plant stress tolerance. In: Emerging technologies and management of crop stress tolerance. Elsevier, Amsterdam, pp 271–291

Vazhappilly CG, Ansari SA, Al-Jaleeli R, Al-Azawi AM, Ramadan WS, Menon V, Hodeify R, Siddiqui SS, Merheb M, Matar R (2019) Role of flavonoids in thrombotic, cardiovascular, and inflammatory diseases. Inflammopharmacology 27:863–869

Volleková A, Košt'álová D, Kettmann V, Tóth J (2003) Antifungal activity of Mahonia aquifolium extract and its major protoberberine alkaloids. Phytother Res 17(7):834–837

Waller GR, Nowacki EK, Waller GR, Nowacki EK (1978) The role of alkaloids in plants. In: Alkaloid biology and metabolism in plants. Springer, Cham, pp 143–181

Weragoda PB (1980) The traditional system of medicine in Sri Lanka. J Ethnopharmacol. https://doi.org/10.1016/0378-8741(80)90033-1

Yeshi K, Crayn D, Ritmejerytė E, Wangchuk P (2022) Plant secondary metabolites produced in response to abiotic stresses has potential application in pharmaceutical product development. Molecules. https://doi.org/10.3390/molecules27010313

Ying HS, Gottron FJ, Choi DW (2000) Assessment of cell viability in primary neuronal cultures. Curr Protoc Neurosci 13(1):7–18

Integrative Approaches for Enhanced Secondary Metabolite Production

17

Pradeep Kumar Shukla, Revanth Kishore Pulamolu, and Pragati Shukla Nee Misra

Abstract

The medicinal properties of plants stem from specific compounds known as plant secondary metabolites, which possess significant potential in treating various human ailments and diseases. However, these metabolites are typically produced in small quantities naturally as a response to different stresses. To address this limitation, researchers explore ways to enhance the production of these valuable metabolites. Biotechnological methods offer numerous advantages, such as consistent, stable, and year-round sustainable production, scalability, and simplified extraction and purification processes. Techniques like hairy root, callus, and suspension cultures are reliable approaches for producing these bioactive compounds in controlled environments. Additionally, ongoing research explores genetic engineering, metabolic engineering, and biotransformation as promising methods to boost biomass and biochemicals. Notably, recent studies highlight the use of elicitors from both biotic and abiotic sources, with nanoparticle elicitation garnering significant attention among researchers. This chapter aims to summarize the latest advancements in the in vitro production of high-value plant secondary metabolites with medicinal significance.

P. K. Shukla
Faculty of Science, Department of Biological Sciences, Sam Higginbottom University of Agriculture Technology and Sciences, Prayagraj, UP, India
e-mail: pradeep.shukla@shiats.edu.in

R. K. Pulamolu
Naini Institute of Agriculture, Sam Higginbottom University of Agriculture Technology and Sciences, Prayagraj, UP, India

P. S. N. Misra (✉)
Centre for Tissue Culture Technology, Jacob Institute of Biotechnology and Bioengineering, Sam Higginbottom University of Agriculture Technology and Sciences, Prayagraj, UP, India
e-mail: pragati.misra@shiats.edu.in

© The Author(s), under exclusive license to Springer Nature Singapore Pte Ltd. 2024

M. Nandave et al. (eds.), *Ethnopharmacology and OMICS Advances in Medicinal Plants Volume 1*, https://doi.org/10.1007/978-981-97-2367-6_17

Keywords

Medicinal plants · Genetic engineering · Biotransformation · Metabolic engineering · Elicitors

17.1 Introduction

Medicinal plants (MPs) serve as an essential source of pharmaceutical drugs, upon which a significant portion of the global population depends on maintaining their health. Many medicinal plants especially endemic species have disappeared, whereas some remain in few numbers and are under risk of extinction (Chupakhin et al. 2020). According to a study released by the Royal Botanic Gardens, Kew, in 2020, about 723 MPs are on the verge of extinction, owing in part to over-collection. Cultivating plant cells and tissues to yield commercially valuable components of pharmaceutical interest has gained substantial attention in recent years (Chen et al. 2016). These plants, long used as remedies for various human diseases, present a rich repository of biologically active compounds, including antimicrobial agents. They stand as vital bio-resources, offering a wide array of traditional medicinal systems, modern medicines, nutraceuticals, food supplements, folk medicines, pharmaceuticals, intermediates, and chemicals crucial for synthetic drugs (Hammer et al. 1999).

Herbs, categorized as plants or plant parts utilized in treating a range of physiological diseases, possess diverse phytochemical and remedial properties (Ahmad et al. 2013). Phytochemicals, characterized as active ingredients exhibiting therapeutic properties akin to medicines or drugs, can be categorized based on their chemical compositions (Arvind 2016). These medicinal plants act as fundamental sources for effective and affordable healthcare solutions for people. Yet, despite their critical roles in benefiting human health, phytochemicals present in these plants cannot be synthesized within the human body (Martinez et al. 2008).

Across various regions, local physicians have historically employed diverse herbal remedies as therapeutic agents (Roustakhiz and Saboki 2016). Dating back to ancient times, people have continuously explored nature, especially plants, in the quest for new drugs, leading to the utilization of numerous medicinal and aromatic plants (MAPs) renowned for their healing properties in treating a spectrum of diseases. Surprisingly, over 80% of the world's population continues to rely on plants or compounds derived from plants to fulfill their primary healthcare needs, emphasizing the sustained significance of MAPs (Laloo et al. 2021). Notably, more than a quarter of drugs prescribed today comprise at least one plant-based compound (Sin et al. 2021), reinforcing the enduring importance of MAPs in healthcare.

17.1.1 Primary Metabolites

Plants engage in synthesizing sugars, amino acids, nucleotides, and fats, constituting primary plant metabolites produced as part of their regular metabolic functions.

These primary metabolites are pervasive across all plants—regardless of species, genera, or families—and play a pivotal role in their sustenance. These uncomplicated molecules serve as building blocks to form crucial polymers vital for plant life. This aspect of plant biochemistry distinguishes the production of primary metabolites from the more intricate molecules, formed through diverse pathways, such as secondary metabolites (Hedin 1983; Bruneton 1999).

17.1.2 Secondary Metabolites

Secondary metabolites, in contrast to primary ones, are synthesized in specific cell types and during distinct developmental stages, rendering their extraction and purification more challenging compared to primary metabolites. These chemical constituents exhibit immense diversity (Bell and Charlwood 1980). Despite the presence of primary metabolites like nucleic acids, proteins, fats, and carbohydrates, secondary metabolites are produced in significantly smaller quantities (Pagare et al. 2015).

Initially perceived as waste materials devoid of function, it is now understood that secondary metabolites play crucial roles in plants, including adaptation, defense, survival, and sustaining their production (Yang et al. 2018; Ramawat and Goyal 2020).

17.1.3 Classification of Secondary Metabolites

Secondary metabolites can be classified into various categories such as terpenoids, phenolics, nitrogenous compounds, and sulfur-containing compounds. These compounds are produced through various metabolic pathways beginning with photosynthesis. Shikimic acid is the initiating source for phenols, tannins, and aromatic alkaloids; acetate–malonate for phenols and alkaloids; pentose for glycosides, polysaccharides, and mevalonic acid; and methyl erythrose phosphate for terpenes, alkaloids, and steroids (Trenchard et al. 2015).

17.1.3.1 Phenolics

Phenolics, an extensive category of phytochemicals, encompass phenolic acids and flavonoids (Castañeda-Loaiza et al. 2020). These compounds exhibit potent antioxidant properties along with anti-inflammatory, anti-obesity, and anticancer activities (Bernal et al. 2011; Ksouri et al. 2012; Khoddami et al. 2013; Złotek et al. 2014; Castañeda-Loaiza et al. 2020).

17.1.3.2 Alkaloids

Alkaloids possess diverse pharmacological properties, including antimalarial (quinine), antiasthma (ephedrine), anticancer (homoharringtonine), and analgesic (morphine) effects (Dixit et al. 2021). Furthermore, there exists a class of sulfur-containing secondary metabolites, such as alliin and defensins, directly or indirectly involved in plant defense mechanisms against microbial diseases. Alkaloids are primarily utilized as drugs (e.g., the source of morphine in poppy plants), terpenoids for their aromatic qualities (e.g., the source of menthol in peppermint plants), and phenolic substances for their antioxidant properties (e.g., the source of rosmarinic acid in rosemary plants) (Baydar 2005).

17.1.3.3 Lignin and Flavonoids

Lignin and flavonoids demonstrate antioxidant and antimicrobial characteristics, displaying potential applications in pharmaceuticals and biomedical fields. Their biocompatibility, UV light absorption capacity, antioxidant, antimicrobial activities, and ability to enhance the mechanical strength of processed biomaterials render them promising for pharmaceutical and biomedical applications. Research indicates the potential utilization of lignin and flavonoids in developing biomedical hydrogels, drug delivery vehicles, or biocomposites for tissue engineering and wound healing (Iiyama and Wallis 1990; Domínguez-Robles et al. 2019, 2020).

17.1.4 Medicinal Properties of Secondary Metabolites

Secondary metabolites (SMs) play a pivotal role not only in the defense signaling pathways of plants but also exhibit various biological activities encompassing antioxidant, anticancer, antimicrobial, antigenotoxic, antihyperlipidemic, antidepressant, anxiolytic, anticonvulsant, anticataractogenic, antinociceptive, anti-inflammatory, antidiabetic, antitussive, aphrodisiac, antiplatelet, antivenin, hypotensive, and more (Singh et al. 2018).

Hyoscyamus niger, classified as an annual or biennial herbaceous plant from the Solanaceae family, holds significant importance in the pharmaceutical industry owing to its rich tropane alkaloids (TAs) content like hyoscyamine and scopolamine, which stand among the oldest medicinal drugs (Dehghan et al. 2012; Pudersell et al. 2012). These compounds have been traditionally used to alleviate various ailments such as stomachaches, earaches, sinusitis, ulcers, kidney and liver conditions, and pain relief and as antispasmodics (John et al. 2010). *H. niger*, due to its TAs, influences the parasympathetic nervous system and finds application in treating ocular, anesthetic, cardiac, and digestive system-related diseases (John et al. 2010). Scopolamine, a component of *H. niger*, has shown beneficial effects in managing acute vertigo and reducing gastric acid secretion during sleep among cancer patients undergoing chemotherapy-induced nausea and ulcers (Clissold and Heel 1985).

Moringa species, belonging to the Moringaceae family and originating from Asia, have extensive use in South Asian folk medicine for addressing various health issues. Each part of the *Moringa* tree serves medicinal purposes, displaying efficacy in lowering blood pressure and lipids, regulating blood sugar, managing body weight, supporting liver health, and exhibiting antioxidant properties (Aekthammarat et al. 2019; Liang et al. 2019; Watanabe et al. 2021).

Lavandula angustifolia Mill, a well-recognized member of the Lamiaceae family, hails from the Mediterranean region and has spread to various parts across the globe. The plant's significance lies in its capacity to produce essential oils extensively used in perfume, cosmetic, flavoring, and pharmaceutical industries (Gonçalves and Romano 2013; Demasi et al. 2018).

Fagonia indica, commonly known as "Sacchi boti," belongs to the Zygophyllaceae family and typically thrives in sandy warm lands present in regions such as Pakistan, India, Afghanistan, Egypt, the UAE, and Africa (Beier 2005). It holds medicinal importance in traditional therapies for treating digestive problems, fever, and urinary illnesses and is renowned for its liquid potion used in the preliminary stages of tumors, particularly breast cancer (Farheen et al. 2015). Furthermore, *F. indica* finds application as a tonic in liver ailments, dropsy, and smallpox. Its diverse uses encompass sedative, anti-malignant, antimicrobial, anti-diabetic, anti-inflammatory, antioxidant, antipyretic, and hepato-protective properties (Alqasoumi et al. 2011; Lan et al. 2020; Saleem et al. 2014).

A wide array of chemical compounds derived from plants have demonstrated strong potential as antiviral agents (Denaro et al. 2020). Ongoing research, such as trials involving traditional Chinese medicine using extracts from *Forsythiae fructus*, is part of the global efforts to find effective treatments for COVID-19 (Maxmen 2020). These medicinal plants, rooted in traditional herbal medicine, could offer multiple benefits, from easing symptoms in COVID-19 patients to serving as a basis for developing powerful antiviral medications. These compounds, notably secondary metabolites, hold immense value in the pharmaceutical industry due to their efficacy in regulating human metabolic processes, addressing diseases through a multi-target approach, and their popularity owing to their safety and minimized side effects (Niazian et al. 2019; Marchev et al. 2020). These reasons contribute to the escalating demand for natural bioactive compounds.

17.2 Enhancement of Secondary Metabolites in Medicinal Plants

Naturally, plants generate a low quantity of secondary compounds in their environment, making downstream processing challenging due to significant variability (Khan et al. 2015). Enhancement of secondary metabolites in medicinal plants is important because, even after selecting a high yielding variety, the products are always not sufficient to meet the commercial needs. After continuous cultivation, the cultivars lose their vigor. Hence, many alternative techniques are employed to improve the content of SMs and obtain efficient yields including conventional and modern strategies (Khani et al. 2012; Isah et al. 2018).

Over recent decades, research has pointed out that various environmental stressors such as drought, extreme temperatures, light exposure, nutrient deficiencies, and soil contamination influence plant physiology in field conditions, stimulating the production of secondary metabolites in both crops and medicinal plants (Dixon and Paiva 1995; Rogerio et al. 2010; Shiringani and Shimelis 2011; Ramakrishna and Ravishankar 2011; Wink 2015; Ghosh et al. 2018; El-Hendawy et al. 2019a, b).

Nitrogen, phosphate, potassium, and sulfur deficiencies have also been linked to inducing stresses that affect the synthesis of specific plant compounds in field conditions (Chalker-Scott and Fnchigami 1989; Dixon and Paiva 1995; Chishaki and Horiguchi 1997; Ramakrishna and Ravishankar 2011; Hassan 2012). Studies have shown that exposure to drought stress during field growth has led to increased production of various secondary metabolites, including terpenes, complex phenols, and alkaloids, as a result of ionic or osmotic stress (de Matos et al. 2014; Niinemets 2015; Quan et al. 2016; Afzal et al. 2017; Piasecka et al. 2017). Oxidative stress imposed by metals also activates a signaling pathway that results in the synthesis of particular secondary metabolites in plants (Anjitha et al. 2021; Yadav et al. 2022).

However, a significant challenge faced in enhancing secondary metabolites in vivo is the variability in yield across seasons and regions for a specific plant species, cultivar, or genotype. This variation often reflects in vitro cultures when attempting to mimic these conditions through media manipulation and cultural adjustments, forming the basis for many in vitro strategies aimed at enhancing secondary metabolites.

The vast medical applications of these secondary metabolites illustrate their critical importance in the pharmaceutical market. In overcoming environmental constraints and productivity challenges, in vitro culture methods (cell, callus, shoot) emerge as the optimal alternative, ensuring a consistent and steady supply of active plant ingredients (Matkowski 2008).

In vitro culture systems operate independently of environmental factors, geographical limitations, and seasonal constraints (Chandran et al. 2020), offering flexibility regardless of the plant's developmental stage (Li et al. 2020). These systems have the capability to produce plants with varying chemical profiles compared to intact plants.

Various culture systems like callus, shoot, roots, hairy roots, and suspension cell cultures find application in the in vitro production of bioactive compounds from medicinal, aromatic, and ornamental plants. Determining the optimal in vitro culture system is crucial to achieving large-scale commercial production of targeted bioactive compounds (Sharma and Agrawal 2018). Furthermore, the ease of extracting metabolites from in vitro-produced tissues adds to the attractiveness of this method for commercial applications (Pavlov et al. 2005).

Elicitation techniques have been widely adopted to enhance the production of essential bioactive compounds within specific parts of plant cells, especially in medicinal plants (Maucieri et al. 2019). When exposed to elicitors, plants trigger the synthesis of bioactive chemicals as a natural defense mechanism. Employing an elicitation strategy ensures increased productivity for industrial-scale production of phytochemicals by initiating the plant's defense response through manipulation of physiological cascades (Zhao et al. 2005; Yang and Stöckigt 2010; Yue et al. 2016).

By practicing precursor feeding, elicitation, and changes in the culture medium conditions, it was seen accumulation of various medicinally useful compounds in both solid and liquid cell, callus, and organ cultures of MAOPs (Mohsen and Sabbatini 2021).

Nanotechnology is a promising upcoming technology, and the involvement of nanoparticles (NPs) as elicitors shows enhancement in SMs in vitro (Khan et al. 2014).

The various modern-day techniques used in the enhancement of medicinally useful secondary metabolites through in vitro techniques are discussed below.

17.2.1 Genetic Engineering

The prerequisite for elevating metabolite production or modifying synthesis lies in identifying and characterizing genes responsible for metabolite synthesis. Such advancements allow for the enhancement of production or transfer of pathways to different organisms (Pickens et al. 2011). Technological breakthroughs in sequencing now enable profiling of numerous genes simultaneously (Sun et al. 2022). Generally, the aim is to increase the level of production of compounds of our interest in MAOPs or to transfer the pathway to other organisms (Satish et al. 2019).

Efforts have been made to decipher genetic information in medicinal and aromatic plants (MAPs). For example, in saffron (*Crocus sativus*), Rubio et al. (2008) isolated four carotenoid cleavage dioxygenase (CCD) genes. Their expression pattern highlighted *Crocus sativus* carotenoid cleavage dioxygenase 1a gene's (CsCCD1a) constant expression, while CsCCD1b was predominantly expressed in stigma tissue. Additionally, cooperative expression of CsCCD4a and CsCCD4b throughout stigma formation resulted in elevated carotene and ionone release. Ahrazem et al. (2010) analyzed CCD4 genomic DNA regions in saffron, identifying the CCD4a promoter sequence as instrumental in initiating GUS expression, specifically in saffron pollen.

Genetic transformation refers to the process involving the insertion and expression of a foreign gene (gene of interest) into the host plant genome (da Silva et al. 2011). The *Agrobacterium*-mediated gene transfer process involves selecting suitable host tissue, acquiring the gene of interest, preparing the strain, infection, selecting transformants, regeneration, and identifying transgenic plants (Chen et al. 2018).

In a specific instance, *A. tumefaciens* strain LBA4404, using a plasmid vector pG35SKNAT1, was utilized to infect *D. lasianthera* protocorms. The experiments indicated that transformation efficiency peaked at an OD600 of 0.6 (bacterial density), 100 mM acetosyringone, with a co-cultivation period of 5 days and a 30-min infection period. PCR analysis confirmed successful transformation by showing KNAT1 gene integration and expression in the *D. lasianthera* genome, promoting multiple shoot formation (Utami et al. 2018). Another study involved *A. tumefaciens*-mediated transformation in *D. catenatum*. Various strains of *A. tumefaciens* were tested for their efficiency in transforming *D. catenatum* protocorms, and the effects of surfactants (Tween-20, Silwet-77, and Triton X-100) and ultrasound were examined. It was found that strain GV3101 at OD600 of 0.6 resulted in the highest transformation. Additionally, the addition of 0.001% Triton X-100 or pretreatment of protocorms with ultrasound significantly enhanced the transformation frequency (Zheng et al. 2012).

As of now, no reports exist on enhancing secondary metabolites through genetic transformation. However, these efficient protocols could potentially be utilized to introduce genes for increased biomass and enhanced secondary metabolite production in various medicinal plants (Saikat et al. 2021).

17.2.2 Biotransformation

Biotransformations refer to the structural alterations of a chemical compound caused by organisms or enzyme systems, resulting in the creation of molecules with relatively higher polarity (Asha and Vidyavathi 2009; Pervaiz et al. 2013). This adaptive mechanism has evolved in microbes to adapt to changes in their environment and holds significant utility in various biotechnological processes. A diverse array of living cells, including bacteria, filamentous fungi, animals, plants, algae, yeast, and actinomycetes, serve as these biological systems (Smitha et al. 2017). Cultivated cells possess the ability to transform supplied compounds into valuable ones, even if those compounds aren't natural intermediates of plant metabolism. This transformation involves various reactions like hydroxylation, oxidation of hydroxyl groups, carbonyl group reduction, hydrogenation of carbon-carbon double bonds, glycosyl conjugation, and hydrolysis, catalyzed by plant enzymes (Murthy et al. 2014).

Biotransformation studies have been extensively used to explore how substances are metabolized in animal models (Kebamo et al. 2015). These studies involving the transformation of substances have been conducted across various biological systems, including fungi, bacteria, plant and animal cell cultures, as well as live animals (Sultana and Saify 2013). The microbial biotransformation process is commonly employed to compare the metabolic pathways of drugs and to increase the quantity of discovered substances in animal models, enabling further assessment of their pharmacological and toxicological properties (Huttel and Hoffmeister 2010). In addition, white biotechnology refers to the utilization of microbial biotransformation to produce specific products of interest (Coelho and Ribeiro 2015). For instance, from the leaves of *Stevia rebaudiana*, the sweet glycoside stevioside yields isosteviol upon hydrolysis. *Aspergillus niger* biotransformed isosteviol into derivatives such as B-OH derivative, ent-7-alpha-hydroxy-16-Ketobeyeran-19-oic, 1-alpha,7-beta-dioxy derivatives, and 1-beta7-alpha-dihydroxy-16-ketobeyenan-19-oic acid. Monoterpenes like pulegone were transformed into iromenthone and (-)-menthone into (+) neomenthol in mentha cell lines. Additionally, *Eucalyptus citriodora*, *Azadirachta indica*, and *Capsicum annum* cell cultures incubated with paclitaxel resulted in biotransformation into baccatin III, deacetyl baccatin III, and an unknown compound by *E. citriodora* cultures (Khirwadkar et al. 2014).

Fungal endophytes often influence the quantity and concentration of final metabolites found in plant tissues. These fungi can directly create or break down certain metabolites, thus impacting the overall metabolite composition of medicinal plants through a process called biotransformation. For instance, specific endophytic fungi like *Flavobacterium* sp. GE 32 and *Arthrinium* sp. GE 17–18 in *Panax* ginseng have the ability to convert ginsenoside Rb1, which has limited bioavailability, into ginsenoside Rg3 and C-K. These transformed products from ginseng roots are believed to offer several health benefits due to their higher bioavailability (Fu 2019; Fu et al. 2016).

Recent advancements in this field have successfully been applied not only in drug synthesis (Louis et al. 2019) but also in environmental applications like pollutant degradation and food processes such as wine brewing (Rho et al. 2020). These advancements present promising opportunities for eco-friendly and efficient solutions to various industrial challenges. While this method stands as one of the most commercially viable approaches, the cost of precursors can limit its economic feasibility in certain cases (Dicosmo and Misawa 1995).

17.2.3 Metabolic Engineering

Plant metabolic engineering as relevant to bioactive secondary metabolites aims at increasing the accumulation of a desired product and/or at reducing accumulation of an undesired metabolite (Farré et al. 2014). This involves leveraging the regulatory and genetic mechanisms within cells to enhance the production of specific compounds (Nielsen 2001). This technology primarily concentrates on modifying microbes or plants to achieve consistent and increased production of secondary metabolites on an industrial scale (Woolston et al. 2013). The approach of metabolic engineering also involves inhibiting competitive pathways to amplify the metabolic flow of intermediates within the targeted biosynthetic pathway, aiming for higher production through various strategies (Fig. 17.1). For instance, inhibiting certain steps in the biosynthetic pathway can lead to the accumulation of preceding intermediates. A prime example is the comprehension and manipulation of the phenyl propanoyl biosynthetic pathway, which is engaged in the synthesis of several plant secondary metabolites, representing a successful and recent application in this field (Nanda et al. 2016). The modification of the tropane alkaloid pattern in *Atropa belladonna* was the initial evidence demonstrating successful metabolic engineering in medicinal plants. This was achieved by boosting the expression of h6h, the gene responsible for hyoscyamine 6-ß hydroxylase, which facilitates the final stages in the creation of scopolamine (Yun et al. 1992).

To enhance the production of the anticancer alkaloid camptothecin in *Ophiorrhiza pumila*, a transgenic root line was developed by Shi et al. (2020), incorporating overexpressed geraniol-10-hydroxylase and secologanin synthase genes. While optimized culture conditions led to reported camptothecin production of up to 2 mg g^{-1} DW in stems and roots (Lee et al. 2020), the engineered hairy root line achieved a yield of 3.5 mg g^{-1} of the compound.

The biosynthesis of pharmaceutically essential terpenoid indole alkaloids (TIAs) in Madagascar periwinkle (*C. roseus*) involves a pathway encompassing more than 35 enzymatic steps in various cellular and subcellular compartments (Sharma et al. 2020). In a recent study, a modular biosynthetic production platform in *Escherichia coli* was constructed, achieving a production titer of 1.16 g L^{-1} psilocybin after

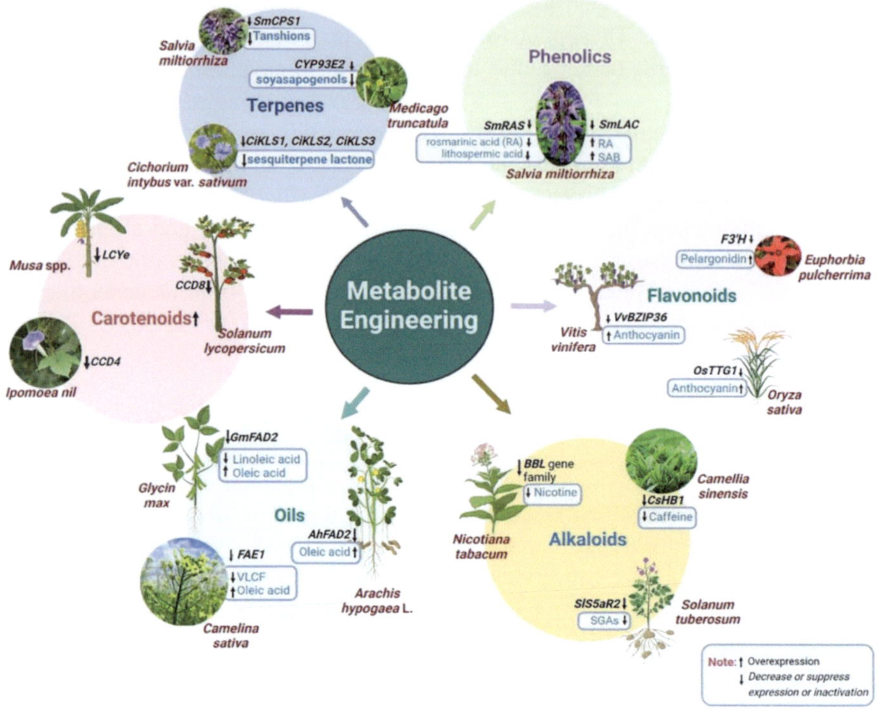

Fig. 17.1 Schematic representation of metabolite engineering in plants (Mipeshwaree Devi et al. 2023)

fermentation optimization through the bioconversion of precursors 4-hydroxyindole, serine, and methionine (Adams et al. 2019). Complete de novo production of psilocybin and related tryptamine derivatives in *Saccharomyces cerevisiae* was accomplished by Milne et al. (2020).

17.2.4 Hairy Root Culture

Hairy root cultures are promising for producing various important secondary metabolites and volatile organic compounds. The process initiates when *Agrobacterium rhizogenes* inserts its T-DNA into the *Ri* plasmid, genetically transforming the plant tissue. *A. rhizogenes*, a gram-negative bacterium similar to *Agrobacterium tumefaciens*, contains Ri plasmids, akin to Ti plasmids, responsible for its virulence (Tõnutare 2015; Kaptan et al. 2018). A notable advantage of the hairy root culture method is that it doesn't necessitate the supplementation of plant growth regulators in the media, as the inserted T-DNA carries genes for endogenous auxin production. Hairy roots, lacking geotropism, exhibit high branching and accelerated growth compared to normal roots. Besides typical root metabolites, they also yield compounds naturally secreted by the plant's aerial parts. Additionally, similar to other cell culture technologies, hairy roots demonstrate physiological and biochemical stability, fast growth, and spontaneous shoot generation and resemble wild-type plant roots in both chemical and morphological aspects. Presently, scientists globally employ approximately 90 *A. rhizogenes* strains across 400 plant species for fundamental and applied research, with the most extensively described strains being the agropine strains ATCC15834 and A4 due to their high virulence. Through continuous efforts, plants have been developed into food and medicinal crops, serving as primary sources of biologically active substances with properties like antioxidant, antiviral, and antimutagenic, among others (Zhao et al. 2016; Radziejewska et al. 2018). Despite their potential, a major obstacle hindering the commercial use of hairy roots lies in the challenges of cultivating them in an industrial system (Giri and Zaheer 2016).

17.2.5 Callus Culture

Callus culture is a pivotal aspect of biotechnology, leveraging plant tissues to generate an undifferentiated mass that can be manipulated to yield beneficial products for pharmaceuticals, food, agriculture, and cosmetics (Eibl 2018). Recently, the development of callus culture, known as pharmaceutical engineering, has gained considerable attention, employing diverse techniques to maximize the production of valuable compounds within specific plant cells (Ogita 2015). Typically, two steps are involved in callus culture for therapeutic compound production: the first aims for maximum callus growth, while the second focuses on secondary metabolite production (Efferth 2019). Optimal conditions encompassing light, temperature, humidity, and nutrients are imperative for callus formation, typically induced by various biotic and abiotic stimuli.

The biosynthesis pathways of secondary metabolites rely on coordination among plant cells, tissues, and organs at intra- and intermolecular levels. Thus, for specific stages of biosynthesis, cell differentiation plays a critical role. In such cases, employing more differentiated tissues or organ cultures and, at times, microplants becomes essential (Cassells and Curry 2001; Duangporn and Siripong 2009). Callus cultures hold significant potential for sustainably producing bioactive compounds from endangered species of medicinal, aromatic, and ornamental plants (MAOPs) (Koufan et al. 2020). However, the level of differentiation, particularly organogenesis, plays a crucial role in influencing the in vitro production of secondary metabolites (Santos et al. 2020). Among the differentiated organ cultures, hairy root culture exhibits higher productivity compared to other culture types (Chandran et al. 2020). Therefore, selecting the most suitable in vitro cultivation method (callus culture, organ culture, or microplants) is a crucial strategic decision in biotechnology that demands special attention. Furthermore, enhancing the biosynthesis of secondary metabolites is often necessary for maximizing the efficiency of in vitro systems (Dzamic et al. 2013; Kim et al. 2015; Zeng et al. 2015).

17.2.6 Suspension Cultures

Plant tissue and cell suspension cultures, enabled by biotechnological techniques, offer a promising platform for producing desired natural products (Fig. 17.2). Medicinal plant cell suspension cultures (MPCSC) comprises free cells or small cell clusters from medicinal plant calluses cultivated in a liquid medium. Typically, callus cultures are transformed into suspension culture systems, employing either batch or continuous fermentation processes to yield these beneficial compounds (Bhatia 2015). These cultures have the potential to produce various secondary metabolites crucial for pharmaceutical purposes (Moscatiello et al. 2013). This technology presents an attractive alternative to whole-plant cultivation for high-value natural compounds like paclitaxel, resveratrol, artemisinin, ginsenosides, and ajmalicine (Ten Hoopen et al. 2002; Baldi and Dixit 2008; Jeong et al. 2008; Li and Tao 2009; Cai et al. 2012a, b). The industrial production process of therapeutic compounds from cells involves selecting the appropriate bioreactor for the fermentation process, maintaining optimal temperature and pH, and supplying the necessary nutrients suitable for the cultured plant.

Compared to conventional methods, MPCSC offers several significant advantages:

(a) Freedom from seasonal, geographical, and environmental limitations
(b) Stability in production, ensuring consistent yield and quality of natural products
(c) Efficient propagation and uniformity, outperforming callus cultures on solidified mediums (Xu et al. 2011a, b)
(d) Potential for synthesizing novel products not typically found in native plants (Ye et al. 2003; Zhang et al. 2011; de Padua et al. 2012)

Culture conditions significantly influence both the quality and quantity of materials derived from MPCSC. Optimizing these conditions, such as carbon and nitrogen sources, growth regulators, pH, temperature, light, and oxygen, plays a crucial role in enhancing the accumulation of desired products. The components within the plant cell culture medium have a direct impact on growth and the production of secondary metabolites (Zhang and Zhong 1997).

17.3 Elicitation in In Vitro Cultures

Secondary metabolites are important compounds synthesized by plant cells in response to environmental stimuli or act as defense mechanisms against invading pathogens. Elicitation, a strategy to enhance secondary metabolite productivity, manipulates cells or tissues by simulating potential biotic or abiotic attacks using agents triggering a defense response (Zhao et al. 2005). Initially, the term "elicitor" referred to compounds that could trigger the production of phytoalexins. However, it is now commonly used to describe substances that stimulate any form of plant defense response (Ebel and Cosio 1994; Hahn 1996; Nürnberger 1999). Ultimately,

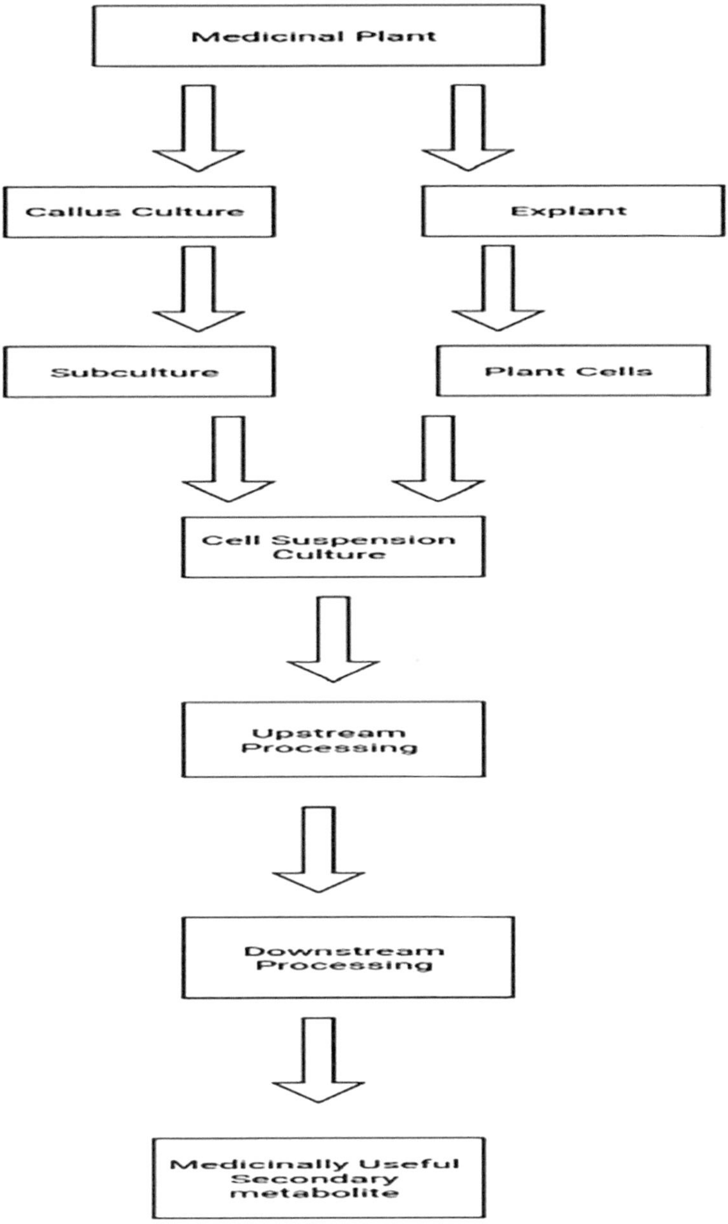

Fig. 17.2 Schematic representation of enhancement of secondary metabolite production through suspension cultures

the activation of defense responses may result in improved resistance (Fig. 17.3). This broader definition of elicitors encompasses both substances originating from pathogens (exogenous elicitors) and compounds released from plants due to pathogen activity (endogenous elicitors) (Boller 1995; Ebel and Cosio 1994). Elicitors incite stress responses in plants, amplifying the synthesis and accumulation of secondary metabolites or even prompting the production of new compounds (Naik and Al-Khayri 2016a, b). They hold pivotal roles in plant growth, development, signal transduction, stress tolerance, and overall productivity (Jan et al. 2021).

Reports suggest that elicitation processes have boosted the biosynthesis of specific secondary metabolites in hairy root and cell suspension cultures, promising large-scale production in the pharmaceutical industry (Srivastava et al. 2017; Singh et al. 2018a, b). While cell suspension cultures primarily serve for elicitation treatment and secondary metabolite production due to their inherent traits, hairy root cultures also exhibit promising characteristics for elicitation experiments (Giri and Zaheer 2016).

Transcription factors like MYB, WRKY, bHLH, and AP2/ERF actively engage in hormonal signaling pathways, augmenting the production of pharmacologically active phytochemicals in diverse medicinal plants (Jan et al. 2021; Zheng et al. 2023). Elicitation aims to modify the metabolic pathways of medicinal plants, prompting the production of bioactive compounds through in vitro plant cultures. This approach holds promise for economically viable secondary metabolite production on a larger scale by utilizing signaling molecules that stimulate pathways leading to increased synthesis and accumulation of phytochemicals (Zhai et al. 2017; Miladinova-Georgieva et al. 2023). Parameters like elicitor concentration, exposure time, elicitor type, and cell culture type significantly impact the accumulation and

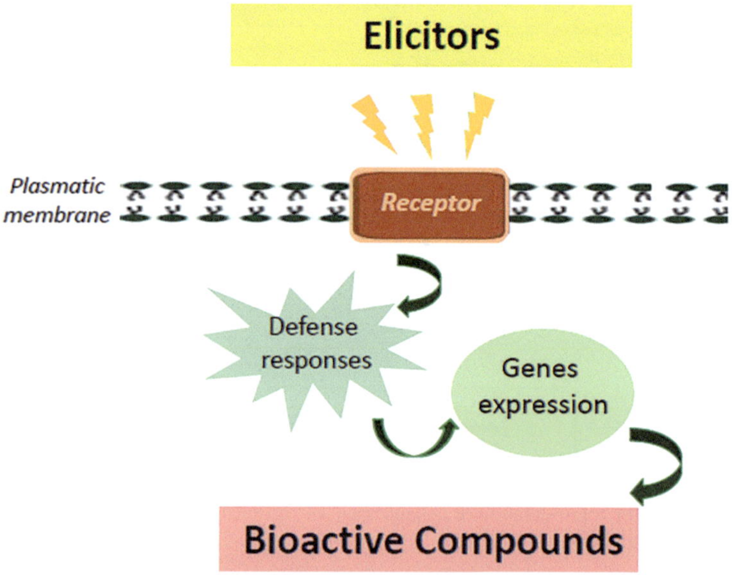

Fig. 17.3 Diagrammatic depiction of the elicitor action producing bioactive compounds (Nieves et al. 2014)

synthesis of metabolites during the elicitation process (Dhiman et al. 2018). Various elicitor treatments, classified as abiotic or biotic, can induce the formation of secondary metabolites in plant cells in vitro, providing a swift production route for bioactive chemicals (Weathers et al. 2010; Donaldson et al. 2004; Naik and Al-Khayri 2016a, b; Welchen et al. 2021).

The researchers investigated the effects of different triggers, including both plant-derived (salicylic acid, jasmonic acid, and methyl jasmonate) and biotic sources (chitosan and yeast extract), on the growth and production of xanthones in Gentianaceae, focusing on two types of hairy root culture samples from *Gentiana dinarica*. These triggers led to the discovery of new xanthone compounds, with variable impacts depending on the root type. For instance, one xanthone, norswertianin-1-O-primeveroside, wasn't significantly affected by SA, JA, or MJ but showed a substantial 24-fold increase when treated with CHI (50 mg L^{-1}) compared to untreated roots (Krstić-Milošević et al. 2017).

Moreover, in a different study, researchers observed increased production of andrographolide in the cell suspension culture of *Andrographis paniculata* by using elicitors derived from *Aspergillus niger* and *Penicillium expansum* (Vakil Moinuddin and Mendhulkar Vijay 2013). An oligosaccharin from fungal cell walls induced specific biochemical changes in plants linked to resistance, triggering the production of phytoalexins, lignin, and ethylene (Ryan 1988). Further investigations revealed that several substances, including yeast cell wall preparations, bacterial antibiotics, oligouronides from plant cell walls, chitosan from fungal cell walls, N-acetyl chitohepatose, and even mechanical injury, rapidly stimulate the production of natural jasmonic acid within plants.

Additionally, the choice of culture medium significantly impacted the elicitation process. In the callus culture of *Erythroxylum coca*, the amounts of cocaine, cinnamoylcocaine, chlorogenic acid (CGA), and 4-coumaroyl quinate (CQA) were notably influenced by the culture medium used (Dicosmo and Misawa 1995). Cocaine production increased nearly tenfold with ARM, GB5, and MMT media, with ARM displaying nearly tenfold higher cocaine levels compared to other media, indicating that medium composition plays a crucial role in the accumulation of tropane alkaloids (TAs).

17.3.1 Biotic Elicitors

Biotic elicitors are extracted from plants and microbes like fungus, bacteria, and yeast (Table 17.1). Moreover, they might be defined composition like a mixture of yeast extract and uniform fungus, inactivated enzymes, chitosan, and polysaccharides such as pectin, cellulose, chitin, and others (Vasconsuelo and Boland 2007). Several biotic elicitors are employed in the enhancement of pharmaceutically viable SMs in medicinal plants, and the major ones are discussed down below.

17.3.1.1 Fungal Elicitors

Fungal elicitors, encompassing both free-living and endophytic fungi, stand out as crucial biotic elicitors extensively employed for synthesizing commercial compounds. This interaction between fungi and plants triggers hypersensitive responses, prompting the activation of defense pathways within the plant, thereby enhancing the production of secondary metabolites more effectively (Zhai et al. 2017). The augmentation of artemisinin, ginsenoside, and anthocyanin, essential bioactive compounds, has been significantly increased using fungal elicitors, namely, *Penicillium oxalicum* and *Trichoderma harzianum*, as demonstrated in studies (Zheng et al. 2016; Biswas et al. 2018). *Aspergillus niger*, another fungal elicitor, is recognized for its role in enhancing the synthesis of secondary metabolites in laboratory cultures (Miclea et al. 2020). Yeast extract, a microbial elicitor, has widespread application in enhancing secondary metabolites in vitro. It has been observed to stimulate the accumulation of various compounds in different plant cultures.

In studies involving *Astragalus membranaceus* and *Aster scaber* hairy root cultures, yeast extract stimulated the accumulation of astragaloside, phenols, and flavonoids (Ghimire et al. 2019; Park et al. 2020). Additionally, yeast extract application in transformed hairy roots of danshen (*Salvia miltiorrhiza*) led to increased tanshinone production, while in hairy root cultures of *Ocimum tenuiflorum (L.)*, it enhanced the production of ursolic acid and eugenol (Wei et al. 2020; Sharan et al. 2019). The combined use of yeast extract and methyl jasmonate (MJ) elicitors resulted in amplified parthenolide accumulation and upregulated TpPTS gene expression in feverfew (*Tanacetum parthenium*) hairy root culture (Pourianezhad et al. 2019).

Furthermore, yeast extract has shown a positive impact on the accumulation of colchicine and thiocolchicoside in cell suspension cultures of *Gloriosa superba* (Mahendran et al. 2018). It has also contributed to the enrichment of schisandra and dibenzocyclooctadiene lignans in micro-shoot cultures of *Schisandra chinensis* (Szopa et al. 2018). These instances underline the diverse applications and effectiveness of yeast extract in enhancing the production of various secondary metabolites across different plant cultures.

17.3.1.2 Bacterial Elicitors

Bacterial elicitation refers to utilizing bacterial cells, components, or extracts to stimulate responses in plants, also aiming to enhance secondary metabolite production under in vitro conditions. The elicitation process is unique for each bacterial species, influencing various secondary metabolites in distinct quantities. This diversity arises from factors like the binding between the elicitor and membrane receptors, activation of G proteins, cytoplasmic acidification, and the generation of reactive oxygen species (Zhao et al. 2005; Biswas et al. 2016). In peppermint shoot cultures, the presence of rhizobacteria (*Bacillus amyloliquefaciens*) significantly mitigated the adverse effects of high salt concentrations (100 mM NaCl). This resulted in increased accumulation of various compounds such as phenols, antioxidants, MDA, menthone, menthol, and pulegone under severe salinity stress, showcasing a beneficial effect on stress tolerance and compound production (Del Rosario

Table 17.1 Various kinds of elicitors and their effect on SMs in in vitro cultures

Medicinal plant	Secondary metabolite	Elicitor	Technique used	Effect on SM	References
Crocus sativus	Carotene and ionone	CsCCD4a and CsCCD4b genes	Genetic engineering	Positive	Rubio et al. (2008)
D. lasianthera	Multiple shoot formation	*A. tumefaciens* strain LBA4404 with a plasmid vector pG35SKNAT1	Genetic transformation	Positive	Utami et al. (2018)
D. catenatum	Protocorn transformation	*A. tumefaciens*	Genetic transformation	Positive	Zheng et al. (2012)
Astragalus membranaceus	Astragaloside, phenols, flavonoids	Yeast extract	Biotic elicitation (fungal)	Positive	Park et al. (2020)
Aster scaber	Astragaloside, phenols, flavonoids	Yeast extract	Biotic elicitation (fungal)	Positive	Ghimire et al. (2019)
Salvia miltiorrhiza	Tanshinone	Yeast extract	Biotic elicitation (fungal)	Positive	Wei et al. (2020)
Ocimum tenuiflorum	Ursolic acid and eugenol	Yeast extract	Biotic elicitation (fungal)	Positive	Sharan et al. (2019)
Tanacetum parthenium	Parthenolide	Yeast extract	Biotic elicitation (fungal)	Positive	Pourianezhad et al. (2019)
Carthamus tinctorius	Red pigment	Micro algae	Biotic elicitation (algae)	Positive	Hanagata et al. (1994)
Artemisia annua L.	Artemisinin, ginsenoside, and anthocyanin	*Penicillium oxalicum* and *Trichoderma harzianum*	Biotic elicitation (fungal)	Positive	Zheng et al. (2016), Biswas et al. (2018)
Gloriosa superba	Colchicine and thiocolchicoside	Yeast extract	Biotic elicitation (fungal)	Positive	Mahendran et al. (2018)
Schisandra chinensis	Schisandra and dibenzocyclooctadiene lignans	Yeast extract	Biotic elicitation (fungal)	Positive	Szopa et al. (2018)

(continued)

Table 17.1 (continued)

Medicinal plant	Secondary metabolite	Elicitor	Technique used	Effect on SM	References
Scutellaria bornmuelleri	Chrysin, wogonin, and baicalein flavonoids	Chitosan + methyl jasmonate + methyl-b-cyclodextrin	Biotic elicitation (chitosan)	Positive	Gharari et al. (2020)
Mentha piperita L.	Phenols, antioxidants, MDA, menthone, menthol, and pulegone	Rhizobacteria (*Bacillus amyloliquefaciens*)	Biotic elicitation (bacterial)	Positive	Del Rosario et al. (2020)
Helicteres isora (L.)	Diosgenin	*Escherichia coli*	Biotic elicitation (bacterial)	Positive	Shaikh et al. (2020)
Dionaea muscipula J. Ellis	Myricetin, caffeic acid, ellagic acid, and plumbagin	*Cronobacter sakazakii*	Biotic elicitation (bacterial)	Positive	Makowski et al. (2020a, b)
Datura stramonium	Hyoscyamine	Tobacco mosaic viruses (PMMoV, TMV, and ToMV)	Biotic elicitation (viral)	Positive	Mihálik et al. (2022)
Hyoscyamus niger	Hyoscyamine and scopolamine	Methyl jasmonate	Biochemical elicitation	Positive	Kang et al. (2004)
Moringa oleifera	Baicalin, rutin, hesperidin, rosmarinic acid, quercitrin, quercetin, and kaempferol	Methyl jasmonate	Biochemical elicitation	Positive	
Withania somnifera	Withanolides	Salicylic acid	Biochemical elicitation	Positive	Sivanandhan et al. (2016)
Stephania venosa	Dicentrine	Salicylic acid	Biochemical elicitation	Positive	Kitisripanya et al. (2013)
Thevetia peruviana	Phenols	Salicylic acid	Biochemical elicitation	Positive	Mendoza et al. (2018)
Gardenia jasminoides	Rutin	Salicylic acid	Biochemical elicitation	Positive	
Taxus cuspidata	Catechin	Methyl jasmonate	Biochemical elicitation	Positive	Bulgakov et al. (2011)

Medicinal plant	Secondary metabolite	Elicitor	Technique used	Effect on SM	References
Lepechinia caulescens (Ortega.)	Terpenes	Methyl jasmonate	Biochemical elicitation	Positive	Martínez et al. (2017)
Centella asiatica	Asiaticoside	Methyl jasmonate	Biochemical elicitation	Positive	Krishnan et al. (2019)
Lavandula angustifolia Mill	Polyphenols	Jasmonic acid	Biochemical elicitation	Positive	Ileana et al. (2020)
Lavandula angustifolia Mill	Polyphenols	Salicylic acid + activated charcoal	Biochemical elicitation	Positive	Ileana et al. (2020)
Bacopa monnieri	Bacoside A	Methyl jasmonate	Biochemical elicitation	Positive	Rajendran et al. (2023)
Bacopa monnieri	Bacoside A	Methyl jasmonate + salicylic acid	Biochemical elicitation	Positive	Rajendran et al. (2023)
C. asiatica	Asiaticoside	Methyl jasmonate + salicylic acid	Biochemical elicitation	Positive	Kim et al. (2007), Sharma et al. (2015), Largia et al. (2015)
Glycyrrhiza glabra	Glycyrrhizin	Methyl jasmonate + salicylic acid	Biochemical elicitation	Positive	Kim et al. (2007), Sharma et al. (2015), Largia et al. (2015)
Ginkgo biloba	Ginkgolide A, B, C, and bilobalide	Methyl jasmonate + salicylic acid	Biochemical elicitation	Positive	Sukito and Tachibana (2016)
Ajuga bracteosa	Phenols and flavonoids	Methyl jasmonate and phenylacetic acid	Biochemical elicitation	Positive	Saeed et al. (2017)
Ajuga bracteosa	Phytoecdysteroid	Methyl jasmonate	Biochemical elicitation	Positive	Saeed et al. (2017)
Hypericum perforatum	Hypericin and pseudohyperian	Salicylic acid	Biochemical elicitation	Positive	Gadzovska et al. (2013)

(continued)

Table 17.1 (continued)

Medicinal plant	Secondary metabolite	Elicitor	Technique used	Effect on SM	References
Bacopa monnieri	Bacoside	Salicylic acid	Biochemical elicitation	Positive	Bunjan et al. (2018), Koul and Mallubhotla (2020)
Bacopa monnieri	Bacoside	Thidiazuron	Biochemical elicitation	Positive	Kharde et al. (2018)
Fagonia indica	Phenols	Methyl jasmonate	Biochemical elicitation	Positive	Khan et al. (2015)
Solanum xanthocarpum	Soladonine	Cholesterol	Biochemical elicitation	Positive	Dornenburg and Knorr (1995)
S. aviculare	Soladonine	Cholesterol	Biochemical elicitation	Positive	Dornenburg and Knorr (1995)
Withania somnifera	Withaferin A	Cholesterol	Biochemical elicitation	Positive	Chitturi et al. (2010)
Datura metel	Hyoscyamine and scopolamine	Salicylic acid	Biochemical elicitation	Positive	Ajungla et al. (2009)
Trigonella foenum-graecum	Trigonelline	Salicylic acid	Biochemical elicitation	Positive	Mathur and Yadav (2011)
Jatropha curcas	Alkanes and fatty acids	Salicylic acid	Biochemical elicitation	Positive	Mahalakshmi et al. (2013)
Salvia miltiorrhiza	Shikimic acid, caffeic acid, *p*-coumaric acid, rosmarinic acid, salvianolic acid B, salvianolic acid A, dihydrotanshinone, cryptotanshinone, tanshinone I, and tanshinone II A	NaCl	Abiotic elicitation (salinity stress)	Positive	Yu et al. (2019)
Stevia rebaudiana Bertoni	Rebaudioside A and stevioside glycosides	Polyethylene glycol (PEG)	Abiotic elicitation (drought stress)	Positive	Ahmad et al. (2020)

Medicinal plant	Secondary metabolite	Elicitor	Technique used	Effect on SM	References
Lepidium sativum	α-Glucosidase and α-amylase	UV-C + melatonin	Abiotic elicitation (UV)	Positive	Ullah et al. (2019)
Lippia gracilis	Carvacrol	Blue light	Abiotic elicitation (light stress)	Positive	Lazzarini et al. (2018)
Plectranthus amboinicus	Carvacrol	Red light	Abiotic elicitation (light stress)	Positive	Silva et al. (2017)
Drosera burmannii	Plumbagin	Blue LED light	Abiotic elicitation (light stress)	Positive	Boonsnongcheep et al. (2019)
Dysphania ambrosioides L.	*p*-Cymene	Blue LED light	Abiotic elicitation (light stress)	Positive	de Carvalho et al. (2020)
Dysphania ambrosioides L.	Ascaridole	White fluorescent light	Abiotic elicitation (light stress)	Positive	de Carvalho et al. (2020)
Withania somnifera	Antioxidants	Red light	Abiotic elicitation (light stress)	Positive	Adil et al. (2019)
Droseraceae peltata and *Droseraceae muscipula*	Proline	Blue–red (ratio 6:1) LED light	Abiotic elicitation (light stress)	Positive	Makowski et al. (2019)
Salvia miltiorrhiza	Salvianolic and rosmarinic acids	Smoke water and smoke-isolated butenolide (HRC)	Abiotic elicitation	Positive	Zhou et al. (2018), Sun et al. (2020)

(continued)

Table 17.1 (continued)

Medicinal plant	Secondary metabolite	Elicitor	Technique used	Effect on SM	References
Fagonia indica	Phenols	Phenyl acetic acid	Abiotic chemical elicitation	Positive	Khan et al. (2015)
Bacopa monnieri	Bacoside	Copper sulfate ($CuSO_4$)	Abiotic chemical elicitation	Positive	Sharma et al. (2015)

et al. 2020). A study by Shaikh et al. (2020) investigated the impacts of bacterial and fungal elicitors, including *Escherichia coli*, *Bacillus subtilis*, *Saccharomyces cerevisiae*, and *Aspergillus niger*, in cell suspension cultures of *Helicteres isora* (L.). The highest diosgenin content (9.1-fold increase) was observed with *E. coli* elicitation (1.5%). Furthermore, these microbial elicitors upregulated the expression levels of regulatory genes, 3-hydroxy 3-methylglutaryl CoA reductase (HMGR), and cycloartenol synthase (CAS), crucial in the diosgenin biosynthetic pathway (Shaikh et al. 2020).

Additionally, bacterial elicitation using *Cronobacter sakazakii* bacteria lysate showed enhanced synthesis of myricetin, caffeic acid, ellagic acid, and plumbagin in liquid tissue culture of Venus flytrap (*Dionaea muscipula* J. Ellis), highlighting the potential of elicitation strategies to boost specific compound production in plant cultures (Makowski et al. 2020a, b).

17.3.1.3 Viral Elicitors

Viruses, as external agents, coordinate with the plant tissues and are responsible for changes in plant secondary metabolism (Lan et al. 2020). A range of SMs are known to show resistance against pathogens (Wink 1988). The utilization of viral infections as a means to stimulate the production of specific secondary metabolites is uncommon, primarily due to their complex mechanism of action compared to typical biotic or abiotic elicitors. Mihálik et al. (2022) conducted a study where they employed three tobacco mosaic viruses (PMMoV, TMV, and ToMV) to artificially infect *Datura stramonium* as a form of in vivo elicitation. Their findings indicated a notable increase in hyoscyamine content compared to both in vitro induction and non-elicited capillary root cultures. This study highlights the potential value and significance of viral elicitation in boosting the production of secondary metabolites.

17.3.2 Biochemical Elicitors

17.3.2.1 Methyl Jasmonate

Methyl jasmonate (MeJA), initially extracted from *Jasminum grandiflorum* essential oil in 1962, serves as a volatile methyl ester of jasmonic acid and plays a crucial role as a major signaling molecule in both abiotic and biotic stresses (Demole et al. 1962; Wang et al. 2021). MeJA stands out as a key elicitor, significantly contributing to the enhancement of secondary metabolite production. Its influence notably impacts alkaloid production, particularly in relation to culture time (Kang et al. 2004; Zayed and Wink 2005).

Hyoscyamus niger, a herbaceous plant belonging to Solanaceae, holds great importance in the pharmaceutical industry due to its rich content of tropane alkaloids (TAs) like hyoscyamine and scopolamine, which have long been used as medications (Dehghan et al. 2012; Pudersell et al. 2012). MeJA plays a pivotal role in augmenting these essential TAs, directly or indirectly affecting key enzymes such as putrescine-N-methyl transferase (PMT) and hyoscyamine-6 beta-hydroxylase (H6H), which are crucial for the biosynthesis of these alkaloids (Kang et al. 2004).

In the context of *Moringa oleifera*, different concentrations of salicylic acid (SA) and methyl jasmonate (MeJA) demonstrated notable effectiveness in enhancing the production of active compounds. Across various callus cultures compared to the control, substantial rises in flavonoid concentrations were observed. Specifically, at 30 mg/l MeJA, the highest baicalin concentration (1728.0 μg/ml) was detected, while the maximum accumulations of rutin, hesperidin, rosmarinic acid, quercitrin, quercetin, and kaempferol (2562.0, 2400.6, 1812.0, 1497.4, 1456.9, and 784.2 μg/ml, respectively) were recorded at 20 mg/l MeJA. In contrast, the untreated control group exhibited lower content levels for baicalin, rutin, hesperidin, quercetin, and kaempferol (722.1, 1358.2, 1182.4, 422.6, and 520.2 μg/ml, respectively) compared to other treatments.

Comparative analysis of the accumulated flavonoids was conducted to ascertain the most effective elicitor and concentration for augmenting these compounds. The results revealed that the application of MeJA at 20 mg/l resulted in the highest total flavonoid content of 11,578.8 μg/ml, showcasing a substantial increase of 94.8% over the control. Additionally, the use of 15 mg/l SA resulted in 10,526.2 μg/ml of flavonoids, indicating a 77% increase rate compared to the control (Abed et al. 2023).

As per Kamińska (2021), jasmonates play a role in signaling pathways that stimulate crucial enzymes involved in biochemical transformations within secondary metabolism. The external application of methyl jasmonate (MeJA) has been observed to yield significant results in various plant systems. For instance, MeJA application led to an increase in catechin production in cell suspension cultures of *Taxus cuspidata* (Bulgakov et al. 2011), a substantial fivefold enhancement in terpene production in *Lepechinia caulescens* (Martínez et al. 2017), and a noteworthy 69-fold rise in Asiaticoside content in *Centella asiatica* (Krishnan et al. 2019). Furthermore, the production of bacoside A, a bioactive compound, was boosted through MeJA application in vitro within shoot cultures of *Bacopa monnieri* (Rajendran et al. 2023).

Additionally, studies suggest that low concentrations of exogenous MeJA have the potential to offer a viable protocol for augmenting bacoside contents in *Bacopa* seedlings (Bunjan et al. 2018; Koul and Mallubhotla 2020). Furthermore, MeJA and phenylacetic acid (PAA) were employed as elicitors to amplify the total phenolic and flavonoid content in the root suspension of *Ajuga bracteosa*. MeJA elicitation over 14 days also led to an enhancement in phytoecdysteroid levels in *A. bracteosa* (Saeed et al. 2017).

17.3.2.2 Salicylic Acid

In various cultures, salicylic acid (SA) has been observed to play a significant role in boosting the production of diverse secondary metabolites in different plant species. For instance, SA supplementation in culture media has led to increased production of withanolides in *Withania somnifera* (Sivanandhan et al. 2016), dicentrine in *Stephania venosa* (Spreng.) (Kitisripanya et al. 2013), phenolic compounds in *Thevetia peruviana* (Schum.) (Mendoza et al. 2018), and rutin in *Gardenia jasminoides* (El-Ashry et al. 2019). Several studies have demonstrated that SA supplementation in culture media stimulates the accumulation of secondary metabolites, such as hyoscyamine and scopolamine in *Datura metel* root cultures (Ajungla et al. 2009), trigonelline in *Trigonella foenum-graecum* cell suspension (Mathur and Yadav 2011), and the synthesis of alkanes and fatty acids in cell cultures of *Jatropha curcas* (Mahalakshmi et al. 2013).

Moreover, the use of elicitors has shown substantial effects on the secretion of polyphenols in lavender explants grown in vitro. In specific cases, the combination of salicylic acid and activated charcoal (SA + AC) in media induced higher total polyphenolic content (TPC) compared to the control condition (Miclea et al. 2020). In *Bacopa monnieri*, the combined application of methyl jasmonate (MeJA) and salicylic acid significantly augmented both biomass and bacoside A content in in vitro shoot cultures.

Studies by Kim et al. (2007), Sharma et al. (2015), and Largia et al. (2015) have reported that the combined use of salicylic acid (SA) and methyl jasmonate (MeJA) effectively enhanced the asiaticoside content in hairy root cultures of *Centella asiatica* after 3 weeks of elicitation. Additionally, this combination showed a positive effect on the production of glycyr–rhizin in *Glycyrrhiza glabra* when MeJA and SA were added. Furthermore, SA and MeJA have been found to significantly improve the production of ginkgolides A, B, and C and bilobalide in *Ginkgo biloba* under in vitro conditions, particularly in immobilized cell cultures (Sukito and Tachibana 2016).

According to Gadzovska et al. (2013), salicylic acid (SA) treatment in suspension culture exhibited a twofold increase in the production of hypericin and pseudohyperian in *Hypericum perforatum* (L.). In the case of *Bacopa monnieri*, SA was employed as an elicitor to augment biomass and bacoside production in in vitro suspension cultures. Koul and Mallubhotla (2020) conducted a study aiming to boost bacoside content in *B. monnieri* through in vitro suspension cultures using various plant growth regulators such as 6-benzyladenine (BA), kinetin, thidiazuron, and chitosan. Among these regulators, samples treated with thidiazuron displayed increased bacoside content (Kharde et al. 2018). Additionally, MeJA, SA, and copper sulfate ($CuSO_4$) were utilized to enhance bacoside content in *B. monnieri* shoot cultures. The findings revealed that $CuSO_4$-treated shoot cultures exhibited the highest yield of bacoside content compared to the other two hormonal elicitors (Sharma et al. 2015).

17.3.2.3 Cholestrol

Cholesterol has been recognized as a compound that can be introduced into the culture medium to stimulate explants, particularly callus and cell suspension, for the production of effective secondary metabolites. Dornenburg and Knorr (1995) observed an increase in soladonine compound content in cell suspensions of *Solanum xanthocarpum* and *S. aviculare* when cholesterol was added to their culture medium. Similarly, Chitturi et al. (2010) noted that the addition of cholesterol to cell cultures of *Withania somnifera* led to an increased content of withaferin A compound compared to untreated control cultures.

17.3.3 Abiotic Elicitors

Abiotic elicitors come from non-biological sources including stresses like deficiency of water, varying temperatures, salt and heavy metal accumulation, exposure to UV light, and others that have significant impact in the amount and nature of secondary metabolites produced in MAOPs (Khare et al. 2020; Mahajan et al. 2020). The stress-triggering factors are known as "eustress factors" (Alvarado et al. 2019). Exposing in vitro cultures to artificial abiotic stresses can be a technique to induce metabolic adjustments and thus increase the production of SMs. Various abiotic stress-inducing factors employed in in vitro techniques are discussed here.

17.3.3.1 Salinity Stress

The utilization of stress-inducing additives such as polyethylene glycol (PEG) and sodium chloride (NaCl) serves as a method to artificially induce drought and salinity stresses, acting as abiotic elicitors to boost secondary metabolite accumulation in various plant species.

For instance, in *Salvia miltiorrhiza* Bunge's cell suspension culture, the application of NaCl elicitation at a concentration of 50 mmol/L significantly increased the content of various secondary metabolites, including shikimic acid, caffeic acid, p-coumaric acid, rosmarinic acid, salvianolic acid B, salvianolic acid A, dihydrotanshinone, cryptotanshinone, tanshinone I, and tanshinone II A (Yu et al. 2019).

17.3.3.2 Drought Stress

Similarly, in *Stevia rebaudiana* Bertoni, exposure to drought stress through PEG 6000 (4%) resulted in a notable rise in rebaudioside A and stevioside glycosides, along with increased levels of total phenolic content, total flavonoid content, total antioxidant capacity, total reducing power, and 1,1-diphenyl-2-picrylhydrazyl (DPPH)-free radical scavenging activity in shoot cultures (Ahmad et al. 2020).

In *Salvia leriifolia* plantlets and calli cultures, varying concentrations of PEG 6000 (5, 7.5, and 10% w/v) application led to increased accumulation of rosmarinic acid, caffeic acid, and salvianolic acid B secondary metabolites. Notably, cultures subjected to PEG 5% showed augmented levels of these metabolites in both culture types (Hosseini et al. 2020).

17.3.3.3 UV Radiation

Ultraviolet (UV) light serves as a significant abiotic elicitor, employed to stimulate phytochemical production in various plant cultures (Xuan et al. 2016). Exposure to UV light stress triggers plant defense mechanisms, thereby encouraging the production of commercially valuable secondary compounds (Yin et al. 2016). Different wavelengths of UV light, such as UV-B and UV-C, have been observed to impact the accumulation and composition of secondary metabolites in plants (Santisree et al. 2020).

For example, in the callus culture of *Lepidium sativum*, researchers assessed the effects of various concentrations of melatonin (ranging from 0 to 50 μM) and different durations of UV-C exposure (ranging from 10 to 150 min) on the biosynthesis of antioxidant and antidiabetic metabolites separately. The results indicated that a concentration of 20 μM of melatonin exhibited higher efficacy compared to UV-C treatment, leading to increased antidiabetic activities, specifically α-glucosidase and α-amylase, in the callus cultures (Ullah et al. 2019).

17.3.3.4 Light Stress

Light spectra variations influence the production of bioactive compounds in different plant species and genotypes. For instance, in *Lippia gracilis*, higher yields of carvacrol were observed under blue light (Lazzarini et al. 2018), while in *Plectranthus amboinicus*, the greatest carvacrol content was obtained under monochromatic red light (Silva et al. 2017). In the case of *Drosera burmannii* and *Drosera indica* shoot cultures, exposure to blue LED light resulted in an increased accumulation of plumbagin (Boonsnongcheep et al. 2019), exhibiting a 2.8-fold and 1.9-fold enhancement, respectively.

Regarding *Dysphania ambrosioides* L. cultures, the effects of various light spectra, including blue (B), red (R), white, and different combinations of B:R (1:1, 2:1, 1:2), were examined. It was found that blue monochromatic LEDs induced higher p-Cymene content but negatively impacted Z-ascaridole levels. In contrast, white fluorescent light facilitated a greater conversion of α-terpinene into ascaridole (de Carvalho et al. 2020).

In experiments involving the in vitro callus cultures of *Withania somnifera*, exposure to different monochromatic light wavelengths (violet, blue, green, yellow, and red) revealed that red light was optimal for achieving maximum biomass accumulation and antioxidant activity in the callus cultures (Adil et al. 2019).

However, in the case of in vitro shoot cultures of *Droseraceae peltata* and *Droseraceae muscipula* subjected to two light regimes—fluorescent light as control and blue–red (6:1 ratio) LED light—it was noted that the blue–red LED light increased proline levels in *D. peltata* and decreased POD activity in both species. Interestingly, this particular light regime was found to be ineffective in eliciting phenolic compounds in both studied plant species (Makowski et al. 2019).

17.3.3.5 Nanoparticles as Elicitors

Nanoparticles have gained significant attention and found extensive applications in medicine, immunology, agriculture, and plant tissue culture. Recent studies, such as those by Jiao et al. (2014), have highlighted nanoparticles as a novel inducer capable of triggering metabolic and physiological responses. Their use in agriculture and plant tissue culture has notably increased for the commercial production of diverse plants and their secondary metabolites, as noted by Asl et al. (2019) and Singh et al. (2022). Nanoparticle-induced stress demonstrates promising potential in stimulating plant defense mechanisms and enhancing the accumulation of bioactive compounds in medicinal plants, as emphasized by Shahhoseini et al. (2020). Moreover, the application of nanoparticles as efficient non-biological elicitors to augment in vitro biosynthesis of secondary metabolites is an upward trend in current research, supported by studies like Mohsen and Sabbatini (2021). Various types of nanoparticles have been utilized in recent studies to enhance secondary metabolite

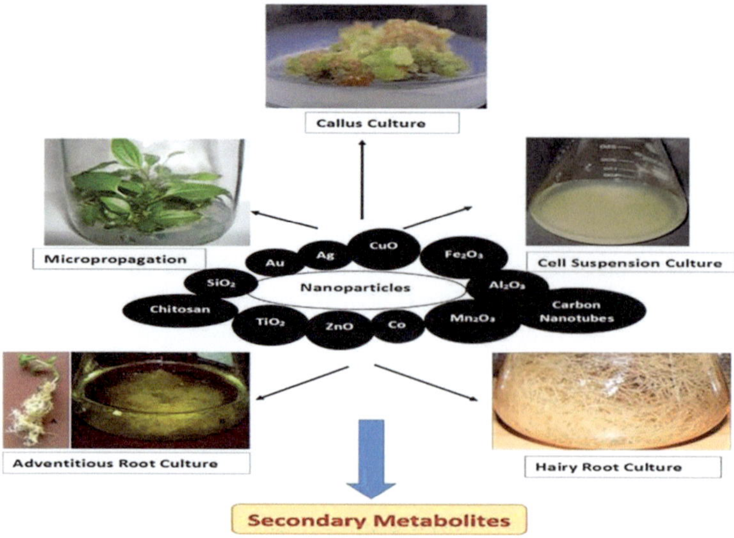

Fig. 17.4 Nanoparticles as elicitors of secondary metabolites under in vitro condition

production in medicinal plants, with specific details presented in Fig. 17.4 and Table 17.2.

17.4 Conclusion

Medicinal plants hold promise as sources of drugs for treating various human illnesses due to the secondary metabolites they produce. However, these metabolites, which have valuable curative properties, are typically generated in insufficient quantities within the plants. Consequently, alternative methods have been developed to increase the production of these medically important compounds. In vitro techniques have emerged as successful and commercially viable strategies to boost the quantity of these metabolites on a large scale.

Methods like callus cultures, hairy root cultures, and suspension cultures, among others, have been extensively used in in vitro tissue culture techniques to enhance the content of secondary metabolites in plant tissues. Although various stress-inducing methods are employed in crop field conditions to obtain higher amounts of these compounds, challenges related to environmental factors and external agents are difficult to manage. Therefore, in vitro culture methods have been instrumental in overcoming these constraints in plant biotechnology. Although any of these methods are not employed on a commercial scale yet, it is noted that suspension culture in particular is the most industrially oriented technique to produce secondary metabolites in large quantities, since it is conducted in huge tanks called bioreactors through a process called bioprocessing.

Table 17.2 Various nanoparticles as elicitors and their effect on secondary metabolite production in in vitro cultures

Medicinal plant	Secondary metabolite	Nanoparticle	Technique used	Effect on SMs	References
Artemisia annua	Artemisia	Silver oxide NPs	Nanotechnology	Positive	Zhang et al. (2013)
Calendula officinalis	Saponin and carotenoid	Silver NP	Nanotechnology	Positive	Ghanati and Bakhtiarian (2014)
Satureja khuzestanica	Essential oils	Multiwalled carbon nanotubes	Nanotechnology	Positive	Ghorbanpour and Hadian (2015)
H. perforatum	Hypericin and hyperforin	Zn and Fe NPs	Nanotechnology	Positive	Sharafi et al. (2013)
Hyoscyamus reticulatus	Hyoscyamine and scopolamine	FeO NPs	Nanotechnology	Positive	Moharrami et al. (2017)
Atropa belladonna	Scopolamine	FeO NPs	Nanotechnology	Positive	Sepehri et al. (2022)
Hyoscyamus reticulatus	Hyoscyamine and scopolamine	ZnO NPs	Nanotechnology	Positive	Asl et al. (2019)
Hyoscyamus reticulatus	Hyoscyamine and scopolamine	Si NPs	Nanotechnology	Positive	Hedayati et al. (2020)
Hyoscyamus reticulatus	Ascorbate peroxidase, catalase, peroxidase Hyoscyamine and scopolamine	Sodium nitroprusside NPs	Nanotechnology	Positive	Khezerluo et al. (2018)
Caralluma tuberculata	Polyphenols	AgNPs	Nanotechnology	Positive	Ali et al. (2019a, b)
Nardostachys jatamansi	Antioxidants	Gold(Au) NPs	Nanotechnology	Positive	Venzhik et al. (2021)

The advent of the technique of "elicitation" has significantly addressed pharmaceutical demands. Elicitors, either biological or non-biological agents, when introduced into in vitro cultures, trigger stress responses in plant tissues, leading to increased production of secondary metabolites beyond normal levels. Over the years, a diverse range of agents, including bacteria, yeast, fungi, algae, plant hormones, and biochemicals, have shown effectiveness as elicitors when added to controlled plant tissue cultures. Additionally, abiological agents like UV radiation, salts, and various chemicals have also demonstrated their efficacy as elicitors. Although several researchers found elicitors as a promising tool, none of them are not tested on an industrial scale as of now.

Currently, there is growing interest in utilizing "nanotechnology" as a tool to enhance secondary metabolites. Recent research conducted by contemporary scientists has shown that nanoparticles made of metals such as Fe, Ag, Au, Cu, Zn, and Ti and non-metals like silicon and carbon, among others, or sometimes combinations of both, can serve as elicitors, effectively enhancing secondary metabolites in various in vitro cultures.

While the use of Nanoparticles as Elictors in tissue cultures is gaining momentum in the field of plant biotechnology, the quest for more effective elicitors and techniques to enhance secondary metabolites continues. Industrialization of these promising in vitro techniques is the need of the hour to meet with the global medicinal demands.

References

Abed AS, Ismail EN, Majeed DM, Jasim Al-Jibouri AM (2023) Increasing amounts of secondary metabolites and medicinal compounds in callus culture of Moringa Oleifera (Lam.) using abiotic elicitors. Iraqi J Sci 64(8):3902–3913

Adams AM, Kaplan NA, Wei Z, Brinton JD, Monnier CS, Enacopol AL, Ramelot TA, Jones JA (2019) In vivo production of psilocybin in *E. coli*. Metab Eng 56:111–119

Adil M, Haider Abbasi B, Ul Haq I (2019) Red light controlled callus morphogenetic patterns and secondary metabolites production in Withania somnifera L. Biotechnol Rep 24:e00380

Aekthammarat D, Pannangpetch P, Tangsucharit P (2019) Moringa oleifera leaf extract lowers high blood pressure by alleviating vascular dysfunction and decreasing oxidative stress in L-NAME hypertensive rats. Phytomedicine 54:9–16

Afzal SF, Yar AK, Ullah RH et al (2017) Impact of drought stress on active secondary metabolite production in Cichorium intybus roots. J Appl Environ Biol Sci 7(7):39–43

Ahmad N, Fazal H, Abbasi BH, Anwar S, Basir A (2013) DPPH free radical scavenging activity and phenotypic difference in hepatoprotective plant (Silybum marianum L.). Toxicol Ind Health 29:460–467

Ahmad MA, Javed R, Adeel M et al (2020) PEG 6000-stimulateddrought stress improves the attributes of *in vitro* growth, steviol glycosides production, and antioxidant activities in Stevia rebaudiana Bertoni. Plants 9:1552

Ahrazem O, Trapero A, Gómez MD, Rubio-Moraga A, Gómez- Gómez L et al (2010) Genomic analysis and gene structure of the plant carotenoid dioxygenase 4 family: a deeper study in Crocus sativus and its allies. Genomics 96(4):239–250

Ajungla L, Patil PP, Barmukh RB, Nikam TD (2009) Influence of biotic and abiotic on accumulation of hyoscyamine and scopolamine in root cultures of Datura metel L. Indian J Biotechnol 8(3):317–322

Ali A, Mohammad S, Khan MA et al (2019a) Silver nanoparticles elicited in vitro callus cultures for accumulation of biomass and secondary metabolites in Caralluma tuberculata. Artif Cells Nanomed Biotechnol 47:715–724

Ali A et al (2019b) Silver nanoparticles elicited *in vitro* callus cultures for accumulation of biomass and secondary metabolites in caralluma tuberculata. Artif Cells Nanomed Biotechnol 47(1):715–724

Alqasoumi SI et al (2011) Anti-inflammatory and wound healing activity of Fagonia schweinfurthii alcoholic extract herbal gel on albino rats. Afr J Pharm Pharmacol 5(17):1996–2001

Alvarado AM, Aguirre-Becerra H, Vazquez-Hernandez MC et al (2019) Influence of elicitors and eustressors on the production of plant secondary metabolites. In: Akhtar M, Swamy MSU (eds) Natural bio-active compounds. Springer, Singapore, pp 333–388

Anjitha KS, Sameena PP, Puthur JT (2021) Functional aspects of plant secondary metabolites in metal stress tolerance and their importance in pharmacology. Plant Stress 2:100038

Arvind KS (2016) Medicinal plants: future source of new drugs. Int J Herb Med 4(4):59–64

Asha S, Vidyavathi M (2009) Cunninghamella–a microbial model for drug metabolism studies–a review. Biotechnol Adv 27(1):16–29
Asl KR, Hosseini B, Sharafi A, Palazon J (2019) Influence of nano-zinc oxide on tropane alkaloid production, H6h gene transcription and antioxidant enzyme activity in Hyoscyamus reticulatus L. hairy roots. Eng Life Sci 19:73–89
Baldi A, Dixit VK (2008) Yield enhancement strategies for artemisinin production by suspension cultures of Artemisia annua. Bioresour Technol 99:4609–4614
Baydar H (2005) Tıbbi, Aromatik ve Keyf Bitkileri Bilimi veTeknolojisi, Süleyman Demirel Üniversitesi Yayınları No: 51
Beier BA (2005) A revision of the desert shrub Fagonia (Zygophyllaceae). Syst Biodivers 3:221–263
Bell EA, Charlwood BV (1980) Secondary plant products. In: Encyclopedia plant physiology, vol 8. Springer, Berlin
Bernal J, Mendiola J, Ibáñez E, Cifuentes A (2011) Advanced analysis of nutraceuticals. J Pharm Biomed Anal 55:758–774
Bhatia S (2015) Plant tissue culture. In: Modern applications of plant biotechnology in pharmaceutical sciences. Elsevier, Amsterdam, pp 31–107
Biswas T, Kalra A, Mathur AK, Lal RK, Singh M, Mathur A (2016) Elicitors influenced differential ginsenoside production and exudation into medium with concurrent Rg3/Rh2 panaxadiol induction in Panax quinquefolius cell suspensions. Appl Microbiol Biotechnol 100:4909–4922
Biswas T, Pandey SS, Maji D, Gupta V, Kalra A, Singh M, Mathur A, Mathur AK (2018) Enhanced expression of ginsenoside biosynthetic genes and in vitro ginsenoside production in elicited Panax sikkimensis (Ban) cell suspensions. Protoplasma 255(1147):1160
Boller T (1995) Chemoperception of microbial signals in plant cells. Annu Rev Plant Physiol Plant Mol Biol 46:189–214
Boonsnongcheep P, Sae-foo W, Banpakoat K et al (2019) Artificial color light sources and precursor feeding enhance plumbagin production of the carnivorous plants Drosera burmannii and Drosera indica. J Photochem Photobiol B Biol 199:111628
Bruneton J (1999) Pharmacognosy, phytochemistry, medicinal plants, 2nd edn. Lavoisier Publishing, London
Bulgakov VP, Tchernoded GK, Veselova MV, Fedoreyev SA, Muzarok TI, Zhuravlev YN (2011) Catechin production in cultured cells of Taxus cuspidata and Taxus baccata. Biotechnol Lett 33(9):1879–1883
Bunjan W, Sujipuli K, Prasarnpun S (2018) Effect of methyl jasmonate elicitation on biomass, gene expression and saponin accumulation in Bacopa monnieri. Int J Biol Sci 13(4):369–377
Cai Z, Kastell A, Knorr D, Smetanska I (2012a) Exudation: an expanding technique for continuous production and release of secondary metabolites from plant cell suspension and hairy root cultures. Plant Cell Rep 31:461–477
Cai Z, Knorr D, Smetanska I (2012b) Enhanced anthocyanins and resveratrol production in Vitis vinifera cell suspension culture by indanoyl-isoleucine, N-linolenoyl-l-glutamine and insect saliva. Enzyme Microb Technol 50:29–34
Cassells AC, Curry RF (2001) Oxidative stress and physiological, epigenetic and genetic variability in plant tissue culture: implications for micropropagators and genetic engineers. Plant Cell Tissue Organ Cult 64:145–157
Castañeda-Loaiza V, Oliveira M, Santos T, Schüler L, Lima AR, Gama F, Salazar M, Neng NR, Nogueira JMF, Varela J et al (2020) Wild vs. cultivated halophytes: nutritional and functional differences. Food Chem 333:127536
Chalker-Scott L, Fnchigami LH (1989) The role of phenolic compounds in plant stress responses. In: Paul HL (ed) Low-temperature stress physiology in crops. CRC Press, Boca Raton, pp 67–79
Chandran H, Meena M, Barupal T, Sharma K (2020) Plant tissue culture as a perpetual source for production of industrially important bioactive compounds. Biotechnol Rep 26:e00450
Chen SL, Yu H, Luo HM, Wu Q, Li CF, Steinmetz A (2016) Conservation and sustainable use of medicinal plants: problems, progress, and prospects. Chin Med 11:1–10

Chen J, Wang L, Chen J, Huang J, Liu F, Guo R, Yang L, Grabon A, Zhao K, Kong F, Liu C, Tian M (2018) Agrobacterium tumefaciens-mediated transformation system for the important medicinal plant Dendrobium catenatum Lindl. In Vitro Cell Dev Biol Plant 54:228–239
Chishaki N, Horiguchi T (1997) Responses of secondary metabolism in plants to nutrient deficiency. Soil Sci Plant Nutr 43(1):987–991
Chitturi D, Venisetty RK, Molmoori RK, Kokate CK, Apte SS (2010) Enhanced bio- production of withaferin A from suspension cultures of Withania somnifera. Ann Biol Res 1(2):77–86
Chupakhin E, Babich O, Prosekov A, Asyakina L, Gureev M, Krasavin M (2020) Plants of the Russian Federation pharmacopeia: an unexhausted natural products research opportunity? Nat Prod Res 35(21):3525–3527
Clissold SP, Heel RC (1985) Transdermal hyoscine (Scopolamine). A preliminary review of its pharmacodynamic properties and therapeutic efficacy. Drugs 29:189–207
Coelho MA, Ribeiro BD (2015) White biotechnology for sustainable chemistry. RCS Publishing, London
da Silva JA, Chin DP, Van PT, Mii M (2011) Transgenic orchids. Sci Hortic 130:673–680
de Carvalho AA, Bertolucci SKV, da Honorato AC et al (2020) Influence of light spectra and elicitors on growth and ascaridole content using in vitro cultures of Dysphania ambrosioides L. Plant Cell Tissue Organ Cult 143:277–290
de Matos NJ, Bertodo LOO, Da Rosa LMG, Von Poser GL, Rech SB (2014) Stress induction of valuable secondary metabolites in Hypericum polyanthemum acclimatized plants. S Afr J Bot 94:182–189
de Padua RM, Meitinger N, Waibel R et al (2012) Biotransformation of 21-O-acetyl- deoxycorticosterone by cell suspension cultures of Digitalis lanata (strain W. 1.4). Steroids 77:1373–1380
Dehghan E, Hakkinen ST, Oksman-Caldentey KM, Ahmadi FS (2012) Production of tropane alkaloids in diploid and tetraploid plants and in vitro hairy root cultures of Egyptian henbane (Hyoscyamus muticus L.). Plant Cell Tiss Org 110:35–44
Del Rosario CC, Chiappero J, Palermo TB et al (2020) Volatile organic compounds from rhizobacteria increase the biosynthesis of secondary metabolites and improve the antioxidant status in Mentha piperita L. grown under salt stress. Agronomy 10:1–16
Demasi S, Caser M, Lonati M, Cioni PL, Pistelli L, Najar B, Scariot V (2018) Latitude and altitude influence secondary metabolite production in peripheral alpine populations of the mediterranean species Lavandula angustifolia Mill. Front Plant Sci 9:983
Demole E, Lederer E, Mercier D (1962) Isolement et détermination de la structure du jasmonate de méthyle, constituant odorant caractéristique de l'essence de jasmin. Helv Chim Acta 45(2):675–685
Denaro M, Smeriglio A, Barreca D, De Francesco C, Occhiuto C, Milano G, Trombetta D (2020) Antiviral activity of plants and their isolated bioactive compounds: an update. Phytother Res 34(4):742–768
Dhiman N, Patial V, Bhattacharya A (2018) The current status and future applications of hairy root cultures. In: Biotechnological approaches for medicinal and aromatic plants. Springer, Singapore, pp 87–155
Dicosmo F, Misawa M (1995) Plant cell and tissue culture: alternatives for metabolite production. Biotechnol Adv 13:425–453
Dixit S, Shukla A, Singh V, Upadhyay SK (2021) Bioprospecting of natural compounds for industrial and medical applications: current scenario and bottleneck. In: Bioprospecting of plant biodiversity for industrial molecules. Wiley, Hoboken, pp 53–71
Dixon RA, Paiva NL (1995) Stress-induced phenylpropanoid metabolism. Plant Cell 7(7):1085
Domínguez-Robles J, Stewart SA, Rendl A, Gonzalez Z, Donnelly RF, Larraneta E (2019) Lignin and cellulose blends as pharmaceutical excipient for tablet manufacturing via direct compression. Biomolecules 9:423
Domínguez-Robles J, Cárcamo-Martínez Á, Stewart SA, Donnelly RF, Larrañeta E, Borrega M (2020) Lignin for pharmaceutical and biomedical applications–could this become a reality? Sustain Chem Pharm 18:100320

Donaldson K, Stone V, Tran C, Kreyling W, Borm PJ (2004) Nanotoxicology. Occup Environ Med 61:727–728
Dornenburg H, Knorr D (1995) Strategies for the improvement of secondary metabolite production in plant cell cultures. Enzym Microb Technol 17(8):674–684
Duangporn P, Siripong P (2009) Effect of auxin and cytokinin on phyllanthusol A production by callus cultures of Phyllanthus Acidus skeels. Am Eurasian J Agric Environ Sci 5:258–263
Dzamic AM, Sokovic MD, Novakovic M, Jadranin M, Ristic M, Teševic V, Marin P (2013) Composition, antifungal and antioxidant properties of Hyssopus Officinalis L. Subsp pilifer (Pant.) Murb. essential oil and deodorized extracts. Ind Crop Prod 51:401–407
Ebel J, Cosio EG (1994) Elicitors of plant defense responses. Int Rev Cytol 148:1–36
Efferth T (2019) Biotechnology applications of plant callus cultures. Engineering 5(1):50–59
Eibl R (2018) Plant cell culture technology in the cosmetics and food industries: current state and future trends. Appl Microbiol Biotechnol 102:8661–8675
El-Ashry AAE-L, Gabr AMM, Arafa NM, El-Bahr MK (2019) Rutin accumulation in Gardenia calli cultures as a response to phenyl alanine and salicylic acid. Bull Natl Res Cent 43(141):1–14
El-Hendawy S, Al-Suhaibani N, Elsayed S et al (2019a) Combining biophysical parameters, spectral indices and multivariate hyperspectral models for estimating yield and water productivity of spring wheat across different agronomic practices. PLoS ONE 14(3):e0212294
El-Hendawy SE, Al-Suhaibani NA, Hassan WM et al (2019b) Evaluation of wave lengths and spectral reflectance indices for high-throughput assessment of growth, water relations and ion contents of wheat irrigated with saline water. Agric Water Manag 212:358–377
Farheen R et al (2015) Triterpenoids and triterpenoid saponins from the aerial parts of Fagonia indica Burm. Phytochem Lett 13:256–261
Farré G, Blancquaert D, Capell T, Van Der Straeten D, Christou P, Zhu C (2014) Engineering complex metabolic pathways in plants. Annu Rev Plant Biol 65:187–223
Fu Y (2019) Biotransformation of ginsenoside Rb1 to Gyp-XVII and minor ginsenoside Rg3 by endophytic bacterium Flavobacterium sp. GE 32 isolated from Panax ginseng. Lett Appl Microbiol 68:134–141
Fu Y, Yin ZH, Wu LP, Yin CR (2016) Biotransformation of ginsenoside Rb1 to ginsenoside C-K by endophytic fungus Arthrinium sp GE 17–18 isolated from Panax ginseng. Lett Appl Microbiol 63:196–201
Gadzovska S, Maury S, Delaunay A, Spasenoski M, Hagege D, Courtois D, Joseph C (2013) The influence of salicylic acid elicitation of shoots, callus, and cell suspension cultures on production of naphtodianthrones and phenylpropanoids in Hypericum perforatum L. Plant Cell Tissue Organ Cult 113:25–39
Ghanati F, Bakhtiarian S (2014) Effect of methyl jasmonate and silver nanoparticles on production of secondary metabolites by Calendula officinalis L (Asteraceae). Trop J Pharm Res 13(11):1783–1789
Gharari Z, Bagheri K, Danafar H, Sharafi A (2020) Enhanced flavonoid production in hairy root cultures of Scutellaria bornmuelleri by elicitor induced over-expression of MYB7 and FNSП2 genes. Plant Physiol Biochem 148:35–44
Ghimire BK, Thiruvengadam M, Chung I-M (2019) Identification of elicitors enhances the polyphenolic compounds and pharmacological potential in hairy root cultures of Aster scaber. S Afr J Bot 125:92–101
Ghorbanpour M, Hadian J (2015) Multi-walled carbon nanotubes stimulate callus induction, secondary metabolites biosynthesis and antioxidant capacity in medicinal plant Satureja khuzestanica grown in vitro. Carbon 94:749–759
Ghosh S, Watson A, Gonzalez-Navarro OE et al (2018) Speed breeding in growth chambers and glasshouses for crop breeding and model plant research. Nat Protoc 13(12):2944–2963
Giri CC, Zaheer M (2016) Chemical elicitors versus secondary metabolite production in vitro using plant cell, tissue and organ cultures: recent trends and a sky eye view appraisal. Plant Cell Tissue Org Cult 126:1–18
Gonçalves S, Romano A (2013) In vitro culture of lavenders (Lavandula spp.) and the production of secondary metabolites. Biotechnol Adv 31:166–174

Hahn MG (1996) Microbial elicitors and their receptors in plants. Annu Rev Phytopathol 34:387–412
Hammer KA, Carson CF, Riley TV (1999) Antimicrobial activity of essential oils and other plant extracts. J Appl Microbiol 86(6):985
Hanagata N, Uehara H, Ito A, Takeuchi T, Karube I (1994) Elicitor for red pigment formation in Carthamus tinctorius cultured cells. J Biotechnol 34:71–77
Hassan A (2012) Effects of mineral nutrients on physiological and biochemical processes related to secondary metabolites production in medicinal herbs. Med Arom Plant Sci Biotechnol 6(1):105–110
Hedayati A, Hosseini B, Palazon J, Maleki R (2020) Improved tropane alkaloid production and changes in gene expression in hairy root cultures of two hyoscyamus species elicited by silicon dioxide nanoparticles. Plant Physiol Biochnol 155:416–428
Hedin PA (1983) Plant resistance to insects. American Chemical Society, Washington
Hosseini NS, Ghasimi Hagh Z, Khoshghalb H (2020) Morphological, antioxidant enzyme activity and secondary metabolites accumulation in response of polyethylene glycol-induced osmotic stress in embryo-derived plantlets and callus cultures of Salvia leriifolia. Plant Cell Tissue Organ Cult 140:143–155
Huttel W, Hoffmeister D (2010) Fungal biotransformations in pharmaceutical sciences. In: Hofrichter M (ed) Industrial applications, the mycota X, 2. Springer, Berlin, pp 293–317
Iiyama K, Wallis AFA (1990) Determination of lignin in herbaceous plants by an improved acetyl bromide procedure. J Sci Food Agric 51:145–161
Ileana M, Suhani A, Zahan M, Bunea A (2020) Effect of jasmonic acid and salicylic acid on growth and biochemical composition of in-vitro-propagated Lavandula angustifolia mill. Agronomy 10:1722
Isah T, Umar S, Mujib A, Sharma MP, Rajasekharan PE, Zafar N, Frukh A (2018) Secondary metabolism of pharmaceuticals in the plant in vitro cultures: strategies, approaches, and limitations to achieving higher yield. Plant Cell Tissue Org Cult 132:239–265
Jan R, Asaf S, Numan M, Lubna L, Kim KM (2021) Plant secondary metabolite biosynthesis and transcriptional regulation in response to biotic and abiotic stress conditions. Agronomy 11:968
Jeong C, Murthy HN, Hahn E, Paek K (2008) Improved production of ginsenosides in suspension cultures of ginseng by medium replenishment strategy. J Biosci Bioeng 105:288–291
Jiao Q, Li L, Mu Q, Zhang Q (2014) Immunomodulation of nanoparticles in nanomedicine applications. Biomed Res Int 2014:e426028
John H, Binder T, Hochstetter H, Thiermann H (2010) LC-ESI MS/MS quantification of atropine and six other antimuscarinic tropane alkaloids in plasma. Anal Bioanal Chem 396:751–763
Kamińska M (2021) Role and activity of jasmonates in plants under in vitro conditions. Plant Cell Tissue Organ Cult 146:3
Kang SM, Jung HY, Kang YM, Yun DJ, Bahk JD, Yang JK, Choi MS (2004) Effects of methyl jasmonate and salicylic acid on the production of tropane alkaloids and the expression of PMT and H6H in adventitious root cultures of Scopolia parviflora. Plant Sci 166:745–751
Kaptan E, Sancar-Bas S, Sancakli A, Bektas S, Bolkent S (2018) The effect of plant lectins on the survival and malignant behaviors of thyroid cancer cells. J Cell Biochem 119:6274–6287
Kebamo S, Tesema S, Geleta B (2015) The role of biotransformation in drug discovery and development. J Drug Metab Toxicol 6:196
Khan I, AbdElsalam NM, Fouad H, Tariq A, Ullah R, Adnan M (2014) Application of ethnobotanical indices on the use of traditional medicines against common diseases. Evid Based Complement Alternat Med 2014:635371
Khan MA, Abbasi BH, Ali H, Ali M, Adil M, Hussain I (2015) Temporal variations in metabolite profiles at different growth phases during somatic embryogenesis of Silybum marianum L. Plant Cell Tissue Organ Cult 120:127–139
Khani S, Barar J, Movafeghi A, Omidi Y (2012) Production of anticancer secondary metabolites: impacts of bioprocess engineering. In: Orhan IE (ed) Biotechnological production of secondary metabolites. Benthan, Potomac, pp 215–240

Kharde AV, Kore SV, Khetmalas MB (2018) Elicitation of Bacoside Content using plant growth regulators in cell suspension culture of Bacopa monnieri (L.) Wettst. Plant Tissue Cult Biotechnol 28(2):191–199
Khare S, Singh NB, Singh A et al (2020) Plant secondary metabolites synthesis and their regulations under biotic and abiotic constraints. J Plant Biol 63:203–216
Khezerluo M, Hosseini B, Amiri J (2018) Sodium nitroprusside stimulated production of tropane alkaloids and antioxidant enzymes activity in hairy root culture of Hyoscyamus reticulatus L. Acta Biol Hung 69:437–448
Khirwadkar P, Dave V, Dashora K (2014) A review on biotransformation. Indian J Res Pharm Biotechnol 2(2):1136
Khoddami A, Wilkes MA, Roberts TH (2013) Techniques for analysis of plant phenolic compounds. Molecules 18:2328–2375
Kim OT, Bang KH, Shin YS, Lee MJ, Jung SJ, Hyun DY, Kim YC, Seong NS, Cha SW, Hwang B (2007) Enhanced production of asiaticoside from hairy root cultures of Centella asiatica (L.) urban elicited by methyl jasmonate. Plant Cell Rep 26(11):1941–1949
Kim BS, Hwang HS, Park H, Kim HH (2015) Effects of selective cleavage of high-mannose-type glycans of Maackia Amurensis leukoagglutinin on sialic acid-binding activity. Biochim Biophys Acta 1850:1815–1821
Kitisripanya T, Komaikul J, Tawinkan N, Atsawinkowit C, Putalun W (2013) Dicentrine production in callus and cell suspension cultures of Stephania venosa. Nat Prod Commun 8(4):443–445
Koufan M, Belkoura I, Mazri MA et al (2020) Determination of antioxidant activity, total phenolics and fatty acids in essential oils and other extracts from callus culture, seeds and leaves of Argania spinosa (L.) skeels. Plant Cell Tissue Organ Cult 141:217–227
Koul A, Mallubhotla S (2020) Elicitation and enhancement of bacoside production using suspension cultures of Bacopa monnieri (L.). 3 Biotech 10:1–4
Krishnan ML, Roy A, Bharadvaja N (2019) Elicitation effect on the production of asiaticoside and asiatic acid in shoot, callus, and cell suspension culture of Centella asiatica. J Appl Pharm Sci 9(6):67–74
Krstić-Milošević D, Janković T, Uzelac B, Vinterhalter D, Vinterhalter B (2017) Effect of elicitors on xanthone accumulation and biomass production in hairy root cultures of *Gentiana dinarica*. Plant Cell Tissue Organ Cult 130:631–640
Ksouri R, Ksouri WM, Jallali I, Debez A, Magné C, Hiroko I, Abdelly C (2012) Medicinal halophytes: potent source of health promoting biomolecules with medical, nutraceutical and food applications. Crit Rev Biotechnol 32:289–326
Laloo D, Kalita JM, Prasad SK (2021) Molecular docking studies of plant-derived bioactive compounds: a comprehensive in silico standardization approach. In: Evidence based validation of traditional medicines. Springer, Singapore, pp 371–404
Lan H, Lai B, Zhao P, Dong X, Wei W, Ye Y, Wu Z (2020) Cucumber mosaic virus infection modulated the phytochemical contents of Passiflora edulis. Microb Pathog 138:103828
Largia MJV, Pothiraj G, Shilpha J, Ramesh M (2015) Methyl jasmonate and salicylic acid synergism enhances bacoside A content in shoot cultures of Bacopa monnieri (L). Plant Cell Tissue Organ Cult 122(1):9–20
Lazzarini LES, Bertolucci SKV, Pacheco FV et al (2018) Quality and intensity of light affect Lippia gracilis Schauer plant growth and volatile compounds in vitro. Plant Cell Tissue Organ Cult 135:367–379
Lee J-Y, Hiyama M, Hikosaka S, Goto E (2020) Effects of concentration and temperature of nutrient solution on growth and camptothecin accumulation of Ophiorrhiza pumila. Plants 9:793
Li Y, Tao W (2009) Paclitaxel-producing fungal endophyte stimulates the accumulation of taxoids in suspension cultures of Taxus cuspidate. Sci Hortic 121:97–102
Li Y, Kong D, Fu Y et al (2020) The effect of developmental and environmental factors on secondary metabolites in medicinal plants. Plant Physiol Biochem 148:80–89
Liang L, Wang C, Li S, Chu X, Sun K (2019) Nutritional compositions of Indian Moringa oleifera seed and antioxidant activity of its polypeptides. Food Sci Nutr 7:1754–1760

Louis B, Sehrish I, Kiran N, Dobgima JF, Elsie LY, Robinson CJ, Yiboh MTN, Pranab R (2019) Biotechnological application of endophytic filamentous bipolaris and curvularia: a review on bioeconomy impact. World J Microbiol Biotechnol 35:69

Mahajan M, Kuiry R, Pal PK (2020) Understanding the consequence of environmental stress for accumulation of secondary metabolites in medicinal andaromatic plants. J Appl Res Med Aromat Plants 18:100255

Mahalakshmi R, Eganathan P, Parida AK (2013) Salicylic acid elicitation on production of secondary metabolite by cell cultures of Jatropha curcas L. Int J Pharm Pharm Sci 5(4):655–659

Mahendran D, Sreeramanan S, Venkatachalam P (2018) Enhanced biosynthesis of colchicine and thiocolchicoside contents in cell suspension cultures of Gloriosa superba L. exposed to ethylene inhibitor and elicitors. Ind Crop Prod 120:123–130

Makowski W, Tokarz B, Banasiuk R et al (2019) Is a blue–red light a good elicitor of phenolic compounds in the family Droseraceae? A comparative study. J Photochem Photobiol B Biol 201:111679

Makowski W, Tokarz KM, Tokarz B et al (2020a) Elicitation-based method for increasing the production of antioxidant and bactericidal phenolic compounds in Dionaea muscipula. J Ellis Tissue Mol 25:1794

Makowski W, Tokarz KM, Tokarz B, Banasiuk R, Witek K, Królicka A (2020b) Elicitation-based method for increasing the production of antioxidant and bactericidal phenolics compounds in Dionaea muscipula. J Ellis Tissue Molecules 25:1794

Marchev AS, Yordanova ZP, Georgiev MI (2020) Green (cell) factories for advanced production of plant secondary metabolites. Crit Rev Biotechnol 40:443–458

Martinez MJA, Lazaro RM, del Olmo LMB, Benito PB (2008) Anti-infectious activity in the anthemideae tribe. Stud Nat Prod Chem 35:45–516

Martínez VMV, Estrada-Soto SE, Arellano-García JJ, Rivera-Leyva JC, Castillo- España P et al (2017) Methyl Jasmonate and salicylic acid enhanced the production of ursolic and oleanolic acid in callus cultures of Lepechinia Caulescens. Pharmacogn Mag 13(4):S886–S889

Mathur L, Yadav RK (2011) Effect of salicylic acid on trigonelline production in Trigonella foenum-graecum L. cell suspension culture. Int Referred Res J 1(17):137–138

Matkowski A (2008) Plant in vitro culture for the production of antioxidants—a review. Biotechnol Adv 26:548–560

Maucieri C, Nicoletto C, Van Os E, Anseeuw D, Van Havermaet R, Junge R (2019) Hydroponic technologies. Aquapon Food Prod Syst 10:77

Maxmen A (2020) More than 80 clinical trials launch to test coronavirus treatments. Nature 578(7795):347–348

Mendoza D, Cuaspud O, Arias JP, Ruiz O, Arias M (2018) Effect of salicylic acid and methyl jasmonate in the production of phenolic compounds in plant cell suspension cultures of Thevetia peruviana. Biotechnol Rep 19:e00273

Miclea I, Suhani A, Zahan M, Bunea A (2020) Effect of jasmonic acid and salicylic acid on growth and biochemical composition of in-vitro-propagated Lavandula angustifolia Mill. Agronomy 10:1722

Mihálik D, Hanˇcinský R, Kanuková Š, Mrkvová M, Kraic J (2022) Elicitation of hyoscyamine production in Datura stramonium L. plants using tobamoviruses. Plants 11:3319

Miladinova-Georgieva K, Geneva M, Stancheva I, Petrova M, Sichanova M, Kirova E (2023) Effects of different elicitors on micropropagation, biomass and secondary metabolite production of Stevia rebaudiana Bertoni—A. Rev Plants 12(1):153

Milne N, Thomsen P, Mølgaard Knudsen N, Rubaszka P, Kristensen M, Borodina I (2020) Metabolic engineering of Saccharomyces cerevisiae for the de novo production of psilocybin and related tryptamine derivatives. Metab Eng 60:25–36

Mipeshwaree Devi A, Khedashwori Devi K, Premi Devi P, Lakshmipriyari Devi M, Das S (2023) Metabolic engineering of plant secondary metabolites: prospects and its technological challenges. Front Plant Sci 14:1171154

Moharrami F, Hosseini B, Sharafi A, Farjaminezhad M (2017) Enhanced production of hyoscyamine and scopolamine from genetically transformed root culture of Hyoscyamus reticulatus L. elicited by iron oxide nanoparticles. In Vitro Cell Dev Biol Plant 53:104–111
Mohsen N, Sabbatini P (2021) Traditional in vitro strategies for sustainable production of bioactive compounds and manipulation of metabolomic profile in medicinal, aromatic and ornamental plants. Planta 254:111
Moscatiello R, Baldan B, Navazio L (2013) Plant cell suspension cultures. In: Yehuda S, Mostofsky DI (eds) Plant mineral nutrients. Humana Press, Totowa, pp 77–93
Murthy HN, Lee EJ, Paek KY (2014) Production of secondary metabolites from cell and organ cultures: strategies and approaches for biomass improvement and metabolite accumulation. Plant Cell Tissue Org Cult 118:1–16
Naik P, Al-Khayri J (2016a) Abiotic and biotic elicitors–role in secondary metabolites production through In vitro culture of medicinal plants. In: Shanker AK, Shanker C (eds) Abiotic and biotic stress in plants-recent advances and future perspectives. IntechOpen, London, pp 247–277
Naik PM, Al-Khayri JM (2016b) Abiotic and biotic elicitors-role in secondary metabolites production through in vitro culture of medicinal plants. In: Abiotic and biotic stress in plants—recent advances and future perspectives. IntechOpen, Rijeka, pp 247–277
Nanda S, Mohanty JN, Mishra R, Joshi RK (2016) Metabolic engineering of phenyl propanoids in plants. In: Jha S (ed) Transgenesis and secondary metabolism: part of the series reference series in phytochemistry. Springer, New York, pp 1–26
Niazian M, Sadat-Noori SA, Tohidfar M et al (2019) Agrobacterium-mediated genetic transformation of ajowan (Trachyspermum ammi (L.) Sprague): an important industrial medicinal plant. Ind Crop Prod 132:29–40
Nielsen J (2001) Metabolic engineering. Appl Microbiol Biotechnol 55:263–283
Nieves B, García-Viguera C, Moreno DA (2014) Elicitation: a tool for enriching the bioactive composition of foods. Molecules 19:13541–13563
Niinemets Ü (2015) Uncovering the hidden facets of drought stress: secondary metabolites make the difference. Tree Physiol 36(2):129–132
Nürnberger T (1999) Signal perception in plant pathogen defense. Cell Mol Life Sci 55:167–182
Ogita S (2015) Plant cell, tissue and organ culture: the most flexible foundations for plant metabolic engineering applications. Nat Prod Commun 10(5):815–820
Pagare S, Bhatia M, Tripathi N, Pagare S, Bansal Y (2015) Secondary metabolites of plants and their role: overview. Curr Trends Biotechnol Pharm 9(3):293–304
Park YJ, Kim JK, Park SU (2020) Yeast extract improved biosynthesis of astragalosides in hairy root cultures of Astragalus membranaceus. Prep Biochem Biotechnol 51(5):467–474
Pavlov A, Popov S, Kovacheva E, Milen Georgiev M, Ilieva M (2005) Volatile and polar compounds in Rosa damascena Mill 1803 cell suspension. J Biotechnol 118:89–97
Pervaiz I, Ahmad S, Madni MA, Ahmad H, Khaliq FH (2013) Microbial biotransformation: a tool for drug designing (review). Prikl Biokhim Mikrobiol 49(5):435–449
Piasecka A, Sawikowska A, Kuczyńska A et al (2017) Drought-related secondary metabolites of barley (Hordeum vulgare L.) leaves and their metabolomic quantitative trait loci. Plant J 89(5):898–913
Pickens LB, Tang Y, Chooi YH (2011) Metabolic engineering for the production of natural products. Annu Rev Chem Biomol Eng 2:211–236
Pourianezhad F, Rahnama H, Mousavi A et al (2019) Effects of combined elicitors on parthenolide production and expression of parthenolide synthase (TpPTS) in Tanacetum parthenium hairy root culture. Plant Biotechnol Rep 13:211–218
Pudersell K, Vardja T, Vardja R, Matto V, Arak E, Raal A (2012) Inorganic ions in the medium modify tropane alkaloids and riboflavin output in Hyoscyamus niger root cultures. Pharmacogn Mag 8:73–77
Quan NT, Anh LH, Khang DT et al (2016) Involvement of secondary metabolites in response to drought stress of rice (Oryza sativa L.). Agriculture 6(2):23

Radziejewska I, Borzym-Kluczyk M, Leszczynska K (2018) Lotus Tetragonolobus, Ulex Europaeus, Maackia Amurensis, and Arachis Hypogaea (peanut) lectins influence the binding of Helicobacter Pylori to gastric carbohydrates. Adv Clin Exp Med 27:807–811

Rajendran J, Muthuramalingam P, Karthick K, Shin H, Choi SH, Ramesh M (2023) Methyl jasmonate and salicylic acid as powerful elicitors for enhancing the production of secondary metabolites in medicinal plants: an updated review. J Plant Biotechnol 153:447–458

Ramakrishna A, Ravishankar GA (2011) Influences of abiotic stress signals on secondary metabolites in plants. Plant Signal Behav 6(11):1720–1731

Ramawat KG, Goyal S (2020) Co-evolution of secondary metabolites during biological competition for survival and advantage: an overview. In: Co-evolution of secondary metabolites. Springer, Berlin, pp 3–17

Rho H, Van Epps V, Kim SH, Doty SL (2020) Endophytes increased fruit quality with higher soluble sugar production in honeycrisp apple (Malus pumila). Microorganisms 8:2076–2607

Rogerio AP, Sá-Nunes A, Faccioli LH (2010) The activity of medicinal plants and secondary metabolites on eosinophilic inflammation. Pharmacol Res 62(4):298–307

Roustakhiz J, Saboki E (2016) Plant tissue culture of Silybum marianum L. and check out elicitor effect on the amount of silymarin. J Nov Appl Sci 5:161–168

Rubio A, Rambla JL, Santaella M, Gómez MD, Orzaez D et al (2008) Cytosolic and plastoglobule-targeted carotenoid dioxygenases from Crocus sativus are both involved in β-ionone release. J Biol Chem 283(36):24816–24825

Ryan CA (1988) Oligosaccharides as recognition signals for the expression of defensive genes in plants. Biotechnol Adv 18:91–120

Saeed S, Ali H, Khan T, Kayani W, Khan MA (2017) Impacts of methyl jasmonate and phenyl acetic acid on biomass accumulation and antioxidant potential in adventitious roots of Ajuga bracteosa Wall ex Benth, a high valued endangered medicinal plant. Physiol Mol Biol Plants 23:229–237

Saikat G, Das A, Mitra M, Chen J-T (2021) Secondary metabolites in orchids: biosynthesis, medicinal uses, and Biotechnology. S Afr J Bot 139:338–351

Saleem S et al (2014) Plants Fagonia cretica L. and Hedera nepalensis K. Koch contain natural compounds with potent dipeptidyl peptidase-4 (DPP-4) inhibitory activity. J Ethnopharmacol 156:26–32

Santisree P, Sanivarapu H, Gundavarapu S et al (2020) Nitric oxide asa signal in inducing secondary metabolites during plant stress. In: Merillon JM, Ramawat K (eds) Co-evolution of secondary metabolites. Reference series in phytochemistry. Springer, Cham, pp 593–621

Santos GS, Sinoti SBP, de Almeida FTC et al (2020) Use of galantamine in the treatment of Alzheimer's disease and strategies to optimize its biosynthesis using the in vitro culture technique. Plant Cell Tissue Organ Cult 143:13–29

Satish L, Rency AS, Muthubharathi BC, Shamili S, Rameshkumar R et al (2019) Transgenic plant cell cultures: a promising approach for secondary metabolite production. In: Natural bio-active compounds. Springer, Singapore, pp 79–122

Sepehri E, Hosseini B, Hedayati A (2022) The effect of iron oxide nano-particles on the production of tropane alkaloids, h6h gene expression and antioxidant enzyme activity in Atropa belladonna hairy roots. Russ J Plant Physiol 69:122

Shahhoseini R, Azizi M, Asili J et al (2020) Effects of zinc oxide nanoelicitors on yield, secondary metabolites, zinc and iron absorption of Feverfew (Tanacetum parthenium (L.) Schultz Bip). Acta Physiol Plant 42:52

Shaikh S, Shriram V, Khare T, Kumar V (2020) Biotic elicitors enhance diosgenin production in Helicteres isora L. suspension cultures via up-regulation of CAS and HMGR genes. Physiol Mol Biol Plants 26:593–604

Sharafi E, Fotokian MH, Loo H (2013) Improvement of hypericin and hyperforin production using zinc and iron nano-oxides as elicitors in cell suspension culture of John's wort (Hypericum perforatum L). J Med Plants By-prod 2(2):177–184

Sharan S, Sarin NB, Mukhopadhyay K (2019) Elicitor-mediated enhanced accumulation of ursolic acid and eugenol in hairy root cultures of Ocimum tenuiflorum L. is age, dose, and duration dependent. S Afr J Bot 124:199–210

Sharma U, Agrawal V (2018) In vitro shoot regeneration and enhanced synthesis of plumbagin in root callus of Plumbago zeylanica L.—an important medicinal herb. In Vitro Cell Dev Biol 54:423–435

Sharma M, Ahuja A, Gupta R, Mallubhotla S (2015) Enhanced bacoside production in shoot cultures of Bacopa monnieri under the influence of abiotic elicitors. Nat Prod Res 29:745–749

Sharma A, Amin D, Sankaranarayanan A, Arora R, Mathur AK (2020) Present status of Catharanthus roseus monoterpenoid indole alkaloids engineering in homo- and hetero-logous systems. Biotechnol Lett 42:11–23

Shi M, Gong H, Cui L, Wang Q, Wang C, Wang Y, Kai G (2020) Targeted metabolic engineering of committed steps improves anti-cancer drug camptothecin production in Ophiorrhiza pumila hairy roots. Ind Crop Prod 148:112277

Shiringani RP, Shimelis HA (2011) Yield response and stability among cowpea genotypes at three planting dates and test environments. Afr J Agric Res 6(14):3259–3263

Silva ST, Bertolucci SKV, da Cunha SHB et al (2017) Effect of light and natural ventilation systems on the growth parameters and carvacrol content in the in vitro cultures of Plectranthus amboinicus (Lour.) Spreng. Plant Cell Tissue Organ Cult 129:501–510

Sin VJ, Anand GS, Koh HL (2021) Botanical medicine and natural products used for erectile dysfunction. Sex Med Rev 9(4):568–592

Singh A, Rani R, Sharma M (2018) Medicinal herbs of Punjab (India). Biol Forum 10:10–27

Singh P, Prasad R, Tewari R, Jaidi M, Kumar S, Rout PK, Rahman L (2018a) Silencing of quinolinic acid phosphoribosyl transferase (QPT) gene for enhanced production of scopolamine in hairy root culture of Duboisia leichhardtii. Sci Rep 8:13939

Singh NR, Rath SK, Behera S, Naik SK, Prasad R, Kumar V (2018b) In vitro secondary metabolite production through fungal elicitation: an approach for sustainability. In: Fungal nanobionics: principles and applications. Springer, Singapore, pp 215–242

Singh A, Sengar RS, Rajput VD, Minkina T, Singh RK (2022) Zinc oxide nanoparticles improve salt tolerance in rice seedlings by improving physiological and biochemical indices. Agriculture 12:1014

Sivanandhan G, Selvaraj N, Ganapathi A, Manickavasagam M (2016) Elicitation approaches for withanolide production in hairy root culture of Withania somnifera (L.) Dunal. Methods Mol Biol 1405:1–18

Smitha MS, Singh S, Singh R (2017) Microbial biotransformation: a process for chemical alterations. J Bacteriol Mycol 4(2):47–51

Srivastava V, Mehrotra S, Verma PK (2017) Biotechnological interventions for production of therapeutic secondary metabolites using hairy root cultures of medicinal plants. In: Current developments in biotechnology and bioengineering: crop modification, nutrition, and food production. Elsevier, Amsterdam, pp 259–282

Sukito A, Tachibana S (2016) Effect of methyl jasmonate and salicylic acid synergism on enhancement of bilobalide and ginkgolide production by immobilized cell cultures of Ginkgo biloba. Bioresour Bioprocess 3:24

Sultana N, Saify ZS (2013) Enzymatic biotransformation of terpenes as bioactive agents. J Enzyme Inhib Med Chem 28(6):1113–1128

Sun H, Xu Z-X, Zhou J, Guo L-P (2020) Karrikins-induced accumulation of Salvianolic acid B is mediated by calcium–calmodulin, brassinolide and jasmonic acid in Salvia miltiorrhiza. S Afr J Bot 132:45–49

Sun Y, Shang L, Zhu QH, Fan L, Guo L (2022) Twenty years of plant genome sequencing: achievements and challenges. Trends Plant Sci 27(4):391–401

Szopa A, Kokotkiewicz A, Krol A, Luczkiewicz M, Ekiert H (2018) Improved production of dibenzocyclooctadiene lignans in the elicited microshoot cultures of Schisandra chinensis (Chinese magnolia vine). Appl Microbiol Biotechnol 102:945–959

Ten Hoopen HJG, Vinke JL, Moreno PRH et al (2002) Influence of temperature on growth and ajmalicine production by Catharanthus roseus suspension cultures. Enzyme Microb Technol 30:56–65

Tõnutare T (2015) Possibilities to affect antioxidant properties of strawberries and some methodical aspects in their determination. Ph.D Thesis, Estonian University of the Life Sciences, Institute of Agricultural and Environmental Sciences, Tartu, Estonia

Trenchard IJ, Siddiqui MS, Thodey K, Smolke CD (2015) De novo production of the key branch point benzylisoquinoline alkaloid reticuline in yeast. Metab Eng 31:74–83

Ullah MA, Tungmunnithum D, Garros L et al (2019) Effect of ultraviolet-C radiation and melatonin stress on biosynthesis of antioxidant and antidiabetic metabolites produced in in vitro callus cultures of Lepidium sativum L. Int J Mol Sci 20:1–19

Utami ESW, Hariyanto S, Manuhara YSW (2018) Agrobacterium tumefaciens-mediated transformation of Dendrobium lasianthera J.J.Sm: an important medicinal orchid. J Genet Eng Biotechnol 16:703–709

Vakil Moinuddin MA, Mendhulkar Vijay D (2013) Enhanced synthesis of andrographolide by Aspergillus niger and Penicillium expansum elicitors in cell suspension culture of Andrographis paniculata (Burm. f.) Nees. Bot Stud 54:49

Vasconsuelo A, Boland R (2007) Molecular aspects of the early stages of elicitation of secondary metabolites in plants. Plant Sci 172:861–875

Venzhik YV, Moshkov IE, Dykman LA (2021) Gold nanoparticles in plant physiology: principal effects and prospects of application. Russ J Plant Physiol 68:401–412

Wang Y, Mostafa S, Zeng W, Jin B (2021) Function and mechanism of jasmonic acid in plant responses to abiotic and biotic stresses. Int J Mol Sci 22(16):8568

Watanabe S, Okoshi H, Yamabe S, Shimada M (2021) Moringa oleifera lam. in diabetes mellitus: a systematic review and meta-analysis. Molecules 26:12

Weathers PJ, Towler MJ, Xu J (2010) Bench to batch: advances in plant cell culture for producing useful products. Appl Microbiol Biotechnol 85:1339–1351

Wei T, Deng K, Gao Y et al (2020) SmKSL overexpression combined with elicitor treatment enhances tanshinone production from Salvia miltiorrhiza hairy roots. Biochem Eng J 158:107562

Welchen E, Canal MV, Gras DE, Gonzalez DH (2021) Cross-talk between mitochondrial function, growth, and stress signaling pathways in plants. J Exp Bot 72:4102–4118

Wink M (1988) Plant breeding: importance of plant secondary metabolites for protection against pathogens and herbivores. Theor Appl Genet 75:225–233

Wink M (2015) Modes of action of herbal medicines and plant secondary metabolites. Medicine 2(3):251–286

Woolston BM, Edgar S, Stephanopoulos G (2013) Metabolic engineering: past and future. Annu Rev Chem Biomol Eng 4:259–288

Xu J, Ge X, Dolan MC (2011a) Towards high-yield production of pharmaceutical proteins with plant cell suspension cultures. Biotechnol Adv 29:278–299

Xu X, Zhang W, Cao X, Xue S (2011b) Abietane diterpenoids synthesized by suspension-cultured cells of Cephalotaxus fortunei. Phytochem Lett 4:52–55

Xuan TD, Khanh TD, Khang DT, Quan NT, Elzaawely AA (2016) Changes in chemical composition, total phenolics and antioxidant activity of Alpinia (Alpinia zerumbet) leaves exposed to UV. Int Lett Nat Sci 55:25–34

Yadav M, Gupta P, Seth CS (2022) Foliar application of α-lipoic acid attenuates cadmium toxicity on photosynthetic pigments and nitrogen metabolism in Solanum lycopersicum L. Acta Physiol Plant 44:112

Yang L, Stöckigt J (2010) Trends for diverse production strategies of plant medicinal alkaloids. Nat Prod Rep 27:1469–1479

Yang L, Wen KS, Ruan X, Zhao YX, Wei F et al (2018) Response of plant secondary metabolites to environmental factors. Molecules 23(4):762

Ye M, Ning L, Zhan J et al (2003) Biotransformation of cinobufagin by cell suspension cultures of Catharanthus roseus and Platycodon grandiflorum. J Mol Catal Benzym 22:89–95

Yin X, Singer SD, Qiao H, Liu Y, Jiao C, Wang H, Li Z, Fei Z, Wang Y, Fan C (2016) Insights into the mechanisms underlying ultraviolet-C induced resveratrol metabolism in grapevine (V. amurensis Rupr.) cv. "Tonghua-3". Front Plant Sci 7:503

Yu Y, Wang T, Wu Y et al (2019) Effect of elicitors on the metabolites in the suspension cell culture of Salvia miltiorrhiza Bunge. Physiol Mol Biol Plants 25:229–242

Yue W, Ming Q-L, Lin B, Rahman K, Zheng C-J, Han T, Qin L-P (2016) Medicinal plant cell suspension cultures: pharmaceutical applications and high-yielding strategies for the desired secondary metabolites. Crit Rev Biotechnol 36:215–232

Yun DJ, Hashimoto T, Yamada Y (1992) Metabolic engineering of medicinal plants: transgenic Atropa belladonna with an improved alkaloid composition. Proc Natl Acad Sci 89:11799–11803

Zayed R, Wink M (2005) Beta-carboline and quinoline alkaloids in root cultures and intact plants of Peganum harmala. Z Naturforsch C J Biosci 60:451–458

Zeng C, Zhu X, Cui Z, Li Y (2015) Antifungal activity of plant extracts against Embellisia astragali, the fungal causal agent of yellow dwarf and root-rot disease of standing milkvetch. Crop Pasture Sci 66:389–395

Zhai X, Jia M, Chen L, Zheng C, Rahman K, Han T, Qin L (2017) The regulatory mechanism of fungal elicitor-induced secondary metabolite biosynthesis in medical plants. Crit Rev Microbiol 43:238–261

Zhang Y, Zhong J (1997) Hyperproduction of ginseng saponin and polysaccharide by high density cultivation of Panax notoginseng cells. Enzyme Microb Technol 21:59–63

Zhang X, Ye M, Dong Y et al (2011) Biotransformation of bufadienolides by cell suspension cultures of Saussurea involucrata. Phytochemistry 72:1779–1785

Zhang B, Zheng LP, Li WY, Wang JW (2013) Stimulation of artemisinin production in Artemisia annua hairy roots by Ag-SiO_2 core-shell nanoparticles. Curr Nanosci 9(3):363–370

Zhao J, Davis LC, Verpoorte R (2005) Elicitor signal transduction leading to production of plant secondary metabolites. Biotechnol Adv 23:283–333

Zhao A, Li L, Li B, Zheng M, Tsao R (2016) Ultrafiltration LC-ESI-MSn screening of 5-lipoxygenase inhibitors from selected Chinese medicinal herbs Saposhnikovia Divaricata, Smilax Glabra, Pueraria Lobata and Carthamus Tinctorius. J Funct Foods 24:244–253

Zheng Q, Zheng YP, Wang GD, Guo WM, Fan WF, Wang C (2012) Sonication assisted Agrobacterium-mediated transformation of the ACC gene to interfere the production of ethylene in spring Dendrobium cv. 'Sanya'. Russ J Plant Physiol 59:266–274

Zheng LP, Tian H, Yuan YF, Wang JW (2016) The influence of endophytic Penicillium oxalicum B4 on growth and artemisinin biosynthesis of in vitro propagated plantlets of Artemisia annua L. Plant Growth Regul 80:93–102

Zheng H, Fu X, Shao J, Tang Y, Yu M, Li L, Huang L, Tang K (2023) Transcriptional regulatory network of high-value active ingredients in medicinal plants. Trends Plant Sci 28:429–446

Zhou J, Xu Z, Ran Z et al (2018) Effects of smoke-water and smoke derived butenolide on accumulation of phenolic acids in cultured hairy roots of Salvia miltiorrhiza Bung. Bangladesh J Bot 47:479–485

Złotek U, Swieca M, Jakubczyk A (2014) Effect of abiotic elicitation on main health-promoting compounds, antioxidant activity and commercial quality of butter lettuce (Lactuca sativa L.). Food Chem 148:253–260

18 Nutraceuticals Derived from Medicinal Plants: A Pediatric Perspective

Yogesh Nikam, Mandar Muley, Satyendra Prasad, and Prakash Itankar

Abstract

Nutraceuticals derived from medicinal plants have gained significant attention for their potential to provide essential nutrients and health benefits to children. Nutraceuticals are predicted to reach a global market value of US$441.7 billion by 2026, with a compound annual growth rate (CAGR) of 7.8% during the forecast period. In 2022, the industry was valued at US$330.6 billion. This chapter aims to present a comprehensive overview of nutraceutical formulations sourced from medicinal plants that cater specifically to the nutritional needs and overall well-being of children. The utilization of local and seasonal natural sources to develop safe and effective products for pediatric nutrition can be an emerging field that requires thorough investigation. This chapter discusses the nutraceutical formulations for the pediatric segment, which provides choice among the functional foods available in the market to meet nutritional requirements as well as for alleviating the metabolic disorders associated with dietary dysregulations. It explores the phytochemical constituents present in plants, such as carbohydrates, proteins, vitamins, minerals, polyphenols, flavonoids, and essential fatty acids, which contribute to growth, immunity, cognitive development, and overall vitality in children. The local approach to nutraceutical products from local flora and fauna is another crucial aspect covered in this chapter. Why the same nutraceutical formulations are not emulated everywhere, why there is no universal formula to tackle the nutritional needs of the children and how to meet the nutraceutical needs of the community using the local resource of medicinal and edible plants is discussed in the present chapter.

Y. Nikam · M. Muley · S. Prasad · P. Itankar (✉)
Department of Pharmaceutical Sciences, Rashtrasant Tukadoji Maharaj Nagpur University, Nagpur, Maharashtra, India

© The Author(s), under exclusive license to Springer Nature Singapore Pte Ltd. 2024

M. Nandave et al. (eds.), *Ethnopharmacology and OMICS Advances in Medicinal Plants Volume 1*, https://doi.org/10.1007/978-981-97-2367-6_18

Keywords

Nutraceuticals · Children · Malnutrition · Medicinal plants · Pediatric

18.1 Introduction

Nutraceuticals can be defined as specially formulated preparations intended to meet specific dietary requirements and provide preventive healthcare benefits. They encompass formulations of nutrients designed to prevent and treat certain diseases while also acting as nutritional supplements. Coined by Dr. Stephen De Felice in 1989, the term "nutraceutical" originates from the fusion of "nutrition" and "pharmaceutical." These substances, whether found in foods or comprising specific food components, offer a range of health advantages, including disease prevention and treatment. The realm of nutritional science has expanded significantly, progressing from addressing nutrient deficiencies to focusing on human health as well as the prevention and management of chronic conditions. The terminology of "nutraceuticals," "food supplements," and "dietary supplements" has emerged subsequent to Dr. De Felice's conceptualization.

Regulatory bodies have not rigidly distinguished between food supplements and nutraceuticals. However, contemporary literature underscores the need to redefine the nutraceutical concept, incorporating considerations of efficacy, safety, and potential toxicity. Food products serve as nourishing substances that sustain life, supply energy, and support growth when consumed. Presently, the extraction of nutrients from these food items is widely acknowledged and practiced. Establishing an epidemiological objective is the first step in differentiating food/dietary supplements from nutraceuticals. This is followed by safety and efficacy studies that clarify the mechanisms of action. One potential distinction between these categories is labeling "food supplements" as agents aimed at rectifying deficiencies in micro- or macronutrients. In contrast, the use of a "nutraceutical" for treating a pathological condition necessitates robust scientific substantiation (Puri et al. 2022).

Nutraceuticals encompass both individual and blended forms of pro- and prebiotic substances, as well as specialized medical foods. Meanwhile, food supplements contain either separate or combined minerals, vitamins, protein supplements, functional foods, and herbal products. By potentially extending or even replacing the reliance on pharmaceuticals for individuals eligible for an alternate non-pharmacological approach to addressing a medical condition, the integration of nutraceuticals into everyday dietary routines could contribute to the preventive management of pathological disorders (Keservani et al. 2017).

Herbal bioactive compounds, such as carotenoids, coumarins, flavonoids, lignans, phthalates, plant sterols, polyphenols, saponins, sulfides, and terpenoids, constitute a significant class of nutraceuticals. Throughout history, we have harnessed the power of herbs for both culinary and medicinal applications. These herbs play a crucial role in maintaining human well-being and elevating the overall quality of life (Sachdeva et al. 2020). In contemplating the nutritional needs of both children and adults, it's essential to keep in mind that resorting to supplementation should be a final consideration for delivering essential nutrients. The foundation of sound nutrition should rest upon a balanced diet, potentially enhanced with fortified and functional foods if deemed necessary (Piekara et al. 2022).

18.2 Global Nutraceutical Market Scenario

According to a recent survey, the nutraceutical market is experiencing significant global expansion, with projections suggesting it could potentially reach a market value of $340 billion by 2024. Between 2016 and 2024, the compound annual growth rate (CAGR) for nutraceuticals is projected to be approximately 7.2%. This notable upsurge in the nutraceutical industry's growth can be attributed to several factors, including an increasing demand for nutraceutical products, growing public awareness about the advantages of nutrition, and a noticeable upward trend in healthcare engagement (Da Costa 2017).

At present, the nutraceutical market is predominantly dominated by Europe, the USA, and Japan, collectively accounting for over 90% of the global market share. Projections indicate that the worldwide nutraceutical market is set to expand from $247 billion in 2019 to approximately $336 billion by 2023, indicating an 8% compound annual growth rate (CAGR). As the global market approaches maturity, the attention of nutraceutical industry stakeholders is shifting toward emerging economies, particularly in the Asia Pacific region, with India taking center stage. The Indian market is expected to grow at a remarkable CAGR of 21% from its modest 2% share of the global nutraceutical market in 2017 to $11 billion by 2023. By 2023, India is also projected to command a market share of at least 3.5% on the global stage (Chopra et al. 2022; Chauhan et al. 2013).

18.3 Overview of Child Nutrition and Nutraceuticals Derived from Medicinal Plants

Approximately 50% of child deaths under 5 results from undernutrition, which increases infection risk severity and hampers recovery. Undernutrition and infection create a dangerous cycle. Insufficient nutrition in a child's first 1000 days leads to stunted growth, affecting cognitive ability and future potential. Despite progress, the 2023 UNICEF-WHO-World Bank Group Joint Malnutrition Estimates reveal over 148 million children under 5 were stunted in 2022, with 45 million experiencing wasting. Global child overweight cases increased from 33 to 37 million from 2000 to 2022. Child malnutrition metrics are vital for tracking development progress, particularly the Sustainable Development Goal 2.2 to "eradicate all manifestations of malnutrition by 2030." In 2022, one-third of children faced stunting in certain regions, while Europe, Central Asia, and Northern America had lower rates. Latin America and the Caribbean showed disparities, with some countries having very high stunting rates, while others were below 2.5%. Stunting affected 22.3% of children under 5 in 2022, improving from 33.0% in 2000. South Asia housed nearly

40% of stunted children, and sub-Saharan Africa was home to another 40% (https://data.unicef.org/topic/nutrition/malnutrition/).

At around 6 months, infants' nutritional needs change rapidly. They should begin solid foods while continuing breastfeeding until at least 2 years old for proper brain and body development. Nutritional requirements for children aged 6–23 months are highest per kilogram of body weight during this period, necessitating varied foods to meet their needs. Unfortunately, global data show that most children aged 6–23 months don't follow recommended nutrition guidelines. Many children don't receive the right timing, frequency, or dietary diversity needed for optimal growth. About 28% of children aged 6–8 months didn't have any solid, semi-solid, or soft foods introduced. Half of children aged 6–23 months didn't meet minimum meal frequency, and a majority (69%) didn't consume foods from at least five of the recommended eight food groups, reflecting minimum dietary diversity (https://data.unicef.org/topic/nutrition/diets/).

Malnutrition is a critical issue in India, posing a significant hurdle to human development. Despite economic growth, malnutrition among children under 5 remains exceptionally high, far surpassing rates in China and sub-Saharan Africa. Of nearly 60 million Indian children, almost half suffer from underweight conditions. Additionally, 45% are stunted, 20% are wasted, 75% are anemic, and 57% lack vitamin A. This malnutrition adversely affects survival, health, education, school dropout rates, and future productivity. Most of this undernourishment occurs during pregnancy and a child's first 2 years, potentially causing irreversible harm to brain development and economic prospects. Persistent undernutrition significantly hinders human development in India, impacting health, education, and economic productivity. This issue jeopardizes India's anticipated demographic dividend and future economic growth. Although malnutrition is common, there are differences in its prevalence between states and socioeconomic classes; females, rural areas, the poorest people, and scheduled tribes and castes are the most affected. Over half of the malnutrition cases in India are found in six states, Bihar, Chhattisgarh, Jharkhand, Madhya Pradesh, Rajasthan, and Uttar Pradesh, with an additional 8–10% concentrated in specific areas of Andhra Pradesh and Maharashtra (https://www.worldbank.org/en/news/feature/2013/05/13/helping-india-combat-persistently-high-rates-of-malnutrition).

The main symptoms of malnutrition are low body weight, slowed growth, and muscle atrophy. Malnutrition causes stunted growth in about 48% of Indian infants under 5 years old. Malnutrition impairs children's immune systems and delays their cognitive and physical development, making them more vulnerable to infectious infections. These children are at greater risk of succumbing to childhood ailments such as diarrhea, as well as diseases like tuberculosis, measles, pneumonia, and malaria (Narayan et al. 2019).

To grasp the concept of dietary supplementation, it's beneficial to revisit the functions of nutrients. Macronutrients play a crucial role in providing energy as well as supporting growth, upkeep, and physical activity. This group includes carbohydrates, fats, proteins, macro-minerals, and water. Carbohydrates transform into glucose and other single sugar molecules. Fats break down to produce glycerol and fatty acids. Amino acids and peptides are produced as proteins breakdown. These macronutrients are used as interchangeable energy sources. Micronutrients include minerals and all forms of vitamins. Water-soluble vitamins include vitamin C, eight members of the vitamin B complex. Fat-soluble vitamins contain vitamins A, D, E, and K. Vital nutrients include minerals, fatty acids, vitamins, amino acids, and several types of carbohydrates. These essential elements must be received through diet because the body is unable to synthesis them. Iron, iodides, fluorine, zinc, chromium, selenium, manganese, molybdenum, and copper are crucial trace minerals. With the exception of iron and zinc, deficiencies in trace minerals are infrequent in industrialized nations (Castiglia 2000).

These required macro- and micronutrients can be obtained from edible medicinal plants; some of the edible plants are mentioned in Table 18.1.

18.3.1 Vitamin D

Urashima et al. investigated how vitamin D supplementation impacted school-age children's seasonal influenza A incidence. They executed a randomized, double-masked, placebo-controlled trial in which they compared the effects of vitamin D supplements (1200 IU/day) against a placebo in schoolchildren. The main objective was to assess the frequency of influenza A cases detected through antigen testing utilizing nasopharyngeal swab specimens. The results indicated that influenza A was observed in 18 out of 167 (10.8%) children in the vitamin D group, whereas 31 out of 167 (18.6%) children in the placebo group were affected (Urashima et al. 2010).

18.3.2 Iodine

In Dunedin, New Zealand, a rigorous study involving 184 children aged 10–13 years was conducted. These kids were split into two groups: one received a daily tablet with 150 μg of iodine, while the other received a placebo for 28 weeks in a double-masked, randomized, placebo-controlled trial. Initially, the children had mild iodine deficiency. After 28 weeks, the iodine-supplemented group saw improvements in their iodine levels, while the placebo group remained deficient. Comparing the two groups, the iodine-supplemented group achieved an overall cognitive score of 0.19 standard deviations higher than the placebo group ($P = 0.011$). The study indicates that in children with modest iodine insufficiency, iodine supplementation improved perceptual thinking. Evidence suggests that in children with modest iodine insufficiency, iodine supplementation improved perceptual thinking (Gordon et al. 2009).

Table 18.1 Plant source of micronutrients and their health benefits

Sr. No.	Micronutrient	Source	Heath benefits	Reference
1.	Vitamin A (retinol)	Amaranth, spinach, sweet potatoes, carrots, pumpkins, yellow maize, mangoes, and papayas	Functions as an antioxidant; contributes to growth, development, and vision health; and aids in the prevention and treatment of cancers and skin disorders	Gilbert (2013)
2.	Vitamin B_1 (thiamine)	Nuts, legumes such as peas, soybeans, and whole grain cereals including brown rice and bran	Assists in the transformation of food into energy and plays a role in neurological functions	Osiezagha et al. (2013)
3.	Vitamin B_3 (niacin)	Peanuts, coffee, beans, barley, wheat, rice, potato, soybean, and maize	Aids in the transformation of food into energy and supports optimal brain function	Hill and Williams (2017)
4.	Vitamin B_6 (pyridoxine)	Legumes, nuts, bananas, and potatoes	Assists in the synthesis of vital proteins and the conversion of protein into energy	Kennedy (2016)
5.	Vitamin B_{12} (cobalamin and cyanocobalamin)	*Boletus* sp., *Macrolepiota procera*, *Pleurotus ostreatus*, *Morchella conica*, *Craterellus cornucopioides*, *Cantharellus cibarius*, *Lentinula edodes*, *Hericium erinaceus*	Assists in the generation of genetic material in cells, supports the creation of red blood cells (RBCs), enhances brain functions, aids in amino acid synthesis, and facilitates the metabolism of biomolecules	Watanabe et al. (2014)
6.	Vitamin C (ascorbic acid)	Herbal teas and tinctures from rose, pine, and bark of trees Fruits of *Emblica officinalis*	Functions as an antioxidant, sustains the well-being of bones, teeth, and gums and assists in wound healing and combating colds	Majumdar et al. (2020)
7.	Vitamin D	*Solanum lycopersicum*, *Solanum tuberosum*, *Cucurbita pepo*, *S. glaucophyllum*, *Nicotiana glauca*, *Cestrum diurnum*, *Medicago sativa*, *Trisetum flavescens*, *Capsicum annuum*	Facilitates the development of bones and teeth and assists in the body's absorption of calcium	Jäpelt and Jakobsen (2013)

(continued)

Table 18.1 (continued)

Sr. No.	Micronutrient	Source	Heath benefits	Reference
8.	Vitamin E	Nuts, leafy vegetables, whole grains, and edible oils	Functions as an antioxidant assist in the creation of blood cells, lungs, muscles, and nerve tissues while also enhancing the immune system	Keen and Hassan (2016)
9.	Folic acid	Leafy vegetables, legumes, citrus fruits	Facilitates the production of genetic material in cells, particularly crucial during the first trimester of pregnancy, contributes to the formation of red blood cells, and protects against heart diseases	Kennedy (2016)
10.	Calcium (Ca)	Orange juice, spinach, soybean, carrots, potato	Aids in fortifying bones and teeth, supporting nerve conduction, and enhancing the function of muscles and glands	Yang et al. (2012)
11.	Iron (Fe)	Cereals, pulses, legumes, fruits, and St John's, leaves of sage, mint, nettle and flowers of chamomile	Assists in generating energy and transporting oxygen to tissues	Geissler and Singh (2011)
12.	Magnesium (Mg)	St. John's wort, sage, chamomile, leaves of peppermint	Facilitates nerve conduction and muscle functioning, contributes to bone formation, and helps prevent premenstrual syndrome	Suliburska and Kaczmarek (2012)
13.	Phosphorous (P)	Some edible oils like corn, and peanut Cereals, lentils, rice, maize flour, cornflakes	Enhancing sturdy bones and teeth, facilitating the creation of genetic material in cells, and supporting the generation and storage of energy	D'Alessandro et al. (2015)
14.	Chromium (Cr)	Aniseed, basil, cinnamon, garlic, laurel, mint, mustard, nutmeg, onion, oregano, parsley, paprika, pepper, pepper, saffron, thyme, vanilla	Assists in the conversion of fats and carbohydrates into energy, working in conjunction with insulin	Sykuła-Zając and Pawlak (2012)

(continued)

Table 18.1 (continued)

Sr. No.	Micronutrient	Source	Heath benefits	Reference
15.	Zinc (Zn)	St John's wort (flowers and leaves), sage (leaves), chamomile (flowers), mint (leaves), and nettle (leaves)	Facilitates cell reproduction, supports growth and development in children, assists in wound healing, and contributes to the production of sperm and testosterone	Suliburska and Kaczmarek (2012)
16.	Copper (Cu)	*Potentilla anserina*, *Mentha arvensis*, *Achillea millefolium*, *Comarum palustre*, *Lysimachia vulgaris*, *Lycopus europaeus*	Aids in hemoglobin formation and collagen production, supports heart functionality, assists in energy generation, and facilitates the absorption of iron from the body	Malinowska and Jankowski (2017)
17.	Iodine (I)	Fruit, berries, nuts, seeds, potatoes	Sustains the proper function of the thyroid gland	Carlsen et al. (2018)

18.3.3 Multiple Micronutrients

This study examined the impact of multiple micronutrient supplements on the growth of children aged 8–14 months with insufficient intake of various micronutrients. Children were split into two groups: one group was given multiple micronutrient supplements that either met or exceeded the recommended dietary requirement, while the other group received a placebo. After around 12.2 months of supplementation, the group receiving micronutrients showed significant gains in body length, with an 8.2 mm difference compared to the placebo group (Rivera et al. 2001).

18.3.4 β-Carotene-Rich Orange-Fleshed Sweet Potatoes

Orange-fleshed sweet potatoes (OFSP) are rich in β-carotene that can combat vitamin A deficiency in developing nations. A study with primary school children showed that daily consumption of 125 g of orange-fleshed sweet potatoes improved their vitamin A status over 2 months. This highlights OFSP's potential as a sustainable, food-based solution for addressing vitamin A deficiency in children in developing countries (Van Jaarsveld et al. 2005).

An investigation offered new insights on the impact of micronutrients on cognitive improvements in school children aged up to 15 years. Thirteen randomized controlled studies, published since 2000, were identified through a comprehensive analysis. These studies focused mainly with assessing the impact of diets fortified with micronutrients on various aspects of cognitive function. Notable micronutrients such as iron, zinc, iodine, and vitamin A were among those investigated. The findings consistently highlighted a positive correlation between micronutrient supplementation and improvements in short-term memory (Khor and Misra 2012).

18.3.5 Vitamin A

This study examined the effectiveness of vitamin A-fortified wheat flour buns (pandesal) in improving vitamin A levels among 6–13-year-old schoolchildren in the Philippines. The fortification of flour was carried out with retinol palmitate. In a 30-week double-masked clinical trial involving 835 children from four rural schools, one group received vitamin A-fortified pandesal, while the other received nonfortified pandesal. The results showed that for children with low initial vitamin A levels, those who consumed the fortified pandesal experienced a significant increase in vitamin A levels compared to the control group. The fortified pandesal also led to a nearly 50% reduction in the proportion of children with insufficient vitamin A levels in the liver. Overall, daily consumption of vitamin A-fortified pandesal improved the vitamin A status of Filipino school-age children with low initial vitamin A levels (Solon et al. 2000).

Supplementing preschoolers with vitamin A can reduce mortality and illness risks from specific diseases like diarrhea, measles, HIV, and malaria, likely due to its immune system effects. Clinical trials have linked vitamin A to increased antibody production, lymphocyte proliferation, gut lining integrity, and T-cell lymphopoiesis. It can enhance vaccine responses in some cases (Villamor and Fawzi 2005).

18.3.6 Zinc

Roy et al. conducted a double-blind, randomized study at Dhaka Hospital in Bangladesh with 179 children aged 3–14 having cholera. They split into two groups: one received 30 mg of zinc daily and the other a placebo. Both took erythromycin. After 3 days, more in the zinc group recovered by Day 2 (49% vs. 32%) and Day 3 (81% vs. 68%). Zinc reduced diarrhea duration by 12% (64.1 vs. 72.8 h) and stool output by 11% (1.6 vs. 1.8 kg/day). Zinc supplementation benefits children with cholera, reducing illness severity and duration (Roy et al. 2008).

Intervention trials have shown that zinc effectively treats acute diarrhea in children in developing nations. A double-blind, placebo-controlled study in malnourished Turkish children with varying zinc levels confirmed this. Within the first 14 days, the study group had shorter diarrhea duration and reduced stool frequency compared to the control group. This indicates zinc supplementation could help alleviate acute diarrhea, especially in children with low zinc levels (Polat et al. 2003).

The study investigated how zinc (Zn) supplementation impacted growth and body development of prepubertal children with sickle cell disease (SCD). Forty-two children (20 girls and 22 boys) aged 4–10 with SCD were randomly split into two groups. One group received 10 mg of zinc in cherry syrup (Zn group), while the other got cherry syrup alone (control group). After 12 months, the Zn group showed significant improvements in various growth parameters, including height, sitting height, knee height, and arm circumference z-scores. In contrast, the control group had decreased height-for-age and weight-for-age z-scores. These findings suggest that zinc supplementation may benefit prepubertal children with SCD by enhancing growth and weight gain (Zemel et al. 2002).

18.3.7 B12

Vitamin B12, also known as cobalamin, plays a vital role in the development of fetuses and children, pregnant women who follow a vegetarian diet, individuals with Crohn's or celiac disease, and individuals who have undergone gastric bypass surgery face an elevated risk of B12 deficiency. Insufficient levels of B12 in the bloodstream have been associated with adverse effects on cognitive function, motor skills, and growth outcomes. Additionally, there seems to be a connection between low cobalamin levels and depression in adults. Certain studies suggest that supplementing with B12 could potentially enhance outcomes in children, although further research is required to explore this area comprehensively (Pepper and Black 2011).

18.3.8 Functional Food

A study assessed the impact of nutraceuticals on growth in short children born SGA. After 6 months of observation, they were split into three groups: growth hormone therapy, a healthy program, and passive observation. The nutritional group received iron, zinc, and vitamin A. Results showed the nutritional group had significant growth acceleration, similar to growth hormone therapy, suggesting nutrient supplementation can boost growth in these children (Zadik et al. 2010).

Examples of marketed nutraceutical products include Revital, Complan, Horlicks, Ensure, Peptamen, NutriMix, Curcumin Boost, Chyawanprash, etc. (Malve and Bhalerao 2023).

18.4 Government Initiatives to Combat Child Malnutrition in India

There are different schemes or services to treat and irradicate malnutrition in India. The schemes are discussed as follows.

18.4.1 Integrated Child Development Scheme (ICDS)

In 1975, the Ministry of Women and Child Development initiated one of India's most comprehensive child development programs. This scheme is jointly funded by the central government of India and UNICEF, with the objective of delivering food and primary healthcare services to children under 6 years old and their mothers. To carry out this mission, India established "Anganwadi centers" in rural areas. The ICDS program was launched in alignment with the National Policy for Children in India, with the goal of combating malnutrition, promoting well-being, and addressing gender disparities. Under this scheme, Anganwadi centers offer immunizations, health assessments, nutritious meals, and informal education (Narayan et al. 2019).

18.4.2 Mid-Day Meal (MDM) Scheme

In 1995, the central government initiated the National Programme of Nutritional Support to Primary Education, commonly referred to as the Mid-Day Meal (MDM) scheme. Its purpose was to boost the nutritional well-being of children, improve school enrollment rates, and enhance attendance. The implementation of the MDM scheme has been effective nationwide. According to the Public Report on Basic Education (PROBE), the mid-day school meal contributed to an improvement in the nutritional status of school children (Sahai 2014).

18.4.3 National Health Mission (NHM)

The National Health Mission was established in 2005 and consists of the National Urban Health Mission (NUHM) and the National Rural Health Mission (NRHM). Its objective is to increase health programs and the delivery of healthcare services. This mission focuses on enhancing maternal, neonatal, child, and adolescent health, thereby preventing diseases. Additionally, it incorporates a "convergence" component, as detailed in the Convergence Action Plan discussion below, to address malnutrition prevention, identification, and management in children. The implementation of this mission involves the collaboration of the Integrated Child Development Services (ICDS) and the Ministry/Department of Health and Family Welfare, Government of India (Narayan et al. 2019).

18.4.4 Food Safety and Standard Authority of India (FSSAI)

The FSSAI, established by the Government of India on 5 September 2008, in accordance with the Food Safety and Standards Act of 2006, serves as a vital resource. It offers technical expertise and scientific counsel to the government for shaping food safety and nutrition policies and programs. Additionally, the FSSAI plays a role in fortifying foods with essential micronutrients and conducts training for school personnel to strengthen mid-day meals, ensuring children receive nutritionally fortified food (Narayan et al. 2019).

Abovementioned nutrients can be obtained from regional flora and fauna; it can help to meet the needs of nutrition in tribal regions. Some of the medicinal plants are mentioned in Table 18.2.

Table 18.2 Nutritionally critical medicinal plants in India

Sr. No.	Nutritional medicinal plants	Use	Reference
1.	*Withania somnifera*	Anti-inflammatory, immunomodulatory, neuroprotective, adaptogenic effect	Mikulska et al. (2023)
2.	*Vigna radiata*	An outstanding reservoir of protein, dietary fiber, minerals, vitamins, and substantial quantities of bioactive compounds	Hou et al. (2019)
3.	*Moringa oleifera*	Anti-inflammatory, antioxidant, anticancer, hepatoprotective, neuroprotective, hypoglycemic, and blood lipid-reducing functions	Kou et al. (2018)
4.	*Phyllanthus emblica*	Anti-aging, immunity booster anticancer, anti-inflammatory, digestive tract, and neurological protective activities	Gul et al. (2022)
5.	*Moringa oleifera*, *Albizia lebbeck*, *Allium cepa*, *Allium sativum*, *Aloe vera*, *Carica papaya*, *Colocasia esculenta*, *Gossypium hirsutum*, *Garcinia cola*, *Ipomoea batatas*, *Jatropha curcas*, *Mangifera indica*, *Mucuna pruriens*, *Musa paradisiaca*, *Ocimum gratissimum*, *Psidium guajava*, *Jatropha tanjorensis*, *Terminalia catappa*, *Gossypium hirsutum*, *Carica papaya*	Treatment of iron deficiency anemia	Devi (2016)
6.	*Moringa oleifera*, *Withania somnifera*, *Punica granatum*, *Litseaglutinosa*, *Panax ginseng*, *Psoralea corylifolia*, *Curcuma aromatica*, *Zingiber officinale*, *Astragalus membranaceus*, *Dioscorea batatas*, *Nigella sativa*, *Podophyllum emodi*	Maintain bone health	Devi (2016)
7.	*Tinospora cordifolia*, *Delonix regia*, *Butea monosperma*, *Andrographis paniculata Nees*, *Abrus precatorius* L., *Butea monosperma* Roxb., *Caesalpinia bonduc ella* L, *Catharanthus roseus* L, *Celosia argentea* L, *Crotalaria prostrata* Rottl., *Lawsonia inermis* L., *Maytenus senegalensis* Lam., *Mimosa pudica* L., *Striga densiflora* Benth., *Tinospora cordifolia* Willd., *Tridax procumbens*	Treatment of malnutrition, treatment of iron, folate, and zinc deficiency	Devi (2016)

(continued)

Table 18.2 (continued)

Sr. No.	Nutritional medicinal plants	Use	Reference
8.	*Centella asiatica*	Improving memory, immunostimulant	Pandey et al. (2011)
9.	*Curcuma longa*	Reducing inflammation, indigestion, antioxidants, and liver problems	
10.	*Foeniculum vulgare*	Stomach bloating, stimulant, digestive spasms	
11.	*Medicago sativa*	Appetite stimulation, nutrition	
12.	*Cassia tora*	Young leaves are used as a vegetable Source of vitamin C	Lilhare et al. (2010), Rathore and Kumar (2020)

18.5 Indian Nutraceutical Market: A Breakthrough Success Story

The nutraceutical market in India is undergoing a substantial transformation, with projections indicating that it will soar to approximately USD 18 billion by the close of 2025, a remarkable increase from USD 4 billion at the conclusion of 2020, which represents more than a fourfold growth over just 5 years. The appeal of the Indian nutraceutical market has significantly piqued the interest of foreign investors. This increase is linked to the relaxation of regulations that permit 100% foreign direct investment (FDI) in the food supplement and nutraceutical production industries. Moreover, these companies now have the opportunity to distribute their products through various channels, including wholesale, retail, and e-commerce platforms. Consequently, FDI in this sector has witnessed a substantial rise, escalating from USD 131.4 million in 2012 to a substantial USD 584.7 million in 2019 (Malve and Bhalerao 2023). With such phenomenal growth, India is poised to be a global leader in nutraceuticals.

In 2014, one of the Equator Prizes was awarded to the Traditional Healer Association, Chhattisgarh local community, India. It is organized by the Equator Initiative within the United Nations Development Programme. The achievements of Traditional Healer Association Chhattisgarh are as follows:

- Mobilization of traditional healers to meet local health needs
- Promotion of medicinal plant gardens and village botanists
- Public education on medicinal plants and traditional healing
- Creating awareness of the conservation of medicinal plants

18.6 Limitations to Using the Same Nutraceutical Formulation Globally

The use of the same nutraceutical formulation globally faces several limitations, including:

1. Cultural Differences: Cultural dietary preferences and eating habits vary significantly across regions and countries. A nutraceutical formulation that works in one part of the world may not be accepted or suitable for consumption in another due to differences in taste, texture, or cultural beliefs surrounding food (Popkin 1999).
2. Nutritional Requirements: Nutritional requirements vary based on factors such as age, gender, and individual health conditions. What may be beneficial for one population group may not meet the specific nutritional needs of another (Meyers et al. 2006).
3. Environmental Factors: Climate and environmental conditions impact the availability of certain ingredients for nutraceutical formulations. Ingredients that are readily available in one region may be scarce or expensive in another (Myers et al. 2017).
4. Regulatory Variations: Regulations governing nutraceuticals and dietary supplements vary widely between countries. Ingredients considered safe and approved in one nation may not meet the regulatory standards of another, leading to restrictions or bans on their sale or use (Shah 2011).
5. Ethical and Social Factors: Ethical concerns, such as fair trade practices and social responsibility, can influence the acceptance and use of nutraceutical formulations. Consumers in different regions may prioritize additional ethical considerations (Crane 2009).
6. Supply Chain Challenges: The global supply chain for nutraceutical ingredients can be complex and subject to disruptions, affecting the availability and consistency of raw materials for formulations.
7. Healthcare Infrastructure: The availability and accessibility of healthcare infrastructure and services can vary widely between countries. Nutraceutical formulations may require monitoring or medical supervision, which may not be feasible in all regions (World Health Organization 2010).

In light of these limitations, it is crucial for stakeholders in the nutraceutical industry to carefully consider local contexts and adapt formulations to meet the specific needs and preferences of diverse populations globally.

18.7 Importance of Local Sources of Nutraceuticals

The importance of local sources of nutraceuticals is significant for several reasons, including promoting sustainability, preserving biodiversity, and addressing region-specific health needs. Here are some key points:

1. Sustainability and Environmental Impact: Utilizing local sources of nutraceuticals often aligns with sustainable agricultural practices, reducing the carbon footprint associated with long-distance transportation. This supports environmental conservation and minimizes the depletion of natural resources (Petersen and Simmonds 2003).
2. Biodiversity Preservation: Relying on local botanical resources encourages the cultivation and conservation of indigenous plant species. It helps protect biodiversity by reducing the demand for exotic and endangered plants that may be overharvested (Cragg and Newman 2013).
3. Cultural and Traditional Knowledge: Local sources of nutraceuticals often align with traditional and cultural knowledge of indigenous communities. This knowledge can be invaluable in understanding the therapeutic properties and safe usage of these substances (Heinrich et al. 1998).
4. Tailoring to Regional Health Needs: Different regions may have unique health concerns and dietary habits. Utilizing local nutraceuticals allows for the development of products that specifically address the health needs of local populations (Hunter and Borelli 2006).
5. Economic Benefits: Supporting local producers of nutraceutical ingredients can boost rural economies, create jobs, and enhance income distribution within communities (Sen 1983).
6. Quality Control and Traceability: Sourcing nutraceuticals locally often allows for better quality control and traceability, as supply chains can be more easily monitored and managed (Cordell and Colvard 2005).
7. Reducing Reliance on Imports: Depending on imported nutraceutical ingredients can make a country vulnerable to supply disruptions. Utilizing local sources can enhance self-sufficiency in meeting nutritional and health needs (Barlow 2013).

Incorporating local sources of nutraceuticals into dietary and healthcare practices not only has health benefits but also contributes to broader sustainability goals and supports the well-being of local communities.

18.8 Conclusion

In conclusion, the exploration of nutraceuticals derived from medicinal plants in the context of pediatrics presents a promising avenue for improving child health and well-being. These natural compounds offer a holistic approach to pediatric healthcare by addressing both nutritional needs and therapeutic benefits. They have the potential to complement traditional medical interventions and mitigate various health challenges faced by children. However, it is essential to exercise caution and thorough research when introducing nutraceuticals into pediatric care. Safety, dosage, and potential interactions with conventional medications must be rigorously studied and monitored. Moreover, the regulatory framework for nutraceuticals in pediatric use should be strengthened to ensure quality, efficacy, and compliance of safety standards.

People's reliance on local flora and wildlife as a source of sustenance has decreased as a result of gimmicky marketing strategies of food industries. As a result, natives are becoming detached from the natural treasures bestowed on them by Mother Nature. Science's intervention and diagnostic tools must be optimized for each geographical location. There should be no universal standard or norm for nutrition; instead, it should be region-specific. Overall, nutraceuticals from medicinal plants offer exciting prospects for enhancing pediatric nutrition and health. Still, their integration into pediatric healthcare must be approached with care and evidence-based practices to reap the full benefits while safeguarding children's well-being.

References

Barlow SM (2013) Toxicological aspects of nutraceuticals. Food Res Int 54(2):1742–1745

Carlsen MH, Andersen LF, Dahl L, Norberg N, Hjartåker A (2018) New iodine food composition database and updated calculations of iodine intake among Norwegians. Nutrients 10(7):930

Castiglia PT (2000) Dietary supplements for children. J Pediatr Health Care 14(6):309–311

Chauhan B, Kumar G, Kalam N, Ansari SH (2013) Current concepts and prospects of herbal nutraceutical: a review. J Adv Pharm Technol Res 4(1):4

Chopra AS, Lordan R, Horbańczuk OK, Atanasov AG, Chopra I, Horbańczuk JO, Jóźwik A, Huang L, Pirgozliev V, Banach M, Battino M (2022) The current use and evolving landscape of nutraceuticals. Pharmacol Res 175:106001

Cordell GA, Colvard MD (2005) Some thoughts on the future of ethnopharmacology. J Ethnopharmacol 100(1-2):5–14

Cragg GM, Newman DJ (2013) Natural products: a continuing source of novel drug leads. Biochim Biophys Acta 1830(6):3670–3695

Crane TA (2009) Of models and meanings: cultural complexity, economic structure, and environmental vitality in Indian home gardens. World Dev 37(8):1469–1479

D'Alessandro C, Piccoli GB, Cupisti A (2015) The "phosphorus pyramid": a visual tool for dietary phosphate management in dialysis and CKD patients. BMC Nephrol 16(1):1–6

Da Costa JP (2017) A current look at nutraceuticals–key concepts and future prospects. Trends Food Sci Technol 62:68–78

Devi BP (2016) Impact of medicinal plants against malnutrition among women: an overview. J Med Plants 4(1):107–110

Geissler C, Singh M (2011) Iron, meat and health. Nutrients 3(3):283–316

Gilbert C (2013) What is vitamin A and why do we need it? Community Eye Health 26(84):65

Gordon RC, Rose MC, Skeaff SA, Gray AR, Morgan KM, Ruffman T (2009) Iodine supplementation improves cognition in mildly iodine-deficient children. Am J Clin Nutr 90(5):1264–1271

Gul M, Liu ZW, Rabail R, Faheem F, Walayat N, Nawaz A, Shabbir MA, Munekata PE, Lorenzo JM, Aadil RM (2022) Functional and nutraceutical significance of amla (Phyllanthus emblica L.): a review. Antioxidants 11(5):816

Heinrich M, Ankli A, Frei B, Weimann C, Sticher O (1998) Medicinal plants in Mexico: healers' consensus and cultural importance. Soc Sci Med 47(11):1859–1871

Hill LJ, Williams AC (2017) Meat intake and the dose of vitamin B3–nicotinamide: cause of the causes of disease transitions, health divides, and health futures. Int J Tryptophan Res 10:1178646917704662

Hou D, Yousaf L, Xue Y, Hu J, Wu J, Hu X, Feng N, Shen Q (2019) Mung bean (Vigna radiata L.): bioactive polyphenols, polysaccharides, peptides, and health benefits. Nutrients 11(6):1238
Hunter D, Borelli T (2006) The poverty of biodiversity and the wealth of human health. Int J Environ Health Res 16(1):1–5
Jäpelt RB, Jakobsen J (2013) Vitamin D in plants: a review of occurrence, analysis, and biosynthesis. Front Plant Sci 4:136
Keen MA, Hassan I (2016) Vitamin E in dermatology. Indian Dermatol Online J 7(4):311
Kennedy DO (2016) B vitamins and the brain: mechanisms, dose and efficacy—a review. Nutrients 8(2):68
Keservani RK, Kesharwani RK, Sharma AK, Gautam SP, Verma SK (2017) Developing new functional food and nutraceutical products. Academic Press, New York
Khor GL, Misra S (2012) Micronutrient interventions on cognitive performance of children aged 5-15 years in developing countries. Asia Pac J Clin Nutr 21(4):476–486
Kou X, Li B, Olayanju JB, Drake JM, Chen N (2018) Nutraceutical or pharmacological potential of Moringa oleifera Lam. Nutrients 10(3):343
Lilhare TK, Koche DK, Kawale MV (2010) Medicinally important wild edible plants of Eastern Vidarbha. Hislopia J 10:73–88
Majumdar A, Shukla SS, Pandey RK (2020) Culinary and herbal resources as nutritional supplements against malnutrition-associated immunity deficiency: the vegetarian review. Future J Pharm Sci 6:1
Malinowska E, Jankowski K (2017) Copper and zinc concentrations of medicinal herbs and soil surrounding ponds on agricultural land. Landsc Ecol Eng 13(1):183–188
Malve H, Bhalerao P (2023) Past, present, and likely future of nutraceuticals in India: evolving role of pharmaceutical physicians. J Pharm Bioallied Sci 15(2):68–74
Meyers LD, Hellwig JP, Otten JJ (eds) (2006) Dietary reference intakes: the essential guide to nutrient requirements. National Academies Press, Washington
Mikulska P, Malinowska M, Ignacyk M, Szustowski P, Nowak J, Pesta K, Szeląg M, Szklanny D, Judasz E, Kaczmarek G, Ejiohuo OP (2023) Ashwagandha (Withania somnifera)—current research on the health-promoting activities: a narrative review. Pharmaceutics 15(4):1057
Myers SS, Smith MR, Guth S, Golden CD, Vaitla B, Mueller ND, Dangour AD, Huybers P (2017) Climate change and global food systems: potential impacts on food security and undernutrition. Annu Rev Public Health 38:259–277
Narayan J, John D, Ramadas N (2019) Malnutrition in India: status and government initiatives. J Public Health Policy 40:126–141
Osiezagha K, Ali S, Freeman C, Barker NC, Jabeen S, Maitra S, Olagbemiro Y, Richie W, Bailey RK (2013) Thiamine deficiency and delirium. Innov Clin Neurosci 10(4):26
Pandey N, Meena RP, Rai SK, Pandey-Rai S (2011) Medicinal plants derived nutraceuticals: a re-emerging health aid. Int J Pharm Bio Sci 2(4):420–441
Pepper MR, Black MM (2011) B12 in fetal development. In: Seminars in cell & developmental biology, vol 22. Academic, New York, pp 619–623
Petersen M, Simmonds MS (2003) Rosmarinic acid. Phytochemistry 62(2):121–125
Piekara A, Krzywonos M, Kopacz M (2022) Dietary supplements intended for children——proposed classification of products available on the market. J Dietary Suppl 19(4):431–442
Polat TB, Uysalol M, Cetinkaya F (2003) Efficacy of zinc supplementation on the severity and duration of diarrhea in malnourished Turkish children. Pediatr Int 45(5):555–559
Popkin BM (1999) Urbanization, lifestyle changes and the nutrition transition. World Dev 27(11):1905–1916
Puri V, Nagpal M, Singh I, Singh M, Dhingra GA, Huanbutta K, Dheer D, Sharma A, Sangnim T (2022) A comprehensive review on nutraceuticals: therapy support and formulation challenges. Nutrients 14(21):4637
Rathore M, Kumar H (2020) Nutritional assessment of cassia tora leaves from different regions in Rajasthan. J Adv Biol Biotechnol 22(4):1–5

Rivera JA, González-Cossío T, Flores M, Romero M, Rivera M, Tellez-Rojo MM, Rosado JL, Brown KH (2001) Multiple micronutrient supplementation increases the growth of Mexican infants. Am J Clin Nutr 74(5):657–663

Roy SK, Hossain MJ, Khatun W, Chakraborty B, Chowdhury S, Begum A, Mah-e-Muneer S, Shafique S, Khanam M, Chowdhury R (2008) Zinc supplementation in children with cholera in Bangladesh: randomized controlled trial. BMJ 336(7638):266–268

Sachdeva V, Roy A, Bharadvaja N (2020) Current prospects of nutraceuticals: a review. Curr Pharm Biotechnol 21(10):884–896

Sahai CS (2014) Mid-day meal scheme: achievements and challenges. Int J Hum Soc Sci Invent 3(10):6–9

Sen A (1983) Development: which way now? Econ J 93(372):745–762

Shah RR (2011) Regulation of nutraceuticals in international markets—a regulatory perspective. Regul Toxicol Pharmacol 59(1):34–41

Solon FS, Klemm RD, Sanchez L, Darnton-Hill I, Craft NE, Christian P, West KP Jr (2000) Efficacy of a vitamin A–fortified wheat-flour bun on the vitamin A status of Filipino schoolchildren. Am J Clin Nutr 72(3):738–744

Suliburska J, Kaczmarek K (2012) Herbal infusions as a source of calcium, magnesium, iron, zinc and copper in human nutrition. Int J Food Sci Nutr 63(2):194–198

Sykuła-Zając A, Pawlak A (2012) Chromium in food products. Biotechnol Food Sci 76(1):27–34

Urashima M, Segawa T, Okazaki M, Kurihara M, Wada Y, Ida H (2010) Randomized trial of vitamin D supplementation to prevent seasonal influenza A in schoolchildren. Am J Clin Nutr 91(5):1255–1260

Van Jaarsveld PJ, Faber M, Tanumihardjo SA, Nestel P, Lombard CJ, Benadé AJ (2005) β-Carotene–rich orange-fleshed sweet potato improves the vitamin A status of primary school children assessed with the modified-relative-dose-response test. Am J Clin Nutr 81(5):1080–1087

Villamor E, Fawzi WW (2005) Effects of vitamin A supplementation on immune responses and correlation with clinical outcomes. Clin Microbiol Rev 18(3):446–464

Watanabe F, Yabuta Y, Bito T, Teng F (2014) Vitamin B12-containing plant food sources for vegetarians. Nutrients 6(5):1861–1873

World Health Organization (2010) World health statistics 2010. World Health Organization, Geneva

Yang J, Punshon T, Guerinot ML, Hirschi KD (2012) Plant calcium content: ready to remodel. Nutrients 4(8):1120–1136

Zadik Z, Sinai T, Zung A, Golander A, Reifen R (2010) "Functional food" for acceleration of growth in short children born small for gestational age. J Pediatr Endocrinol Metab 23(5):435–441

Zemel BS, Kawchak DA, Fung EB, Ohene-Frempong K, Stallings VA (2002) Effect of zinc supplementation on growth and body composition in children with sickle cell disease. Am J Clin Nutr 75(2):300–307

Nutraceutical Formulations from Medicinal Plants: A Potential Therapeutic Agent

19

Rohan Sarkar, Swarajya Laxmi Nayak, Manish Kumar Suthar, and Manish Das

Abstract

The rising health concerns within contemporary population have significantly contributed to the growing acceptance of various nutraceutical products and functional foods. Recent market trends and innovative research endeavours have further intensified the focus on nutraceutical, driving the development of wide range of products that provides diverse health benefits. Leveraging the power of medicinal and herbal plants as valuable resources for nutraceutical development has gained momentum. These plants have been integral to traditional knowledge systems throughout history. Scientific studies over several decades revealed presence of various bioactive compounds mostly polyphenolics that are responsible for different health-promoting effects. When it comes to nutraceutical formulations, herbal extracts find utility in various forms like powders, pills, tablets, capsules and blended juices, tailored to specific applications. Additionally, herbal-infused products like soap, toothpaste, beverages and body oil have seamlessly integrated into our daily routines. Extensive research works are on track for development, improvement, modification and quality control of the herbal nutraceutical products. In light of this landscape, the present article was formulated addressing the role of phytochemicals as potential nutraceutical agents, different herbal formulations and their therapeutic potential along with the market landscape, the regulations, challenges and future prospects in a lucid way as possible.

Keywords

Medicinal plants · Phytochemicals · Nutraceutical products · Health curing agent

---

Rohan Sarkar, Swarajya Laxmi Nayak and Manish Kumar Suthar contributed equally with all other contributors.

R. Sarkar (✉) · S. L. Nayak · M. K. Suthar · M. Das
ICAR-Directorate of Medicinal and Aromatic Plants Research, Boriavi, Anand, Gujarat, India

© The Author(s), under exclusive license to Springer Nature Singapore Pte Ltd. 2024

M. Nandave et al. (eds.), *Ethnopharmacology and OMICS Advances in Medicinal Plants Volume 1*, https://doi.org/10.1007/978-981-97-2367-6_19

19.1 Introduction

A nutraceutical product is defined as a substance having specific physiological or biological benefit that can provide protection or cure against chronic health ailments. It is an amalgamation of "nutrition" and "pharmaceutical". Nutraceuticals are the best alternatives that tend to erase the boundary between food and medicine. Due to increase in awareness regarding importance of balance diet among the people, they are so much cautious on what they are eating. Nutrition is the indispensable factor in overall well-being of health. People who are frequently suffering from diseases and are in deficit of appropriate nutrition often shift towards nutraceuticals because they are of natural origin and are formulated based on the essential components need for a balanced diet. Unlike commercial synthetic drugs, the nutraceuticals are preventive or prophylactic in nature. The chemical substances present in the products provide protection and/or treat the illness (Conover 2002). Nutraceuticals provide a different approach, i.e. curing health issues based on nutrition and overall well-being of the body. These nutraceutical products or functional foods can be categorized based on their sources, chemical constituents and pharmacological properties. The most common nutraceuticals are herbal or natural products, nutrient supplements and dietary supplements among which medicinal plant-based formulations play a significant role as potential therapeutic agents.

Plant-derived chemicals or phytochemicals are a large group of compounds that help to prevent occurrence of diseases. The specific chemical nature of the compounds is primarily responsible for such kind of activity. Basically, these chemicals have the ability to interact with harmful components of our body and can regulate the secretion of hormones, enzymes and also their binding to the specific receptors. Many compounds boost the production as well as activities of several enzymes that can block the carcinogens and other toxic metabolites inside body (Pandey et al. 2011). Although mode of action of many phytochemicals is still under investigation, their health beneficial effects are undeniable in conformity to the age-old traditional medicine system. A large number of medicinal plants are in use since long back that have important role in disease prevention. As, for example, *Andrographis paniculata* (kalmegh) helps in bile flow, digestion; *Withania somnifera* (ashwagandha) has immunomodulatory, antimicrobial activity; *Tinospora cordifolia* (giloy) shows antiviral, antimicrobial and immunomodulatory activity; *Commiphora wightii* (guggul) has cholesterol-reducing property; *Gymnema sylvestre* (gudmar) shows antidiabetic activity; and many more. Phytochemicals like polyphenolics, alkaloids and saponins are present in the plant matrices that provide specific bioactivity. These medicinal plant extracts are an excellent source for development of potential nutraceutical products having strong therapeutic action.

The market of nutraceutical is comprised of two primary segments, i.e. dietary supplements and functional foods. These type of products are prepared from herbal extracts by formulating the extract into usable form as per convenience. As the plant extract can't be used as such in its concentrated form, it must be converted into suitable products so that the maximum benefit of these phytochemicals can be obtained by the consumers. Herbal products come in different forms that can be used both internally and externally. It includes liquid decoction, teas, tablets and capsules, oils, ointments, etc. For example, ointments prepared from aloe vera gels are used as skin care products, herbal teas form chamomile act against sleeplessness and anxiety; and tablets and capsules of ashwagandha, giloy extracts, are used as immunity booster, increasing body strength and curing various diseases. Mixtures of different plant extracts are also popular among users. Products like mixed herbal *churnas*, chyavanprash, triphala juice and mixed giloy-tulsi juice are highly popular among consumers due to diverse health benefit effects. Development of suitable nutraceutical formulations is of utmost importance to enhance the quality of the extracts. These herbal nutraceutical products can definitely be a game changer in curing health diseases and be the best alternative to synthetic drugs, provided appropriate knowledge and extensive research work is required to formulate efficient products based on need and demand.

In this context, this present article is developed that gives an overview about different phytochemicals, their chemical nature and potential role as nutraceuticals and various herbal formulations with their use as therapeutic agents. This chapter also discusses about present market scenario, regulatory frameworks and quality control aspects associated with herbal products together with the sustainability and future aspects of this sector in a detailed manner.

19.2 Phytochemicals as Potential Nutraceuticals

A plethora of compounds come into the nutraceutical category as they fit the criteria of providing nutrition as well as health-promoting and health-protecting activities inside body. From regular house hold commodities (spices, condiments, blended oil, herbs) up to different supplements and functional foods come under the broad umbrella of nutraceuticals. These nutraceuticals can be broadly categorized into various forms like polyphenols and antioxidants, probiotics, prebiotics, polyunsaturated fatty acids and dietary enzymes. Each of the group has different chemical and/or biological compounds having diverse health beneficial effects. Due to difference in their chemical nature, the mode of action is also different that ultimately provide specific biological activity. The natural compounds obtained from various plant sources (phytochemicals) are rich in polyphenols and their derivatives. These compounds provide several heath beneficial effects including antioxidant, anticarcinogenic, antidiabetic and anti-inflammatory activities that make them potential candidates for development of nutraceutical products.

Polyphenols are large group of chemical compounds that occur naturally within plant system. These compounds have wide range of chemical structures from simple phenolic acids to complex polymeric networks. The basic structural unit in polyphenols is the phenolic ring, mostly classified as phenolic acids, phenolic alcohols, their esters and glycosides. Based on the nature of phenolic ring, polyphenols can be divided into many classes as phenolic acids, flavonoids, phenolic alcohols, stilbenes and lignans. These polyphenols are also mentioned as phytochemicals as the primary source of these chemical compounds are different plant matrices. The phytochemicals are the major bioactive compounds that are involved in protecting human health against various chronic diseases and ailments. Among phenolic acids, hydroxybenzoic acid and hydroxycinnamic acid are the two major classes of phenolic acids available naturally. Derivatives of benzoic acid and cinnamic acid have skeleton of C_1-C_6 and C_1-C_3 backbone, respectively. Among different hydroxybenzoic acid derivatives, gallic acid is found to be present as most predominantly in human diet. Similarly, different cinnamic acid derivatives such as rosmarinic acid, ferulic acid, chlorogenic acid, *p*-coumaric acid, caffeic acid, etc. are abundantly present in human food having significant role in human health (Hussain et al. 2005). Flavonoids are made of two benzene rings that are connected with three carbon chain form a pyran ring. These flavonoids are further classified into different groups based on position of two benzene ring attached with central pyran ring and/or different substituents on the both rings (hydroxylation configuration). This includes flavanol, flavanone, flavones, isoflavones and anthocyanidins. The backbone of flavonoids is basically C_6-C_3-C_6 that contains two units of phenolic structure (C_6) along with one unit of pyran ring (C_3) that connects the two units. In most of the flavonoids, the second phenolic ring is attached with C_2 of pyran ring, but attachments in C_3 and C_4 are also observed. Different sugar moieties are attached with these flavonoids, thus remaining as glycosides in plant matrices. Flavanols are mostly found in mono and polymeric form rather in glycosylated form. Catechins and epicatechins are major compounds in this class. In flavones, there exists a double bond between C_3 and C_2. For isoflavone the attachment of phenolic ring is with C_3 edge of pyran moiety. Compounds like genistein and glycitein fall under this group. On the other hand, flavanones contain an oxygen molecule at C_4 along with saturated pyran ring. Naringin and hesperidin are the two primary compounds come under flavanone group (O'Leary et al. 2004; Sadik et al. 2003). Anthocyanins are one of the major class flavonoids that impart colour (red, blue, purple) to different food commodities. It contains sugar moiety at C_3 of pyran ring and hydroxyl group at 5, 7-position of one benzene ring and $3'$, $4'$and $5'$position of another benzene ring. Unlike flavonoids and anthocyanins, the stilbenes contain C_6-C_2-C_6 type of skeleton in their structure. It has an ethylene moiety at central position connecting two substituted phenyl group on both end of the double bond. The stilbene exists in two geometric isomer forms, i.e. cis-stilbene (two substituted phenyl ring at the same side of the double bond) and trans-stilbene (two substituted phenyl ring at opposite side of the double bond). Resveratrol, phytoalexins and piceatannol are the major candidates under this phenolic group. Lignans are such phenolic compounds that exist in dimeric form with 2,3-dibenzylbutane structure. The basic skeleton is C_6-C_3 that remain as dimers coupled at 8–8′ positions. Some of the lignans belong to dibenzocyclooctadiene, furan and dibenzyl butane subgroups. Podophyllotoxin, pinoresinol, niranthin and nirtetralin are the major lignans found in natural sources.

Different herbal extracts mostly contain these phenolics and flavonoids having potential nutraceutical properties. The primary mode of action of these compounds is scavenging of free radicals from the system, acting as antioxidants. These compounds react with various free radicals and reactive oxygen species in the body and form stable molecules, thus reducing the harmful effects of these reactive molecules. Several studies suggested that many of the diseases including cardiovascular disease, chronic inflammations, cancer, degenerative diseases, etc. are directly linked with oxidative stress generated from different free radicals like reactive oxygen species (ROS) and reactive nitrogen species (RNS). The phytochemicals or polyphenols have ability to neutralize the free radicals by donating hydrogen atom or an electron. The characteristic chemical structure of these compounds, i.e. conjugated double bonds in combination with hydroxylation patterns over the benzene ring, is considered as the primal factor for imparting the antioxidant properties. These polyphenols subdue the effect of free radicals either by inhibiting their formation or by deactivating their precursors and active species. These antioxidants can act as chain breakers of lipid peroxidation reaction (stopping the chain reaction) by scavenging the generated free radicals directly. This is mediated through neutralizing as well as stabilizing the radicals by donating electron to the reactive species (Rice-Evans et al. 1996). Apart from the radical scavenging property, polyphenols can also impart reducing and metal ion chelating activities. Their strong therapeutic potential lies in these properties too. These molecules form stable chelate with oxophilic transition metals like Fe^{2+} that leads to formation of stable five or six membered rings, ultimately reducing the Fenton reaction kinetics directly. This prevents the effect of reactive hydroxyl radicals and reduces chances of oxidation (Pietta 2000). Most of the polyphenols are included in single metal binding site group, as they contain one potential site for chelation. Curcuminoids, stilbenes, lignans, flavanols, isoflavonoids, and anthocyanins fall under this group. On the other hand, flavanones and flavones come under two binding sites group whereas, flavanols and tannins are considered as three binding sites group. Basically, this categorization primarily depends on the type and number substitutions in benzene ring with hydroxyl and/or oxo group (Lakey-Beitia et al. 2021). Figure 19.1 illustrates different groups of phytochemicals or polyphenols having antioxidant activities.

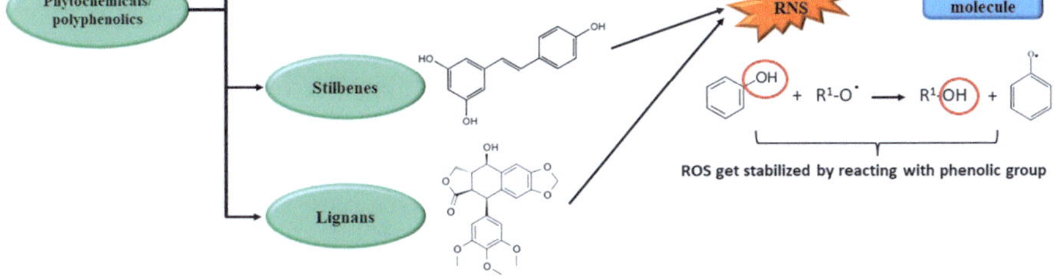

Fig. 19.1 Phytochemicals or polyphenols having antioxidant activities that can stabilize ROS and/or RNS generated inside human system (Source: Luo et al. 2021[110])

Besides antioxidant activity, the polyphenols impart good killing and inhibitory effects over inflammatory cells. This is mediated by influencing the effect of cytokines and their receptors and/or by regulating their secretion inside the system. Some of the phenolic compounds like rutin and hesperidin are reported to exhibit anti-inflammatory activity as comparable with that of standard synthetic drugs (Soni et al. 2013; Ding et al. 2017). The antioxidant and anti-inflammatory activity of polyphenols makes them a potential agent to reduce risk of cancer. They can retard the proliferation of tumour, its harmful effects over cells and apoptosis also. Resveratrol is reported to have significant inhibitory effect over tumour cell proliferation that ultimately reduces the occurrence of cancer and other related diseases (Lee and Lee 2021). Other phenolic compounds have also been identified as potential anticarcinogenic agents like quercetin in prevention of oesophageal cancer, silymarin to reduce the growth of liver cancer cells, epigallocatechin gallate against breast cancer (Liu and Ho 2018; Bimonte et al. 2020), etc. Apart from anticarcinogenic activity, some polyphenols show pro-oxidant activity also that is highly dose dependent. At high doses, this can cause DNA damage and even apoptosis. The pro-oxidant activity of resveratrol leads to accelerate the aging process and shorten the life span of yeast used in brewery sector. Similarly, at large dose, gallic acid tends to accelerate the oxidative loss of amine and thiol groups that largely affects the structural properties and bioactivities of proteins.

Polyphenols also provide good antimicrobial effects including bacteria, fungus, parasites, etc. This attribute is especially for flavonoids having profound antibacterial activity against some of the bacterial strains that become resistant to multiple synthetic antibiotics. These polyphenols can also show synergistic effect with antibiotics exhibiting excellent antibacterial properties. Some studies suggested the antidiabetic potential of polyphenolics. They have ability to regulate the insulin pathway and increase its sensitivity in the peripheral tissues. Many of these compounds display strong inhibition towards alpha glucosidase and alpha amylase that regulate the absorption of glucose in intestine, thus maintaining blood sugar balance inside the body (Dominguez Avila et al. 2017). A report demonstrated promising therapeutic as well as preventive potential of quercetin against diabetes in both in vitro and in vivo conditions (M Eid and S Haddad 2017). On the other hand, flavanol-type polyphenolic compounds including proanthocyanidins and catechins provide antihypertensive effect by improving the endothelial cell function effectively that helps to reduce the sensitivity of low-density lipoproteins, ultimately increasing vasodilation. Apart from flavanols, phenolic acids, anthocyanins, resveratrol and tannins have positive effect over vasodilation and other aspects related to controlling blood pressure level. Polyphenols exhibit anti-obesity properties too. They can inhibit proliferation of adipocyte tissues, stimulating their apoptosis and promoting fat oxidation as well as lipolysis. Other mechanisms involve inhibition of pancreatic lipase and pancreatic protein lipase activity, reducing adipocyte differentiation and inducing secretion of lipase to accelerate lipolysis. Compounds like catechin and gallocatechin can play significant role for controlling obesity (Sung et al. 2010). A number of polyphenolic compounds have the potential to protect the cardiovascular system (anti-atherosclerotic property) via different mechanisms such as maintaining the balance of lipoproteins in body, i.e. decreasing the low-density lipoprotein, increasing high-density lipoprotein and preventing their oxidation. Ellagic acid, epigallocatechin gallate and resveratrol have profound anti-atherosclerotic role that improves the function of endothelial barrier (Bahramsoltani et al. 2019).

Other than polyphenolics, medicinal plant extracts contain several other health-promoting principles such as triterpenoids especially steroids, saponins, alkaloids, lactones, etc. The steroids are large group of compounds, chemically triterpene in nature. Phytosteroids have various health beneficial effects. These compounds can reduce the level of cholesterol from plasma and have anticarcinogenic, antidiabetic and anti-inflammatory activities (Miras-Moreno et al. 2016). Phytosterols present in *T. cordifolia* like β-sitosterol, stigmasterol and ecdysterone have ability to reduce blood cholesterol level. On the other hand, steroids in *W. somnifera* such as withanolides, withafein and withanone have strong anti-inflammatory activity. All these compounds are potential candidate for development of functional food and/or medicines. Saponins are also steroidal in nature where the steroid portion (aglycone) is attached with sugar moiety, remaining as steroidal glycosides. The steroidal portion is either tetracyclic triterpene or pentacyclic triterpene in nature. The sugar moiety has a variety of pentoses or hexoses sugars. Saponins possess a wide range of therapeutic potential such as antimicrobial activity, antiviral activity, anti-obesity activity, etc. Several plant extracts like *Glycyrrhiza glabra* (mulethi), *Chlorophytum borivilianum* (safed musli) and *Panax* sp. (ginseng) are rich source of saponins that are used for development of potential therapeutic drugs. Researches revealed a great potential of these compounds against infectious diseases like SARS-CoV without having any adverse health effects (Sharma et al. 2021). Saponins in *Centella asiatica* (Indian pennywort) (asiatic acid, madecassic acid) together with their respective aglycones (asiaticoside and madecassoside) provide magnitude of health-promoting effects like anticarcinogenic, antimicrobial, neuroprotective, anti-inflammation, wound healing, etc. Other than steroidal compounds, alkaloids are also an abundantly available secondary metabolite present in plant system that are basic in nature with at least one nitrogen atom in its structure. These alkaloids act as strong protecting agents against chronic inflammations and neurodegenerative disorders. Alkaloids present in *T. cordifolia* such as berberine, palmatine and jatrorrhizine provide antioxidant as well as anticarcinogenic effect. Another alkaloid, vasicine, obtained from *Justicia adhatoda* (vasaka) leaf provides pronounced bronchodilatory activity (Avula et al. 2008) that can be extensively used in cough syrup as substitute of synthetic drug, bromhexine. The alkaloids present in *Papaver somniferum* (opium poppy), i.e. morphine, codeine and papaverine, are mainly responsible for its diverse applications in pharmaceutical industries.

Altogether the diverse group of bioactive compounds present in medicinal plant extracts provide health-promoting activities in different ways that suggest their immense potential in nutraceutical sector. If these plant extracts are formulated into useful and convenient form, their acceptability would get increased among the consumers. The nutraceutical potential of the phytochemicals can best be utilized through development of suitable products that can be a promising substitute for synthetic medicines.

19.3 Herbal/Natural Formulations

Herbal formulations refer to products that contain one or more herbs, either in their natural state or processed, blended in specific quantities. These formulations are designed to offer distinct nutritional or cosmetic advantages and are intended for use in diagnosing, treating or mitigating diseases in humans or animals, as well as potentially influencing the physiological makeup of individuals (Singh et al. 2022). Herbal formulations have been an integral component of traditional medical practices for centuries. These formulations are created using a variety of techniques, including extraction, distillation, fractionation, purification, concentration, expression and even fermentation of diverse herbs (Nirmal et al. 2022). Presented below are several examples of herbal compositions.

19.3.1 Herbal Infusions

Infusion is the method of drawing out chemical compounds or flavours from plant materials using solvents like water, oil or alcohol (Yadav et al. 2022; Zioga et al. 2022). This is achieved by immersing the plant material in the solvent for a prolonged duration. Herbal infusions can be produced using a diverse array of plants and plant components, including leaves, flowers and bark. Several well-known herbs, namely, *Matricaria chamomilla* (chamomile), *Melissa officinalis* (lemon balm), *Echinacea angustifolia* (narrow-leaved purple coneflower), *Mentha x piperita* (peppermint) and *Aloysia citrodora* (Greek lemon verbena), are usually prepared as infusions, due to their appealing sensory attributes (Mateus et al. 2023).

19.3.2 Herbal Decoction

It is a method of extracting herbal or plant chemicals by boiling various plant parts such as stems, bark, seeds, berries, roots and rhizomes to dissolve the chemicals present in these plant materials (Alamgir and Alamgir 2017). The procedure involves boiling a specific quantity of dried, ground and powdered plant material in a specified volume of water. The mixture is stirred while being heated, followed by a cooling period and eventual straining. This entire process just takes a few minutes, typically around 15 minutes (Shireen and Zarena 2022). Unlike infusions, decoctions tend to encompass a higher concentration of oil-soluble substances (Zhang et al. 2018).

19.3.3 Herbal Tinctures

Herbal tinctures are created through a process where leaves, bark, berries and various other plant components are soaked in alcohol for an extended period (Azwanida 2015). In a mixture of alcohol and water, many plant elements dissolve more effectively than pure water (Romero et al. 2005). Infusion and decoction techniques are favoured for extracting water-soluble chemicals, such as anti-inflammatory plant steroids. On the other hand, tinctures are more suitable for extracting chemicals that dissolve effectively in alcohol, including anti-bacterial alkaloids. This may explain why a tincture is traditionally used to treat a range of bacterial infections (Alamgir and Alamgir 2017).

19.3.4 Herbal Capsules

Capsules are solid forms of medication that encase specific substances often enclosed within a gelatine container (Winfield et al. 2009). Hard gelatine capsules are used for dry powered or granulated herbal components, whereas soft gelatine capsules are used for herbal oils and herbal compounds that are either dissolved or suspended in oil (Bayor et al. 2011). The gelatine shell of capsules readily ruptures and dissolves upon oral administration. In comparison to tablets, capsules generally facilitate a more rapid release of drugs (Hoffmann 2003). The stability of herbal capsule is relatively better than the aqueous preparations such as infusions and decoctions. In herbal capsules, botanical components are meticulously packed into the capsule shells. Alternatively, herbal extracts can also be employed, derived from the respective herbal material(s), with appropriate fillers (Johnson et al. 2010).

19.3.5 Herbal Tablet

A tablet is a round, oval or square-shaped compact medicine. Herbal tablets are specifically designed for oral consumption with various herbal materials incorporated to elicit particular therapeutic effects (Mishra et al. 2012). Like the herbal capsules, incorporation of the herbal material may be done with the finely powdered or extracts from the plant materials using appropriate solvents suitable for oral consumption (Nitin et al. 2010). The formulation may incorporate certain additives, such as binders, and lubricants to facilitate the tableting process. Disintegrants are also included to ensure proper breakup of the tablet within the gastrointestinal tract. Additionally, to mask the unpleasant taste of certain active ingredients, sweeteners or flavours can be integrated, while pigments may be employed to enhance the visual appeal of uncoated tablets (Hoffmann 2003).

19.3.6 Herbal Powders

Herbal powders are one kind of compositions that can be purchased as single herbal powder or as a mix of different herbal powders. The efficacy of the formulation is depended upon the specific blend of powders and accuracy in the combination ratio. The powders can be consumed in a variety of ways. They can be consumed whole in the specified quantities, mixed with boiling water and shipped as tea or even blended with honey and swallowed as a little pill (Bruton-Seal and Seal 2014). They may be finely sifted herbal materials from various parts of plants meant for a particular therapeutic effect (Bouwman-Boer et al. 2015).

19.3.7 Herbal Teas

These are the infusion, typically consumed in the form of tea. It's important to note that prepared infusions should be promptly consumed after preparation, since they don't possess a prolonged shelf life due to the involvement of water in the extraction process (Packer et al. 2004). These teas are conveniently available as tea bags suitable for hot infusion (Bruton-Seal and Seal 2014) or as powdered herbal materials (typically pulverized leaves) that are boiled in hot water for a brief duration, strained and then enjoyed as a brewed tea (Bouwman-Boer et al. 2015). The stability of the powdered plant material used in the preparation depends on the type as well as the moisture content. Teas that are stored in hermetically sealed containers can retain their quality for up to a year, whereas those kept in tea bags may have a relatively shorter shelf life.

19.3.8 Herbal Ointments

It is a semisolid dosage form designed for application to the skin, rectum or nasal mucosa that contains less than 20% water and more than 50% hydrocarbons, waxes or polyethylene glycols (Winfield et al. 2009). Ointments are used for a variety of functions, including providing protection, antiseptics, acting as emollients and antipruritic (Aulton 2007). However, it is important to note that herbal ointments are not suitable for treating deep wounds (Bascom 2002). In herbal ointments, the plant material(s) are typically integrated into the base, either finely sifted or extracted (Al-Achi 2008). Herbal ointment of *Azadirachta indica* (neem) and *Curcuma longa* (turmeric) extract is the most common herbal ointment formulations widely used (Sawant and Tajane 2016).

19.3.9 Herbal Cream

Creams are semi-solid emulsions that are mixtures of oil and water (hydrophilic) (Winfield et al. 2009). Herbal creams normally contain the herbal material in either finely sifted form or incorporated as an extract. Some herbalists tend to confuse between creams and ointments. Herbal creams are defined by their hydrophilic base, while the base is entirely hydrophobic for ointment (Zdoryk et al. 2014). They may have a relative shorter shelf life compared to ointments due to the presence of water in the base. The herbal cream is useful for moisturizing, nourishing, whitening and treating a variety of skin disorder.

19.3.10 Herbal Oil

These are formulations where herbal components are either suspended or dissolved within an oily medium. Among these oils, it's common to find herbal ingredients such as leaves infused with essential oils. These oils, often referred to as infused or macerated oils, should not be conflated with essential oils, which are fragrant oils extracted through distillation of plant materials (Bruton-Seal and Seal 2014). These preparations are normally meant for external or topical use as liniments (Winfield et al. 2009). Nonetheless, some of these formulations might be intended for oral consumption as well (Bruton-Seal and Seal 2014). Common herbal oils are peppermint oil, clove oil, etc. The stability of various essential oils is predominantly influenced by the type of oil chosen for the extraction process (Turek and Stintzing 2013).

19.3.11 Herbal Shampoo

A surfactant-based formulation, herbal shampoo comes in various forms: liquid, solid, or powder. The creation of herbal shampoo typically involves the combination of hair-nurturing herbs with surfactants like sodium laurate sulphate, alongside co-surfactants such as cocamido propyl betaine, all suspended within water (Gubitosa et al. 2019). Among the assortment of options, recognizable herbal shampoos include those incorporating ingredients like *Acacia concinna* (shikakai), *Sapindus mukorossi* (reetha) and *Lawsonia inermis* (henna) each contributing to a distinctive hair care experience (Lodha 2019).

19.3.12 Herbal Soap

The soap is essentially the salt derived from a fatty acid often made through the saponification process involving a fatty acid and caustic soda or an appropriate base (Jones 2011). Herbal soaps have the herbal materials incorporated in the detergent base. These botanical ingredients usually exhibit antifungal and antibacterial properties upon contact with the skin, enhancing the cleansing ability (Grosso 2013). Herbal soaps are predominantly designed to address microbial skin conditions like dandruff, eczema, ringworm and boils (Panda 2000). To extend the shelf life of soaps, the incorporation of preservatives or antioxidants can be added.

19.4 Role of Different Herbal Products in Curing Ailments

In recent times, there's a growing demand for natural, plant-derived medicines due to their lower risk of adverse reactions compared to synthetic alternatives. Herbal products are gaining prominence in managing chronic medical conditions like breast cancer, liver diseases, HIV, asthma and rheumatological disorders. According to World Health Organization (WHO), around three-quarters of the global population incorporate herbs and traditional medicines into their health routines for various ailments. Traditional Indian medicine systems like Ayurveda and Unani are closely connected to Indian medicinal plants, which carry a heritage of contributing to well-being through their therapeutic qualities and holistic healing approach. Indian medicinal plants play a vital role in promoting overall wellness in various ways.

19.4.1 Skin Ailments

Skin diseases are common health issues, and medicinal plants offer a valuable source of active compounds, providing safe and cost-effective treatments. Mohammad et al. (2019) developed an Ayurvedic formulation including ingredients like *Cassia fistula*, *Saccharum officinarum*, *Embelia ribes*, *Terminalia chebula*, *Phyllanthus emblica*, *Operculina turpethum*, *Syzygium aromaticum*, *Elettaria cardamomum* and *Piper nigrum* together with honey targeting skin disorders. Similarly, Mazzio and Soliman (2008) created a topical ointment containing extracts of various plants, namely, black walnut hull, turmeric rhizome, wormwood, garlic, chamomile, licorice root, *Aloe vera* and niacin for skin infections. Medicinal plants from the euphorbiaceae family, such as *Phyllanthus emblica*, *Acalypha indica*, *Jatropha curcas* and *Euphorbia thymifolia*, have been traditionally used for diverse skin issues (Tirant et al. 2018). Reddy et al. (2007) formulated an ointment using herbal extracts of *Wrightia tinctoria* and *Cocos nucifera* to regress dermal vessel tortuosity. Khanuja et al. (2007) developed an anti-dematophytic cream with a synergistic combination of garlic extract and essential oil of spearmint, cinnamon and geraniol, while Mariana (2005) created an herbal blend of mallow, rosemary and eucalyptus extracts for skin problems. Lee (2005) designed an acne treatment solution with herbal extracts such as dandelion, senecio, elephantopi, lonicera, angelica, scute and witch hazel. Arora et al. (2003) patented a composition for psoriasis using plant extracts of *Argemone mexicana* and *Cuminum cyminum*. Ramakrishna et al. (2002) invented an herbal formulation using *Tagetes erecta*, *Ocimum sanctum*, *Moringa oleifera*, *Tridax procumbens*, *Aloe vera* and *Gum olibanum* for skin disorders, whereas, Farooqi et al. (2002) developed a skin care formulation with sandalwood oil and rose water for skin cleansing, restoration and moisturization.

19.4.2 Pulmonary and Respiratory Disorders

Respiratory disorders are a major global health problem, which impact all age groups (Smruti 2021). Herbal solutions have gained attention in this sector. Kayani et al. (2014) created a formulation, employing *Althea officinalis*, *Saxifraga hirculus*,

Sisymbrium erysimoides, *Solanum virginianum* and *Inula obtusifolia*, addressing 25 respiratory disorders. Gibbs (2009) utilized aerial parts of *Adhatoda vasica* for bronchitis and asthma. Bandyopadhyay et al. (2002) developed a *Piper betel* and *Murraya koenigii* blend for treating respiratory conditions of humans. Similarly, Brindavanam et al. (2003) offered a liquid composition, potent against bronchial asthma, incorporating *Solanum virginianum*, *Justicia adhatoda*, *Piper longum*, *Elettaria cardamomum*, *Syzygium aromaticum* and *Mesua ferrea*.

19.4.3 Cancer Treatment

Medicinal herbs and compounds gain recognition for cancer treatment. Clinical studies highlight their positive impact when combined with conventional therapies (Yin et al. 2013). Qi et al. (2010) formulated an herbal formulation using astragalus, turmeric, ginseng, boosting immune function and sensitizing chemoradiotherapy. On the other hand, Rao et al. (2003) introduced a composition from *Cedrus deodara* and wikstromol, inhibiting various cancer growths.

19.4.4 Cardiovascular Diseases

Cardiovascular diseases are a major global concern, having surged in recent decades. Due to prevalent side effects of current treatments, alternative methods like medicinal plants and natural products gain preference (Hesari et al. 2021). Malik et al. (2017) tested herbal formulation of *Cichorium intybus*, *Gymnema sylvestre*, *Nigella sativa* and *Trigonella foenum graecum* extracts for blood pressure reduction. Chou (2009) introduced an herbal composition with *Rhodiola rosea*, *Ginkgo biloba*, *Panax notoginseng* and *Ligusticum chuanxiong*, promoting brain and cardiovascular health. Chen (2003) invented an herbal composition with *Ligusticum chuanxiong*, *Panax ginseng*, *Panax notoginseng* and *Rheum palmatum* extracts for treating cerebral vascular diseases. Lee et al. (2002) proposed a composition (*Angelica sinensis* 30–35%, *Radix ligustici* 35–65%, *Radix ginseng* 7–13%) to enhance blood circulation, successfully demonstrating therapeutic efficacy in human.

19.4.5 Treatment of Diabetes

Diabetes mellitus, a widespread metabolic disorder, poses a significant global health threat. Over 463 million individuals are affected worldwide, projected to rise to 578 and 700 million by 2030 and 2045 (Saeedi et al. 2019). Popoola and Obembe (2013) suggested *Moringa oleifera* Lam. leaves, combined with "Osan Orombo" (*Citrus* sp.), for diabetes treatment. Brindavanam et al. (2002) proposed an herbal composition containing extracts of *Momordica charantia*, *Gymnema sylvestre* and *Pterocarpus marsupium*. Palpu et al. (2007) developed an antidiabetic formulation using *Glycine max*, *Curcuma longa* and *Zingiber officinale* extracts, exhibiting significant activity at 50–200 mg/kg doses.

19.4.6 Treatment of Depression

Depression is a significant public health concern, profoundly impacting individuals and society, with around 121 million affected globally (Zhang et al. 2012). Du et al. (2020) highlighted a traditional Chinese medicine formula comprising herbs like Chaihu, Gancao, Fuling, Suanzaoren, Yujin, Baizhu, Yuanzhi, Shichangpu and Banxia, effective in reducing poststroke depression in rats. Kung and Kong (2006) introduced a pharmaceutical formulation using *Psoralea corylifolia* L. and *Poria cocos* (Schw) Wolf extract to treat depression. Oral treatment at 100–250 mg/kg/day with appropriate extract proportions showed potent and rapid antidepressant effects in animal models.

19.4.7 Treatment of Obesity

Obesity is recognized as an important metabolic disorder, affecting a substantial global population. The World Health Organization (WHO) ranks it among the top 10 preventable diseases (Wilborn et al. 2005). Hasani-Ranjbar et al. (2009) found that extract of several plants including *Cissus quadrangularis*, *Sambucus nigra*, *Asparagus officinalis*, *Garcinia atroviridis* and *Zingiber officinale* led to notable body weight reduction. Kamali et al. (2012) introduced Itrifal Saghir, a polyherbal preparation with *Phyllanthus emblica*, *Terminalia chebula* and *Terminalia bellirica*, marketed in Iran for obesity treatment. Palpu et al. (2005) proposed a synergistic composition containing *Gentiana kurroo*, *Murraya koenigii*, *Allium sativum*, *Zingiber officinale*, *Amorphophallus campanulatus* extracts and additives, offering potential against atherosclerosis and hyperlipidemia treatment.

19.4.8 Treating Infections

Phyto-constituents from medicinal herbs are extensively researched for antiviral and antibacterial potential. An isoquinoline alkaloid from *Cephaelis ipecacuanha* treats *Escherichia histolytica* infections (Ganjhu et al. 2015). Ma et al. (2018) reported the traditional Chinese medicine formula called San Wu Huangqin Decoction (SWHD), composed of *Sophora flavescens*, *S. baicalensis* and *Rehmannia glutinosa*, that can be utilized to address influenza. Notable herbs like *Panax ginseng*, *Panax quinquefolius*, *Camellia sinensis* and *Allium sativum* prevent viral respiratory infections (Yarnell 2018). Ghosal (2007) patented an oral poly-herbal formulation for viral infections. It included *Withania somnifera*, *Mangifera indica* extracts and purified Shilajit. It maintained treatment post initial viral load reduction, causing minimal side effects. Khanuja et al. (2003) developed an antimicrobial formulation comprising thymol from *Trachyspermum ammi* and mint oil from a hybrid of *Mentha spicata* and *Mentha arvensis*, along with conventional additives. Mint oil concentration ranged from 0.1 to 0.5% w/w and thymol from 20 to 50% w/w. Evaluation against bacteria, particularly *mycobacterium* and *Escherichia* groups, highlighted its efficacy.

19.4.9 Treatment of Infertility

Herbs have long been recognized as aphrodisiacs for impotence treatment. Hsieh et al. (2016) presented an oral herb formula containing *Astragalus membranaceus*, *Lepidium meyenii*, *Ophiocordyceps sinensis*, *Panax quiquefolium*, *Piper nigrum*, *Rhodiola rosea* and *Serpentes cnidium monnieri*. A poly-herbal formulation by Hussain et al. (2018) combined various plant parts of *Chlorophytum borivilianum*, *Hygrophila spinosa*, *Mucuna pruriens*, *Mimosa pudica*, *Acacia Senegal*, *Astragalus membranaceus*, *Plantago ovata*, *Bombax ceiba*, *Eurycoma longifolia* and rocky candy in a ratio of 1:1:1:1:1:1:1:1:1:8 to enhance fertility. Palpu et al. (2007) introduced herbal bidis and cigarettes to boost sexual desire, treat impotence and aid digestion. Ingredients included *Matricaria chamomilla*, *Hygrophila auriculata* and *Aquilaria agallocha*, aiming to enhance vitality and digestion.

Herbal medicine offers significant benefits in treating gynecological disorders. Aziz et al. (2018) highlighted the use of *Allium sativum* mixed with turmeric powder to stimulate uterine muscles. Katiyar et al. (2002) developed an enriched herbal capsule containing ashoka, lodhra and various other components to manage gynecological disorders and prevent anaemia from excessive menstrual bleeding. The formulation also contained plant coagulates from spinach, amaranth, berseem and cowpea, aiming to address menstrual issues and associated anaemia.

19.4.10 Gastrointestinal Disorders

Medicinal plants effectively treat a range of gastrointestinal disorders, from vomiting to peptic ulcers. Triphala, composed of *Terminalia chebula*, *Terminalia bellirica* and *Phyllanthus emblica*, benefits the entire gastrointestinal tract (GI) tract (Tarasiuk et al. 2018). Palpu et al. (2008) developed a formulation for treating ulcers containing extracts of *Ulteria salicifolia*, *Asparagus racemosus*, *Foeniculum vulgare* and *Ficus glomerata*. Rao et al. (2005) introduced an antiulcer composition with *Asparagus racemosus*, *Glycyrrhiza glabra* and more, showing superior activity to omeprazole. Another method used extracts of *Ulteria salicifolia*, *Asparagus racemosus* and more to relieve acidity and heal ulcers, proving more effective than commercial antiulcer agents.

19.4.11 Immunostimulants

Ancient herbal formulations have been utilized to boost the immune system, which safeguards the body against infections. Gupta et al. (2006) noted that ashwagandha *churna* holds potential to stimulate cell-mediated immunity. Shoemake (2006) created a natural blend of powdered herbs to restore the immune system against infectious diseases and allergies. This formulation included ingredients like carrot powder, pau d'arco bark, ginkgo leaf, celery leaf, red clover blooms, burdock root, goldenseal root, poke root, garlic bulb powder and beet root powder encapsulated in gel capsules. Managoli (2008) introduced an herbal composition for immune stimulation, comprising extracts of *Glycine max*, *Asparagus racemosus*, *Boerhavia diffusa*, *Withania somnifera*, *Saraca indica* and *Coleus forskohlii*. Its therapeutic effects encompassed anticancer, antimetastatic, immunomodulatory and adaptogenic properties.

19.4.12 Therapeutic Application of Indian Medicinal Plants

Herbal formulations have a long history of use worldwide and are appreciated for their relatively low adverse effects compared to modern medications (Musthaba et al. 2010; Srivastava et al. 2012). They are widely utilized as remedies for various diseases by a significant portion of the global population. In India, these formulations have gained widespread acceptance as therapeutic agents for various conditions. Several herbal formulations have been extensively studied for their pharmacological effects in various research laboratories across India. A few examples are provided in the summarized Table 19.1.

Table 19.1 Therapeutic uses of different medicinal plants

Plants used in herbal formulation	Form of application	Therapeutic use	Reference
Aloe ferox and *Withania somnifera*	Aqueous extracts	Anti-viral effect (herpes simplex virus)	Kambizi et al. (2007)
Basil extract	Cream	Anti-aging potential	Rasul and Akhtar (2011)
Brahmi and Sarpagandha	Capsule	Management of hypertension	Mishra and Tubaki (2019)
Ocimum basilicum (basil)	Ethanolic extract	Cardioprotective effects	Fathiazad et al. (2012)
Eclipta alba (bhringraj), *Andrographis paniculata* (kalmegh), *Boerhaavia diffusa* (punarnava), *Picrorhiza kurroa* (kutaki), *Cichorium intybus* (kasani), *Phyllanthus niruri* (Bhumyamlaki), *Tephrosia purpurea* (Sharpunkha), *Zingiber officinale* (Shunthi)	Liquid dosage forms	Hepatoprotective effect	Pundir et al. (2009)
Andrographis peniculata, *Zingiber officinalis* and *Citrus Limon*	Hand wash	Anti-microbial activity	Rukari et al. (2022)
Lemon grass and ginger extract	Ice cream	Develop reduced sugar herbal ice cream	Chamchan et al. (2017)

19.5 Market Scenario and Future Perspective

About 80% of the world's population depends on traditional herbal systems for primary healthcare. As people seek comprehensive wellness solutions, traditional medicines offer diverse options that resonate with their preferences and beliefs, contributing to their market expansion. The changing consumer inclination from chemical-based drugs towards Ayurvedic products, as they aid in increasing resistance to diseases, eliminating impurities, preventing inflammation while improving overall health, etc., is primarily driving the Ayurvedic products market. The global herbal medicine market size was valued at USD 151.91 billion in 2021. The herbal market is growing with a CAGR of 5.34% in the period of 2021 to 2028 and is expected to reach USD 426.43 billion by 2028. The Indian market is also expanding in proportion to the global one. The Indian Ayurvedic market was valued at INR 626 billion in 2022 that is expected to grow at CAGR of 19.3% during 2023–2028.

The market for traditional medicines is growing due to various factors. Modern lifestyles and healthcare practices often focus on treating specific symptoms rather than addressing the underlying causes of health issues. Natural remedies and alternative medicine emphasize a holistic approach that considers the interconnectedness of the body, mind and environment. Many conventional pharmaceuticals and treatments can come with potential side effects and long-term consequences. Natural remedies, derived from plants and herbs, are often perceived as gentler alternatives with fewer adverse effects. Across cultures, traditional healing practices have utilized plants, herbs and other natural elements for centuries. As people seek to reconnect with their cultural heritage and ancestral wisdom, interest in these practices has grown. The availability of information through the Internet has empowered individuals to take a proactive role in their health. Natural remedies often provide a sense of agency, allowing people to manage their health more independently.

Moreover, there is a growing emphasis on preventing illness rather than just treating it when it occurs. Natural remedies and alternative medicine often focus on supporting the body's innate healing mechanisms and boosting the immune system. Alternative medicine approaches often consider an individual's unique constitution, lifestyle and preferences. This personalized approach resonates with individuals seeking healthcare that feels tailored to their needs. Perception of natural purity is also a key factor for the adoption of herbal products. Many consumers perceive natural remedies as pure and in harmony with nature. This perception aligns with a broader cultural movement towards eco-friendly and sustainable living. Some individuals use natural remedies as complementary treatments alongside conventional medicine. Integrative medicine combines the strengths of both approaches for a more holistic treatment plan. Further, as scientific research continues to explore the potential benefits of natural remedies, more evidence is emerging to support their efficacy. This research contributes to the credibility of alternative medicine practices. High-profile endorsements, media coverage of natural remedies and the rise of wellness influencers have contributed to raising awareness and interest in these approaches.

19.6 Regulatory Landscape, Quality Control and Sustainability

The regulatory landscape is framed in accordance with quality control that helps to ensure the safety, efficacy and standardization of products. In India, various regulatory authorities and guidelines oversee the cultivation, harvesting, processing and distribution of the medicinal plants. Various regulatory frameworks have been designed for quality control and quality assurance of Indian medicinal plants that includes different organizations, agencies and practices. It is discussed in brief as follows:

Central Drugs Standard Control Organization (CDSCO): CDSCO is the primary regulatory authority in India that regulates the import, manufacture, sale and distribution of drugs, including herbal products. It ensures that herbal medicines adhere to safety, efficacy and quality standards.

Ayurvedic Pharmacopoeia of India (API): API is an authoritative document that outlines quality standards for traditional Ayurvedic medicines. It provides guidelines on the identity, purity, strength and safety of herbal ingredients and formulations.

Traditional Knowledge Digital Library (TKDL): TKDL is a valuable resource that documents traditional knowledge related to medicinal plants and their uses. It helps prevent biopiracy by preventing the misappropriation of traditional knowledge for commercial purposes.

Quality Control Laboratories: Herbal products and raw materials are tested in quality control laboratories to ensure they meet specified standards. These laboratories analyse parameters such as identity, purity, strength, and safety.

Herbal Drugs Consultative Committee (HDCC): HDCC advises the government on matters related to the regulation and development of herbal drugs. It plays a role in formulating policies and guidelines for the herbal industry.

Testing and Certification Agencies: Various government and private organizations provide testing and certification services for herbal products, verifying their adherence to quality and safety standards.

Labelling and Packaging Regulations: Herbal products must follow specific labelling and packaging requirements, including accurate ingredient listing, usage instructions, and safety information.

Phytochemical Research: Research institutions and organizations conduct studies to identify and quantify the bioactive compounds present in medicinal plants. This contributes to understanding their therapeutic potential and standardization.

Good Agricultural and Collection Practices (GACP): GACP guidelines provide recommendations for the cultivation, collection and processing of medicinal plants to ensure their quality and sustainability. These guidelines cover practices related to soil, water, cultivation methods, harvesting and post-harvest handling.

Good Manufacturing Practices (GMP): GMP standards ensure that herbal products are manufactured in a controlled environment with proper quality control measures. They cover aspects such as facility design, equipment maintenance, hygiene, documentation, and quality assurance.

International Collaboration: India collaborates with international organizations, such as the World Health Organization (WHO), to harmonize regulations and quality control standards for herbal products.

Overall, the regulatory landscape and quality control measures aim to ensure the safety, efficacy and authenticity of herbal products. Consumers can make informed choices when purchasing these products, and the industry benefits from maintaining high-quality standards and preventing adulteration or mislabelling.

Another important aspect to maintain the quality and viability of this sector in long term is sustainability. It also promotes fair practices and environmental stewardship. Sustainability and ethical sourcing safeguard these valuable resources for future generations and also contribute to the well-being of local communities and ecosystems. By adopting responsible practices, businesses and consumers can contribute to the preservation of biodiversity and the cultural heritage associated with traditional healing practices. There are some challenges while maintaining the sustainability of this herbal sector like overharvesting and unplanned collection of plants from natural habitat, deforestation, habitat destruction, weak regulatory frameworks and limited enforcement, overexploitation of wildlife species, etc. These issues can be overcome successfully by adopting suitable and effective strategies and implementing them in appropriate manner. Figure 19.2 displays different challenges and appropriate strategies that can be adapted towards maintaining the sustainability of herbal industry as whole. Encouraging farmers to cultivate the

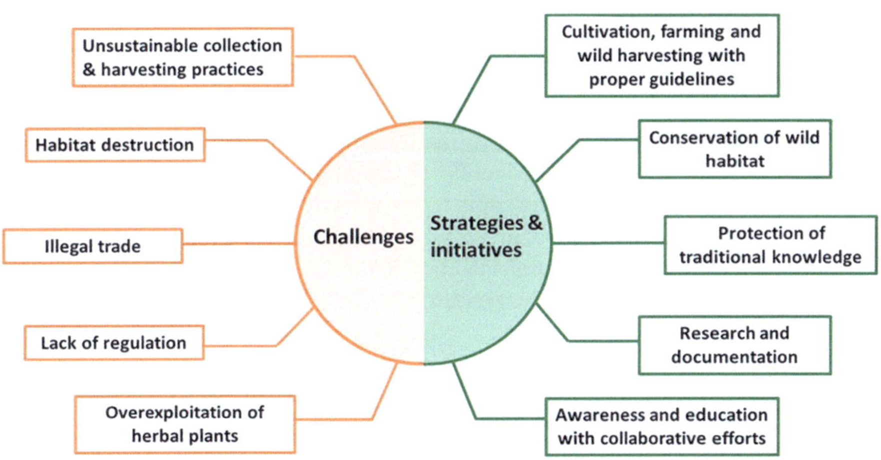

Fig. 19.2 Overview of several challenges and appropriate strategies that can be adapted towards accomplishing the sustainability of herbal sector

medicinal plants on demand and need basis, developing and promoting guidelines for sustainable wild harvesting, such as limited collection, selective harvesting and rotation, can be very fruitful to reduce the loss of wild populations. Conservation of wild habitat is another option to accomplish the same. Involving local communities in the management and decision-making process ensures that their traditional knowledge is respected and that they get benefitted economically from sustainable practices. Recognizing and protecting the intellectual property rights is also an important aspect in this matter. Various awareness programs and education campaign need to be organized in collaboration between governments, NGOs, industry stakeholders, and local communities that will foster coordinated efforts for sustainable sourcing and conservation. Government of India has constituted the National Medicinal Plants Board (NMPB) that promotes the cultivation, conservation and sustainable use of medicinal plants through various schemes and programs. Biological Diversity Act and associated rules also regulate the access to genetic resources and ensure that benefits from their utilization are shared with local communities. By implementing proper strategies at different levels and making people aware about the importance of sustainable approach, the viability of this herbal sector can be maintained that will be useful for both producers and consumers.

19.7 Technological Innovations

Technological innovations in medicinal plants have revolutionized the way we cultivate, process and utilize these valuable resources. Cutting-edge advancements are enhancing the efficacy, quality and sustainability of herbal products. One notable innovation is precision farming, which employs sensors, data analytics and remote monitoring to optimize cultivation conditions. This results in higher yields, improved quality, and reduced environmental impact. Similarly, genetic modification techniques enable the enhancement of specific medicinal properties in plants, making them more potent and effective.

Extraction methods have also evolved. Supercritical fluid extraction and nanotechnology-based processes yield highly concentrated and bioavailable herbal extracts. This enhances the bioactivity and absorption of medicinal compounds, maximizing their therapeutic benefits. Technological interventions has also improved quality control. Spectroscopic techniques and DNA barcoding ensure the authenticity and purity of herbal ingredients, preventing adulteration. Blockchain and traceability systems enable transparency in the supply chain, assuring consumers of the product's origin and ethical sourcing. Virtual reality (VR) and augmented reality (AR) have found applications in education and research, allowing scientists and practitioners to explore virtual herbal gardens and learn about plant properties.

These technological innovations are reshaping the medicinal plant industry, offering solutions for sustainable cultivation, potent extracts, quality assurance and educational enhancements. These advancements are contributing to the growth and credibility of the herbal products market while preserving the essence of traditional practices.

19.8 Challenges and Future Outlook

The medicinal plant market faces various challenges and holds both promising opportunities and uncertainties for the future. Here's a concise overview of the challenges and future outlook of the medicinal plant market.

19.8.1 Challenges

Sustainability: Overharvesting, habitat loss and unsustainable practices threaten the availability of medicinal plants, leading to potential resource depletion and biodiversity loss.

Quality Control: Ensuring consistent quality, authenticity and safety of herbal products remains a challenge, with issues like adulteration and mislabelling affecting consumer trust.

Regulation: Navigating complex regulatory landscapes across different regions poses challenges for businesses operating in the global market, impacting product development and trade.

Cultural Preservation: Balancing the commercial demand for medicinal plants with the preservation of traditional knowledge and cultural practices is a delicate challenge.

Climate Change: Changing climatic conditions can affect the growth patterns and medicinal properties of plants, impacting their reliability as therapeutic sources.

Market Fragmentation: The diversity of herbal products and traditional practices can lead to a fragmented market, making standardization and communication challenging.

19.8.2 Future Outlook

Growing Demand: The trend towards natural remedies and holistic wellness is expected to drive continued demand for medicinal plants, creating growth opportunities.

Innovation: Advancements in technology and research will lead to more effective cultivation methods, extraction techniques and quality control measures.

Sustainability Focus: Industry stakeholders are increasingly prioritizing ethical sourcing, sustainable practices and conservation efforts to ensure long-term availability.

Collaboration: Partnerships between governments, NGOs, communities and businesses will foster responsible sourcing, knowledge sharing and biodiversity conservation.

Global Integration: The global herbal market will likely see increased international collaboration, harmonization of regulations and standardized quality benchmarks.

Consumer Awareness: Heightened consumer awareness about the benefits of herbal products and their holistic approach to health will drive market expansion.

Education: Enhanced education about medicinal plants' uses, benefits and proper sourcing will improve consumer trust and ethical consumption.

Traditional Wisdom: Integrating traditional knowledge into modern research and product development will enhance both efficacy and cultural preservation.

While challenges persist, the medicinal plant market holds a positive future driven by increased demand, technological innovation, sustainability efforts and a greater emphasis on collaboration and consumer education. The industry's success will depend on navigating challenges responsibly while embracing opportunities for growth and positive impact.

19.9 Conclusion

Nutraceutical formulations from medicinal plants provide a novel avenue for overall well-being of health including immunity enhancement, disease prevention and its cure. This article focuses on different phytochemicals with their nutraceutical potential, various herbal formulations, their therapeutic applications along with the market dynamics, future outlook, regulatory framework and sustainability considerations related to the herbal nutraceutical formulations. With the availability of plethora of plant species present on the earth, there is always a possibility for exploration of the plants to develop different herbal formulations having a wide array of health promoting activities. These products show promise in addressing obesity, diabetes, cancer, heart disease, neurodegenerative conditions and many more. In essence, nutraceuticals play an indispensable role in disease management worldwide. Although researchers are actively involved in developing different advanced nutraceutical formulations, public awareness and acceptance is crucial to expand the scope. With proper quality control measurements and regulatory guidelines, these products can be disseminated among the consumers successfully so that the best benefits can be obtained by the end users for a healthy disease-free lifestyle.

References

Al-Achi A (2008) An introduction to botanical medicines: history, science, uses, and dangers. Bloomsbury Publishing, New York

Alamgir ANM, Alamgir ANM (2017) Herbal drugs: their collection, preservation, and preparation; evaluation, quality control, and standardization of herbal drugs. In: Therapeutic use of medicinal plants and their extracts: pharmacognosy, vol 1. Springer, Cham, pp 453–495

Arora S, Gupta L, Srivastava V, Sanganabhatla N and Saraf D (2003) Herbal composition for treating various disorders including psoriasis, a process for preparation thereof and method for treatment of such disorders. U.S. patent application 10/340,195

Aulton ME (2007) Aulton's pharmaceutics: the design and manufacture of medicines, 3rd edn. Elsevier, Amsterdam, pp 298–299

Avula B, Begum S, Ahmed S, Choudhary MI, Khan IA (2008) Quantitative determination of vasicine and vasicinone in Adhatoda vasica by high performance capillary electrophoresis. Pharmazie 63(1):20–22

Aziz MA, Khan AH, Ullah H, Adnan M, Hashem A, Abd Allah EF (2018) Traditional phytomedicines for gynecological problems used by tribal communities of Mohmand agency near the Pak-afghan border area. Rev Bras 28:503–511
Azwanida NN (2015) A review on the extraction methods use in medicinal plants, principle, strength and limitation. Med Aromat Plants 4(196):2167–0412
Bahramsoltani R, Ebrahimi F, Farzaei MH, Baratpourmoghaddam A, Ahmadi P, Rostamiasrabadi P, Rasouli Amirabadi AH, Rahimi R (2019) Dietary polyphenols for atherosclerosis: a comprehensive review and future perspectives. Crit Rev Food Sci Nutr 59(1):114–132
Bandyopadhyay S, Roy KC, Roy M, Pal BC, Bhadra R, Das K and Bhattacharaya S (2002) Herbal formulation of a combination of Piper betel and Murrya koenigii extracts for blocking 5 lipoxygenase activity. U.S. Patent 6,413,553
Bascom A (2002) Incorporating herbal medicine into clinical practice. FA Davis, Philadelphia
Bayor MT, Johnson R, Gbedema SY (2011) The oral capsule-the most appropriate dosage form for croton membranaceus. Int J Pharm Sci Res 2(1):55–62
Bimonte S, Cascella M, Barbieri A, Arra C, Cuomo A (2020) Current shreds of evidence on the anticancer role of EGCG in triple negative breast cancer: an update of the current state of knowledge. Infect Agent Cancer 15:1–6
Bouwman-Boer Y, Fenton-May V, Le Brun P (2015) Practical pharmaceutics. An international guideline for the preparation, care and use of medicinal products. Springer, Cham
Brindavanam N, Katiyar C and Rao Y (2002) Novel herbal composition for diabetes patients and a process for producing the same. U.S. patent application 09/845,723
Brindavanam N, Katiyar C and Rao Y (2003) Novel herbal composition for the management of bronchial asthma and a process of manufacturing the same. U.S. patent application 09/881,042
Bruton-Seal J, Seal M (2014) Hedgerow medicine: harvest and make your own herbal remedies. Merlin Unwin Books Limited, Ludlow
Chamchan R, Sinchaipanit P, Disnil S, Jittinandana S, Nitithamyong A, On-nom N (2017) Formulation of reduced sugar herbal ice cream using lemongrass or ginger extract. Br Food J 119(10):2172–2182
Chen S (2003) Herbal composition PHY828 and its use US0143289
Chou WH (2009) Compositions and methods for promoting brain and cardiovascular health, preventing and treating brain and cardiovascular disorders. U.S. patent application 12/457,155
Conover EA (2002) Over-the-counter products: nonprescription medications, nutraceuticals, and herbal agents. Clin Obstet Gynecol 45(1):89–98
Ding HW, Huang AL, Zhang YL, Li B, Huang C, Ma TT, Meng XM, Li J (2017) Design, synthesis and biological evaluation of hesperetin derivatives as potent anti-inflammatory agent. Fitoterapia 121:212–222
Dominguez Avila JA, Rodrigo Garcia J, Gonzalez Aguilar GA, De la Rosa LA (2017) The antidiabetic mechanisms of polyphenols related to increased glucagon-like peptide-1 (GLP1) and insulin signaling. Molecules 22(6):903
Du Y, Ruan J, Zhang L, Fu F (2020) Jieyu anshen granule, a Chinese herbal formulation, exerts effects on poststroke depression in rats. Evid Based Complementary Altern Med 2020:7469068
Farooqi AHA, Sharma S and Kumar S, (2002) Herbal skin care formulation and a process for the preparation thereof. US Patent 6,368,639
Fathiazad F, Matlobi A, Khorrami A, Hamedeyazdan S, Soraya H, Hammami M, Maleki-Dizaji N, Garjani A (2012) Phytochemical screening and evaluation of cardioprotective activity of ethanolic extract of Ocimum basilicum L.(basil) against isoproterenol induced myocardial infarction in rats. DARU J Pharm Sci 20:1–10
Ganjhu RK, Mudgal PP, Maity H, Dowarha D, Devadiga S, Nag S, Arunkumar G (2015) Herbal plants and plant preparations as remedial approach for viral diseases. Virus 26:225–236
Ghosal S (2007) Polyherbal compositions and methods for treating viral infections. U.S. Patent 7,250,181
Gibbs BF (2009) Differential modulation of IgE-dependent activation of human basophils by ambroxol and related secretolytic analogues. Int J Immunopathol Pharmacol 22(4):919–927
Grosso A (2013) The everything soap making book: learn how to make soap at home with recipes. Adams Media, Massachusetts

Gubitosa J, Rizzi V, Fini P, Cosma P (2019) Hair care cosmetics: from traditional shampoo to solid clay and herbal shampoo, a review. Cosmetics 6(1):13

Eid HM, Haddad SP (2017) The antidiabetic potential of quercetin: underlying mechanisms. Curr Med Chem 24(4):355–364

Hasani-Ranjbar S, Nayebi N, Larijani B, Abdollahi M (2009) A systematic review of the efficacy and safety of herbal medicines used in the treatment of obesity. World J Gastroenterol 15(25):3073

Hesari M, Mohammadi P, Khademi F, Shackebaei D, Momtaz S, Moasefi N, Farzaei MH, Abdollahi M (2021) Current advances in the use of nanophytomedicine therapies for human cardiovascular diseases. Int J Nanomedicine 16:3293–3315

Hoffmann D (2003) Medical herbalism: the science and practice of herbal medicine. Simon and Schuster, New York

Hsieh CH, Tsai HC, Hsu GL, Chen CC, Hsu CY (2016) Herb formula enhances treatment of impotent patients after penile venous stripping: a randomised clinical trials. Andrologia 48(7):754–760

Hussain T, Gupta S, Adhami VM, Mukhtar H (2005) Green tea constituent epigallocatechin-3-gallate selectively inhibits COX-2 without affecting COX-1 expression in human prostate carcinoma cells. Int J Cancer 113(4):660–669

Hussain SA, Hameed A, Nasir F, Wu Y, Suleria HAR, Song Y (2018) Evaluation of the spermatogenic activity of polyherbal formulation in oligospermic males. BioMed Res Int 2018:2070895

Johnson R, Bayor MT, Adotey J (2010) Formulation and evaluation of Bridelia ferruginea and Canthium glabriflorum herbal capsules. J Herb Med 1:18–22

Jones M (2011) The complete guide to creating oils, soaps, creams, and herbal gels for your mind and body: 101 natural body care recipes. Atlantic Publishing Company, London

Kamali SH, Khalaj AR, Hasani-Ranjbar S, Esfehani MM, Kamalinejad M, Soheil O, Kamali SA (2012) Efficacy of 'Itrifal Saghir', a combination of three medicinal plants in the treatment of obesity; a randomized controlled trial. DARU J Pharm Sci 20:1–8

Kambizi LGBM, Goosen BM, Taylor MB, Afolayan AJ (2007) Anti-viral effects of aqueous extracts of Aloe ferox and Withania somnifera on herpes simplex virus type 1 in cell culture: research in action. S Afr J Sci 103(9):359–360

Katiyar CK, Duggal RK and Rao BVJ, Dabur Research Foundation (2002) Herbal composition and method of manufacturing such composition for the management of gynecological disorders. U.S. Patent 6,455,077

Kayani S, Ahmad M, Zafar M, Sultana S, Khan MPZ, Ashraf MA, Hussain J, Yaseen G (2014) Ethnobotanical uses of medicinal plants for respiratory disorders among the inhabitants of Gallies–Abbottabad, Northern Pakistan. J Ethnopharmacol 156:47–60

Khanuja SPS, Srivastava S, Shasney A K, Darokar MP, Kumar TRS, Agarwal KK, Ahmed A, Patra NK, Sinha P, Dhawan S, Saikia D and Kumar S (2003) US Patent No 6514541

Khanuja SPS, Chaturvedi P, Singh AK, Shasany AK, Agarwal VK, Gupta VK, Gupta SC, Tripathy AK, Pal A, Saikia D and Darokar MP (2007) Anti-dermatophytic preparation and use thereof. U.S. Patent 7,291,349

Kung H and Kong L (2006) Anti-depression formulations. US20060147569

Lakey-Beitia J, Burillo AM, La Penna G, Hegde ML, Rao KS (2021) Polyphenols as potential metal chelation compounds against Alzheimer's disease. J Alzheimers Dis 82(1):S335–S357

Lee W (2005) Treatment of acne by topical herbal extract solution. U.S. patent application 10/688,227

Lee SH, Lee YJ (2021) Synergistic anticancer activity of resveratrol in combination with docetaxel in prostate carcinoma cells. Nutr Res Pract 15(1):12–25

Lee FY, Wu H and Chang MC (2002) Chinese herbal composition for improving blood circulation and the method of preparing the same. U.S. Patent 6,447,814

Liu HT, Ho YS (2018) Anticancer effect of curcumin on breast cancer and stem cells. Food Sci Human Wellness 7(2):134–137

Lodha G (2019) Formulation and evaluation of polyherbal shampoo to promote hair growth and provide antidandruff action. J Drug Deliv Ther 9(4-A):296–300

Luo J, Si H, Jia Z, Liu D (2021) Dietary anti-aging polyphenols and potential mechanisms. Antioxidants 10(2):283

Ma Q, Yu Q, Xing X, Liu S, Shi C, Luo J (2018) San Wu Huangqin decoction, a Chinese herbal formula, inhibits influenza a/PR/8/34 (H1N1) virus infection in vitro and in vivo. Viruses 10(3):117

Malik A, Mehmood MH, Channa H, Akhtar MS, Gilani AH (2017) Pharmacological basis for the medicinal use of polyherbal formulation and its ingredients in cardiovascular disorders using rodents. BMC Complement Altern Med 17:1–12

Managoli NB (2008) Herbal composition for treatment and maintenance of hormone dependent conditions and circulatory conditions, and immunostimulation. U.S. Patent 7,368,134

Mateus ARS, Crisafulli C, Vilhena M, Barros SC, Pena A, Sanches Silva A (2023) The bright and dark sides of herbal infusions: assessment of antioxidant capacity and determination of tropane alkaloids. Toxins 15(4):245

Mazzio EA and Soliman KF (2008). Topical treatment for dyshidrosis (pompholyx) and dry skin disorders. U.S. Patent 7,357,950

Miras-Moreno B, Sabater-Jara AB, Pedreño MA, Almagro L (2016) Bioactivity of phytosterols and their production in plant in vitro cultures. J Agric Food Chem 64(38):7049–7058

Mishra D, Tubaki BR (2019) Effect of Brahmi vati and Sarpagandha Ghana vati in management of essential hypertension–A randomized, double blind, clinical study. J Ayurveda Integr Med 10(4):269–276

Mishra US, Murthy PN, Sahoo KS, Sahu KC (2012) Formulation and evaluation of herbal tablet containing methanolic extract of Calophyllum inophyllum. Int J Pharm 2(1):181–186

Mohammad H, Prabhu K, Rao MRK, Sundaram RL, Shil S, Vijayalakshmi N, Dinakar S (2019) The GC MS study of one ayurvedic medicine, aragwadharishtam. Res J Pharm Technol 12(3):1111–1114

Musthaba M, Baboota S, Athar TMD, Thajudeen KY, Ahmed S, Ali J (2010) Patented herbal formulations and their therapeutic applications. Recent Pat Drug Deliv Formul 4(3):231–244

Nirmal P, Singh R, Kumar N, Sharma S (2022) Phytopharmaceutical regulated new class: an industrial initiative of ayurvedic drugs towards the advancement of India system of medicine. World J Adv Res Rev 15(3):407–419.23

Nitin BG, Surendra GG, Shailesh SC (2010) Design, development and evaluation of oral herbal formulations of Piper nigrum and Nyctanthes arbortristis. Int J Pharmtech Res 2(1):171–176

O'Leary KA, de Pascual-Tereasa S, Needs PW, Bao YP, O'Brien NM, Williamson G (2004) Effect of flavonoids and vitamin E on cyclooxygenase-2 (COX-2) transcription. Mutat Res 551(1–2):245–254

Packer L, Wachtel-Galor S, Ong CN, Halliwell B (2004) Herbal and traditional medicine: biomolecular and clinical aspects. CRC Press, Boca Raton

Palpu P, Mehrotra S, Rao CV, Ojha SK, Raghavan G, Rao GMM and Padmavathi S (2005) Synergistic composition for treating hyperlipidemia. US20050136138

Palpu P, Rao C, Govindarajan R, Rawat A, Srivastava S, Deb B and Subramanian S, Council of Scientific, Industrial Research CSIR and Godfrey Phillips India Ltd (2007) Functional aphrodisiac rolled herbal bidis and cigarettes. U.S. Patent Application 11/379,449

Palpu P, Rao CV, Kishore K, Gupta YK, Kartik R and Govindrajan R (2008) Herbal formulation as memory enhancer in alzheimer condition. U.S. Patent 7,429,397

Panda H (2000) Herbal soaps and detergents handbook. Natl Institute Industrial Res 1:194–196

Pandey N, Meena RP, Rai SK, Pandey-Rai S (2011) Medicinal plants derived nutraceuticals: a re-emerging health aid. Int J Pharm Bio Sci 2(4):420–441

Pietta PG (2000) Flavonoids as antioxidants. J Nat Prod 63(7):1035–1042

Popoola JO, Obembe OO (2013) Local knowledge, use pattern and geographical distribution of Moringa oleifera lam.(Moringaceae) in Nigeria. J Ethnopharmacol 150(2):682–691

Pundir R, Singh G, Pandey AA, Saraf SA (2009) Demand of herbal hepatoprotective formulations in Lucknow-a survey. Pharm Res 1(1):23–33

Qi F, Li A, Inagaki Y, Gao J, Li J, Kokudo N, Li XK, Tang W (2010) Chinese herbal medicines as adjuvant treatment during chemoor radio-therapy for cancer. Biosci Trends 4(6):297

Ramakrishna S, Sitaramam BS, Diwan PVR and Raghavan KV (2002) Herbal formulation useful for treatment of skin disorders. U.S. Patent 6,383,495
Rao JM, Srinivas PV, Yadav JS, Raghavan KV, Saxena AK, Shanmugavek M, Kampasi H and Qazi GN (2003) Herbal chemical composition for the treatment of cancer. U.S. Patent 6,649,650
Rao JM, Sampathkumar U, Sastry BS, Yadav JS, Raghavan, K.V., Palit G, Bhalla DN, Rai D, Varier PM, Muraleedharan TS and Muraleedharan K (2005) Novel herbal composition for the treatment of gastric ulcer. US0089583
Rasul A, Akhtar N (2011) Formulation and in vivo evaluation for anti-aging effects of an emulsion containing basil extract using non-invasive biophysical techniques. Daru 19(5):344
Reddy NB, Kumar S, Reddy V, Torgalkar A and Murugan NR, Apptec Inc (2007) Compositions for safe and effective regression of dermal vessel tortuosity. U.S. Patent Application 11/288,923
Rice-Evans CA, Miller NJ, Paganga G (1996) Structure-antioxidant activity relationships of flavonoids and phenolic acids. Free Radic Biol Med 20(7):933–956
Romero CD, Chopin SF, Buck G, Martinez E, Garcia M, Bixby L (2005) Antibacterial properties of common herbal remedies of the southwest. J Ethnopharmacol 99(2):253–257
Rukari T, Pawar N, Gawas S, Bhoite J, Jadhav P, Madkholkar S, Jagtap V (2022) To study invitro antimicrobial activity of polyherbal handwash formulation. Asian J Pharm Technol 12(3):237–241
Sadik CD, Sies H, Schewe T (2003) Inhibition of 15-lipoxygenases by flavonoids: structure–activity relations and mode of action. Biochem Pharmacol 65(5):773–781
Saeedi P, Petersohn I, Salpea P, Malanda B, Karuranga S, Unwin N, Colagiuri S, Guariguata L, Motala AA, Ogurtsova K, Shaw JE (2019) Global and regional diabetes prevalence estimates for 2019 and projections for 2030 and 2045: results from the international diabetes federation diabetes atlas. Diabetes Res Clin Pract 157:107843
Sawant SE, Tajane MD (2016) Formulation and evaluation of herbal ointment containing neem and turmeric extract. J Sci Innov Res 5(4):149–151
Sharma P, Tyagi A, Bhansali P, Pareek S, Singh V, Ilyas A, Mishra R, Poddar NK (2021) Saponins: extraction, bio-medicinal properties and way forward to anti-viral representatives. Food Chem Toxicol 150:112075
Shireen S, Zarena AS (2022) A generalized review on extraction of biomolecules. J Pharm Nutr Sci 12:175–187
Shoemake R (2006) Immune system reconstructor composition. U.S. Patent 7,112,343
Singh B, Malik A, Tiwari G, Noor G, Hema G, Ruchi Tiwari AC (2022) Standardization procedure of herbal medicines and biodynamic agriculture. NeuroQuantology 20(6):9644
Smruti P (2021) A review on natural remedies used for the treatment of respiratory disorders. Int J Pharm 8:104–111
Soni H, Malik J, Singhai AK, Sharma S (2013) Antimicrobial and antiinflammatory activity of the hydrogels containing rutin delivery. Asian J Chem 25(15):8371
Srivastava S, Lal VK, Pant KK (2012) Polyherbal formulations based on Indian medicinal plants as antidiabetic phytotherapeutics. Phytopharmacology 2(1):1–15
Sung HY, Hong CG, Suh YS, Cho HC, Park JH, Bae JH, Park WK, Han J, Song DK (2010) Role of (−)-epigallocatechin-3-gallate in cell viability, lipogenesis, and retinol-binding protein 4 expression in adipocytes. Naunyn Schmiedeberg's Arch Pharmacol 382:303–310
Suresh Gupta M, Shivaprasad HN, Kharya MD, Rana AC (2006) Immunomodulatory activity of the ayurvedic formulation "Ashwagandha Churna". Pharm Biol 44(4):263–265
Tarasiuk A, Mosińska P, Fichna J (2018) Triphala: current applications and new perspectives on the treatment of functional gastrointestinal disorders. Chin Med 13:1–11
Tirant M, Lotti T, Gianfaldoni S, Tchernev G, Wollina U, Bayer P (2018) Integrative dermatology–the use of herbals and nutritional supplements to treat dermatological conditions. Open Access Maced J Med Sci 6(1):185
Turek C, Stintzing FC (2013) Stability of essential oils: a review. Compr Rev Food Sci Food Saf 12(1):40–53

Wilborn C, Beckham J, Campbell B, Harvey T, Galbreath M, La Bounty P, Nassar E, Wismann J, Kreider R (2005) Obesity: prevalence, theories, medical consequences, management, and research directions. J Int Soc Sports Nutr 2(2):4

Winfield AJ, Rees AJ, Smith I (2009) Pharmaceutical practice. Churchill Livingstone, London

Yadav R, Choubey A, Mishra MA (2022) Industrial pharmacognosy. Booksclinic Publishing, Bilaspur

Yarnell E (2018) Herbs for viral respiratory infections. Altern Complement Ther 24(1):35–43

Yin SY, Wei WC, Jian FY, Yang NS (2013) Therapeutic applications of herbal medicines for cancer patients. Evid Based Complement Alternat Med 2013:1

Zdoryk OA, Khokhlova KO, Georgiyants VA, Vyshnevska LI (2014) Investigation of physical and chemical stability of ointment with herbals. Int J Pharm Compound 18:248–252

Zhang Y, Han M, Liu Z, Wang J, He Q, Liu J (2012) Chinese herbal formula xiao yao san for treatment of depression: a systematic review of randomized controlled trials. Evid Based Complement Alternat Med 2012:931636

Zhang QW, Lin LG, Ye WC (2018) Techniques for extraction and isolation of natural products: A comprehensive review. Chin Med 13:1–26

Zioga M, Chroni A, Evageliou V (2022) Utilisation of pectins extracted from orange peels by non conventional methods in the formation of edible films in the presence of herbal infusions. Polysaccharides 3(3):574–588

Nanobiotechnology for Medicinal Plants: A Comprehensive Review

20

Shreya Agrawal, Neha Bhandari, and Pooja Saklani

Abstract

Nanotechnology deals with the synthesis and manipulation of particles of nanosize (1–100 nm). They have unique properties owing to their small size and large surface area and can be used in different industries. Their properties at the nanoscale offer advantages not seen in their bulk counterparts, making them a versatile asset. Top-down or bottom-up approach is used for their synthesis. There are different methods to synthesize, viz., physical, chemical, or biological. Nanoparticles can be metallic or nonmetallic. The synthesized nanoparticles can be characterized by visualizing the color change and using techniques like UV-visible spectroscopy, Fourier-transform infrared spectroscopy (FTIR), scanning electron microscopy (SEM), energy-dispersive X-ray spectroscopy (EDX), transmission electron microscopy (TEM), selected area electron diffraction (SAED), dynamic light scattering (DLS), X-ray diffractometry (XRD), zeta potential analysis, atomic force microscopy (AFM), etc. Silver nanoparticles (AgNPs) are notable metallic nanoparticles that have large number of applications. The nanoparticles can be synthesized biologically by using microorganisms and plants. Different plant parts like barks, fruits, stems, leaves, seeds, flowers, rhizomes, roots, and fruit peels are used for the synthesis of nanoparticles. This plant-mediated or green synthesis method has several advantages like eco-friendly, high yield, low toxicity, and energy-efficient. The plants contain secondary metabolites which act as reducing agent for the nanoparticles' synthesis. The silver nanoparticles hold the potential to revolutionize various industrial sectors and environmental solutions. However, it's important to acknowledge that their multifaceted nature presents both opportunities and challenges, as highlighted by various studies revealing potential adverse effects on mankind.

S. Agrawal · N. Bhandari · P. Saklani (✉)
Department of Biotechnology, Hemvati Nandan Bahuguna Garhwal University (A Central University), Srinagar, Uttarakhand, India

© The Author(s), under exclusive license to Springer Nature Singapore Pte Ltd. 2024

M. Nandave et al. (eds.), *Ethnopharmacology and OMICS Advances in Medicinal Plants Volume 1*, https://doi.org/10.1007/978-981-97-2367-6_20

Therefore, silver nanoparticles embody a dichotomy of potential benefits and risks. Their diverse applications hold immense promise, but the intricacies of their interaction with biological systems require careful consideration to harness their advantages while mitigating potential harms.

Keywords
Nanoparticles · Green synthesis · Silver nanoparticles · Secondary metabolites · FTIR · SEM · TEM

20.1 Introduction

Nanotechnology has rapidly emerged as a thriving industry, specializing in the production, manipulation, and imaging of nanoscale structures, with a specific emphasis on nanoparticles ranging in diameter from 1 to 100 nm (Kumar et al. 2017). The inception of nanotechnology dates back to 1959 when Feynman first proposed the concept. These nanoparticles exhibit distinctive properties attributed to their size and shape, featuring a substantial surface-to-volume ratio due to their diminutive dimensions. Consequently, they find diverse applications across various industries, including but not limited to bioremediation, biosensors, diagnostics, electronics, pharmaceuticals, smart drug delivery systems, and inhibition of biofilm formation in certain bacteria (Balachandar et al. 2019).

Nanoparticle synthesis commonly employs two principal methods known as "top-down" and "bottom-up" (Forough and Farhadi 2010). The synthesis of nanoparticles can be achieved through physical, chemical, and biological means. Metallic nanoparticles encompass materials such as copper, gold, silver, magnetic elements (cobalt, nickel), and semiconductors like cadmium, lead, and zinc. Conversely, nonmetallic nanoparticles predominantly consist of carbon-based materials like carbon nanotubes and nanofibers. Extensive exploration has been conducted on metallic nanoparticles, driven by their unique electrical, optical, and catalytic properties (Bharathi et al. 2018).

Among metallic nanoparticles, silver nanoparticles (AgNPs) have garnered significant attention due to their distinctive physical, chemical, and biological attributes. AgNPs exhibit versatility in various applications, including targeted drug delivery, wound healing, antibacterial agents, and electronics (Jain et al. 2008; Sharma et al. 2009; Marimuthu et al. 2011; Bindhu and Umadevi 2015; Velusamy et al. 2015; Verma and Mehata 2016).

Traditional medicine has relied on medicinal plants and their components for centuries, owing to the presence of a diverse array of beneficial elements, particularly bioactive compounds. These bioactive compounds, encompassing alkaloids, phenolics, and tannins, play a crucial role by acting as reducing agent and converting metal ions to nanoparticles following the biosynthesis pathways that are environment friendly (Munir et al. 2021). Various plant components, including leaves, fruits, stems, roots, seeds, rhizome, flowers, bark, and peels of fruits, are utilized to produce nanoparticles, through the process known as green synthesis of nanoparticles. This method offers distinct advantages in terms of cost-effectiveness, environmental friendliness, and the ready availability of raw materials (Nathani 2021).

20.2 Methods of Synthesis of Nanoparticles

Nanoparticle formulation employs two primary approaches: the top-down and bottom-up methods (Fig. 20.1). In the top-down method, the bulky material is reduced to the particles of nanosize. The reduction in size is accomplished through processes such as lithography, laser ablation, thermal decomposition, etching, sputtering, and mechanical milling (Abou El-Nour et al. 2010). Whereas, the "bottom-up" method is preferred for the nanoparticle synthesis. This method involves a homogeneous system where catalysts, such as reducing agents and enzymes, facilitate the synthesis of nanoparticles. The synthesis process is controlled by the properties of the catalyst, reaction media, and conditions (including solvents, stabilizers, and temperature) (Abou El-Nour et al. 2010).

Nanoparticle synthesis techniques fall into three main categories: physical, chemical, and biological approaches. Of these, the biological synthesis of AgNPs has attracted considerable attention, thanks to its advantageous properties, including high yield, solubility, and stability. This distinguishes it from the more labor-intensive and riskier physical and chemical methods (Husen and Siddiqi 2014; Ahmed et al. 2016a, b). Various synthesis methods for nanoparticles are outlined in Table 20.1.

20.3 Plant-Mediated/Green Synthesis of Silver Nanoparticles

The biosynthesis of silver nanoparticles using plant materials plays a crucial role in bioreduction processes. The diverse bioactive molecules present in plant extracts, including amino acids, enzymes, proteins, polysaccharides, vitamins, and organic

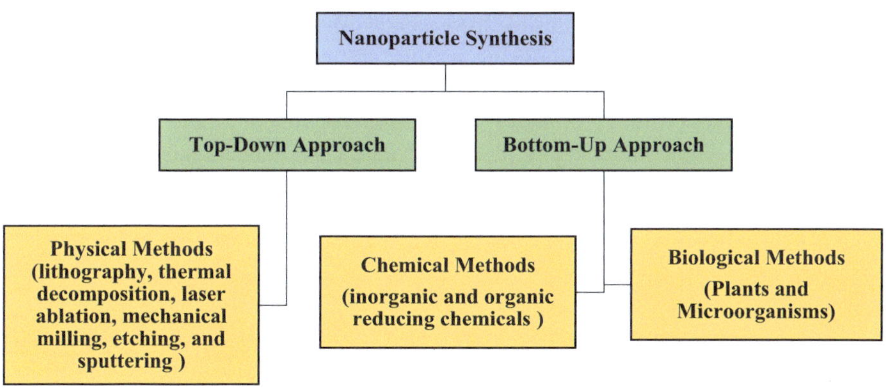

Fig. 20.1 General types of synthesis of metal nanoparticles

Table 20.1 Different methods used to synthesize nanoparticles

Methods	Description	Advantages	Disadvantages	References
Physical methods	Evaporation-condensation, laser ablation	Fast synthesis, no harmful reagents	Contamination of solvent, distribution not uniform, low yield, high consumption of energy	Sylvestre et al. (2004), Elsupikhe et al. (2015)
Chemical methods	Utilizes inorganic and organic reducing chemicals to reduce silver ions to nanoparticles	High yield, controllable particle size	Costly, use of toxic chemicals, potential aggregation	Agnihotri et al. (2014), Patil et al. (2012), Zhang et al. (2016)
Biological methods	Employs biological entities such as plants or microorganisms to synthesize nanoparticles	Eco-friendly, high yield, low toxicity, energy-efficient	Challenges in growth and maintenance of cell culture, variable particle sizes	Fayaz et al. (2010), Logeswari et al. (2015), Saravanan et al. (2018), Guilger-Casagrande and Lima (2019), Pallela et al. (2018), Afreen et al. (2020)

acids, have potential metal ion reducing capacity (Mohammadlou et al. 2016). Plant extracts also play an important a role in stabilizing the nanoparticles (Bar et al. 2009; Marambio-Jones and Hoek 2010; Velayutham et al. 2013). The significance of utilizing plants for nanoparticle synthesis stems from their ready availability and the abundance of active agents distributed throughout the plant, encompassing roots, latex, stem, seeds, and leaves (Kharissova et al. 2013).

The rapid generation of metal nanoparticles utilizing plant extracts has been demonstrated at room temperature, with the specific timeframe contingent on the particular plant extract employed. Various critical factors influence the nanoparticle formation process, including the concentration of the plant extract, temperature, pH, type of metal salt, and contact time (Mittal et al. 2013). Consequently, plant extracts have emerged as ideal candidates for the green synthesis of silver nanoparticles due to their rapid growth, non-pathogenic nature, and environmentally friendly, one-step reaction protocols that are highly cost-effective. This approach provides a sustainable and environment friendly alternative for nanoparticle synthesis and harnesses the rich biochemistry of plants to create valuable nanomaterials.

Various researches have explored the plant-mediated synthesis of silver nanoparticles, and a selection of these studies is outlined in Table 20.2.

Table 20.2 Plants and their parts utilized for the synthesis of AgNPs

Plants	Part used	Structure	Nanoparticle size	References
Vitis Vinifera	Grape extract	Spherical shape	18–20 nm	Rao and Paria (2013)
Helicteres isora	Root extract	Crystalline and spherical	30–40 nm	Bhakya et al. (2016)
Azadirachta indica	Leaf extract	Spherical and polydispersed	43 nm	Gavhane et al. (2012)
Coriandrum sativum	Leaf extract	Spherical	8–75 nm	Sathyavathi et al. (2010)
Elaeagnus latifolia	Leaf extract	Spherical	30–50 nm	Phanjom et al. (2012)
Solanum nigrum and *Clitoria ternatea*	Leaves extract	Spherical	23 nm	Jassal et al. (2016)
Cranberry	Fruit powder extract	Spherical	2.8 ± 2.1, 1.4 ± 0.8, 8.6 ± 2.5 nm	Ashour et al. (2015)
Phyllanthus niruri	Leaves extract	Crystalline face centered cubic (FCC) and spherical	30–60 nm	Suresh et al. (2015)
Tagetes erecta	Flower broth	Irregular, hexagonal, and spherical	10–90 nm	Padalia et al. (2015)
Bryophyllum and *Asiatic pennywort*	Leaves extract	Crystalline FCC and spherical	18–21 nm	Saikia et al. (2015)
Zingiber officinale	Rhizome broth extract	Crystalline and spherical	3.1 nm	Shalaby et al. (2015)
Curry leaf	Leaf extract	Crystalline FCC	10–25	Christensen et al. (2011)
Aloe vera	Leaves extract	Spherical, rectangular, cubical, and triangular	70	Medda et al. (2015)
Calotropis procera	Latex serum extract	Spherical	12.33	Mohamed et al. (2014)
Catharanthus roseus	Root extract	Crystalline FCC and spherical	35–55	Rajagopal et al. (2015)
Rauvolfia tetraphylla	Leaf Extract	Crystalline	26–37 nm	Kalaiarasi et al. (2013)
Turbinaria ornata (turner)	Seaweed aqueous extract	Spherical and crystalline	20–32 nm	Deepak et al. (2017)
Catharanthus roseus	Leaves extract	Crystalline FCC and spherical	20 nm	Al-Shmgani et al. (2017)
Ananas comosus	Juice	Spherical	12 nm	Ahmad and Sharma (2012)

20.4 Role of Secondary Metabolites in the Synthesis of Silver Nanoparticles

Plant extracts serve as abundant sources of diverse primary and secondary metabolite compounds, encompassing myriad molecules like proteins, phenolic acids, flavonoids, alkaloids, terpenoids, amino acids, antioxidants, and organic acids that actively participate in redox reactions and play a crucial role in reducing metal ions to form metallic nanoparticles (Aromal and Philip 2012). While the precise mechanisms underlying the green synthesis of nanoparticles are not fully elucidated, it is suggested that silver ions initially bind to protein surfaces through electrostatic interactions within the plant extract (Marslin et al. 2018). Subsequently, proteins facilitate the reduction of these ions, altering their result in structure and forming silver nuclei. These nuclei grow as silver ions continue to undergo reduction, accumulating around them (Li et al. 2007).

Researchers encounter challenges in conclusively elucidating the compounds involved in green synthesis processes. Nevertheless, FT-IR analyses offer valuable insights into potential biomolecules participating in these processes. Studies that focus on isolated compounds, such as flavonoids and terpenoids, have successfully demonstrated their roles in green synthesis. Fourier-transform infrared spectroscopy (FT-IR) plays a crucial role in identifying potential biomolecules engaged in metal ion reduction and capping. Table 20.3 enumerates the secondary metabolites present in the silver nanoparticles of the respective plant species.

20.5 Characterization of Silver Nanoparticles

The characterization of silver nanoparticles (AgNPs) is essential to confirm their formation and assess their suitability for intended applications. Physicochemical characteristics play a pivotal role in influencing the actions, biodistribution, safety, and efficacy of nanoparticles. Thorough characterization of AgNPs using various analytical techniques is crucial to evaluate their functional properties.

One of the fundamental methods involves visually observing the color change of the solution containing plant extract and silver nitrate from yellow to brown, indicating the production of AgNPs. This color change is a result of the excitation of surface plasmon resonance in AgNPs. To further confirm the formation of AgNPs, multiple methods are employed, including UV-visible spectroscopy, Fourier-transform infrared spectroscopy (FTIR), scanning electron microscopy (SEM), energy-dispersive X-ray spectroscopy (EDX), transmission electron microscopy (TEM), selected area electron diffraction (SAED), dynamic light scattering (DLS), X-ray diffractometry (XRD), zeta potential analysis, and atomic force microscopy (AFM). These techniques collectively provide valuable insights into the properties of AgNPs.

Table 20.3 List of secondary metabolites present in the AgNPs synthesized from different plant species

Plant species	Metabolites identified	References
Azadirachta indica	Flavonoids, terpenoids	Ahmed et al. (2016a, b)
Coleus aromaticus	Flavonoids	Vanaja and Annadurai (2013)
Hibiscus rosa-sinensis	Carboxylate ion groups	Nayak et al. (2015)
Helianthus annuus	Flavonoids, proteins, fatty acids, triglycerides, phenolics, tocopherols	Thakore et al. (2014)
Dioscorea bulbifera	Flavonoids, polyphenol, diosgenin, ascorbic acid	Ghosh et al. (2012)
Glycyrrhiza glabra	Flavonoids and terpenoid	Dinesh et al. (2012)
Achyranthes aspera	Polyols	Elumalai et al. (2016)
Syzygium cumini	Polyphenols	Kumar et al. (2010)
Decalepis hamiltonii	Polyols, phenols	Rashmi and Sanjay (2017)
Desmodium triflorum	Ascorbic acid	Ahmad et al. (2011)
Anacardium occidentale	Proteins, polyols	Mukunthan and Balaji (2012)
Aegle marmelos	Tannin	Jagajjanani Rao and Paria (2013)
Trianthema decandra	Saponin, catechins, hydroxyflavones	Geethalakshmi and Sarada (2013)
Solanum xanthocarpum	Alkaloids, phenolic, sugars	Sengottaiyan et al. (2016)
Zingiber officinale	Flavonoid, alkaloids	Velmurugan et al. (2014)
Mimusops elengi	Polyphenols	Kiran et al. (2014)
Lantana camara	Flavonoids	Ajitha et al. (2015)
Nigella arvensis	Flavonoids, alkaloids	Chahardoli et al. (2017)
Acalypha indica	Quercetin	Krishnaraj et al. (2010)
Citrus sinensis	Flavonoids, ascorbic acid, volatile oils	Kaviya et al. (2011)
Ocimum sanctum	Eugenols, linalool, terpenes	Ramteke et al. (2013)
Rosmarinus officinalis	Polyphenols	Ghaedi et al. (2015)
Gardenia jasminoides	Rutin, gallic acid, chlorogenic acid	Fenfen et al. (2014)
Withania somnifera	Catechin p-coumaric acid, luteolin-7-glucoside, withanolides	Marslin et al. (2015)
Lippia citriodora	Verbascoside, isoverbascoside, chrysoeriol-7-O-diglucoronide, luteonin-7-O-diglucoronide	Cruz et al. (2010)

(continued)

Table 20.3 (continued)

Plant species	Metabolites identified	References
Sesuvium portulacastrum	Flavones, proteins, terpenoids	Nabikhan et al. (2010)
Mentha piperita	Alkaloids, flavones, steroids, polysaccharides, amino acids, oximes, proteins, menthol	MubarakAli et al. (2011)
Iresine herbstii	Phenolic compound	Dipankar and Murugan (2012)
Morinda pubescens	Flavonoids, triterpenoids, polyphenols	Kumar et al. (2015)
Carica papaya	Proteins, alcohols, phenolics	Banala et al. (2015)

20.5.1 UV-Visible Spectrophotometry

UV-vis spectroscopy is a technique employed for monitoring the synthesis and characterization of silver nanoparticles (AgNPs) without the need for extensive calibration procedures. This method relies on the interaction between AgNPs and specific wavelengths of light, taking advantage of their unique properties that make them highly responsive to particular light wavelengths. This responsiveness is attributed to the phenomenon of surface plasmon resonance (SPR), where the conduction and valence bands of AgNPs are closely spaced, allowing electrons to collectively oscillate in resonance with incident light waves. Consequently, this collective electron oscillation results in the characteristic SPR absorption band observed in UV-vis spectra. Detection of nanoparticles typically occurs within the wavelength range of 400–800 nm (Fig. 20.2a). The intensity and position of the SPR peak, appearing at wavelengths between 380 and 450 nm, can be utilized to assess the quality of the synthesized nanoparticles.

Previous research by Smitha et al. (2008) suggests that a narrow and low wavelength absorption peak indicates a small size for the nanoparticles, while a broad peak at a higher wavelength suggests larger size or aggregated AgNPs. UV-vis spectroscopy offers advantages such as speed, ease of use, sensitivity, and specificity for different types of nanoparticles. Additionally, it provides insights into the stability of AgNPs over time. Consistency in the SPR peak at the same wavelength, as observed using UV-Vis spectroscopy, indicates that AgNPs prepared through biological methods can maintain their properties for extended periods, often exceeding a year (Zhang et al. 2016).

20.5.2 FTIR Spectroscopy

FTIR analysis plays a pivotal role in the identification of biomolecules responsible for the reduction of silver ions ($Ag+$) and the stabilization of the synthesized silver nanoparticles (AgNPs) (Anandalakshmi et al. 2016). This technique involves scanning samples with infrared light to observe their chemical bonds, as different chemical groups absorb infrared radiation at distinct wavelengths (Palithya et al. 2022).

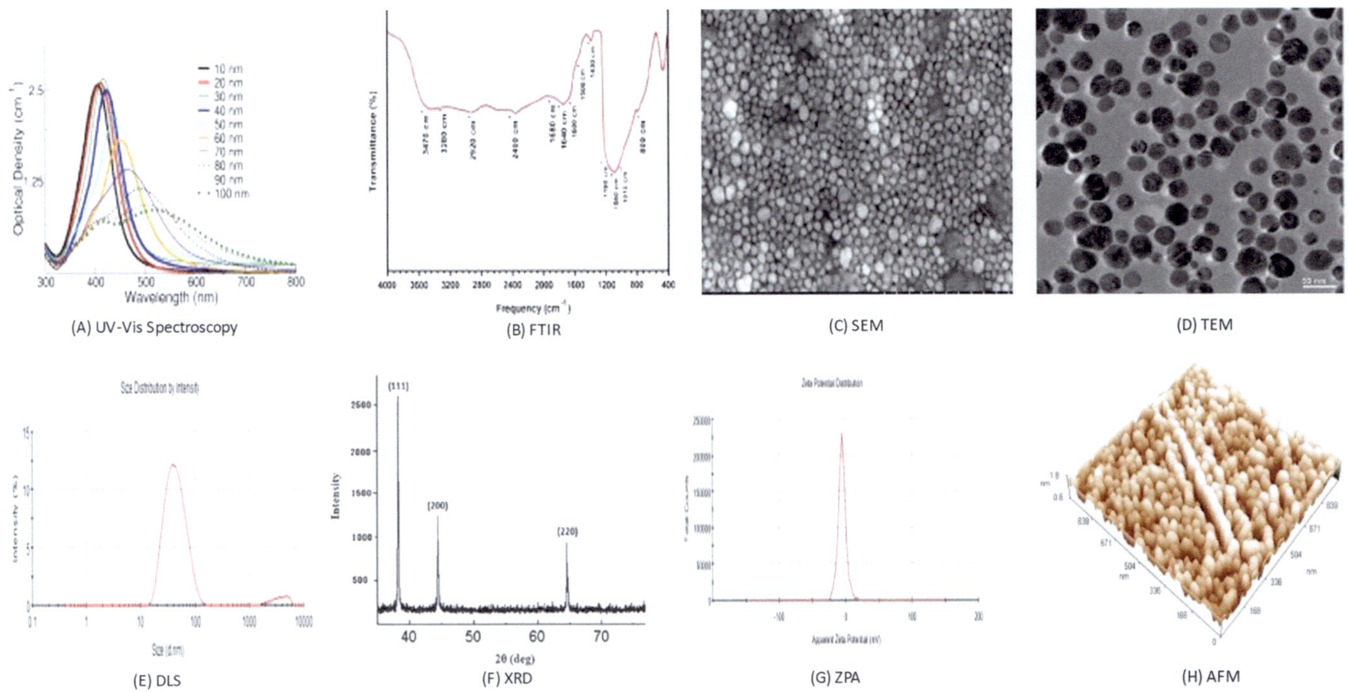

Fig. 20.2 Characterization methods of silver nanoparticles, viz., (**a**) UV-visible spectroscopy, (**b**) FTIR, (**c**) SEM, (**d**) TEM, (**e**) DLS, (**f**) XRD, (**g**) ZPA, and (**h**) AFM

Peaks in FTIR spectra indicate the presence of secondary metabolites in plants, including flavonoids, terpenoids, phenols, glycosides, and tannins. These compounds are often associated with functional groups such as aldehydes, ketones, and carboxylic acids, contributing to the capping of AgNPs (Fig. 20.2b). Moreover, the heightened levels of antioxidants and flavonoids in plant extracts, such as ascorbic acid and gallic acid, play a role in reducing metal salts during the synthesis process. Analyzing the FTIR spectra of medicinal plant extracts and comparing them with those of biosynthesized AgNPs allows for the identification of functional groups responsible for coating and effectively stabilizing the nanoparticles.

FTIR spectroscopy provides an economical and valuable means of uncovering the role of biological molecules in the green synthesis and stabilization of AgNPs (Akintelu et al. 2020). It offers qualitative analysis by revealing peaks in the absorption spectrum associated with specific chemical bonds, aiding in the identification of biomolecules involved in the process.

20.5.3 Scanning Electron Microscopy with Energy-Dispersive X-Ray Spectroscopy (SEM- EDX)

Scanning electron microscopy (SEM) is a powerful electron microscopy technique widely used to investigate the surface characteristics of nanoparticles, providing essential information about their size, shape, and distribution (Fig. 20.2c) (Sadeghi and Gholamhoseinpoor 2015). Field emission SEM (FESEM) enhances SEM by utilizing a potent electric field to accelerate electrons, producing high-resolution images (Chandraker et al. 2021). In SEM, an electric current passes through electromagnetic coils and lenses, generating a focused electron beam that interacts with the sample's surface, producing secondary electrons. These secondary electrons are then employed to construct detailed representations of the sample's surface morphology. SEM is especially effective for imaging electrically conductive nanoparticles like silver and gold. SEM can be complemented with energy-dispersive X-ray spectroscopy (EDX) to perform elemental analysis. EDX can confirm the presence of elements, such as silver in AgNPs. By combining SEM with EDX, researchers can gain a comprehensive understanding of nanoparticle characteristics, aiding in quality assessment and agglomeration analysis (Sadeghi and Gholamhoseinpoor 2015).

20.5.4 Transmission Electron Microscopy (TEM)

Transmission electron microscopy (TEM) is a valuable tool for characterizing nanoparticles, revealing about their size, size distribution, morphology, and crystalline nature (Fig. 20.2d). TEM provides improved spatial resolution compared with SEM and enables a more in-depth analysis of nanoparticles. However, TEM has certain advantages and disadvantages. Sample preparation for TEM can be laborious and requires a high vacuum environment, and the examined area is typically small, which may not represent the entire sample. Furthermore, in the case of biomaterials, the electron beam can potentially damage the sample. Despite these drawbacks, TEM provides atomic-level precision and is widely used for visualizing nanoparticles (Chandraker et al. 2021). Additionally, TEM can be coupled with selected area electron diffraction (SAED). SAED is a technique that can determine particle size, shape, distribution, and the crystalline nature of nanoparticles (Chandraker et al. 2021). For example, it was used to confirm the face-centered cubic (FCC) crystalline structure of silver in plant-mediated synthesized AgNPs.

20.5.5 Dynamic Light Scattering (DLS)

DLS is a valuable technique for nanoparticle characterization, offering insights into surface charge, size, and size distribution (Fig. 20.2e) (Saxena et al. 2010). This method capitalizes on the interaction between the Brownian motion of nanoparticles and light passing through a colloidal solution. DLS reveals the hydrodynamic diameter of nanoparticles, influenced by surface charges and capping agents (Bamal et al. 2021). DLS works well for particles ranging from 1 to 500 nm, though agglomerated particles pose challenges (Akbari et al. 2011). DLS can analyze polydisperse mixtures, such as AgNPs from *Phoenix dactylifera* seeds, with a 203 nm hydrodynamic diameter (Khader et al. 2022).

20.5.6 X-Ray Diffraction (XRD)

XRD is a versatile and powerful technique for the qualitative and quantitative analysis of nanoparticles, providing critical details about their crystal structure and characteristics. It enables the identification of materials by comparing diffraction patterns to reference databases like the Joint Committee on Powder Diffraction Standards (JCPDS) library (Zhang et al. 2016). XRD can readily assess material purity by detecting the presence of impurities. The fundamental principle behind XRD is Bragg's law, which helps determine Bragg reflections specific to AgNPs. Notably, the (111), (200), (220), (311), (331), and (222) crystallographic planes are distinctive for AgNPs (Fig. 20.2f) (Mashwani et al. 2015). XRD patterns reveal the nanoparticle size, with sharp peaks indicating the formation of nanosized particles (Alaqad and Saleh 2016). The Debye-Scherrer formula is often used to determine average crystalline size:

$$D = \frac{K\lambda}{\beta \cos\theta}$$

where D is size, K is a constant, λ is X-ray wavelength, β is line broadening, and θ is Bragg's angle. XRD is broadly applicable, examining molecular and crystal structures in various materials, from inorganic catalysts to biomolecules. The configuration of diffraction peaks is crucial for material property assessment, with each material producing a distinct diffraction pattern.

20.5.7 Zeta Potential Analysis

Zeta potential is a critical parameter in assessing nanoparticle dispersion, effective surface electric charge, and stability (Zhang et al. 2002). It is calculated based on factors like peak area and peak number, which are indicative of the particle charge (Fig. 20.2g). Previous research (Baldassarre et al. 2015) has shown that nanoparticles with a high positive or negative charge tend to repel each other, resulting in stable particles with a reduced tendency to aggregate. Typically, suspensions are considered stable when they exhibit a zeta potential within the range of 20–30 mV. Conversely, dispersions with a lower zeta potential can lead to AgNP aggregation due to van der Waals attraction (Jastrzębska et al. 2015).

20.5.8 Atomic Force Microscopy (AFM)

AFM has found valuable applications in the examination of nanoparticle dispersion, aggregation, size, shape, and structure (Fig. 20.2h). The advantage of AFM for the research related to nanoparticle is the capacity to prepare 3D images. AFM is comparatively cheaper and less spacious than SEM and TEM. AFM enables the real-time study of nanoparticle interactions with supported lipid bilayers, a capability not achievable with existing electron microscopy techniques.

20.6 Applications of Silver Nanoparticles

20.6.1 Antibacterial Activity of AgNPs

Silver nanoparticles possess the capability to continuously release silver ions, which can be considered the underlying mechanism for their antimicrobial action. This phenomenon is attributed to the affinity of silver ions for sulfur proteins. Due to electrostatic attraction, they can get adhered to the cell wall and cytoplasm of the bacteria and augment the cytoplasmic membrane permeability, thereby disrupting the envelope of the bacteria. Furthermore, when free silver ions are internalized by cells, they can deactivate respiratory enzymes. Given that sulfur and phosphorus are integral constituents of DNA, the interaction between silver ions and these elements can lead to complications during DNA replication and cell reproduction and potentially result in the termination of microorganisms. Additionally, the inhibitory effect of silver ions extends to protein synthesis, achieved through the denaturation of ribosomes within the cytoplasm. Apart from their ability of releasing silver ion, AgNPs exhibit the inherent capacity to directly eliminate bacteria. These nanoparticles can amass within crevices formed on the cell wall after anchoring on the cell surface. These accumulated silver nanoparticles prompt denaturation of the cell membrane. Moreover, silver nanoparticles possess the ability to permeate bacterial cell walls due to their nanoscale dimensions, inducing alterations in the cell membrane structure. This membrane denaturation can lead to the rupture of organelles and even result in cell lysis. Furthermore, silver nanoparticles have the potential to engage in bacterial signal transduction processes. These processes, reliant on the phosphorylation of protein substrates, can be influenced by nanoparticles that dephosphorylate tyrosine residues on peptide substrates. This interference with signal transduction pathways can trigger cell apoptosis and halt cell proliferation (Yin et al. 2020). Gram-negative bacteria exhibit higher susceptibility to the effects of silver nanoparticles. This disparity is attributed to the narrower cellular walls found in gram-negative strains compared to their gram-positive counterparts. This differentiation in the antibacterial impact of silver nanoparticles on gram-negative versus gram-positive bacteria underscores the significance of nanoparticle uptake in influencing the antibacterial outcome. It is widely acknowledged that silver nanoparticles with sizes smaller than 10 nm possess the ability to directly modify cell permeability, enabling their entry into bacterial cells and subsequent induction of cell damage (Meikle et al. 2020).

20.6.2 Antifungal Activity of AgNPs

The recognition of fungi as significant pathogens among critically ill patients, especially within nosocomial environments, has been steadily growing. While the antibacterial attributes of Ag-NPs are well-established, the antifungal potential of this material remains an area that necessitates further exploration. The evaluation of antifungal effects of Ag-NPs against 44 strains encompassing six fungal species, incorporating clinical isolates and ATCC strains of *Trichophyton mentagrophytes* and *Candida albicans*, has been conducted. The outcomes displayed inhibitory concentrations (IC80) ranging from 1 to 7 μg/mL. In the context of *C. albicans*, the antifungal impact of Ag-NPs was linked to the disruption of the cell membrane structure, alongside the inhibition of the normal budding process attributed to the compromised integrity of the membrane (Kim et al. 2008a, b). In recent times, driven by advancements in nanotechnology and the quest for novel antimicrobial agents, there has been an escalating interest in employing nanoparticles to address skin microbial infections. A study has been conducted to investigate the in vitro inhibitory efficacy of silver nanoparticles (AgNP) against clinical isolates of *M. furfur* to visualize the interaction between AgNP and *Malassezia*, assess potential synergism with ketoconazole (KTZ), and formulate an antimicrobial gel combining AgNP and KTZ. The synthesized AgNP exhibited a fungicidal effect with minimal inhibitory concentration values that were consistently low. These nanoparticles were found to be randomly distributed across the yeast surface, contributing to their antimicrobial action (Mussin et al. 2019). Researchers have also investigated the antifungal attributes of plastic catheters coated with Ag-NPs (approximately 100 nm thick), revealing near-complete growth inhibition for *C. albicans*. Silver nanoparticles (AgNPs) that were synthesized using strawberry waste have also shown antifungal activity against *F. oxysporum*, which is a plant fungus. This serves as an additional illustration of the significance of AgNPs in promoting both plant and human health (Khan et al. 2021). In summation, Ag-NPs have demonstrated potential as biocidal agents against fungal strains, presenting prospects for the mitigation of fungal infections and safeguarding human well-being.

20.6.3 Antiviral Activity of AgNPs

The development of antiviral agents is crucial due to the prevalence and increasing prominence of viral-mediated diseases worldwide. An essential component of antiviral therapy involves understanding the mechanisms underlying the antiviral activity of AgNPs. These nanoparticles exhibit notable interactions with viruses and bacteria, with their efficacy influenced by specific shapes and size ranges. A study explored the antiviral potential of nano-silver incorporated into polysulfone ultrafiltration membranes against the MS2 bacteriophage revealing that the enhanced hydrophilicity of the membrane played a significant role in its antiviral activity. Silver nanoparticles (AgNPs) have demonstrated effective inhibitory actions against viruses such as hepatitis B virus (HBV) and human immunodeficiency virus (HIV) (Zhang et al. 2016). The efficacy of tannic acid-modified AgNPs (TA-AgNPs) was examined in mouse models afflicted with vaginal HSV-2 infections which demonstrated that these nanoparticles prompted the generation of cytokines and chemokines crucial for an antiviral response (Orlowski et al. 2014).

Multiple research endeavors have elucidated that the primary mode of action for AgNPs involves direct physical interactions with either the unbound viral particle or the viral particle that has adhered to a cell. In this manner, AgNPs possess the capability to exhibit virucidal effects, encompassing the deactivation of infectious viral particles or the modification of virion morphology. Additionally, AgNPs can impede the initial stages of viral replication, which encompasses the hindrance of virus binding to the host cell and the processes associated with virus penetration.

20.6.4 Anti-Inflammatory Activity of AgNPs

Inflammation is a pathological state that constitutes a component of our immune system's reaction to detrimental stimuli, including irritants, damaged cells, and pathogens. Lately, a multitude of distressing chronic illnesses have been established to have connections with inflammation, such as rheumatoid arthritis, atherosclerosis, diabetes, Alzheimer's disease, inflammatory bowel disease, and cancer. Silver nanoparticles are renowned for their anti-inflammatory effects. The synthesis of AgNPs through an extraction method involving petroleum ether, along with a minor quantity of ethyl acetate, has revealed robust inhibition of cyclooxygenase-2 (COX-2), an enzyme associated with inflammation. By incorporating the natural extracts with anti-inflammatory properties during the synthesis of silver nanoparticles using petroleum ether extraction, the resultant nanoparticles' anti-inflammatory potential becomes augmented (Abdelhafez et al. 2020). Studies suggest that rats treated with nanocrystalline silver, either through intra-colonic administration or oral ingestion, experience a significant decrease in colonic inflammation. The recuperation process facilitated by AgNPs leads to faster recovery and improved cosmetic results, showcasing a dose-dependent trend. Recently, researchers extracted silver nanoparticles from a soft coral named *Nephthea* sp., which inherently possesses anti-inflammatory attributes. An investigation indicated that the AgNPs display anti-inflammatory effects in a rat arthritis model by diminishing edema and levels of interleukin-6 (IL-6) and high-sensitive C-reactive within paw tissues, in contrast to the untreated arthritic rats (Khashan et al. 2023). Collectively, these findings substantiate that AgNPs hold potential as a promising anti-inflammatory effect.

20.6.5 Anticancer Activity of AgNPs

Cancer is an intricate ailment that impacts not only individual organs but also disrupts the overall functioning of the body. Investigations into novel technologies aimed at mitigating systemic and bystander side effects, along with enhancements to existing medications, have spurred the creation of therapeutics grounded in nanotechnology. AgNPs hold the potential to offer an immediate remedy for treating these fatal diseases and mitigating or eradicating associated challenges. Doxil (doxorubicin encapsulated in liposomes) and Caelyx are prescribed for treating ovarian cancer and AIDS-related Kaposi's sarcoma and, in conjunction with bortezomib, for multiple myeloma therapy. Furthermore, Caelyx and Myocet find application in patients diagnosed with metastatic breast cancer. These drugs stand as pioneering organic nano-formulations, having secured FDA approval for cancer treatment (Rodríguez et al. 2022). As demonstrated by several studies, AgNPs show remarkable control over the proliferation of human liver (HepG2) and breast (MCF-7) cancer cells, even at significantly low concentrations. The cytotoxic impact of AgNPs on HepG2 and MCF-7 cancer cells was found to be contingent on the dose as assessed via MTT and NRU assays. Exposure to AgNPs led to a 136% increase in the production of reactive oxygen species (ROS) in HepG2 cells and a 142% increase in MCF-7 cells (Elsupikhe et al. 2015). Numerous investigations have highlighted the potential anticancer properties of AgNPs, encompassing those derived from diverse plant extracts, notably against breast (MCF-7), lung (A549), cervical (HeLa), colon (HT-29, Caco-2), prostate (PC-3), and VCaP cancer cell lines (Kathiravan et al. 2014; Venugopal et al. 2017; Jeyaraj et al. 2013). The half-maximal inhibitory concentration (IC50) of AgNPs demonstrated comparability to that of established standard anticancer medications solidifying the validation of AgNPs' potential utility as anticancer agents.

20.6.6 Antiangiogenic Activity of AgNPs

Angiogenesis is a significant process integral to normal growth and wound healing. However, an imbalance in the growth factors driving this process can lead to the acceleration of diverse diseases, encompassing malignancies, ocular issues, and inflammatory conditions. Intervening in the pathway of angiogenesis to inhibit it presents a promising approach to impede the advancement of these diseases. Researchers have suggested that biologically derived AgNPs exhibited dose-dependent antiangiogenic effects, as evidenced by the parameters derived from CAM images. Additionally, quantitative real-time polymerase chain reaction (qRT-PCR) analysis unveiled the downregulation of iNOS and VEGF genes. Furthermore, a two-dimensional wound healing assay demonstrated the inhibition of migration and motility of DU145 cells throughout the 72-hour evaluation period (Kitimu et al. 2022). In recent study, in vitro evaluations were conducted to assess wound healing potential using a scratch wound healing assay, invasion capabilities through a transwell Matrigel invasion assay, and apoptosis levels via DAPI and annexin V-FITC assays, all conducted on HeLa cells. The synthesized biogenic AgNPs have demonstrated antiangiogenic properties, as substantiated by an in vivo chick chorioallantois membrane assay. AgNPs proficiently hindered the development of fresh blood microvessels induced by VEGF in the mouse Matrigel plug assay. To decipher the fundamental mechanism behind the angiogenesis-inhibitory impact of Ag-NPs, our findings illustrated that AgNPs had the capacity to impede the activation of the PI3K-Akt pathway (Gurunathan et al. 2009).

20.6.7 Air Disinfection by AgNPs

Upon inhalation, the particulate matter in the air penetrates the bronchi, obstructing airway passages and giving rise to significant health concerns. These issues encompass respiratory ailments, exacerbation of conditions like asthma and bronchitis, as well as cardiovascular morbidity and mortality. The findings indicated the efficacy of Ag-deposited activated carbon fiber (ACF) filters in eliminating bioaerosols. The assessment of antibacterial activity regarding Ag-coated ACF filters revealed complete inhibition of *Bacillus subtilis* and *E. coli* bacteria within 10 and 60 min, respectively. Notably, the deposition of silver did not alter the physical attributes of ACF filters, such as pressure drop and filtration efficiency (Tran and Le 2013; Watson et al. 2022). Silver nanoparticles being a good antibacterial and antiviral agent becomes best choice for coating air filters with it. The evident decrease in viable bacterial cells within silver-treated filters underscores the necessity of antimicrobial filter technology for future applications.

20.6.8 Biological Wastewater and Groundwater Disinfection by AgNPs

Water pollution has become a global concern due to the escalating use of chemical compounds. As heavy metals, a variety of organic and inorganic dyes are added as effluents to water sources. These dyes are primarily derived from well-known carcinogens like benzidine and naphthalene, which when they enter biological systems can transform into carcinogens due to microbial degradation (Marimuthu et al.

2020). To make up for the losses and prevent further strain on water bodies, the contaminated water must be treated prior to discharge into main water resources. Several researches have demonstrated the catalytic action of AgNPs synthesized by plant extract against methylene blue (MB) which is a great example of hazardous dyes causing water pollution (Saha et al. 2017). Hybrids combining silver nanoparticles (AgNPs) with cellulose are suggested to offer a dual advantage: they can be environmentally safe if thoughtfully designed and are readily producible from recycled materials at minimal costs while also exhibiting reusability potential. These advanced hybrids, rooted in AgNP-based constituents, are poised for rapid deployment in environmental contexts, a development that holds the potential for profound economic and ecological ramifications (Fiorati et al. 2020).

20.6.9 Role of AgNPs in Food Processing and Packaging

Foodborne illnesses pose a significant threat to consumer health, particularly among vulnerable populations and in developing nations. The persistent pursuit of effective antimicrobial solutions for the food industry has ushered in the realm of nanotechnology. Among these solutions, silver nanoparticles (Ag-NPs) stand out for their exceptional antimicrobial efficacy, thus presenting significant promise for applications in food processing and packaging. Many studies outline the primary applications of Ag-NPs as antimicrobial agents in food preservation, alongside an exploration of prevailing regulations concerning these materials, and their influence on the oral and gut microbiota is also highlighted (Zorraquín-Peña et al. 2020). As feed additives, Ag-NPs have demonstrated efficacy in diminishing potentially pathogenic microorganisms, including *E. coli* and *Clostridium perfringens*. This potential reduction in pathogenic organisms could offer a viable alternative to antibiotic use in livestock. Furthermore, certain Ag-NPs have exhibited notable antiparasitic activity. Research conducted on impact of AgNP-coated packing material on the shelf-life of perishable food such as tomatoes depicted those tomatoes stored within AgNP-coated packing material remained protected from microbial spoilage, retaining their freshness for up to 30 days. In contrast, tomatoes wrapped in regular polythene bags displayed partial spoilage, accompanied by surface wrinkles (Kumari et al. 2021). AgNPs enhance the quality of food packaging materials and hence are preferred for this purpose (Fig. 20.3).

20.7 Challenges Associated with Silver Nanoparticles

Silver nanoparticles have gained attention for their wide range of applications, particularly in biomedicine. They show promise in diagnostics, drug delivery, cell imaging, and implantation. However, their multifaceted nature presents challenges, including potential adverse effects on health and the environment.

A comprehensive toxicology study focused on silver nanoparticles found that even low exposure levels (10–50 $\mu g\ ml^{-1}$) induced oxidative stress and disrupted

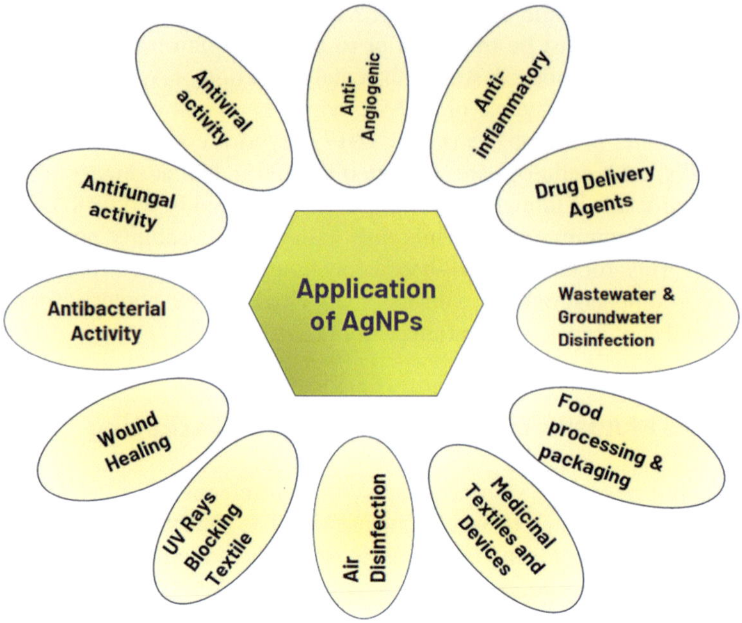

Fig. 20.3 Applications of silver nanoparticles in various fields of biology

mitochondrial function in rat liver cells. Higher doses (>1.0 mg L^{-1}) led to increased cytotoxicity and morphological changes in cells (McAuliffe and Perry 2007). Silver nanoparticles also interfered with mitochondrial activities and altered cell membrane permeability in in vitro mouse germ line stem cells.

Extended exposure through ingestion or inhalation raised concerns about chronic disorders like argyria or argyrosis in humans, affecting skin microvessels and eye health. Various in vitro cell culture studies demonstrated the cytotoxic effects of silver nanoparticles on different cell types, primarily attributed to the generation of reactive oxygen species (ROS) (Carrola et al. 2016; Kim and Shin 2014; Sabella et al. 2014; Li et al. 2018; Sahu et al. 2014; Shi et al. 2014; Jiang et al. 2018). Exposure to silver nanoparticles led to glutathione depletion, increased ROS levels, lipid peroxidation, upregulated ROS-responsive genes, DNA damage, apoptosis, and necrosis (Haider and Kang 2015).

Due to their high surface area to volume ratio, silver nanoparticles can easily enter the body through ingestion, inhalation, or skin contact, potentially distributing throughout the body (Sung et al. 2008). They were observed to cross the blood-testes barrier and accumulate in the testes, impacting the male reproductive system and affecting peripheral blood mononuclear cell proliferation and cytokine expression (Kim et al. 2008a, b).

The AgNPs have some toxic effects on the environment. They can be released into wastewater from various sources, including consumer products and medical applications, that can persist in wastewater treatment plants and transform into more toxic forms, raising concerns about their release into aquatic environments. Aquatic ecosystems, fish in particular, are susceptible to AgNP exposure. AgNPs can also enter soil through various pathways and can alter the abundance and diversity of soil bacteria, potentially disrupting nutrient cycling and soil health. The nanoparticles can get accumulated in plant tissues, resulting in phytotoxic effects, reduced biomass, and altered nutrient uptake, hence posing a risk to both agricultural and natural ecosystems. The bioaccumulation potential of biogenic AgNPs is also a growing concern.

In conclusion, silver nanoparticles offer both benefits and risks. While they hold promise for various applications, their interactions with biological systems necessitate careful consideration to harness their advantages while minimizing potential harms.

20.8 Future Prospects and Conclusion

Researchers are increasingly turning to green methods for synthesizing nanoparticles due to their environmental friendliness, potential health benefits, and cost-effectiveness. Further studies are required to understand the exact mechanism involved in the green synthesis of AgNPs for better control and reproducibility. Optimizing the physicochemical characteristics of AgNPs can lead to safer nanoparticles with minimal toxicity while maximizing their biological efficacy. More studies have to be conducted on AgNPs to analyze their clinical applications. For instance, research should focus on identifying specific targets for AgNPs creating a targeted drug delivery system. AgNPs are studied to find out their efficacy as antimicrobial agent against several microbes this may lead to development of potent antimicrobial agent. Extensive in vivo studies have to be conducted to develop AgNPs as a clinically viable treatment option for chronic diseases like cancer. Future studies should also explore the impact of AgNPs on biological systems, the response of living organisms to various exposure routes, facilitating measures to protect organisms and the environment.

In conclusion, green synthesis of AgNPs using plant extracts is a promising and environment-friendly approach with broad potential applications. While synthesis optimization and toxicity assessment of AgNPs remain challenging, continued research and a deeper understanding of the process will pave the way for safer and more effective AgNPs in various fields, including medicine and industry.

References

Abdelhafez OH, Ali TFS, Fahim JR, Desoukey SY, Ahmed S, Behery FA, Abdelmohsen UR (2020) Anti-inflammatory potential of green synthesized silver nanoparticles of the soft coral *Nephthea sp.* supported by metabolomics analysis and docking studies. Int J Nanomed 15:5345–5360

Abou El-Nour KM, Eftaiha AA, Al-Warthan A, Ammar RA (2010) Synthesis and applications of silver nanoparticles. Arab J Chem 3(3):135–140

Afreen A, Ahmed R, Mehboob S, Tariq M, Alghamdi HA, Zahid AA, Ali I, Malik K, Hasan A (2020) Phytochemical-assisted biosynthesis of silver nanoparticles from *Ajuga bracteosa* for biomedical applications. Mater Res Express 7(7):075404

Agnihotri S, Mukherji S, Mukherji S (2014) Size-controlled silver nanoparticles synthesized over the range 5–100 nm using the same protocol and their antibacterial efficacy. RSC Adv 4(8):3974–3983

Ahmad N, Sharma S, Singh VN, Shamsi SF, Fatma A, Mehta BR (2011) Biosynthesis of silver nanoparticles from *Desmodium triflorum*: A novel approach towards weed utilization. Biotechnol Res Int 2011:1–8

Ahmad N, Sharma S (2012) Green synthesis of silver nanoparticles using extracts of *Ananas comosus*. Green Sustain Chem 3:376

Ahmed S, Ahmad M, Swami BL, Ikram S (2016a) A review on plants extract mediated synthesis of silver nanoparticles for antimicrobial applications: a green expertise. J Adv Res 7(1):17–28

Ahmed S, Ahmad M, Swami BL, Ikram S (2016b) Green synthesis of silver nanoparticles using *Azadirachta indica* aqueous leaf extract. J Radiat Res Appl Sci 9(1):1–7

Ajitha B, Reddy YAK, Shameer S, Rajesh KM, Suneetha Y, Reddy PS (2015) *Lantana camara* leaf extract mediated silver nanoparticles: antibacterial, green catalyst. J Photochem Photobiol B Biol 149:84–92

Akbari B, Tavandashti MP, Zandrahimi M (2011) Particle size characterization of nanoparticles–a practical approach. Iran J Mater Sci Eng 8(2):48–56

Akintelu SA, Agboola OS, Fagbohun MA (2020) Biogenic synthesis of silver nanoparticles using leaf extract of *Alchornea cordifolia* and its antimicrobial activity. Result Chem 2:100027

Alaqad K, Saleh TA (2016) Gold and silver nanoparticles: synthesis methods, characterization routes and applications towards drugs. J Environ Anal Toxicol 6(4):525–2161

Al-Shmgani HS, Mohammed WH, Sulaiman GM, Saadoon AH (2017) Biosynthesis of silver nanoparticles from *Catharanthus roseus* leaf extract and assessing their antioxidant, antimicrobial, and wound-healing activities. Artif Cells Nanomed Biotechnol 45(6):1234–1240

Anandalakshmi K, Venugobal J, Ramasamy VJAN (2016) Characterization of silver nanoparticles by green synthesis method using *Pedalium murex* leaf extract and their antibacterial activity. Appl Nanosci 6:399–408

Aromal SA, Philip D (2012) Green synthesis of gold nanoparticles using *Trigonella foenum-graecum* and its size-dependent catalytic activity. Spectrochim Acta A Mol Biomol Spectrosc 97:1–5

Ashour AA, Raafat D, El-Gowelli HM, El-Kamel AH (2015) Green synthesis of silver nanoparticles using cranberry powder aqueous extract: characterization and antimicrobial properties. Int J Nanomedicine 2022:7207–7221

Balachandar R, Gurumoorthy P, Karmegam N, Barabadi H, Subbaiya R, Anand K, Boomi P, Saravanan M (2019) Plant-mediated synthesis, characterization and bactericidal potential of emerging silver nanoparticles using stem extract of *Phyllanthus pinnatus*: a recent advance in phytonanotechnology. J Clust Sci 30:1481–1488

Baldassarre F, Cacciola M, Ciccarella G (2015) A predictive model of iron oxide nanoparticles flocculation tuning Z-potential in aqueous environment for biological application. J Nanopart Res 17:1–21

Bamal D, Singh A, Chaudhary G, Kumar M, Singh M, Rani N, Mundlia P, Sehrawat AR (2021) Silver nanoparticles biosynthesis, characterization, antimicrobial activities, applications, cytotoxicity and safety issues: an updated review. Nano 11(8):2086

Banala RR, Nagati VB, Karnati PR (2015) Green synthesis and characterization of *Carica papaya* leaf extract coated silver nanoparticles through X-ray diffraction, electron microscopy and evaluation of bactericidal properties. Saudi J Biol Sci 22(5):637–644

Bar H, Bhui DK, Sahoo GP, Sarkar P, Pyne S, Misra A (2009) Green synthesis of silver nanoparticles using seed extract of *Jatropha curcas*. Colloids Surf A Physicochem Eng Asp 348(1–3):212–216

Bhakya S, Muthukrishnan S, Sukumaran M, Muthukumar M (2016) Biogenic synthesis of silver nanoparticles and their antioxidant and antibacterial activity. Appl Nanosci 6:755–766

Bharathi D, Diviya Josebin M, Vasantharaj S, Bhuvaneshwari V (2018) Biosynthesis of silver nanoparticles using stem bark extracts of *Diospyros montana* and their antioxidant and antibacterial activities. J Nanostructure Chem 8:83–92

Bindhu MR, Umadevi M (2015) Antibacterial and catalytic activities of green synthesized silver nanoparticles. Spectrochim Acta A Mol Biomol Spectrosc 135:373–378

Carrola J, Bastos V, Jarak I, Oliveira-Silva R, Malheiro E, Daniel-da-Silva AL, Duarte IF (2016) Metabolomics of silver nanoparticles toxicity in HaCaT cells: structure–activity relationships and role of ionic silver and oxidative stress. Nanotoxicology 10(8):1105–1117

Chahardoli A, Karimi N, Fattahi A (2017) Biosynthesis, characterization, antimicrobial and cytotoxic effects of silver nanoparticles using *Nigella arvensis* seed extract. Ir J Pharm Res 16(3):1167

Chandraker SK, Ghosh MK, Lal M, Shukla R (2021) A review on plant-mediated synthesis of silver nanoparticles, their characterization and applications. Nano Express 2(2):022008

Christensen L, Vivekanandhan S, Misra M, Kumar Mohanty A (2011) Biosynthesis of silver nanoparticles using *Murraya koenigii* (curry leaf): an investigation on the effect of broth concentration in reduction mechanism and particle size. Adv Mater Lett 2(6):429–434

Cruz D, Falé PL, Mourato A, Vaz PD, Serralheiro ML, Lino ARL (2010) Preparation and physicochemical characterization of ag nanoparticles biosynthesized by *Lippia citriodora* (lemon verbena). Colloids Surf B: Biointerfaces 81(1):67–73

Deepak P, Sowmiya R, Ramkumar R, Balasubramani G, Aiswarya D, Perumal P (2017) Structural characterization and evaluation of mosquito-larvicidal property of silver nanoparticles synthesized from the seaweed, *Turbinaria ornata* (turner) J. Agardh 1848. Artif Cells Nanomed Biotechnol 45(5):990–998

Dinesh S, Karthikeyan S, Arumugam P (2012) Biosynthesis of silver nanoparticles from *Glycyrrhiza glabra* root extract. Arch Appl Sci Res 4(1):178–187

Dipankar C, Murugan S (2012) The green synthesis, characterization and evaluation of the biological activities of silver nanoparticles synthesized from *Iresine herbstii* leaf aqueous extracts. Colloids Surf B: Biointerfaces 98:112–119

Elsupikhe RF, Shameli K, Ahmad MB, Ibrahim NA, Zainudin N (2015) Green sonochemical synthesis of silver nanoparticles at varying concentrations of κ-carrageenan. Nanoscale Res Lett 10:1–8

Elumalai D, Ashok K, Suresh A, Hemavathi M (2016) Green synthesis of silver nanoparticle using *Achyranthes aspera* and its larvicidal activity against three major mosquito vectors. Eng Agric Environ Food 9(1):1–8

Fayaz AM, Balaji K, Girilal M, Yadav R, Kalaichelvan PT, Venketesan R (2010) Biogenic synthesis of silver nanoparticles and their synergistic effect with antibiotics: a study against gram-positive and gram-negative bacteria. Nanomedicine 6(1):103–109

Fenfen LÜ, Yixian GAO, Huang J, Daohua S, Qingbiao LI (2014) Roles of biomolecules in the biosynthesis of silver nanoparticles: case of *Gardenia jasminoides* extract. Chin J Chem Eng 22(6):706–712

Fiorati A, Bellingeri A, Punta C, Corsi I, Venditti I (2020) Silver nanoparticles for water pollution monitoring and treatments: ecosafety challenge and cellulose-based hybrids solution. Polymers 12(8):1635

Forough M, Farhadi K (2010) Biological and green synthesis of silver nanoparticles. Turk J Eng Environ Sci 34(4):281–287

Gavhane AJ, Padmanabhan P, Kamble SP, Jangle SN (2012) Synthesis of silver nanoparticles using extract of neem leaf and triphala and evaluation of their antimicrobial activities. Int J Pharm Bio Sci 3(3):88–100

Geethalakshmi R, Sarada DVL (2013) Characterization and antimicrobial activity of gold and silver nanoparticles synthesized using saponin isolated from *Trianthema decandra* L. Ind Crop Prod 51:107–115

Ghaedi M, Yousefinejad M, Safarpoor M, Khafri HZ, Purkait MK (2015) *Rosmarinus officinalis* leaf extract mediated green synthesis of silver nanoparticles and investigation of its antimicrobial properties. J Ind Eng Chem 31:167–172

Ghosh S, Patil S, Ahire M, Kitture R, Kale S, Pardesi K, Cameotra SS, Bellare J, Dhavale DD, Jabgunde A, Chopade BA (2012) Synthesis of silver nanoparticles using *Dioscorea bulbifera* tuber extract and evaluation of its synergistic potential in combination with antimicrobial agents. Int J Nanomedicine 7:483–496

Guilger-Casagrande M, Lima RD (2019) Synthesis of silver nanoparticles mediated by fungi: a review. Front Bioeng Biotechnol 7:287

Gurunathan S, Lee KJ, Kalishwaralal K, Sheikpranbabu S, Vaidyanathan R, Eom SH (2009) Antiangiogenic properties of silver nanoparticles. Biomaterials 30(31):6341–6350

Haider A, Kang IK (2015) Preparation of silver nanoparticles and their industrial and biomedical applications: a comprehensive review. Adv Mater Sci Eng 2015:1–16

Husen A, Siddiqi KS (2014) Phytosynthesis of nanoparticles: concept, controversy and application. Nanoscale Res Lett 9:1–24

Jain PK, Huang X, El-Sayed IH, El-Sayed MA (2008) Noble metals on the nanoscale: optical and photothermal properties and some applications in imaging, sensing, biology, and medicine. Acc Chem Res 41(12):1578–1586

Jagajjanani Rao K, Paria S (2013) Green synthesis of silver nanoparticles from aqueous *Aegle marmelos* leaf extract. Mater Res Bull 48(2):628–634

Jassal V, Shanker U, Kaith BS (2016) *Aegle marmelos* mediated green synthesis of different nanostructured metal hexacyanoferrates: activity against photodegradation of harmful organic dyes. Scientifica 2016:2715026

Jastrzębska AM, Karwowska E, Olszyna AR, Kunicki A (2015) Influence of bacteria adsorption on zeta potential of Al_2O_3 and Al_2O_3/Ag nanoparticles in electrolyte and drinking water environment studied by means of zeta potential. Surf Coat Technol 271:225–233

Jeyaraj M, Rajesh M, Arun R, MubarakAli D, Sathishkumar G, Sivanandhan G, Ganapathi A (2013) An investigation on the cytotoxicity and caspase-mediated apoptotic effect of biologically synthesized silver nanoparticles using *Podophyllum hexandrum* on human cervical carcinoma cells. Colloids Surf B: Biointerfaces 102:708–717

Jiang X, Lu C, Tang M, Yang Z, Jia W, Ma Y, Wang H (2018) Nanotoxicity of silver nanoparticles on HEK293T cells: a combined study using biomechanical and biological techniques. ACS Omega 3(6):6770–6778

Kalaiarasi R, Prasannaraj G, Venkatachalam P (2013) A rapid biological synthesis of silver nanoparticles using leaf broth of *Rauvolfia tetraphylla* and their promising antibacterial activity. Indo Am J Pharm Res 3(10):8052–8062

Kathiravan V, Ravi S, Ashokkumar S (2014) Synthesis of silver nanoparticles from *Melia dubia* leaf extract and their in vitro anticancer activity. Spectrochim Acta A Mol Biomol Spectrosc 130:116–121

Kaviya S, Santhanalakshmi J, Viswanathan B, Muthumary J, Srinivasan K (2011) Biosynthesis of silver nanoparticles using *Citrus sinensis* peel extract and its antibacterial activity. Spectrochim Acta A Mol Biomol Spectrosc 79(3):594–598

Khader SZA, Ahmed SSZ, Mahboob MR, Prabaharan SB, Lakshmanan SO, Kumar KR, David D (2022) In vitro anti-inflammatory, anti-arthritic and anti-proliferative activity of green synthesized silver nanoparticles-*Phoenix dactylifera* (Rothan dates). Braz J Pharm Sci 58:e18594

Khan M, Khan AU, Bogdanchikova N, Garibo D (2021) Antibacterial and antifungal studies of biosynthesized silver nanoparticles against plant parasitic nematode *Meloidogyne incognita*, plant pathogens *Ralstonia solanacearum* and *Fusarium oxysporum*. Molecules 26(9):2462

Kharissova OV, Dias HR, Kharisov BI, Pérez BO, Pérez VMJ (2013) The greener synthesis of nanoparticles. Trends Biotechnol 31(4):240–248

Khashan AA, Dawood Y, Khalaf YH (2023) Green chemistry and anti-inflammatory activity of silver nanoparticles using an aqueous curcumin extract. Result Chem 5:100913

Kim MJ, Shin S (2014) Toxic effects of silver nanoparticles and nanowires on erythrocyte rheology. Food Chem Toxicol 67:80–86

Kim KJ, Sung WS, Moon SK, Choi JS, Kim JG, Lee DG (2008a) Antifungal effect of silver nanoparticles on dermatophytes. J Microbiol Biotechnol 18(8):1482–1484

Kim YS, Kim JS, Cho HS, Rha DS, Kim JM, Park JD, Yu IJ (2008b) Twenty-eight-day oral toxicity, genotoxicity, and gender-related tissue distribution of silver nanoparticles in Sprague-Dawley rats. Inhal Toxicol 20(6):575–583

Kiran Kumar HA, Mandal BK, Mohan Kumar K, Maddinedi SB, Sai Kumar T, Madhiyazhagan P, Ghosh AR (2014) Antimicrobial and antioxidant activities of *Mimusops elengi* seed extract mediated isotropic silver nanoparticles. Spectrochim Acta A Mol Biomol Spectrosc 130:13–18

Kitimu SR, Kirira P, Abdille AA, Sokei J, Ochwang ID, Mwitari P, Maina N (2022) Anti-angiogenic and anti-metastatic effects of biogenic silver nanoparticles synthesized using *Azadirachta indica*. Adv Biosci Biotechnol 13(04):188–206

Krishnaraj C, Jagan EG, Rajasekar S, Selvakumar P, Kalaichelvan PT, Mohan NJCSBB (2010) Synthesis of silver nanoparticles using *Acalypha indica* leaf extracts and its antibacterial activity against water borne pathogens. Colloids Surf B: Biointerfaces 76(1):50–56

Kumar V, Yadav SC, Yadav SK (2010) *Syzygium cumini* leaf and seed extract mediated biosynthesis of silver nanoparticles and their characterization. J Chem Technol Biotechnol 85(10):1301–1309

Kumar KR, Nattuthurai N, Gopinath P, Mariappan T (2015) Synthesis of eco-friendly silver nanoparticles from *Morinda tinctoria* leaf extract and its larvicidal activity against Culex quinquefasciatus. Parasitol Res 114:411–417

Kumar R, Ghoshal G, Jain A, Goyal M (2017) Rapid green synthesis of silver nanoparticles (AgNPs) using (*Prunus persica*) plants extract: exploring its antimicrobial and catalytic activities. J Nanomed Nanotechnol 8(4):1–8

Kumari SC, Padma PN, Anuradha K (2021) Green silver nanoparticles embedded in cellulosic network for fresh food packaging. J Pure Appl Microbiol 15(3):1236–1244

Li S, Shen Y, Xie A, Yu X, Qiu L, Zhang L, Zhang Q (2007) Green synthesis of silver nanoparticles using *Capsicum annuum* L. extract. Green Chem 9(8):852–858

Li L, Cui J, Liu Z, Zhou X, Li Z, Yu Y, Wu Y (2018) Silver nanoparticles induce SH-SY5Y cell apoptosis via endoplasmic reticulum-and mitochondrial pathways that lengthen endoplasmic reticulum-mitochondria contact sites and alter inositol-3-phosphate receptor function. Toxicol Lett 285:156–167

Logeswari P, Silambarasan S, Abraham J (2015) Synthesis of silver nanoparticles using plants extract and analysis of their antimicrobial property. J Saudi Chem Soc 19(3):311–317

Marambio-Jones C, Hoek EM (2010) A review of the antibacterial effects of silver nanomaterials and potential implications for human health and the environment. J Nanopart Res 12:1531–1551

Marimuthu S, Rahuman AA, Rajakumar G, Santhoshkumar T, Kirthi AV, Jayaseelan C, Bagavan A, Zahir AA, Elango G, Kamaraj C (2011) Evaluation of green synthesized silver nanoparticles against parasites. Parasitol Res 108:1541–1549

Marimuthu S, Antonisamy AJ, Malayandi S, Rajendran K, Tsai PC, Pugazhendhi A, Ponnusamy VK (2020) Silver nanoparticles in dye effluent treatment: a review on synthesis, treatment methods, mechanisms, photocatalytic degradation, toxic effects and mitigation of toxicity. J Photochem Photobiol B Biol 205:111823

Marslin G, Selvakesavan RK, Franklin G, Sarmento B, Dias AC (2015) Antimicrobial activity of cream incorporated with silver nanoparticles biosynthesized from *Withania somnifera*. Int J Nanomedicine 10:5955–5963

Marslin G, Siram K, Maqbool Q, Selvakesavan RK, Kruszka D, Kachlicki P, Franklin G (2018) Secondary metabolites in the green synthesis of metallic nanoparticles. Materials 11(6):940

Mashwani ZUR, Khan T, Khan MA, Nadhman A (2015) Synthesis in plants and plant extracts of silver nanoparticles with potent antimicrobial properties: current status and future prospects. Appl Microbiol Biotechnol 99:9923–9934

McAuliffe ME, Perry MJ (2007) Are nanoparticles potential male reproductive toxicants? A literature review. Nanotoxicology 1(3):204–210

Medda S, Hajra A, Dey U, Bose P, Mondal NK (2015) Biosynthesis of silver nanoparticles from Aloe vera leaf extract and antifungal activity against *Rhizopus* sp. and *Aspergillus* sp. Appl Nanosci 5:875–880

Meikle TG, Dyett BP, Strachan JB, White J, Drummond CJ, Conn CE (2020) Preparation, characterization, and antimicrobial activity of cubosome encapsulated metal nanocrystals. ACS Appl Mater Interfaces 12(6):6944–6954

Mittal AK, Chisti Y, Banerjee UC (2013) Synthesis of metallic nanoparticles using plant extracts. Biotechnol Adv 31(2):346–356

Mohamed NH, Ismail MA, Abdel-Mageed WM, Shoreit AAM (2014) Antimicrobial activity of latex silver nanoparticles using *Calotropis procera*. Asian Pac J Trop Biomed 4(11):876–883

Mohammadlou M, Maghsoudi H, Jafarizadeh-Malmiri HJIFRJ (2016) A review on green silver nanoparticles based on plants: synthesis, potential applications and eco-friendly approach. Int Food Res J 23(2):446

MubarakAli D, Thajuddin N, Jeganathan K, Gunasekaran M (2011) Plant extract mediated synthesis of silver and gold nanoparticles and its antibacterial activity against clinically isolated pathogens. Colloids Surf B: Biointerfaces 85(2):360–365

Mukunthan KS, Balaji S (2012) Cashew apple juice (*Anacardium occidentale* L.) speeds up the synthesis of silver nanoparticles. Int J Green Nanotechnol 4(2):71–79

Munir H, Bilal M, Mulla SI, Abbas Khan H, Iqbal HM (2021) Plant-mediated green synthesis of nanoparticles. In: Advances in green synthesis: avenues and sustainability. Springer, Cham, pp 75–89

Mussin JE, Roldán MV, Rojas F, Sosa MDLÁ, Pellegri N, Giusiano G (2019) Antifungal activity of silver nanoparticles in combination with ketoconazole against *Malassezia furfur*. AMB Express 9(1):1–9

Nabikhan A, Kandasamy K, Raj A, Alikunhi NM (2010) Synthesis of antimicrobial silver nanoparticles by callus and leaf extracts from saltmarsh plant, *Sesuvium portulacastrum* L. Colloids Surf B: Biointerfaces 79(2):488–493

Nathani S (2021) Synthesis of metallic nanoparticles using plant derivatives, vol 166. AIJR Abstracts, Dhaurahra

Nayak D, Ashe S, Rauta PR, Nayak B (2015) Biosynthesis, characterisation and antimicrobial activity of silver nanoparticles using *Hibiscus rosa-sinensis* petals extracts. IET Nanobiotechnol 9(5):288–293

Orlowski P, Tomaszewska E, Gniadek M, Baska P, Nowakowska J, Sokolowska J, Krzyzowska M (2014) Tannic acid modified silver nanoparticles show antiviral activity in herpes simplex virus type 2 infection. PLoS One 9(8):e104113

Padalia H, Moteriya P, Chanda S (2015) Green synthesis of silver nanoparticles from marigold flower and its synergistic antimicrobial potential. Arab J Chem 8(5):732–741

Palithya S, Gaddam SA, Kotakadi VS, Penchalaneni J, Golla N, Krishna SBN, Naidu CV (2022) Green synthesis of silver nanoparticles using flower extracts of *Aerva lanata* and their biomedical applications. Part Sci Technol 40(1):84–96

Pallela PNVK, Ummey S, Ruddaraju LK, Pammi SVN, Yoon SG (2018) Ultra small, mono dispersed green synthesized silver nanoparticles using aqueous extract of *Sida cordifolia* plant and investigation of antibacterial activity. Microb Pathog 124:63–69

Patil RS, Kokate MR, Jambhale CL, Pawar SM, Han SH, Kolekar SS (2012) One-pot synthesis of PVA-capped silver nanoparticles their characterization and biomedical application. Adv Nat Sci Nanosci Nanotechnol 3(1):015013

Phanjom P, Sultana A, Sarma H, Ramchiary J, Goswami K, Baishya P (2012) Plant-mediated synthesis of silver nanoparticles using *Elaeagnus latifolia* leaf extract. Dig J Nanomater Biostruct 7(3):1117–1123

Rajagopal T, Jemimah IAA, Ponmanickam P, Ayyanar M (2015) Synthesis of silver nanoparticles using *Catharanthus roseus* root extract and its larvicidal effects. J Environ Biol 36(6):1283

Ramteke C, Chakrabarti T, Sarangi BK, Pandey RA (2013) Synthesis of silver nanoparticles from the aqueous extract of leaves of *Ocimum sanctum* for enhanced antibacterial activity. J Chem 2013:2013

Rao KJ, Paria S (2013) Green synthesis of silver nanoparticles from aqueous *Aegle marmelos* leaf extract. Mater Res Bull 48(2):628–634

Rashmi V, Sanjay KR (2017) Green synthesis, characterisation and bioactivity of plant-mediated silver nanoparticles using *Decalepis hamiltonii* root extract. IET Nanobiotechnol 11(3):247–254

Rodríguez F, Caruana P, De la Fuente N, Español P, Gamez M, Balart J, Céspedes MV (2022) Nano-based approved pharmaceuticals for cancer treatment: present and future challenges. Biomol Ther 12(6):784

Sabella S, Carney RP, Brunetti V, Malvindi MA, Al-Juffali N, Vecchio G, Pompa PP (2014) A general mechanism for intracellular toxicity of metal-containing nanoparticles. Nanoscale 6(12):7052–7061

Sadeghi B, Gholamhoseinpoor F (2015) A study on the stability and green synthesis of silver nanoparticles using *Ziziphora tenuior* (Zt) extract at room temperature. Spectrochim Acta A Mol Biomol Spectrosc 134:310–315

Saha J, Begum A, Mukherjee A, Kumar S (2017) A novel green synthesis of silver nanoparticles and their catalytic action in reduction of methylene blue dye. Sustain Environ Res 27(5):245–250

Sahu SC, Zheng J, Graham L, Chen L, Ihrie J, Yourick JJ, Sprando RL (2014) Comparative cytotoxicity of nanosilver in human liver HepG2 and colon Caco2 cells in culture. J Appl Toxicol 34(11):1155–1166

Saikia D, Gogoi KP, Phukan P, Bhuyan N, Borchetia S, Saikia J (2015) Green synthesis of silver nanoparticles using *Asiatic pennywort* and *Bryophyllum* leaves extract and their antimicrobial activity. Adv Mater Lett 6(3):260–264

Saravanan M, Barik SK, MubarakAli D, Prakash P, Pugazhendhi A (2018) Synthesis of silver nanoparticles from *Bacillus brevis* (NCIM 2533) and their antibacterial activity against pathogenic bacteria. Microb Pathog 116:221–226

Sathyavathi R, Krishna MB, Rao SV, Saritha R, Rao DN (2010) Biosynthesis of silver nanoparticles using *Coriandrum sativum* leaf extract and their application in nonlinear optics. Adv Sci Lett 3(2):138–143

Saxena A, Tripathi RM, Singh RP (2010) Biological synthesis of silver nanoparticles by using onion (*Allium cepa*) extract and their antibacterial activity. Dig J Nanomater Bios 5(2):427–432

Sengottaiyan A, Mythili R, Selvankumar T, Aravinthan A, Kamala-Kannan S, Manoharan K, Thiyagarajan P, Govarthanan M, Kim JH (2016) Green synthesis of silver nanoparticles using *Solanum indicum* L. and their antibacterial, splenocyte cytotoxic potentials. Res Chem Intermed 42:3095–3103

Shalaby TI, Mahmoud OA, El Batouti GA, Ibrahim EE (2015) Green synthesis of silver nanoparticles: synthesis, characterization and antibacterial activity. Nanosci Nanotechnol 5(2):23–29

Sharma VK, Yngard RA, Lin Y (2009) Silver nanoparticles: green synthesis and their antimicrobial activities. Adv Colloid Interf Sci 145(1–2):83–96

Shi J, Sun X, Lin Y, Zou X, Li Z, Liao Y, Zhang H (2014) Endothelial cell injury and dysfunction induced by silver nanoparticles through oxidative stress via IKK/NF-κB pathways. Biomaterials 35(24):6657–6666

Smitha SL, Nissamudeen KM, Philip D, Gopchandran KG (2008) Studies on surface plasmon resonance and photoluminescence of silver nanoparticles. Spectrochim Acta A Mol Biomol Spectrosc 71(1):186–190

Sung JH, Ji JH, Yoon JU, Kim DS, Song MY, Jeong J, Yu IJ (2008) Lung function changes in Sprague-Dawley rats after prolonged inhalation exposure to silver nanoparticles. Inhal Toxicol 20(6):567–574

Suresh U, Murugan K, Benelli G, Nicoletti M, Barnard DR, Panneerselvam C, Kumar PM, Subramaniam J, Dinesh D, Chandramohan B (2015) Tackling the growing threat of dengue: *Phyllanthus niruri*-mediated synthesis of silver nanoparticles and their mosquitocidal properties against the dengue vector *Aedes aegypti* (Diptera: Culicidae). Parasitol Res 114:1551–1562

Sylvestre JP, Kabashin AV, Sacher E, Meunier M, Luong JH (2004) Stabilization and size control of gold nanoparticles during laser ablation in aqueous cyclodextrins. J Am Chem Soc 126(23):7176–7177

Thakore S, Rathore PS, Jadeja RN, Thounaojam M, Devkar RV (2014) Sunflower oil mediated biomimetic synthesis and cytotoxicity of monodisperse hexagonal silver nanoparticles. Mater Sci Eng C 44:209–215

Tran QH, Le AT (2013) Silver nanoparticles: synthesis, properties, toxicology, applications and perspectives. Adv Nat Sci Nanosci Nanotechnol 4(3):033001

Vanaja M, Annadurai G (2013) *Coleus aromaticus* leaf extract mediated synthesis of silver nanoparticles and its bactericidal activity. Appl Nanosci 3:217–223

Velayutham K, Rahuman AA, Rajakumar G, Roopan SM, Elango G, Kamaraj C, Marimuthu S, Santhoshkumar T, Iyappan M, Siva C (2013) Larvicidal activity of green synthesized silver nanoparticles using bark aqueous extract of *Ficus racemosa* against *Culex quinquefasciatus* and *Culex gelidus*. Asian Pac J Trop Med 6(2):95–101

Velmurugan P, Anbalagan K, Manosathyadevan M, Lee KJ, Cho M, Lee SM, Park JH, Oh SG, Bang KS, Oh BT (2014) Green synthesis of silver and gold nanoparticles using *Zingiber officinale* root extract and antibacterial activity of silver nanoparticles against food pathogens. Bioprocess Biosyst Eng 37:1935–1943

Velusamy P, Das J, Pachaiappan R, Vaseeharan B, Pandian K (2015) Greener approach for synthesis of antibacterial silver nanoparticles using aqueous solution of neem gum (*Azadirachta indica* L.). Ind Crop Prod 66:103–109

Venugopal K, Rather HA, Rajagopal K, Shanthi MP, Sheriff K, Illiyas M, Maaza M (2017) Synthesis of silver nanoparticles (ag NPs) for anticancer activities (MCF 7 breast and A549 lung cell lines) of the crude extract of *Syzygium aromaticum*. J Photochem Photobiol B Biol 167:282–289

Verma A, Mehata MS (2016) Controllable synthesis of silver nanoparticles using neem leaves and their antimicrobial activity. J Radiat Res Appl Sci 9(1):109–115

Watson R, Oldfield M, Bryant JA, Riordan L, Hill HJ, Watts JA, de Cogan F (2022) Efficacy of antimicrobial and anti-viral coated air filters to prevent the spread of airborne pathogens. Sci Rep 12(1):2803

Yin IX, Zhang J, Zhao IS, Mei ML, Li Q, Chu CH (2020) The antibacterial mechanism of silver nanoparticles and its application in dentistry. Int J Nanomedicine 15:2555–2562

Zhang Y, Kohler N, Zhang M (2002) Surface modification of superparamagnetic magnetite nanoparticles and their intracellular uptake. Biomaterials 23(7):1553–1561

Zhang XF, Liu ZG, Shen W, Gurunathan S (2016) Silver nanoparticles: synthesis, characterization, properties, applications, and therapeutic approaches. Int J Mol Sci 17(9):1534

Zorraquín-Peña I, Cueva C, Bartolomé B, Moreno-Arribas MV (2020) Silver nanoparticles against foodborne bacteria. Effects at intestinal level and health limitations. Microorganisms 8(1):132

Genetic Enhancement Strategies for Medicinal Plants: Exploring Mutagenesis

21

Manish Sharma, Shailaja Punetha, B. K. Rajkumar, Vineet Kaswan, and Pooja Oli

Abstract

Mutagenesis, a powerful technique in plant breeding, holds significant promise for enhancing the genetic diversity and therapeutic potential of medicinal plants. Medicinal plants are valued for their secondary metabolites, which serve as sources for valuable drugs. However, challenges in cultivation, environmental changes, and limited genetic diversity hinder their potential. Induced mutagenesis emerges as a solution to address these challenges, offering targeted genetic alterations through physical and chemical agents. This chapter reviews mutagenesis approaches in medicinal plants, including random mutagenesis through irradiation and chemical treatments. It also delves into the evolving landscape of targeted mutagenesis, exemplified by the revolutionary CRISPR/Cas9 system. The success stories of mutagenesis in various medicinal plants, such as black henbane, opium poppy, and isabgol, demonstrate its potential to boost desired traits like alkaloid content and yield. Despite its promise, mutagenesis techniques have limitations, including off-target effects and regulatory considerations. As the demand for medicinal plants rises due to climate change and population growth, integrating traditional breeding practices with advanced

M. Sharma
Pulses Research Station, Sardarkrushinagar Dantiwada Agricultural University, Sardarkrushinagar, Gujarat, India

S. Punetha (✉) · P. Oli
G.B. Pant National Institute of Himalayan Environment, Kosi-Katarmal, Uttarakhand, India

B. K. Rajkumar
Main Cotton Research Station, Navsari Agricultural University, Surat, Gujarat, India

V. Kaswan
College of Basic Science and Humanities, Sardarkrushinagar Dantiwada Agricultural University, Sardarkrushinagar, Gujarat, India

© The Author(s), under exclusive license to Springer Nature Singapore Pte Ltd. 2024

M. Nandave et al. (eds.), *Ethnopharmacology and OMICS Advances in Medicinal Plants Volume 1*, https://doi.org/10.1007/978-981-97-2367-6_21

mutagenesis tools offers a pathway to harness the full potential of these valuable resources.

Keywords

CRISPR-Cas9 · Targeted mutagenesis · Secondary metabolites · Bioactive compounds

Abbreviations

ODM	Oligonucleotide-directed mutagenesis
SDN	Site-directed nuclease
DSB	Double-strand break
GMO	Genetically modified organisms

21.1 Introduction

Plants are called medicinal as they are the source of several valuable drugs formulated from secondary metabolites present in them. These secondary metabolites may be extracted from the stem, root, bark, seeds, and leaves or whole of the plant. Medicinal plants play a vital role in disease prevention and treatment. Though such is the importance of medicinal plants, only a few of them are being cultivated commercially, while most of them are found in the wild (Ramawat 2019). With the advent of climate change detrimental to wild habitats and ever-increasing population pressure, demand for medicinal plants will peak, which can be fulfilled by the cultivation of medicinal plants with improved genetic makeup. Most farmers still use wild types or local varieties of medicinal plants as improved varieties are either not available or not known to them. Breeding medicinal plants is a complicated phenomenon as the production of secondary metabolites is known to alter with changing environments and the growth cycle of plants. Therefore, medicinal plants are often found difficult to breed using conventional breeding methods (Wang et al. 2020). Genetic variations can also be induced by mutagenesis and is one of the effective tools for crop improvement practices. It is the deliberate application of physical or chemical substances to intentionally cause genetic alterations or mutations in the DNA of a plant, which can be accomplished using a variety of techniques, including irradiation, chemical mutagens, or the application of certain enzymes (Udage 2021). Traditionally, mutations are induced in plants by exposing propagules to physical as well as chemical agents with mutagenic properties, bringing heritable changes in DNA sequences, and, consequently, changing the phenotype of the treated plants.

Further technological advances in molecular biology have revolutionized mutation breeding, making it highly target-specific and efficient (Tadele 2016). In order to introduce novel mutations into crop plants, there are a number of new techniques known as "new breeding techniques" (NBTs) available for targeted mutagenesis at specific locations in the plant genome (Holme et al. 2019). Mutagenesis in medicinal plants is beneficial for a variety of reasons, as detailed below:

- Mutagenesis in medical plants can result in beneficial genetic changes that improve attributes such as productivity, quality, and pest and disease resistance.
- Mutagenesis may foster the development of entirely novel molecules. New bioactive compounds with potential therapeutic uses may be found as an outcome. Such substances can be employed to create novel pharmaceuticals or herbal remedies for the treatment of a variety of ailments.
- The amount of bioactive chemicals in medicinal plants can be increased through mutagenesis, which is especially advantageous because higher concentrations of these substances frequently correspond with more decisive pharmacological action.
- Due to the diversity of medicinal plants and their ecological conditions, the breeding of these plants tends to be exceptionally complex. Mutagenesis offers a simple breeding method for these complex requirements.
- Due to their lengthy life cycles, traditional breeding techniques for medicinal plants are frequently time-consuming. On the other hand, mutagenesis can hasten the breeding process by causing genetic mutations in a brief period. As a result, improved medicinal plant varieties can evolve more quickly.
- With the advent of climate change and all of the challenges that accompany it, plant types that are adapted to certain climatic circumstances, such as drought or heat tolerance, can be developed through mutagenesis.

21.2 Mutagens and Mutagenesis Approaches

Plant mutation breeding is not a novel topic or recent approach. Long back in 1927, Muller demonstrated that X-ray treatment could generate genetic deviants in Drosophila that were indistinguishable from typically occurring phenotypes. Blakeslee, Stadler, and Goodspeed have shown the ability to obtain very high mutation rates in agricultural plants such as tobacco, maize, and barley using X-rays. Mutation breeding has since been implemented in both plants and animals. Numerous induced mutagenesis methods are available for medicinal plants. The bulk of mutations occurring are random and can be induced by physical mutagens, like fast neutron, UV, X-ray, and gamma radiations, and chemical mutagens, including ethyl methanesulfonate (EMS) N-methyl-N-nitrosourea (MNU), sodium azide, hydrogen fluoride (H.F.), etc. (Koornneef 2002). Nowadays, site-directed (targeted) mutagenesis has become an essential molecular crop improvement tool, and it is based on nucleases that create a specific double-stranded break at desired locations in the genome to modify a gene of interest within a genome directly. Techniques like transcription activator-like effector nuclease (TALEN), zinc finger nuclease (ZFN), and Clustered Regularly Interspaced Short Palindrome Repeats (CRISPR) alter the

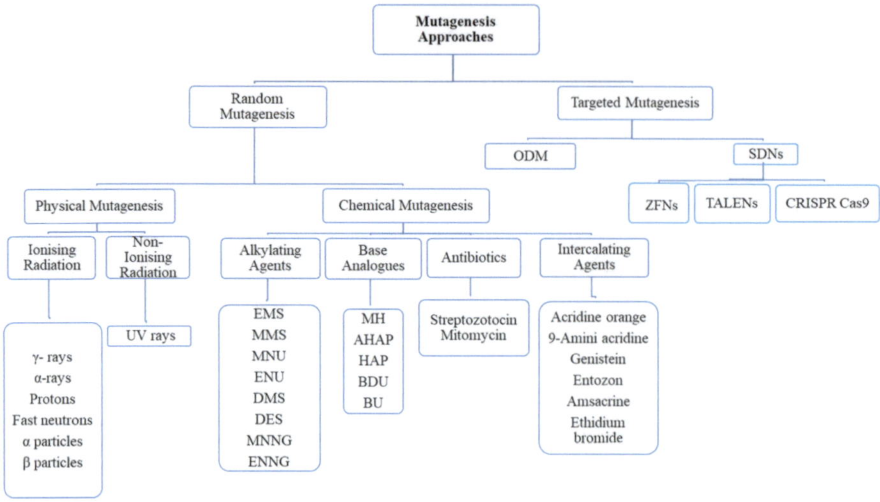

Fig. 21.1 Mutagenesis approaches and commonly used mutagens. *EMS* ethyl methanesulphonate, *MMS* methyl methanesulphonate, *MNU* 1-methyl-1-nitrosourea, *ENU* 1-ethyl-1-nitrosourea, *DMS* dimethyl sulfate, *DES* diethyl sulfate, *MNNG* 1-methyl-2-nitro-1-nitrosoguanidine, *ENNG* 1-ethyl-2-nitro-1-nitrosoguanidine, *MH* maleic hydrazide, *AHAP* 2-amino-6 N-hydroxylaminopurine, *HAP* 6-N-hydroxylaminopurine, *BDU* bromodeoxyuridine, *B.U.* 5-bromo-uracil, *ODM* oligonucleotide-directed mutagenesis, *SDN* site-directed nuclease, *ZFN* zinc finger nucleases, *TALEN* transcription activator-like (TALEN), *CRISPR* clustered regularly interspaced short palindromic repeat-associated

DNA sequence at a precise location, creating site-specific insertion or deletions (Holme et al. 2019). Different mutagenesis approaches and mutagens commonly used are detailed in Fig. 21.1.

21.2.1 Random Mutagenesis in Medicinal Plants

The two most common strategies for inducing random mutations in plants are irradiation and chemical mutagen treatment (Leitao 2012). Physical mutagenesis has emerged as one of the most successful methods for acquiring new genetic resources and correcting some intrinsic flaws in a variety of medicinal plants. Lal et al. (2020) reported that in medicinal plants, viz., Roselle (*Hibiscus Sabdariffa* L.), psyllium/isabgol (*Plantago ovata* F.), black henbane (*Hyoscyamus niger* L.), pyrethrum (*Chrysanthemum cinerariifolium*), kewanch (*Mucuna pruriens* (L.) DC.), and opium poppy (*Papaver somniferum* L.) mutation breeding can rapidly rectify inherent defects compared to other selective breeding methods.

21.2.1.1 Physical Mutagenesis in Medicinal Plants

It has been well established that traditional radiation mutation breeding techniques, which primarily use X-ray and gamma-ray applications to induce a large number of genomic mutations and hasten the production of mutant traits through energy deposition directly or indirectly onto DNA, are helpful for crop variation (Ma et al. 2021). This method allows for the initiation of desired characteristics that are either not exhibited in natural form or have been lost during the period of evolution. In this section, we will go through some of the accomplishments made possible by radiation breeding in medicinal plants.

Hyoscyamus niger, often known as black henbane (BH), has been utilized as a medicine for many years. Alkaloids such as hyoscyamine, atropine, tropane, and scopolamine can be found in the plant's leaves, seeds, and roots. It has bronchodilating, antisecretory, spasmolytic, relaxing on the urinary bladder, psychedelic, hypnotic, dilates pupil, sedative, and anti-diarrheal properties (Alizadeh et al. 2014). Gamma irradiation of 15–40 kR was employed to increase herbage yield and active constituent production. In the M2 generation of gamma irradiation (Co60 at 40 kR) offspring of *Hyoscyamus niger*, a prolific yellow flowered mutant with high tropane alkaloid content was obtained. The mutant strain Y-17 was released as var. Aela for commercial cultivation due to its better alkaloid biosynthesis and yield (Sharma et al. 1989). Aekla, a different black henbane mutant that has undergone gamma radiation (20 kR), has been made available for commercial production in India (Sharma et al. 1997). It has resulted from a biological mega-event, the unbranched (Ub) mutant of *H. niger* which differs from the average plant in as many as 15 morpho-chemical traits that are all altered at the same time.

Kewanch (*M. pruriens* L. DC.), a genus comprising roughly 33 species of climbing vines and shrubs, is a member of the Fabaceae family and is a widely recognized as medicinal plant of India, and it has long been used in traditional Ayurvedic medicine system to treat disorders such as Parkinson's disease. Pharmacological research on this plant is being done for a number of purposes, including its potential as an antidiabetic, aphrodisiac, anticancer, anti-epileptic, and antibacterial agent (Sathiyanarayanan and Arulmozhi 2007). CSIR-CIMAP is engaged in extensive mutation breeding to create the novel mutant variety with desired characters. Mutant variety CIM-Sfurity is a chemotype having an L-dopa concentration of 4–5% and high seed production of 17–25 ql/ha. The key characteristics that set this variety apart from others are its small size and highly thick inflorescence axis, the light brown color of the pods, and pods from the mutant's basal section (Dhawan et al. 2018). Another mutant variety is CIM-Nirom. In field experiments, the variety regularly produced a lot of seeds (33 ql/ha) having L-dopa content of 4.50% and seed yield of 33 q/ha that has been released for commercial cultivation. Due to its lintless pods, it is perfect for orchards, where it can be grown as a climber crop and is simple to harvest (Lal et al. 2015).

The opium poppy (*Papaver somniferum*) is the world's most ancient medicinal plants that serves as an effective model system for studies on secondary metabolism. Currently, poppies are a widespread crop farmed both legally and illegally in South America, Asia, Europe, and Oceania. Numerous benzylisoquinoline alkaloids (BIAs), including morphine, codeine, thebaine, papaverine, and noscapine, are derived from it and employed by the pharmaceutical industry and in the clandestine production of heroin (Labanca et al. 2018). The alkaloids in poppies used for food applications must be absent or barely present. Mutagenesis is a valuable technique to weaken, inactivate, or prevent the production of any secondary metabolite. By genetically transforming latex-producing "opium poppies" into latex-free "seed poppies" through mutation breeding, the LL-34 straw mutant was created. It has very few alkaloids and very little opium. Once the plant reaches the lancing stage, latex production stops in the mutant LL-34 because it lacks opium. Insignificant amounts of the primary opium alkaloids—nicotine, codeine, thebaine, papaverine, and morphine—can also be found in the straw (capsule hull). More importantly, seeds have a vegetable oil content of 52–59%, which is mainly unsaturated and advantageous for the dietary control of lipid abnormality-related diabetes and coronary heart disease. Aside from being a potential contribution to the production of edible seeds and seed oil, the mutant L.L. 34, currently known as variety Sujata, is also a low-cost, long-term answer to ending opium-related social abuses worldwide (Sharma et al. 2002).

An annual herb called Isabel (*Plantago ovata* Forsk) has a number of pharmacological and therapeutic characteristics. It is commonly used in a variety of medicines due to its medical properties, such as mucilage, super disintegrants, gelling agent, and suspending agent, as well as its pharmacological activities, such as anti-diarrheal, anti-constipation, wound healer, hypocholesterolemic, and hypoglycemia (Sutradhar et al. 2022). Mutation breeding has been fruitful and resulted in the development of varieties Niharika and Mayuri by CSIR-CIMAP Lucknow, Uttar Pradesh (India). Using mutagenesis of 20 kR radiation to reorganize the genetic background, M 20–22 genotype having more panicles per plant and high seed yield was ultimately made available as the commercial variety Niharika (Lal et al. 1998) for north Indian conditions. Another isabgol mutation breeding effort was launched in an inbred line to promote heterogeneity in its genetic structure and select suitable polygenic mutants with earliness, high seed, and husk output. The most promising mutant, CIMAP/S 5/FEA 5, was obtained at a dosage of 20 kR. The elite strain CIMAP/S 5/FEA 5 consistently beats the control variety Niharika in terms of seed yield per unit area. The superior strain was given the name "Mayuri" (Lal et al. 2007).

21.2.1.2 Chemical Mutagenesis in Medicinal Plants

Even though there are many mutagenic chemicals, only a few have been studied and utilized in plant experimental mutagenesis and plant mutation breeding (Leitao 2012). Among them, too, only a tiny subset of alkylating chemicals has found widespread use in medicinal plant development. Here, we will discuss the use of chemical agents in creating variation among different medicinal plants.

Chicory (*C. intybus*) is a medicinal plant that can withstand high temperature throughout the vegetative as well as reproductive stages of growth. Important phytochemicals are found throughout the plant, but the root contains the majority of them. Chicory is known as a multipurpose medicinal herb because of its antihepatotoxic, antiulcerogenic, anticancer, and anti-inflammatory effects, among others (Street et al. 2013). Khan et al. (2009) investigated the genotoxicity of methyl methanesulfonate on chicory gametic cells. Progenies were produced from seeds treated with four different concentrations of MMS aqueous solutions: 0.04, 0.06, 0.08, and

0.10%. Stickiness, univalents, multivalents, and premature separation of chromosomes were discovered during metaphase; bridges, laggards, and polyads were discovered during anaphase and telophase. Such aberrations were dosage-dependent and increased with increasing mutagen doses. MMS can, therefore, be employed as a possible mutagen to induce genetic variability in this plant.

Catharanthus roseus (L.) *G. Don* has a long history of being used as a folk cure in many nations, including South Africa, China, India, Mexico, and Malaysia, where it is used to treat diabetes problems (Tiong et al. 2013). It is one of the therapeutic plants that has undergone chemical mutagenesis. When exposed to the powerful mutagen ethyl methane sulfonate (EMS), mutant cell lines with fast growth and high indole alkaline content were produced (Xiusheng et al. 2004). Additionally, several studies have revealed that combining radiation and chemical therapy can raise the frequency of mutations in medicinal plants (Wang et al. 2020).

21.3 Limitations of Random Mutagenesis

In order to make any progress, breeding objectives are particular and target-oriented, which is why random mutagenesis is no longer the method of choice for plant breeders (Chaudhary et al. 2019). Further, randomly induced mutation suffers from the non-reproducibility of results. It follows that even following the same mutation dose and procedures might not produce the same mutation events because induced mutations are random occurrences. Working with substantial population numbers and applying statistical probability are two ways to reduce this uncertainty (Mba et al. 2010). Many medicinal plants have dormancy in their seeds, so prior to receiving treatments for induced mutagenesis, it's critical to break out of dormancy. Before treating seeds, preliminary seed viability tests to determine whether or not the seeds are dormant are typically advised to prevent other underlying causes from confounding the estimations of radiosensitivity (Kodym and Afza 2003). Radioactivity may be harmful to health because it can cause cancer and mutations. Thus, only qualified and authorized personnel should operate radioactive sources. Although both physical and chemical mutagenesis have advantages and limitations in plant breeding, it may be able to fulfill the full potential of these techniques by selecting optimum combinations and doses. Furthermore, to improve breeding efficiency, particular emphasis should be paid to the combination of mutant breeding with a helping hand from molecular markers and tissue culture technologies.

21.4 Targeted Mutagenesis in Medicinal Plants

The traditional mutation breeding approaches are time-consuming and create imprecise mutations, producing results with poor predictability and problems, and unexpected variations are now paving the way for targeted gene editing technology to precisely modify the locus information of a genome and achieve targeted gene deletion, insertion, or replacement (Holme et al. 2019). New breeding technique (NBT) mutation tools consist of a number of novel technologies for the introduction of discrepancy into crop plants. In this section, we will concentrate on the two strategies used to generate mutations at predetermined sites in the plant genome, namely, oligonucleotide-directed mutagenesis (ODM) and, in particular, the site-directed nuclease (SDN) tools.

Both ODM and SDN tools are powerful molecular biology techniques used to introduce specific and intentional changes into a gene of interest. The basic principles following the precision of the ODM and SDN mutation tools are very different. ODM utilizes a short, single-stranded oligonucleotide ranging in length from 20 to 100 nucleotides that are designed identical to a similar sequence in the plant genome to introduce the desired mutation into a stretch of DNA. The oligonucleotides interact with the complementary DNA sequence in the genome, causing mismatches that are corrected by the cell's DNA repair machinery. As a result, the desired modification is made at a precise location in the genome. SDNs are tools that may be programmed to recognize and cut at specified places in a genome, resulting in a double-strand break (DSB) at the region of interest. By using the cell's DSB repair mechanisms, the DSB makes it possible to create many forms of mutations (Voytas 2013). SDN1 and SDN2 are frequently used to refer to the various mutational types produced by the two main repair processes, homologous recombination and non-homologous end joining.

The three primary SDN approaches are ZFN (Bibikova et al. 2003), TALENs (Bogdanove and Voytas 2011), and CRISPR/Cas9 (Jinek et al. 2012). Both zinc finger nucleases (ZFNs) and TAL effector nucleases (TALENs) can be used to mutate genomes at specific loci. Still, these systems need the presence of two distinct DNA binding proteins, each of which has a C-terminal FokI nuclease module flanking a sequence of interest. However, because of their complex design and labor-intensive synthesis of particular DNA-binding proteins for each target gene, these technologies have not been widely considered by plant scientists. CRISPR technology has emerged as one of the new plant breeding techniques (NPBTs) alongside ZFNs and TALENs. By allowing targeted cleavage of genomic DNA under the control of a programmable short noncoding RNA, the CRISPR/Cas system enables both homology-directed repair (HDR) and non-homologous end joining (NHEJ) methods to modify genes. Contrarily, the CRISPR/Cas system is flexible and only needs a single Cas9 nuclease that may be modified through sgRNA engineering (Belhaj et al. 2013). All of these methods allow for the introduction of plant genome alterations that are identical from those introduced through traditional breeding and chemical and physical mutagenesis but with more specificity and precision.

21.4.1 CRISPR/Cas9-Induced Mutagenesis in Medicinal Plants

CRISPR/Cas9 is a precision-based genome editing technique, and its use in therapeutic plants is relatively new. Li et al. used CRISPR/ Cas9 in *Salvia miltiorrhiza*, a traditional Chinese medicinal herb, to accurately knock down the committed diterpene synthase gene (SmCPS1), which is essential in tanshinone production. *S. miltiorrhiza* has been a suitable resource for medicinal plant genetics and epigenetics study due to its effective genetic transformation system, short life cycle and simple micropropagation procedures, and effective genetic transformation system. The inclusion of lipid-soluble chemicals known as tanshinones and water-soluble phenolic acids such as lithospermic acid, salvianolic acid, and rosmarinic acid contributes significantly to its pharmacological effectiveness. Using CRISPR/Cas9, 3 homozygous and 8 chimeric mutants were obtained from 26 independent transgenic hairy root lines by *Agrobacterium rhizogenes*-mediated transformation. Tanshinones, particularly cryptotanshinone, tanshinone I, and tanshinone IIA, were utterly absent in homozygous mutants without having any effect on other phenolic acid metabolites, according to a metabolomic investigation based on LC-qTOF-MS and Q-TRAP-LC-MS/MS. This study paved the way for large-scale genome editing in *S. miltiorrhiza*, which is crucial for elucidating secondary metabolite pathways, improving quality, and increasing yields. It also demonstrates that CRISPR/Cas9-mediated transformation by *Agrobacterium rhizogenes* is a fast and effective technique for modifying the genome in *S. miltiorrhiza*.

Apart from *S. miltiorrhiza*, CRISPR/Cas9-mediated targeted mutagenesis has been accomplished in opium poppy (*Papaver somniferum* L.) for 3-hydroxyl-N-methylcoclaurine 4-O-methyltransferase gene in BIAs biosynthetic pathway (Alagoz et al. 2016), in comfrey (*Symphytum officinale* L. Boraginaceae) for homospermidine synthase gene (HSS) in P.A. biosynthetic pathway (Zakaria et al. 2021) and *Cannabis sativa* for phytoene desaturase gene (Zhang et al. 2021). These illustrations show how CRISPR-Cas9 technology can alter the transcriptional expression of genes from the source by performing targeted gene deletion, insertion, replacement, and chromosomal recombination at the DNA level (Guo et al. 2022). In order to improve quality and breed potential genotypes for medicinal plants, the CRISPR-Cas9 system must be applied to investigations of the function of genes and the regulation of metabolic networks in medicinal plants.

21.4.1.1 Limitations of Target Mutagenesis

NBT mutations have few limitations when compared to traditional mutagenesis. The primary distinction between regular mutation breeding and NBT mutations is that NBT mutation tools must be supplied into the cells, which necessitates the use of well-established laboratory methods that are relatively expensive. One frequently expressed concern regarding these tools is off-target mutations, i.e., mutations will be created in regions in the genome where the tools were not designed to mutate. Both ODM and all SDN tools can induce these off-target mutations, but due to a less specific binding capacity, the CRISPR/Cas9 tool has a higher frequency (Zischewski et al. 2017). Despite the off-target mutations created by NBT mutagenesis, the SDN tools generate non-planned mutations at significantly lower frequencies than conventional mutation breeding. The existence of a protospacer adjacent motif (PAM) sequence is necessary for the Cas9 protein to bind and cut DNA, which limits the number of target sites available for genome editing (Holme et al. 2019).

Furthermore, CRISPR/Cas9 can only be used on a few model plants that have complete genomic information and efficient genetic transformation mechanisms. Delivering CRISPR/Cas9 components into plant cells can be difficult, especially in crops that are resistant to transformation. Crops edited with CRISPR/Cas9 technology are deemed genetically modified organisms (GMOs) in some countries and are subject to regulatory limitations, and organisms transformed using these new procedures are not excluded from current EU GMO legislation (Holme et al. 2019). Despite these limitations, CRISPR/Cas9 technology has enormous crop improvement potential. With significant advances in sequencing technologies, genetic transformation techniques and improved regeneration protocols, and many other related techniques, CRISPR-Cas9 is expected to transform medicinal plant breeding in the near future.

21.5 Conclusion

The monetary value of medicinal plants far exceeds that of any other crop plant due to the tremendous benefits they provide to both people and the environment. However, due to the widespread neglect of medicinal plant research and development, there is a very little information about their genetic makeup, heterozygosity, growth cycle, and self-incompatibility, which has slowed breeding attempts. The breeding of these plants tends to be incredibly complex due to the diversity of medicinal plants and the ecological conditions in which they flourish (Wang et al. 2020). Therefore, a focused effort should be made to resolve this issue. Mutation breeding using either physical, chemical, or enzymatic agents has been expansively used by medicinal plant breeders, as we have seen throughout this chapter. Mutagenesis in medicinal plants has come a long way with advancements in science and has improved its efficiency and robustness. Given the diversity of medicinal plant species, the complexities of environmental conditions, and the global distribution of medicinal plants, it is critical that research efforts be bolstered in order to generate new and improved therapeutic plant varieties, and mutagenesis can be one such practical approach for achieving the same (Ramawat and Arora 2021). While contemporary biotechnology will result in innovative advancements, traditional mutation breeding methods will still need to be integrated with current NBT mutation tools to advance the development and application of medicinal plant resources.

References

Alagoz Y, Gurkok T, Zhang B, Unver T (2016) Manipulating the biosynthesis of bioactive compound alkaloids for next-generation metabolic engineering in opium poppy using CRISPR-Cas 9 genome editing technology. Sci Rep 6(1):30910

Alizadeh A, Moshiri M, Alizadeh J, Balali-Mood M (2014) Black henbane and its toxicity–a descriptive review. Avicenna J Phytomed 4(5):297

Belhaj K, Chaparro-Garcia A, Kamoun S, Nekrasov V (2013) Plant genome editing made easy: targeted mutagenesis in model and crop plants using the CRISPR/Cas system. Plant Methods 9(1):1–10

Bibikova M, Beumer K, Trautman JK, Carroll D (2003) Enhancing gene targeting with designed zinc finger nucleases. Science 300(5620):764–764

Bogdanove AJ, Voytas DF (2011) TAL effectors: customizable proteins for DNA targeting. Science 333(6051):1843–1846

Chaudhary J, Deshmukh R, Sonah H (2019) Mutagenesis approaches and their role in crop improvement. Plan Theory 8(11):467

Dhawan SS, Singh SK, Lal RK, Mishra A, Gupta P, Singh R, Singh M, Shankar K, Singh S (2018) Registration of an advance breeding line of Kewanch with early and high yield of seeds and L-DOPA: CIM–Sfurti. J Med Aromatic Plant Sci 40:67–71

Guo M, Chen H, Dong S, Zhang Z, Luo H (2022) CRISPR-Cas gene editing technology and its application prospect in medicinal plants. Chin Med 17(1):33

Holme IB, Gregersen PL, Brinch-Pedersen H (2019) Induced genetic variation in crop plants by random or targeted mutagenesis: convergence and differences. Front Plant Sci 10:1468

Jinek M, Chylinski K, Fonfara I, Hauer M, Doudna JA, Charpentier E (2012) A programmable dual-RNA–guided DNA endonuclease in adaptive bacterial immunity. Science 337(6096):816–821

Khan Z, Gupta H, Ansari MYK, Chaudhary S (2009) Methyl methanesulphonate induced chromosomal variations in a medicinal plant Cichorium intybus L. during microsporogenesis. Biol Med 1(2):66–69

Kodym A, Afza R (2003) Physical and chemical mutagenesis. Methods Mol Biol 236:189–203

Koornneef M (2002) Classical mutagenesis in higher plants. Mol Plant Biol 1:1–11

Labanca F, Ovesna J, Milella L (2018) Papaver somniferum L. taxonomy uses and new insight in poppy alkaloid pathways. Phytochem Rev 17:853–871

Lal RK, Sharma JR, Misra HO (1998) Development of a new variety Niharika of Isabgol (*Plantago ovata*). J Med Aromat Plant Sci 20:421–422

Lal RK, Singh N, Misra HO, Sharma JR, Bahl JR, Shasany AK and Khanuja SPS (2007) A distinct high yielding, early maturing Psyllium (*Plantago ovata* F.) variety Mayuri as described and illustrated with developmentally expressed pigment marker. Patent number, U.S. PP17505P3

Lal RK, Singh S, Chandra R, Sarkar S, Singh S, Dhawan OP, Gupta AK, Singh VR, Kumar B, Gupta MM, Dhawan SS (2015) Registration of new variety CIM–Nirom: a hairless pods, high seed and L-dopa yielding variety of Kewanch (*Mucuna pruriens* L.). J Med Aromat Plant Sci 37:57–60

Lal RK, Chanotiya CS, Gupta P (2020) Induced mutation breeding for qualitative and quantitative traits and varietal development in medicinal and aromatic crops at CSIR-CIMAP, Lucknow (India): past and recent accomplishment. Int J Radiat Biol 96(12):1513–1527

Leitao JM (2012) Chemical mutagenesis. In: Plant mutation breeding and biotechnology. CABI, Wallingford, pp 135–158

Ma L, Kong F, Sun K, Wang T, Guo T (2021) From classical radiation to modern radiation: past, present, and future of radiation mutation breeding. Front Public Health 9:768071

Mba C, Afza R, Bado S, Jain SM (2010) Induced mutagenesis in plants using physical and chemical agents. In: Plant cell culture: essential methods, vol 20. Wiley, New York, pp 111–130

Muller HJ (1927) Artificial transmutation of the gene. Science 66(1699):84–87

Ramawat KG (2019) An introduction to biodiversity and chemotaxonomy. In: Biodiversity and chemotaxonomy. Springer, Cham, pp 1–14

Ramawat KG, Arora J (2021) Medicinal plants domestication, cultivation, improvement, and alternative technologies for the production of high value therapeutics: an overview. In: Medicinal plants: domestication, biotechnology and regional importance. Springer, Cham, pp 1–29

Sathiyanarayanan L, Arulmozhi S (2007) Mucuna pruriens Linn.-A comprehensive review. Pharmacogn Rev 1(1):157–162

Sharma JR, Lal RK, Misra HO, Gupta MM, Ram RS (1989) Potential of gamma radiation enhancing the biosynthesis of tropane alkaloids in black henbane (*Hyoscyamus niger* L.). Euphytica 40:253–258

Sharma JR, Lal RK, Misra HO, Gupta MM (1997) Register of new cultivars and genetoypes: aekla henbane. J Med Aromat Plant Sci 19(2):443–445
Sharma JR, Lal RK, Gupta AP, Misra HO (2002) Origin and utility of the first non-narcotic var. Sujata of opium poppy. J Med Aromat Plant Sci 24:148–148
Street RA, Sidana J, Prinsloo G (2013) Cichorium intybus: traditional uses, phytochemistry, pharmacology, and toxicology. Evid Based Complement Alternat Med 2013:579319
Sutradhar K, Tahsin MR, Ashrafi S, Ferdousy S, Akter T, Aktar F, Chowdhury JA, Chowdhury AA, Kabir S, Amran MS (2022) Phytochemical, biochemical and pharmacological properties of *Plantago ovata* (Ispaghula husk)–a review. Dhaka Univ J Pharm Sci 21(2):231–243
Tadele Z (2016) Mutagenesis and TILLING to dissect gene function in plants. Curr Genomics 17(6):499–508
Tiong SH, Looi CY, Hazni H, Arya A, Paydar M, Wong WF, Cheah SC, Mustafa MR, Awang K (2013) Anti-diabetic and antioxidant properties of alkaloids from *Catharanthus roseus* (L.) G. Don. Molecules 18(8):9770–9784
Udage AC (2021) Introduction to plant mutation breeding: different approaches and mutagenic agents. J Agric Sci 16(3):466
Voytas DF (2013) Plant genome engineering with sequence-specific nucleases. Annu Rev Plant Biol 64:327–350
Wang W, Xu J, Fang H, Li Z, Li M (2020) Advances and challenges in medicinal plant breeding. Plant Sci 298:110573
Xiusheng Z, Rongtao Z, Lan C, Yong W, Shufang WN (2004) Mutation in *Catharanthus roseus* induced by EMS. *Zhong cao Yao*=. Chin Tradit Herb Drug 35(11):1293–1296
Zakaria MM, Schemmerling B, Ober D (2021) CRISPR/Cas9-mediated genome editing in comfrey (*Symphytum officinale*) hairy roots results in the complete eradication of pyrrolizidine alkaloids. Molecules 26(6):1498
Zhang X, Xu G, Cheng C, Lei L, Sun J, Xu Y, Deng C, Dai Z, Yang Z, Chen X, Liu C (2021) Establishment of an agrobacterium-mediated genetic transformation and CRISPR/Cas9-mediated targeted mutagenesis in hemp (*Cannabis sativa* L.). Plant Biotechnol J 19(10):1979–1987
Zischewski J, Fischer R, Bortesi L (2017) Detection of on-target and off-target mutations generated by CRISPR/Cas9 and other sequence-specific nucleases. Biotechnol Adv 35(1):95–104

Intellectual Property Rights Related to Medicinal Plants

22

Irmanida Batubara, Shadila Fira Asoka, I Ketut Mudite Adnyane, and Wisnu Ananta Kusuma

Abstract

Natural ingredients now have quite a high prestige because people are starting to adopt a "back to nature" lifestyle. This lifestyle also influences how to treat illnesses, so people choose medicinal plants available in the surrounding area. This attracts the attention of individuals/companies to utilize these natural resources. Unfortunately, there are many cases of exploitation of medicinal plants, so the existence of these plants has decreased drastically. In addition, there are threats of biopiracy and violations of customary law associated with the massive utilization of these medicinal plants. Therefore, legal protection and rights are needed for this matter. Intellectual property rights protect medicinal plants, indigenous peoples, individuals, or companies from losses resulting from massive exploitation. The positive trend in society shows that awareness about intellectual property rights related to medicinal plants is increasing. There is still a need to understand IPR pertaining to medicinal plants, starting from the cultivation process, extraction, standardization, and formula to become a product. This chapter will discuss and summarize intellectual property rights related to medicinal plants.

I Batubara (✉) · S. F. Asoka
Tropical Biopharmaca Research Center—IPB University, Bogor, Indonesia

Department of Chemistry, Faculty of Mathematics and Natural Sciences, IPB University, Bogor, Indonesia
e-mail: ime@apps.ipb.ac.id

I. K. M. Adnyane
School of Veterinary Medicine and Biomedical Sciences, IPB University, Bogor, Indonesia

W. A. Kusuma
Tropical Biopharmaca Research Center—IPB University, Bogor, Indonesia

Department of Computer Sciences, Faculty of Mathematics and Natural Sciences, IPB University, Bogor, Indonesia

© The Author(s), under exclusive license to Springer Nature Singapore Pte Ltd. 2024
M. Nandave et al. (eds.), *Ethnopharmacology and OMICS Advances in Medicinal Plants Volume 1*, https://doi.org/10.1007/978-981-97-2367-6_22

Keywords

Geographic indication · Health · Patent · Traditional knowledge

22.1 Introduction

Medicinal plants have been influential in human history as a source of natural medicine that has been used for thousands of years. As an integral part of biodiversity, medicinal plants have great potential to solve various human health problems. The natural ingredients in the form of bioactive compounds contained therein are very beneficial for human health and well-being. Protection of biodiversity and traditional knowledge, including intellectual property rights, is becoming increasingly important to ensure sustainability. Especially at this time, there is much exploitation of plants because there is no clarity and awareness of rules or laws that protect plants. Many cases have been of exploitation, claiming, and violation of plants from upstream to downstream. These cases often result in natural damage, even extinction of a variety. Another case is the claim of traditional medicine so that it can be commercialized but does not have permission from the local community. These cases can be minimized by the rules and protection of rights in the form of intellectual property rights.

Intellectual property (IP) protection for medicinal plants has an essential role in maintaining the sustainability of natural resources, advancing science, and encouraging innovation in medicine and pharmaceuticals. The importance of IP protection for medicinal plants is becoming increasingly evident due to several key factors: (1) Research and development in therapeutic plant science has resulted in the discovery of unique compounds with significant medicinal potential and economic value. IP protection ensures that inventors of these compounds receive legal recognition and benefit from the downstream results of their research. (2) IP protection also plays a role in encouraging natural resource conservation. With an increase in the economic value of medicinal plants, the community will be more motivated to protect the natural habitat of these plants in order to remain sustainable. IP protection can also promote collaborative research and innovation in developing plant-based medicines.

22.2 Intellectual Property Right

Intellectual property is the work of thought from a person/group. Intellectual property rights are divided into traditional/private and non-traditional/communal. According to the World Intellectual Property Organization (WIPO), conventional property rights are intellectual abilities that have roots, are developed, and are still traditional. Conventional property rights are divided into copyright and industrial property rights. Traditional intellectual property comprises patents, copyrights, industrial design rights, trademarks, trade secrets, geographical indications, and plant variety rights (Fig. 22.1). Copyrights include the rights over books and other writings for a minimum of 50 years after the author's death (Baby and Suriyapraakash

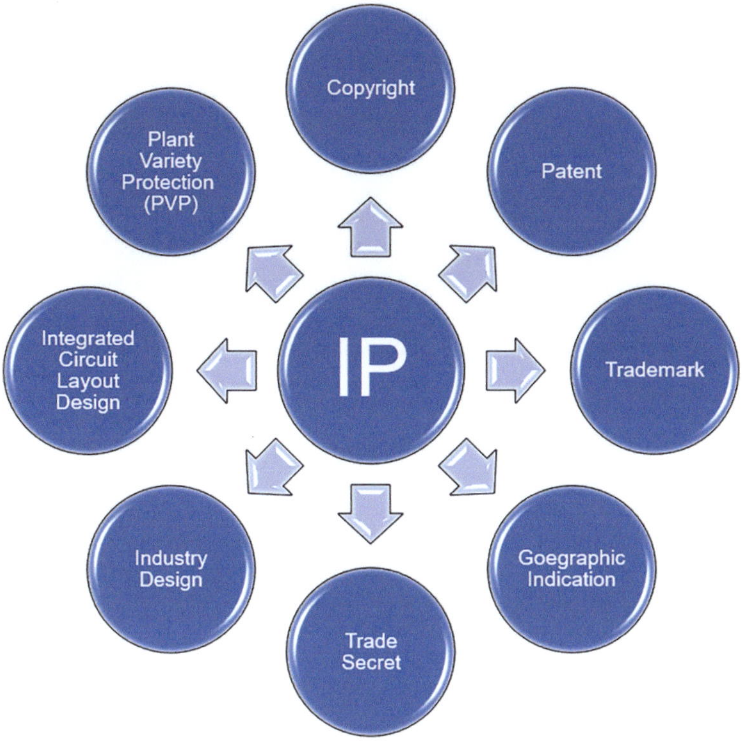

Fig. 22.1 Types of traditional intellectual property rights

2021). Industrial property is intangible rights connected with agriculture, commerce, and industry. According to the Paris Convention, Article 1(3), it is going to be interpreted broadly to encompass not just classic business and industry but also the extractive industries and all manufactured or natural goods. Patents are legal documents issued by the local government to protect an invention from being commercialized where the patent was issued (Baby and Suriyapraakash 2021). According to World Intellectual Property, a trade secret is an IPR on confidential information that may be sold or licensed. Industrial design differs from industrial property because industrial design refers more to the aesthetic aspects of invention, such as shapes, colors, patterns, and appearance. A trademark is a symbol, word, or phrase distinguishing the goods or services offered by a specific person or organization. Signing for specific commodities from a particular geographical origin with a reputation is a geographical indication. Plants with new varieties can be propagated and protected with plant variety protection IPR. Communal property rights refer to the intellectual property rights held collectively by a community or a traditional society divided into genetic resources, traditional knowledge, and folklore (Putri et al. 2022). Types of communal rights can be seen in Fig. 22.2. In addition, many products and things derived from plants are still vulnerable to exploitation and piracy. An example is the method of extracting plants and synthesizing active

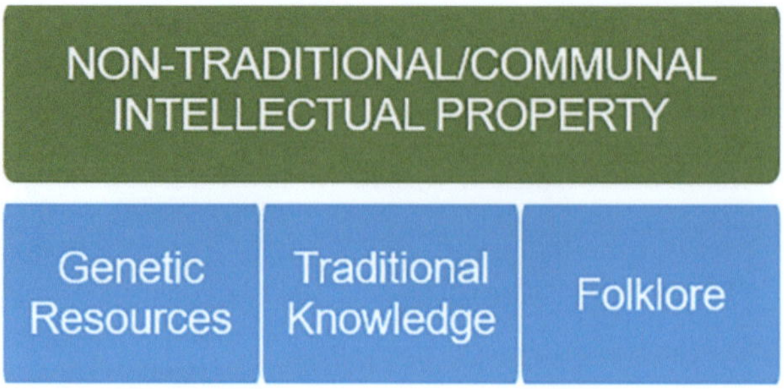

Fig. 22.2 Types of communal intellectual property rights

compounds in plants. Publications and books discussing IPR related explicitly to medicinal plants are still limited.

Based on the requirements for geographical indications, institutions that represent communities in a geographical area that process goods and products from natural resources, industrial products and handicrafts, and geographical indications in Indonesia are also a collective right owned by Indonesian law. Property rights and communal systems regulated in the TRIPS Agreement, such as genetic resources, traditional knowledge, and traditional cultural expressions. Applications for geographical indications can also come from district/city or provincial governments. Apart from the geographical location and tropical climate which are natural resources for herbal medicine products, herbal medicine products are one of the commodities that have the potential to be protected through the Indonesian Geographical Indication system because traditional knowledge regarding herbal medicine products has been used for generations since our ancestors. Indonesian people are very familiar with ethnomedicine which is practiced by various ethnic groups from various tribes in Indonesia. With the COVID-19 pandemic, diagnosis and treatment, as well as several recent findings regarding children suffering from acute kidney failure due to long-term consumption of synthetic medicinal products, ethnomedicine is essentially a national or state asset that must be protected, preserved, and developed to provide benefits, positive for society.

There is a lengthy history of medicinal plant study, particularly when it comes to determining a plant's quality. Using taste, smell, and appearance as bodily senses, the initial strategies were organoleptic. These eventually gave rise to increasingly sophisticated instrumental approaches. While many nations have their own ancient medical practices, China is currently at the forefront of publications devoted to the examination of medicinal plants and parts of their pharmacopoeia. These publications include monographs that provide guidance on the kind of analysis that ought to be carried out. This usually means that manufacturers need to have access to evermore sophisticated instruments. Numerous facets of analytical analysis have advanced, most notably the hyphenation and development of chromatographic and spectroscopic techniques. Metabolomics has been made possible by the ability to process data using multivariate analysis software. This has increased our understanding of the many chemical compound variations found in medicinal plants and given us more confidence regarding the quality of the plants and medicines as well as their suitability for clinical research. Thanks to technological advancements, it is now possible to classify and analyze plants efficiently and identify toxins and contaminants at low concentrations. However, technological advances cannot provide us with all the answers we need to deliver high-quality herbal medicines, and the more traditional techniques of assessing quality remain as important today (Fitzgerald et al. 2020).

Based on current pharmacological review, several of the traditional combination formulae seem to have a solid scientific foundation. During the production (decoction) of a prescribed herbal formula, significant chemical changes occurred. For instance, the chemical interactions between the substances resulted in the considerable reduction of some hazardous chemicals and the generation of new active molecules. Several combination formulations had pharmacological outcomes that were noticeably superior than those of the individual herbal medications included in the formula. These results imply that the regulatory and drug screening processes now in place will not be suitable for the development of a botanical treatment including a class of phytochemicals. A vital part of treating illness is the synergistic interaction of chemical substances. Assuming we take a holistic and dynamic view of a diseased state, we may see that numerous biological systems within the human body interact with one another and that these interactions alter as the disease gets better or worse. It might not make sense or be physically possible to treat such an illness with a single pharmaceutical substance. Combination formulas may become the therapeutics of choice in the future due to the synergistic effect and dynamic adjustment achieved by the multiple ingredients that will restore the balance of an imbalanced or diseased human body (Jia et al. 2004).

National borders and territories have different legal features pertaining to the protection of intellectual property rights. The public's right of access to these inventions and the creators' moral and financial rights are synchronized by national federal laws protecting intellectual property. Article 27 of the United Nations Universal Declaration of Human Rights (UDHR 1948) advocates for the rights of participation and protection in the progress of culture, the arts, and science. International treaties provide consistency in the protection of intellectual property rights across country legal systems. The World Intellectual Property Organization (WIPO) was established by the WIPO Convention and went into effect in 1970. The WIPO Convention is one of the 26 treaties that this WIPO oversees. The most extensive framework for patents, trademarks, industrial designs, utility models, service marks, trade names, geographical indications, and the prevention of unfair competition was established in 1883 with the adoption of the Paris Convention.

One of the current concerns is the role of medicinal plants, which is still debated whether they include intellectual property. On the other hand, there are already many products derived from medicinal plants that have patents. The preparations and product formulas already have copyrights, patents, and trademarks. Trade secrets and trademarks are the most popular forms of IPR for herbal medicines. According to Overwalle (2006), medicinal plants can have a patent protection type IPR but with specific criteria. Genetically modified medicinal plants, plant cells, plant seeds, modification methods related to medicinal plants, chemical compounds isolated from medicinal plants, and embedded chemical compounds are patentable. Several things are still related to medicinal plants but cannot be IPR. Wild plants and traditionally bred medicinal plant varieties are not patentable. However, there are still many things that have not been discussed regarding the IPR regarding medicinal plants. Therefore, this paper will discuss IPR related to medicinal plants and its derivatives.

22.3 Medicinal Plant

Plants worldwide are available in various forms, including multiple forms of leaves, stems, flowers, roots, and others, which are then grouped into multiple families (Ros 1999). The diversity of plants provides benefits for a variety of human life, such as food, clothing, building materials, and medicinal materials. Plants used for treatment are known as medicinal plants. The availability of plants around humans is the reason humans use these plants. Some indigenous peoples still use plants as medicine, known as traditional medicinal plants.

On the other hand, with the development of science and technology, research on developing plants as ingredients for treatment continues to grow, which encourages the discovery of plant species previously unused for certain diseases to become materials for preventing and treating diseases. Furthermore, the discussion will be divided into two parts, namely, traditional medicinal plants and medicinal plants, as a result of new findings using research.

22.3.1 Traditional Medicinal Plant

Indigenous people still use traditional medicinal customs in an area, and knowledge about this treatment is called indigenous knowledge (Alikodra and Sudarmonowati 2022). Traditional medicine still uses plants that grow in the surrounding area and only requires simple preparations. Many conventional treatments using medicinal plants are still used today. Several books and evidence from the past have recorded the types, parts, and ways traditional communities use medicinal plants. Most of the widely known traditional medicine comes from Asia. In China, traditional medicine is known as traditional Chinese medicine (TCM); in Japan, it is known as Kampo; in India, it is known as Ayurveda; in the Greco-Arabian system of medicine, it is known as Unani (Nikhat and Fazil 2020); in Tibet it is known as Tibetan medicine; and in Indonesia it is known as herbal medicine and others.

- **Traditional Chinese medicine,** sometimes referred to as yin-yangism, emphasizes the dynamic processes of material structure, which is comparable to the humoral theory of ancient Greece and Rome. The Huangdi Neijing (Canon of the Yellow Emperor's Inner) and the Compendium of Materia Medica, which contains a variety of herbal medicines, are two texts that have referenced TCM medication, gua sha, acupuncture, cupping therapy, massage (tui na), die-da (bone strengthening), qigong (exercise), and food therapy. Records from these books regarding the use of plants have been made in the form of databases such as the Traditional Chinese Medicine Integrated Database (TCMID), the Traditional Chinese Medicine Information Database (TCM-ID), and others (Li et al. 2022).

- **Kampo** is a traditional Japanese medicine. The root of Kampo is traditional Chinese medicine, which also uses medicinal plants. The oldest written record focusing on the therapeutic uses of plants is the Shennong Ben Cao Jing (first century BC), classifying 365 plant species. Furthermore, Japanese doctors developed their own views based on own knowledge to produce herbal concoctions in standard proportions. Currently, Kampo medicines are produced by various manufacturers using the ministry's standardized methodology and are also prescribed by doctors. The new Kampo drugs were evaluated using modern techniques to assess their mechanism of action.
- **Ayurveda** and Unani are used in India. Ayurveda is a system of medicine based on history in the Indian subcontinent that uses plants (including their oils), meditation, yoga, and massage. Most Ayurvedic medicine comes from plants, the roots, leaves, fruit, bark, or seeds. The principle of Ayurveda is the balance of dosha, which consists of three elements: vāta, pitta, and kapha. The ancient works on Ayurvedic medicine are from the three primary early texts of Ayurveda (Charaka Samhita, Sushruta Samhita, and Bhela Samhita).
- **Unani** medicine is based on at least five books, i.e., Al-Qanoon fil Tib (The Law in Medicine) of Ibn Sina (Latinized as Avicenna, 980–1035 CE), Kitab al-Hawi (The Comprehensive Book of Medicine) of Zakariya Razi (Latinized as Rhazes, 865–925 CE), Kitab al-Kulliyat (The Complete Book on Medicine) by Ibn Rushd (Latinized as Averroes,1126–1198 CE), Kitab al-Mansoori (Book dedicated to Caliph Mansoor) by Zakariya Razi (865–925 CE), and Kitab al-Mukhtarat fil-Tib (The Book on Choice of Medicine) by Ibn Hubal Baghdadi (1121–1213 CE) (Nikhat and Fazil 2020). The roots of Unani medicine are civilization in ancient Greece, Hippocrates (460–377 BC) regarding mizaj (temperament) and akhlat (humor). Unani medicine started with the concepts related to microbes that are not good for humans, so the plants that are widely used have antimicrobial properties (Riaz et al. 2020). In India, Unani medicine is integrated into the national healthcare system and is officially given the drug name Unani (Subbarayappa 2001).
- **Jamu** is a traditional medicine in Indonesia that is believed to have originated from the Mataram Kingdom era, c. 1300 years ago. Evidence that Indonesia has been using herbal medicine for a long time is the discovery of a stone mortar and pestle with a long cylindrical stone mortar used in making traditional herbal medicine, images on reliefs at Borobudur Temple depicting the activities of making potions, selling drinks, doctors, and masseurs treating their clients, also on the Madhawapura inscription which mentions a particular profession of herb mixer and combiner (herbalist), which is called Caraki. Ayurveda heavily influences Jamu; however, the use of Jamu varies by region in Indonesia. Jamu recipes are often unwritten, especially in remote areas. Records related to traditional medicine can be seen in a medical book from Mataram (1700), which contains 3000 herbal recipes, and in the Javanese classic literature Serat Centhini (1814),

which describes several recipes for herbal medicine. One of the local knowledge used by indigenous people in Indonesia is the selection of plants based on symptoms or the shape of diseased human organs. For example, to treat jaundice (the body becomes yellowish), people use parts of plants that are yellow (e.g., the rhizome of turmeric (*Curcuma domestica*), temulawak (*Curcuma xanthorrhiza*), Fig. 22.3).

The use of medicinal plants that have long been used traditionally by the community in an area is based on several factors, such as the availability of plants, the environment, the diseases encountered, and the knowledge and habits that become a culture (Fig. 22.4). Plants available around a community will cause the community to use them. The types of diseases that have been cured traditionally also determine the selection of plants. The community will be encouraged to use the plants around them to treat diseases. The environment and community knowledge from its predecessors also influence the types of plants chosen as medicinal plants. All practices of using plants for treatment in a particular area eventually become the culture and knowledge of the local community, known as traditional knowledge. Traditional knowledge mechanisms and rules must be clarified so indigenous peoples can benefit from them, and there will be no more biopiracy problems.

Traditional communities often use a mixture of plants to make ingredients to prevent and cure disease, such as in TCM, Kampo, and Jamu (Martin et al. 2020; Afendi et al. 2013). This mixture is hoped to have a synergistic effect that accelerates healing. In Indonesia, traditional medicinal ingredients contain four efficacies, namely, antimicrobial, anti-inflammatory, analgesic, and specific efficacy for targeted treatment (Afendi et al. 2010, 2013). Antimicrobials are always added because Indonesia, a tropical country with high humidity, makes it easy for harmful microbes to grow, and antimicrobials need to be added to its traditional medicinal ingredients. Anti-inflammatory are added to the ingredients because most diseases are related to inflammation, while analgesics are added to relieve pain. An illustration of the components of Indonesian herbal medicine can be seen in Fig. 22.5.

Plants that grow in different environments will synthesize other metabolites, especially the differences in their secondary metabolites, which can lead to differences in the use of a plant species from one region to another. In addition,

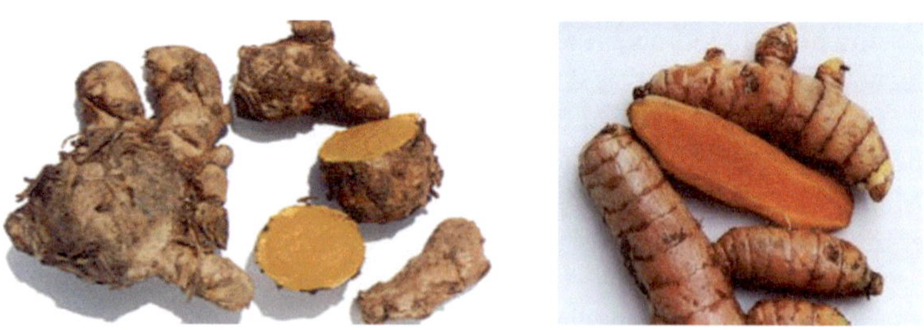

Fig. 22.3 The rhizome of *Curcuma domestica* (left) and *Curcuma xanthorrhiza* (right) used to treat jaundice

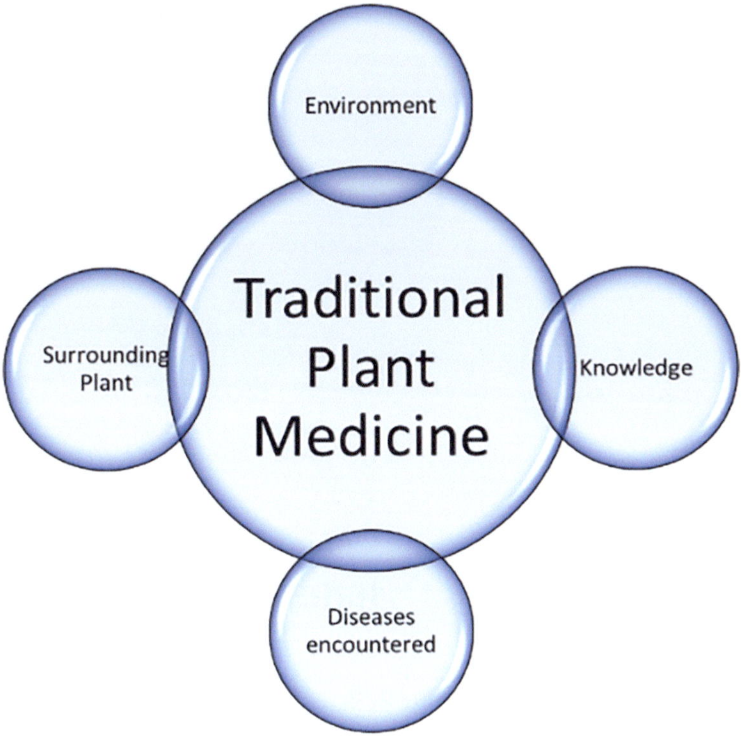

Fig. 22.4 The influencing selection of traditional medicinal plants

metabolites synthesized explicitly by plants that grow in certain areas can produce unique plants and be a separate geographical indication, which can be applied to communal intellectual property rights, namely, geographical indication.

22.3.2 Medicinal Plants as a Result of New Findings Using Research

Many plants are used traditionally, but the number of available plants is much more so that both plants that have been used traditionally and plants that have not been used need to search for their benefits, including as raw materials for medicine. Efforts to discover the benefits of plant biodiversity have been carried out systematically and organized to produce a product that can be commercialized through a bioprospecting process. Bioprospecting is searching, classifying, and systematically investigating valuable products such as new chemical compounds, active ingredients, genes, proteins, and other genetic information for commercial purposes with actual and potential economic value found in living resources such as medicinal plants (Alikodra and Sudarmonowati 2022). Bioprospection has scientifically proven local community knowledge using available science and technology.

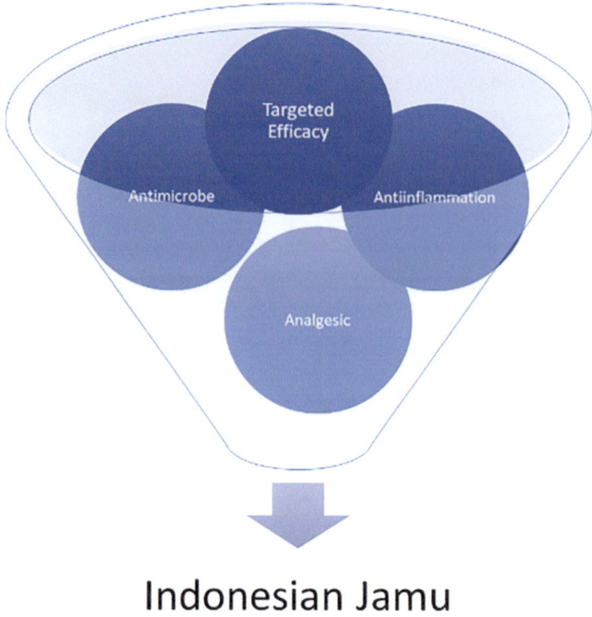

Fig. 22.5 Four components of Indonesian Jamu (Afendi et al. 2013)

Chemical compounds in plants are also isolated; their structure and efficacy are determined, thus confirming the ability of these compounds and plants. Ultimately, these plants are developed into products that can be commercialized, such as drugs, nutraceuticals, cosmetics, etc. Compounds obtained are often structurally modified through chemical and biochemical synthesis to produce new compounds expected to have better efficacy than the original compounds.

Creation of items with scientific backing can be patentable and made commercially through technology transfer while guaranteeing profit sharing and access for all parties involved. Benefits are anticipated to be shared by scientific community inventors, grassroots biodiversity management committees, and tribal knowledge producers and holders. In searching for new plants as new medicinal plants, many approaches can be used, such as based on chemotaxonomy, in silico, omics, etc. Based on the compounds found in a plant known to have efficacy, the same can be found in other plants. This approach is chemotaxonomic. Chemotaxonomy, or chemical taxonomy, is used to classify plants based on their chemical content (Singh 2016). Plant metabolites, especially secondary metabolites, are produced through specific biosynthetic pathways limited to taxonomically related organisms, making them useful in classification.

The in silico approach has accelerated drug discovery in less time, less effort, and lower costs (Shaker et al. 2021). This approach can discover new drug compounds using computational methods. In silico predictions about interactions between plant compounds and human proteins can be made using data-intensive science and machine learning methods. In silico research can be used to verify proteins that are the target of plant compounds so that it will help select plant-based drug candidates. In addition to determining drug candidates, the in silico approach provides a better understanding of traditional medicine at the omics level. These in silico results were further validated by docking simulations between predicted compound-protein interactions and would be better validated in vitro or in vivo in the laboratory. We can also explore supporting chemical or biological properties in predicted interactions, such as the similarity between the target compound and the known ligand of the predicted protein (Wijaya et al. 2021). When the active compounds have been proven, the plants containing these compounds can be used as raw materials for medicine.

The discipline of omics comprises areas as diverse as proteomics and metabolomics. This technology is also considered to provide results in a shorter time because it uses a variety of advanced analysis, data storage, and bioinformatics technologies. Proteomics is used to study the structure, function, and interactions of proteins with proteins or other ligands such as bioactive compounds. Metabolomics identifies and quantifies specific metabolites in samples, which are helpful in determining the concentration of active compounds, plant fingerprinting, and profiling.

Some research has been published related to finding potential plants or selecting the possible genotypes of plants to become medicinal plants. The genotypes of *Orthosiphon aristatus* (Blume) Miq. leaves were chosen using chemometrics (Batubara et al. 2020). Three varieties of Indonesian ginger (gajah ginger, emprit ginger, and red ginger, Fig. 22.6) have been selected for their potential as anti-skin aging (Asoka et al. 2022). Selecting fermented or fresh garlic as an antigout agent has also been conducted using metabolomic and docking analysis (Lestari et al. 2022).

The Tropical Biopharmaca Research Center of IPB University, Indonesia, established the Ijah Analytics system to connect all information on medicinal plants. Ijah Analytics (http://apps.cs.ipb.ac.id/ijah) is the system for predicting interactions between compounds and protein targets using machine learning and visualizing the results as a plant-compound-protein-disease network (Fig. 22.7). This web-based application was developed to analyze and predict herbal formulas based on their composition. At least three relationships were developed, namely the relationship between plants and the compounds they contain, the relationship between chemical

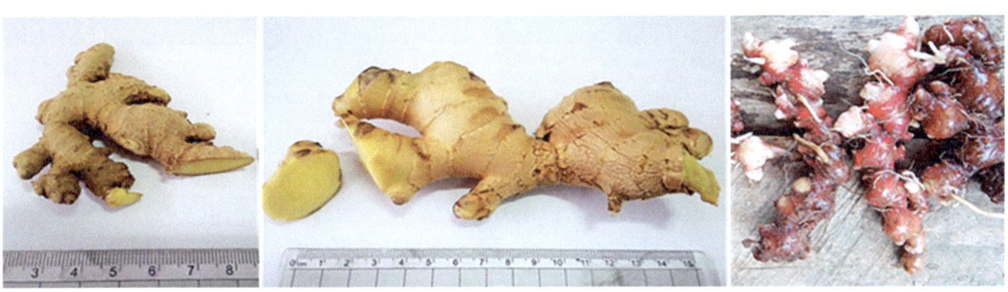

Fig. 22.6 Three varieties of Indonesian ginger; emprit ginger (left), gajah ginger (middle), and red ginger (right)

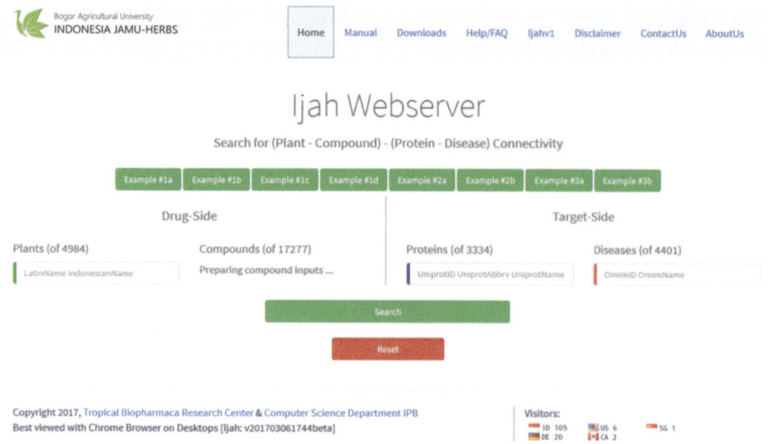

Fig. 22.7 Ijah Analytics interface

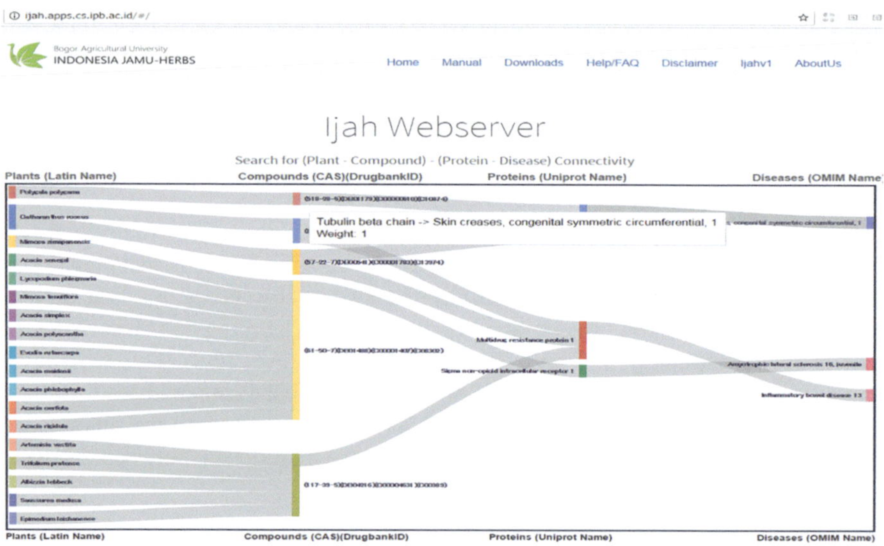

Fig. 22.8 An Ijah Analytics Prediction Result in the form of a plant-compound-protein-disease network

compounds and proteins found in humans, and the relationship between proteins and diseases. One example of a relationship can be seen in Fig. 22.8.

Using this webserver, new types of plants that have the potential as raw materials for natural medicines can be determined (Fig. 22.7). Mixing several types of plants into ingredients to target specific diseases can also be determined.

22.4 Application IPR on Medicinal Plants

Intellectual property rights protect against various problems that threaten intellectual property, including the massive exploitation of available natural resources, commonly known as biopiracy (Alikodra and Sudarmonowati 2022). Therefore, bioprospection is needed. The search for biological, chemical, and genetic material that could be useful for upcoming pharmaceutical goods is known as bioprospecting (Peterson 2017), which covers a wide range of matters, from the cultivation process to production for commercial purposes. Unfortunately, few are aware of the importance of bioprospecting to our natural resources.

Articles that discuss IPR relating to the legality of medicinal plants in detail still need to be explored. Table 22.1 summarizes some examples of intellectual property rights related to medicinal plants. However, some articles discuss the IPR of the medical cannabis industry from upstream to downstream. Wyse and Luria (2021) suggest that the IPR trend regarding medical cannabis will increase until 2020. The most common types of IPR registered for medical cannabis are utility patents, breeders' proper plates, and plant patents (USA only). The distribution of registered IPRs is uneven for the upstream, midstream, and downstream of the medical cannabis industry. IPR does not protect the upstream portions of the medicinal cannabis supply chain, particularly when it comes to cannabis production. In contrast, midstream parts of the medical cannabis supply chain have more than 300 patents. Not many IPRs have been registered for downstream parts, but it is hoped that there will be more because several patents are already being processed for legalization.

The Colgate-Palmolive Company submitted a tooth cleaning product formulation containing ginger extract which the Australian government has patented with application number AU 2010326133 B2. The patent discusses that ginger extract in dental cleaning products is suitable for treating and preventing oral diseases. The patented formula is 0.01–5% by weight of the combination of extracts in toothpaste, tooth gels, and tooth powders and 0.01–2% by weight of the combination of extracts in rinses and washes. The European Patent Office has granted a patent to the Hebei Yiling Medicine Research Institute for the use of ephedra in treating bronchitis, along with the preparation technique. The patent has been filed under the worldwide publication number W.O. 2011/000149. Several patented pharmaceutical composition formulas contain ephedra with the present invention, preferably, more preferably, and alternatively levels. Patented formulation preparation methods include weighing raw materials, extraction under reflux, concentration, and mixing in premix. Formulation of various medicinal plants to provide a hypoglycemic effective composition consists of extracts patented by Holomed Pharmaceuticals, Ltd. with US patent number 6,551,627 B1. This patent includes information on concentrations of extracts of medicinal plants such as *Anemarrhena asphodeloides*, *Morus alba*, *Orthosiphon aristatus*, *Ophiopogon japonicus*, *Pterocarpus marsupium*, or *Trichosanthes kirilowii*.

Table 22.1 Intellectual property rights related to medicinal plant

No	Region	Title	Plant variety	Publication number	References
1	European patent	Pharmaceutical composition containing *Ephedrae* for treating bronchitis and preparation method therefor	*Ephedrae herba*, *Mori cortex*, *Citri reticulate pericarpium*, *Rhei radix et rhizoma*, *Gypsum fibrosum*, *Armeniacae semen amarum*, *Fritillariae thunbergii bulbus*, *Platycodonis radix*, *Forsythiae fructus*, *Peucedani radix*, *Arctii fructus*, *Glycyrrhizae radix et rhizoma*, *Scutellariae radix*, *Pinelliae rhizoma*, *Lonicerae flos*	WO 2011/000149 (06.01.2011 Gazette 2011/01)	Wu et al. (2009)
2	US patent	Medicinal herbal compounds for the prevention and treatment of diabetes	*Pterocarpus marsupium*, *Morus alba*, *Orthosiphon aristatus*, *Opiophogon japonicus*, *Rosa rugosa*, *Commelina communis*, *Trichosanthis kirilowii*, and *Anemarrhena asphodeloides*	US 6,551,627 B1	Yoon and Jun (2003)
3	US patent	Process for the preparation of plant extracts for treating skin disorders and enhancing the healing of wounds	*Plectranthus amboinicus* extracts	US 9,872,879 B2	Wu et al. (2007)
4	US patent	Anti-cholesterolemic compounds and methods of use	*Emblica officinalis*	US 9,376,369 B2	Chatterjee (2016)
5	US patent	Emulsion containing a plant extract, method for producing said emulsion and for obtaining said plant extract	–	US 7,482,383 B2	Scheffler (2001)
6	US patent	Phytochemical extraction system and methods to extract phytochemicals from plants including plants of the family Cannabaceae sensu stricto	*Cannabis sativa*	US 11,465,072 B2	Lopa (2022)

7	US patent	Extraction of pharmaceutically active components from plant material	*Cannabis sativa*	U.S. 2021/0128657 A1	Whittle et al. (2021)
8	Australian patent	Oral compositions containing extracts of *Zingiber officinale* and related methods	*Zingiber officinale*	AU 2010326133 B2	Trivedi and Gittins (2009)
9	World intellectual property organization	A bioactive fraction from *Zingiber officinale* and a process for the preparation thereof	*Zingiber officinale*	W.O. 2007/113855 A1	Dharmesh and Siddaraju (2007)
10	US patent	Process of preparation of biodegradable films from semi-refined kappa carrageenan	*Eucheuma cottonii*	US 7,067,568 B1	Ghosh et al. (2006)
11	Indonesia patent	Formulation of medicinal plant extracts as antigout	*Garcinia mangostana L & Zingiber officinale*	P00201507749	Harmely and Dira (2015)
12	Indonesia patent	Essential oil from the craration plant (*Lantana camara* L) for external wound medicine	*Lantana camara* L	IDP000081479	Hariyanto (2022)
13	Indonesia patent	Utilization of *Acalypha indica* medicinal plant as a growth promoter in tilapia (*Oreochromis niloticus*)	*Acalypha indica*	S00202109070	Manoppo et al. (2021)
14	World intellectual property organization	Plant-based coffee composition, as well as preparation method and application thereof	Coffee	CN113812498	Xu et al. (2024)
15	World intellectual property organization	Medicinal plant extract preserving agent using grape extract and method for preserving medicinal plant extract	*Vitis vinifera*	1,020,020,090,732	Lee and Lee (2002)

Trichopus zeylanicus was validated scientifically, leading to the filing and awarding of three patents. The patent application titled "A process for the isolation of a glycolipid fraction from *T. zeylanicus* possessing adaptogenic activity" was granted Indian patent IN183071 in September 1999. Indian patent IN187975, dated August 2002, was awarded for the patent application entitled "A process for preparation of novel immunoenhancing, antifatigue, antistress and hepatoprotective herbal drug." "A process for the preparation of a novel herbal medicinal composition for cancer treatment from *Janakia arayalpathra* root and *T. zeylanicus* leaf" is the title of the third patent, IN193609, which was granted on September 22, 2006, and covers a combination of drugs that includes *Janakia arayalpathra* root and *T. zeylanicus* leaf. Subsequently, TBGRI produced a new polyherbal Ayurvedic medication in granular form called "Jeevani" by utilizing Ayurvedic pharmacologic procedures along with current scientific validation methods. The word "elixir of life" is what "Jeevani" signifies. The ingredients in "Jeevani" were *Trichopus zeylanicus* ssp. *travancoricus* Burkill. Ex Narayanan, *Evolvulus alsinoides* (Linn.) L., *Withania somnifera* (L.) Dunal., and *Piper longum* L. Clinical trial of "Jeevani" was conducted on more than a hundred subjects with different backgrounds. Besides modern drug efficacy tests, the results were evaluated based on Ayurvedic pharmacology. Subsequently, the technical knowledge for producing the Jeevani drug was transferred to an Ayurvedic drug manufacturing company for 7 years for a license fee of Rs.10 lakhs and a 2% annual royalty on an ex-factory sales price.

The community needs to be aware of the importance of IPR because many cases harm both the community and stakeholders because they underestimate IPR. Willcox et al. (2015) summarizes several cases of IPR negligence relating to medicinal plants. *Prunus africana*'s bark is used to treat benign prostate hyperplasia. As a result, the bark of *P. africana* was exploited, causing a significant decrease in the number of *P. africana* plants. There are considerations for patenting and protecting *P. africana* plants using monitoring harvest rates; local communities could have royalties and sustainable cultivation. However, *P. africana* is still alive, and it will be a lesson in the future that IPR is needed to protect plants from massive exploitation. Another case relates to the local indigenous people with standard patents for *Hoodia gordonii*. One of the NGOs patented the active compound from the *H. gordonii* plant so that they received royalty and agreed to share it with Indigenous Peoples. The patent was sold to one of the companies in cooperation with another company. Indigenous peoples should have benefited from selling these patents, but unfortunately, until they have changed hands again, they still have to wait 7 years for their rights to be fulfilled. Another case concerns *Pelargonium sidoides* extract, whose patent was contested by Indigenous Peoples because it was considered to violate traditional knowledge known as biopiracy. The case study conducted by Willcox et al. (2015) concluded that IPR is not easy to apply to the development of remedies from medicinal plants. Still, it is worth fighting for because many positive things can be felt.

22.5 Conclusion

Intellectual property is vulnerable to infringement because it is considered a trivial matter. Intellectual property can also relate to medicinal plants, starting from upstream to downstream processes, which are not spared from specific intellectual property violations called biopiracy. Laws and regulations protecting property rights related to medicinal plants need more attention and encouragement. Public awareness of intellectual property rights shows a positive trend from year to year. Many types of IPR, from traditional to communal, relate to medicinal plants. Traditional IPRs are mainly in the form of patents, which generally contain methods ranging from cultivation, extraction, and standardization to purification of active compounds. The assessment process for communal IPR considerations is now more straightforward because there is help from websites that can automatically search for compounds/mixtures that can treat a disease. It is hoped that with this convenience, more intellectual property rights related to medicinal plants can be protected by IPR.

References

Afendi FM et al (2010) System biology approach for elucidating the relationship between Indonesian herbal plants and the efficacy of Jamu. In: IEEE International Conference on Data Mining Workshops, Sydney, NSW, Australia. IEEE, Sydney, pp 661–668. https://doi.org/10.1109/ICDMW.2010.105

Afendi FM, Ono N, Nakamura Y, Nakamura K, Darusman LK, Kibinge N, Morita AH, Tanaka K, Horai H, Altaf M, Kanaya S (2013) Data mining methods for omics and knowledge of crude medicinal plants toward big data biology. Comput Struct Biotechnol J 4(5):e201301010. https://doi.org/10.5936/csbj.201301010

Alikodra HS, Sudarmonowati E (2022) Potensi Bioprospeksi Indonesia bagi Pembangunan Ekonomi NKRI (In Indonesia). IPB Press, Bogor, ID

Asoka SF, Batubara I, Lestari AR, Wahyuni WT, Wahyudi ST (2022) Compounds in Indonesian ginger rhizome extracts and their potential for anti-skin aging based on molecular docking. Cosmetics 9:128. https://doi.org/10.3390/cosmetics9060128

Baby BT, Suriyapraakash TNK (2021) Intellectual property rights: bioprospecting, biopiracy, and protection of traditional knowledge—an Indian perspective. IntechOpen, London

Batubara I, Komariah K, Sandrawati A et al (2020) Genotype selection for phytochemical content and pharmacological activities in ethanol extracts of fifteen types of Orthosiphon aristatus (Blume) Miq. Leaves using chemometric analysis. Sci Rep 10:20945. https://doi.org/10.1038/s41598-020-77991-2

Chatterjee S (2016) Anti-cholesterolemic compounds and method of use. (United States patent US 9,376,369 B2). US patent and trademark office

Dharmesh SM, Siddaraju MN (2007) A bioactive fraction from Zingiber officinale and a process for the preparation thereof (WO 2007/113855 A1). World Intellectual Property Organization, Geneva

Fitzgerald M, Heinrich M, Booker A (2020 Jan) Medicinal plant analysis: a historical and regional discussion of emergent complex techniques. Front Pharmacol 9(10):1480. https://doi.org/10.3389/fphar.2019.01480. PMID: 31998121; PMCID: PMC6962180

Ghosh PK, Siddhanta AK, Prasad K, Meena R, Bhattacharya A (2006) Council of Scientific and Industrial Research CSIR, Process of preparation of biodegradable films from semi refined kappa carrageenan. U.S. Patent 7,067,568

Hariyanto Y (2022) Minyak Esensial dari Tanaman Krasi (*Lantana camara*) untuk Obat Luka Luar. Indonesian Paten IDP000081479

Harmely F, Dira (2015) Formulasi Ekstrak Tanaman Obat Tradisional Anti Gout, Indonesian Patent P00201507749

Jia W, Gao W, Yan Y, Wang J, Xu Z, Zheng W, Xiao P (2004) The rediscovery of ancient Chinese herbal formulas. Phytother Res 18:681–686. https://doi.org/10.1002/ptr.1506

Lee DH, Lee SJ (2002) Medicinal plant extract preserving agent using grape extract and method for preserving medicinal plant extract. Korean Patent KR1020020090732

Lestari AR, Batubara I, Wahyudi ST, Ilmiawati A, Achmadi SS (2022) Bioactive compounds in garlic (Allium sativum) and black garlic as antigout agents, using computer simulation. Life 12:1131. https://doi.org/10.3390/life12081131

Li X, Ren J, Zhang W, Zhang Z, Yu J, Wu J, Sun H, Zhou S, Yan K, Yan X, Wang W (2022) LTM-TCM: a comprehensive database for the linking of traditional Chinese medicine with modern medicine at molecular and phenotypic levels. Pharmacol Res 178:106185. https://doi.org/10.1016/j.phrs.2022.106185

Lopa FA. (2022) Phytochemical extraction system and methods to extract phytochemicals from plants including plants of the family Cannabaceae Sensu Stricto (United States Patent US 11,465,072 B2). US Patent and Trademark Office

Manoppo H, Kalesaran OJ, Lantu S (2021) Pemanfaatan tanaman obat Acalypha indica sebagai promotor pertumbuhan pada ikan nila (Oreochromis niloticus). Indonesian paten S00202109070

Martin F, Michael H, Anthony B (2020) Medicinal plant analysis: a historical and regional discussion of emergent complex techniques. Front Pharmacol 10:1480. https://doi.org/10.3389/fphar.2019.01480

Nikhat S, Fazil M (2020) Overview of Covid-19; its prevention and management in the light of Unani medicine. Sci Total Environ 728:138859

Peterson K (2017) Benefit sharing for all?: bioprospecting NGOs, intellectual property rights, new governmentalities. Intellectual Property Routledge:263–276

Putri YM, Putri RM, Tisnanta HS (2022) Communal rights as the hegemony in third world regime: an Indonesian perspective. Indones J Int Law 19(2):1–28. https://doi.org/10.17304/ijil.vol19.2.5

Riaz U, Ali SA, Omar MA, Abdulaziz MA, Samir I, Mohammed B (2020) A review on ethnomedicinal plants used in traditional medicine in the Kingdom of Saudi Arabia. Saudi J Biol Sci 27(10):2706–2718. https://doi.org/10.1016/j.sjbs.2020.06.020

Ros IA (1999) Medicinal plants of the world: chemical constituents, traditional, and modern medicinal uses. Humana Press, Totowa, NJ

Scheffler A (2001) Emulsion containing a plant extract, method for producing said emulsion and for obtaining said plant extract (United States patent US 7,482,383 B2). US Patent and Trademark Office

Shaker B, Ahmad S, Lee J, Jung C, Na D (2021) In silico methods and tools for drug discovery. Comput Biol Med 137:104851. https://doi.org/10.1016/j.compbiomed.2021.10485

Singh R (2016) Chemotaxonomy: a tool for plant classification. J Med Plants Stud 4(2):90–93

Subbarayappa BV (2001) The roots of ancient medicine: an historical outline. J Biosci. https://doi.org/10.1007/BF02703637

Trivedi HM, Gittins EK (2009) Oral compositions containing extracts of Zingiber officinale and related methods. (Australiant Patent AU 2010326133 B2). Australian Patent Office

Whittle BA, Hill CA, Flockhart IR, Downs DV, Gibson P, Wheatley GW (2021) Extraction of pharmaceutically active components from plant materials (United States patent US 2021/0128657 A1). Patent Application Publication

Wijaya SH, Afendi FM, Batubara I, Huang M, Ono N, Kanaya S, Altaf-Ul-Amin M (2021) Identification of targeted proteins by jamu formulas for different efficacies using a machine learning approach. Life 11:866. https://doi.org/10.3390/life11080866

Willcox M, Diallo D, Sanogo R, Giani S, Graz B, Falquet J, Bodeker G (2015) Intellectual property rights, benefit-sharing and development of improved traditional medicines: a new approaches. J Ethnopharmacol 176:281–285
Wu RY, Chung YS, Wu YY, Siu ML, Hsiao CW (2007) Process for the preparation pf plant extracts for treating skin disorders and enhancing healing of wounds. (United States patent US 9,872,879 B2). US patent and trademark office
Wu Y, Xu H, Song H, Li W, Li X, Wang H, Zhang H, An J, Li Y, Wang C (2009) Pharmaceutical composition containing ephedrae for treating bronchitis and preparation method therefor (European Patent EP 2 450 047 B1). European Patent Office
Wyse J, Luria G (2021) Trends in intellectual property rights protection for medical cannabis and related products. J Cannabis Res 3(1):1–22. https://doi.org/10.1186/s42238-020-00057-7
Xu W, Shaoquan W, Rongbo Z, Cizhou L, Yu W, Fengyuan Z, Baoguang X, Jie D. (2024) Plant-based Coffee Composition, as well as Preparation Method and Application Thereof. CN113812498
Yoon JW, Jun HS (2003) Medicinal herbal compounds for the prevention and treatment of diabetes. (United States patent US 6,551,627 B1). US patent and trademark office

MIX
Papier aus verantwortungsvollen Quellen
Paper from responsible sources
FSC® C105338

If you have any concerns about our products,
you can contact us on
ProductSafety@springernature.com

In case Publisher is established outside the EU,
the EU authorized representative is:
Springer Nature Customer Service Center GmbH
Europaplatz 3, 69115 Heidelberg, Germany

Printed by Libri Plureos GmbH
in Hamburg, Germany